Lecture Notes in Networks and Systems

Volume 394

The series "Lecture Notes in Networks and Systems" publishes the latest developments in Networks and Systems—quickly, informally and with high quality. Original research reported in proceedings and post-proceedings represents the core of LNNS.

Volumes published in LNNS embrace all aspects and subfields of, as well as new challenges in, Networks and Systems.

The series contains proceedings and edited volumes in systems and networks, spanning the areas of Cyber-Physical Systems, Autonomous Systems, Sensor Networks, Control Systems, Energy Systems, Automotive Systems, Biological Systems, Vehicular Networking and Connected Vehicles, Aerospace Systems, Automation, Manufacturing, Smart Grids, Nonlinear Systems, Power Systems, Robotics, Social Systems, Economic Systems and other. Of particular value to both the contributors and the readership are the short publication timeframe and the world-wide distribution and exposure which enable both a wide and rapid dissemination of research output.

The series covers the theory, applications, and perspectives on the state of the art and future developments relevant to systems and networks, decision making, control, complex processes and related areas, as embedded in the fields of interdisciplinary and applied sciences, engineering, computer science, physics, economics, social, and life sciences, as well as the paradigms and methodologies behind them.

Indexed by SCOPUS, INSPEC, WTI Frankfurt eG, zbMATH, SCImago.

All books published in the series are submitted for consideration in Web of Science.

For proposals from Asia please contact Aninda Bose (aninda.bose@springer.com).

More information about this series at https://link.springer.com/bookseries/15179

Ali Kashif Bashir · Giancarlo Fortino ·
Ashish Khanna · Deepak Gupta
Editors

Proceedings of International Conference on Computing and Communication Networks

ICCCN 2021

Springer

Editors
Ali Kashif Bashir
Manchester Metropolitan University
Manchester, UK

Giancarlo Fortino
University of Calabria
Rende, Italy

Ashish Khanna
Maharaja Agrasen Institute of Technology
New Delhi, Delhi, India

Deepak Gupta
Maharaja Agrasen Institute of Technology
New Delhi, Delhi, India

ISSN 2367-3370 ISSN 2367-3389 (electronic)
Lecture Notes in Networks and Systems
ISBN 978-981-19-0603-9 ISBN 978-981-19-0604-6 (eBook)
https://doi.org/10.1007/978-981-19-0604-6

This Springer imprint is published by the registered company Springer Nature Singapore Pte Ltd.
The registered company address is: 152 Beach Road, #21-01/04 Gateway East, Singapore 189721, Singapore

ICCCN-2021 Steering Committee

General Chair

Prof. Dr. Omer Rana, Cardiff University, UK

Honorary Chairs

Prof. Dr. Vincenzo Piuri, University of Milan, Italy
Prof. Dr. Janusz Kacprzyk, FIEEE, Polish Academy of Sciences, Poland
Prof. Dr. Manu Malek, EiC Computer and Electrical Engineering, Stevens Institute of Technology, USA
Dr. Jon G. Hall, EiC Expert Systems (WILEY), The Open University, UK

Conference Chair

Dr. Ali Kashif Bashir, Manchester Metropolitan University, UK
Prof. Dr. Darren Dancey, Manchester Metropolitan University, UK

Technical Program Chair

Prof. Dr. Joel J. P. C. Rodrigues, Senac Faculty of Ceará, Fortaleza—CE, Brazil
Prof. Dr. Giancarlo Fortino, Università della Calabria, Italy

Convener

Dr. Pancham Shukla, London Metropolitan University, London, UK
Dr. Ashish Khanna, Maharaja Agrasen Institute of Technology, India
Dr. Deepak Gupta, Maharaja Agrasen Institute of Technology, India
Dr. Utku Kose, Suleyman Demirel University, Isparta, Turkey

Publication Chair

Prof. Dr. Valentina Emilia Balas, Aurel Vlaicu University of Arad, Romania
Prof. Dr. Zdzislaw Polokoswski, Jan Wyzykowski University, Polkowice
Prof. Dr. Vijay Singh Rathore, IIS (Deemed to be University), Jaipur
Dr. Hamid Reza Boveiri, Sama College, Islamic Azad University, Shoushtar Branch, Iran

Publicity Chair

Prof. Dr. Fadi Al-Turjman, Near East University, Nicosia, Turkey

Co-convener

Mr. Moolchand Sharma, Maharaja Agrasen Institute of Technology, India

ICCCN-2021 Advisory Committee and Technical Program Committee

Prof. Dr. Vincenzo Piuri, University of Milan, Italy
Prof. Dr. Oscar Castillo, Tijuana Institute Technology, Tijuana, Mexico
Prof. Dr. Valentina Emilia Balas, Aurel Vlaicu University of Arad, Romania
Prof. Dr. Aboul Ella Hassanien, Cairo University, Egypt
Prof. Dr. Joel J. P. C. Rodrigues, Federal University of Piaui, Brazil
Prof. Dr. Vineet Kansal, ProVC, AKTU; Director, IET, Lucknow, India
Prof. Dr. Giancarlo Fortino, Università della Calabria, Italy
Prof. Dr. João Manuel R. S. Tavares, Universidade do Porto (FEUP), Portugal
Prof. Dr. Victor Hugo C. de Albuquerque, Federal University of Ceará, Brazil
Prof. Dr. Yu-Dong Zhang, University of Leicester, LE1 7RH, UK
Prof. Dr. Neeraj Kumar, Thapar Institute of Engineering and Technology, Patiala, Punjab, India
Prof. Dr. Zdzislaw Polkowski, Jan Wyzykowski University, Poland
Prof. Dr. George A. Tsihrintzis, University of Piraeus, Greece
Prof. Dr. Arun Sharma, Indira Gandhi Delhi Technical University for Women, India
Prof. Dr. Abhishek Swaroop, Bhagwan Parshuram Institute of Technology, India
Prof. Dr. Giorgos Karagiannidis, Aristotle University of Thessaloniki, Greece
Prof. Dr. Sheng-Lung Peng, National Dong Hwa University, Taiwan
Dr. Dijana Oreski, University of Zagreb, Varazdin, Croatia
Dr. Ahmed Zobaa, Brunel University London, UK
Dr. Vicente García Díaz, University of Oviedo, Spain
Prof. Dr. A. K. Singh, National Institute of Technology Kurukshetra, India
Prof. Dr. Anil Kumar Ahlawat, KIET Group of Institutes, India
Dr. Jafar A. Alzubi, Al-Balqa Applied University, Salt—Jordan
Prof. Dr. Alex Norta, Tallinn University of Technology, Estonia
Dr. Utku Kose, Suleyman Demirel University, Isparta, Turkey
Prof. Dr. Isabel de la Torre Díez, University of Valladolid, Spain
Dr. Oana Geman, Universitatea Stefan cel Mare din Suceava, 720229 Suceava, Romania
Dr. Keping Yu, Waseda University, Tokyo, Japan
Dr. Varun G. Menon, SCMS School of Engineering and Technology, Kochi, India

Dr. Mohammad Shojafar, University of Surrey, UK
Dr. Muhammad Habib ur Rehman, King's College London, UK
Dr. Irfan Mehmood, University of Bradford, UK
Dr. Muddesar Iqbal, London South Bank University, UK
Dr. Shahid Mumtaz, Institute de Telecommunication, Portugal
Dr. Waleed Ejaz, Lakehead University, Canada
Dr. Anish Jindal, University of Essex, UK
Dr. Suresh Chavhan, Vellore Institute of Technology, Vellore, India
Dr. D Jude Hemanth, Karunya University, Coimbatore, India
Dr. GaganGeet Singh Aujla, Durham University, UK
Dr. Sachin Kumar, South Ural State University, Chelyabinsk, Russian Federation
Dr. Prayag Tiwari, Aalto University, Finland
Dr. Pradeep Malik, KIIT University, India
Dr. Akash Kumar Bhoi, Sikkim Manipal Institute of Technology, India
Dr. Hamid Reza Boveiri, Sama College, IAU, Shoushtar Branch, Shoushtar, Iran
Dr. Sahil Garg, École de technologie supérieure, Université du Québec, Montreal, Canada
Dr. Gulshan Shrivastava, Sharda University, Greater Noida, India
Dr. Gabriella Casalino, University of Bari, Italy
Dr. Aditya Khamparia, Babasaheb Bhimrao Ambedkar University, India
Dr. Amit Kumar Jaiswal, University of Leeds, UK
Dr. Qianqian Xie, University of Manchester, Manchester, UK
Dr. Yousaf Bin Zikria, Yeungnam University, South Korea
Dr. Francesco Piccialli, University of Naples Federico II, Italy
Dr. Ashiq Anjum, University of Leicester, UK
Prof. Dr. Nuno M. Garcia, University of Beira Interior, Covilhã, Portugal
Prof. Dr. Kashif Saleem, Universiti Teknologi Malaysia, Riyadh, Saudi Arabia
Dr. Le Hoang Son, University of Danang, Vietnam
Dr. Jaafar Alghazo, Virginia Military Institute, Lexington, VA
Dr. Jalil Piran, Sejong University, South Korea
Dr. Hari Mohan Pandey, Edge Hill University, UK
Prof. Dr. P. Sanjeevikumar, Aarhus University, Herning, Denmark
Dr. Kemal Polat, Abant Izzet Baysal University, Turkey
Dr. Juhriyansyah Dalle, Universitas Lambung Mangkurat, Indonesia
Dr. Ahmed Elngar, Beni-Suef University, Egypt
Dr. Prajoy Podder, Institute of Information and Communication Technology, BUET, Dhaka
Dr. M. Rubaiyat Hossain Mondal, Institute of Information and Communication Technology, BUET, Dhaka
Prof. Dr. Arun Kumar Sangaiah, VIT University, India
Prof. Dr. Robert Hsu, National Chung Cheng University, Taiwan
Dr. Sarada Prasad Gochhayat, Old Dominion University, USA
Dr. Daniel Nogueira, University of Minho, Brazil
Prof. Dr. Gwanggil Jeon, Incheon National University, South Korea
Dr. Khan Muhammad, Sejong University, South Korea

Dr. Yenumula B. Reddy, Grambling State University, Louisiana
Dr. Chandran Venkatesan, KPR Institute of Engineering and Technology, India
Dr. Alireza Jolfaei, Macquarie University, Australia
Dr. Souvik Ganguli, Thapar Institute of Engineering and Technology, India
Dr. Flah Aymen, National School of Engineering of Gabes, Tunisia
Prof. Dr. Placido Rogerio Pinheiro, University of Fortaleza, Brazil
Dr. Seifedine Kadry, Noroff University College, Norway
Dr. Daniela Clara Moraru, University of Luxembourg, Luxembourg
Dr. Gautam Srivastava, Brandon University, Canada
Dr. Vassilis C. Gerogiannis, University of Thessaly, Greece
Prof. Dr. B. S. Manoj, Indian Institute of Space Science and Technology, India
Dr. Mª Luz Castro Pena, Universidade da Coruña, Spain
Dr. Ilya Levin, Tel Aviv University, Israel
Dr. Muhibul Haque Bhuyan, Southeast University, Bangladesh
Prof. Dr. Med Salim BOUHLEL, Lab SETIT, Sfax University, Tunisia
Dr. Mamoun Alazab, Charles Darwin University, Australia
Dr. Lalit Garg, University of Malta, Msida, Malta
Dr. Arij Naser Abougreen, University of Tripoli, Libya
Dr. Sherif Mohamed Ismail, Egyptian German Academy
Prof. Dr. Vijay Singh Rathore, IIS Deemed to be University, India
Dr. Aslanbek Naziev, Ryazan State University named after S. A. Esenin, Russia
Dr. Mwaffaq Otoom, Yarmouk University, Jordan
Dr. Ahmed A. Ewees, Damietta University, Egypt
Dr. Iwan Adhicandra, University of Sydney, Australia
Prof. Dr. Meng Li, Hefei University of Technology, China
Dr. Korhan Cengiz, Trakya University, Turkey
Dr. Muhammad Bilal, Hankuk University of Foreign Studies, South Korea
Dr. R. R. Venkatesha Prasad, TU Delft, The Netherlands
Dr. Özge Korkmaz, Malatya Turgut Özal University, Turkey
Dr. Alexander Fedorov, Rostov State University of Economics, Russia
Prof. Dr. Alfredo Grieco, Politecnico di Bari, Italy
Prof. Dr. Quoc-Viet Pham, Pusan National University, South Korea
Dr. Enkeleda Lula, University Haxhi Zeka, Peja, Kosovo
Dr. Fides del Castillo, De La Salle University, Philippines
Dr. Dr Houda Chihi, Innov'COM Lab of Sup'COM Tunisia
Prof. Dr. Tu Nguyen, Kennesaw State University, Kennesaw, USA
Prof. Dr. Christos Douligeris, University of Piraeus, Greece
Dr. Dr. Surbhi Bhatia, King Faisal University, Saudi Arabia
Dr. Feras M. Awaysheh, Tartu University, Delta Research Center, Estonia
Dr. Assunta Di Vaio, University of Naples "Parthenope" (Italy)
Dr. Mehdi Gheisari, Southern University of Science and Technology, Shenzhen, Guangdong Province, P. R. China

Preface

We hereby are delighted to announce that Manchester Metropolitan University, Manchester, UK, and Universal Inovators have hosted the eagerly awaited and much coveted International Conference on Computing and Communication Networks (ICCCN-2021) in virtual mode during November 19–20, 2021. The conference has attracted many high-quality submissions and stimulates the cutting-edge research discussions among many academic pioneering researchers, scientists, industrial engineers, students with the reception of papers including more than 1150 authors from different parts of the world. The committee of professionals dedicated toward the conference is striving to achieve a high-quality technical program with tracks on Networks and Computing Technologies, Advances in Artificial Intelligence and Machine Learning, Security and Privacy, Emerging Topics in 5G/6G Communication Systems, Cyber Physical Systems, Emerging Trends in Data Analytics, Cyber Security for Industry 4.0 and Smart and Sustainable Environmental Systems. All the tracks chosen in the conference are interrelated and are very famous among present-day research community. Therefore, a lot of research is happening in the above-mentioned tracks and their related sub-areas. More than 210 full-length papers have been received, among which the contributions are focused on theoretical, computer simulation-based research, and laboratory-scale experiments. Among these manuscripts, 53 papers have been included in the Springer proceedings after a thorough two-stage review and editing process. All the manuscripts submitted to ICCCN-2021 were peer-reviewed by at least two independent reviewers, who were provided with a detailed review pro forma. The comments from the reviewers were communicated to the authors, who incorporated the suggestions in their revised manuscripts. The recommendations from two reviewers were taken into consideration while selecting a manuscript for inclusion in the proceedings. The exhaustiveness of the review process is evident, given the large number of articles received addressing a wide range of research areas. The stringent review process ensured that each published manuscript met the rigorous academic and scientific standards. It is an exalting experience to finally see these elite contributions materialize into a volume as ICCCN-2021 proceedings by Springer entitled "International Conference on Computing and Communication Networks."

ICCCN-2021 invited three keynote speakers, who are eminent researchers in the field of computer science and engineering, from different parts of the world. In addition to the plenary sessions on each day of the conference, five concurrent technical sessions are held every day to assure the oral presentation of around 53 accepted papers. Keynote speakers and session chair(s) for each of the concurrent sessions have been leading researchers from the thematic area of the session. A technical exhibition is held during all the 2 days of the conference, which has put on display the latest technologies, expositions, ideas and presentations. The research part of the conference was organized in a total of 15 special sessions and 2 workshops. These special sessions and workshops provided the opportunity for researchers conducting research in specific areas to present their results in a more focused environment.

An international conference of such magnitude and release of the ICCCN-2021 proceedings by Springer have been the remarkable outcomes of the untiring efforts of the entire organizing team. The success of an event undoubtedly involves the painstaking efforts of several contributors at different stages, dictated by their devotion and sincerity. Fortunately, since the beginning of its journey, ICCCN-2021 has received support and contributions from every corner. We thank them all who have wished the best for ICCCN-2021 and contributed by any means toward its success. The edited proceedings volume by Springer would not have been possible without the perseverance of all the steering, advisory and technical program committee members.

All the contributing authors owe thanks from the organizers of ICCCN-2021 for their interest and exceptional articles. We would also like to thank the authors of the papers for adhering to the time schedule and for incorporating the review comments. We wish to extend our heartfelt acknowledgment to the authors, peer-reviewers, committee members and production staff whose diligent work put shape to the ICCCN-2021 proceedings. We especially want to thank our dedicated team of peer-reviewers who volunteered for the arduous and tedious step of quality checking and critique on the submitted manuscripts. The management, faculties, administrative and support staff of the university have always been extending their services whenever needed, for which we remain thankful to them.

Lastly, we would like to thank Springer for accepting our proposal for publishing the ICCCN-2021 proceedings. Help received from Mr. Aninda Bose, the acquisition senior editor, in the process has been very useful.

Manchester, UK

Ashish Khanna
Deepak Gupta
Organizers and Conveners
ICCCN-2021

Contents

Editors and Contributors

About the Editors

Ali Kashif Bashir is Associate Professor at the Department of Computing and Mathematics, Manchester Metropolitan University, United Kingdom. He is also with Visual Research Intelligent Center, University of Electronics Science and Technology of China (UESTC) as Honorary Professor and Chief Adviser; with University of Science and Technology, Islamabad (NUST) as Adjunct Professor, and with University of Guelph, Canada, as Special Graduate Faculty. He is a senior member of IEEE, member of IEEE Industrial Electronic Society, member of ACM, and distinguished speaker of ACM. He received his Ph.D. in computer science and engineering from Korea University, South Korea. He has authored over 200 research articles; and received over 3 Million USD funding as PI and Co-PI from research bodies of South Korea, Japan, EU, UK and Middle East. His research interests include internet of things, wireless networks, distributed systems, network/cyber security, network function virtualization, machine learning, etc. He is serving as the editor-in-chief of the IEEE FUTURE DIRECTIONS NEWSLETTER. He is also serving as the area editor and associate editor in several top ranking journals. He is leading many conferences as a chair (program, publicity, and track) and had organized workshops in flagship conferences like IEEE Infocom, IEEE Globecom, and IEEE Mobicom.

Giancarlo Fortino (SM'12) is Full Professor of Computer Engineering at the Department of Informatics, Modeling, Electronics, and Systems of the University of Calabria (Unical), Italy. He received a Ph.D. in Computer Engineering from Unical in 2000. He is also Distinguished Professor at Wuhan University of Technology and Huazhong Agricultural University (China), high-end expert at HUST (China), Senior Research Fellow at the Italian ICAR-CNR Institute, CAS PIFI Visiting Scientist at SIAT—Shenzhen, and Distinguished Lecturer for IEEE Sensors Council. He is Web of Science Highly Cited Researcher 2020. His research interests include wearable computing systems, e-Health, Internet of Things, and agent-based computing. He is currently the scientific responsible of the Digital Health group of the Italian CINI

National Laboratory at Unical. He is author of 500+ papers in international journals, conferences, and books. He is (founding) Series Editor of IEEE Press Book Series on Human-Machine Systems and EiC of Springer Internet of Things series and AE of many international journals such as IEEE TAC, IEEE THMS, IEEE IoTJ, IEEE SJ, IEEE JBHI, IEEE SMCM, IEEE OJEMB, IEEE OJCS, Information Fusion, JNCA, EAAI, etc. He organized as chair many international workshops and conferences (100+), was involved in a huge number of int'l conferences/workshops (500+) as IPC member, is/was guest-editor of many special issues (60+).

Ashish Khanna [M'19, SM'20] received his Ph.D. degree from National Institute of Technology, Kurukshetra, in March 2017. He has completed his M.Tech. in 2009 and B.Tech. from GGSIPU, Delhi, in 2004. He has completed his Post-Doc (PDF) from Internet of Things Lab at Inatel, Brazil, in 2018. He has around 152 accepted and published research papers and book chapters in reputed SCI, Scopus journals, conferences and reputed book series including 82 papers accepted in SCI indexed Journals. He also has 4 published Patents to his credit. Additionally, He has authored, edited and editing around 40 books. He has also served as a keynote speaker, resource person and given several invited. He has worked as Guest Editor in several reputed journals. He is also serving as Series Editor in publishing houses like De Gruyter (Germany) of "Intelligent Biomedical Data Analysis" series, Elsevier of "Intelligent Biomedical Data Analysis" and CRC Press of "Intelligent Techniques in Distributed Systems: Principles and Applications". He is also acting as a consulting editor for Elsevier. His research interest includes, Distributed Systems and its variants (MANET, FANET, VANET, IoT), machine learning. He is the recipient of 2021 IEEE System Council Best Paper Award. He is serving as convener in some springer international conferences series like ICICC, ICDAM, and DoSCI. He is a senior IEEE member (SMIEEE) and ACM member too. He has played key role in promoting and initiating several startups and is also a startup consultant.

Deepak Gupta [M'19, SM'20] received a B.Tech. degree in 2006 from the Guru Gobind Singh Indraprastha University, Delhi, India. He received M.E. degree in 2010 from Delhi Technological University, India and Ph.D. degree in 2017 from Dr. A. P. J. Abdul Kalam Technical University (AKTU), Lucknow, India. He has completed his Post-doc from National Institute of Telecommunications (Inatel), Brazil, in 2018. He has co-authored more than 192 journal articles and 44 conference articles. He has authored/edited 50 books, published by IEEE-Wiley, Elsevier, Springer, Wiley, CRC Press, DeGruyter, and Katsons. He has filled four Indian patents. He is convener of ICICC, ICDAM, and DoSCI Springer conferences series. Currently, he is Associate Editor of Expert Systems (Wiley), and Intelligent Decision Technologies (IOS Press). He is the recipient of 2021 IEEE System Council Best Paper Award. He has been featured in the list of top 2% scientist/researcher database in the world [Table-S7-singleyr-2019]. He is also working towards promoting startups and also serving as a startup consultant. He is also a series editor of "Elsevier Biomedical Engineering" at Academic Press, Elsevier, "Intelligent Biomedical Data Analysis" at De Gruyter, Germany, "Explainable AI (XAI) for Engineering Applications" at CRC Press.

Contributors

Ahmed Subhi Abdalkafor Career Development Center, University Of Anbar, Anbar, Iraq

Dhafar Hamed Abd Department of Computer Science, Al-Maarif University College, Anbar, Iraq

Elsaeed E. AbdElrazek Computer Department, Damietta University, Damietta, Egypt

Husam Ali Abdulmohsin Computer Science Department, Faculty of Science, University of Baghdad, Baghdad, Iraq

Omar Azzawi Abdulteef Ministry of Education, Anbar Directorate, Planning Department, Anbar, Iraq

Belal Al-Khateeb Computer Science Department, College of Computer Science and Information Technology, University of Anbar, Anbar, Iraq

Atta E. Alalfy Computer Department, Mansoura University, Mansoura, Egypt

Abdulrahman Alamer Department of Information Technology and Security, Jazan University, Jazan, Saudi Arabia

Sameer I. Ali Al-Janabi Collage of Islamic Science, University of Anbar, Anbar, Iraq

Salah A. Aliesawi College of Computer Science and Information Technology, University Of Anbar, Anbar, Iraq

J. Amudha Department of Computer Science and Engineering, Amrita School of Engineering, Amrita Vishwa Vidyapeetham, Bengaluru, India

Ankita Chitkara University Institute of Engineering and Technology, Chitkara University, Rajpura, Punjab, India

Rajakumar Arul Department of CSE, Vellore Institute of Science and Technology, Chennai, India

Ege Baran Ayhan Department of Automotive Engineering, Atılım University, Ankara, Turkey

Tan Nguyen Bao Viettel Solutions, Hanoi, Vietnam

Mahmoud H. Barghout Faculty of Engineering, Menoufia University, Shebin El-Kom, Egypt

Ali Kashif Bashir Department of Computing and Mathematics, Manchester Metropolitan University, Manchester, UK

Rohit Beniwal Department of Computer Science and Engineering, Delhi Technological University, New Delhi, India

Duthil Benjamin EIGSI, La Rochelle, France;
IT, Image, Interaction Laboratory (L3I), University of La Rochelle, La Rochelle, France

Sylvia Bhattacharya Electrical Engineering Technology, Kennesaw State University, Marietta, GA, USA

A. Bhuvaneswari Vellore Institute of Technology, Chennai, Tamil Nadu, India

Mehmet Turhan Bilgin Department of Computer Engineering, Atılım University, Ankara, Turkey

Anil Carie School of Computer Science, VIT-AP, Amaravati, India

Satish Chand Jawaharlal Nehru University, New Delhi, India

Adil Chekati Faculty of Sciences of Tunis (FST), University of Tunis ElManar, Tunis, Tunisia

Meenu Chopra Vivekananda Institute of Professional Studies, Guru Gobind Singh Indraprastha University, Delhi, India

Sümeyye Çiçek Department of Internal Medicine, University of Health Sciences, Bursa, Turkey

Duong Le Dai Viettel Solutions, Hanoi, Vietnam

Praneeth Kumar Reddy Dendi Department of Computer Science, Kennesaw State University, Marietta, GA, USA

Sandeep Kumar Dey Faculty of Management and Economics, Tomas Bata University in Zlin, Zlin, Czech Republic

Wanyu Dong School of Software, Dalian University of Technology, Dalian, China

Vinh Luong Duc Viettel Solutions, Hanoi, Vietnam

Duygu Duru Department of Chemical Engineering, Atılım University, Ankara, Turkey

Chitra Ekambaram Department of ECE, SRM Institute of Science and Technology, Deemed to be University, Kattankulathur, Chennnai, India

Amany S. Elsharawy Computer Department, Damietta University, Damietta, Egypt

Ahmed A. Ewees Computer Department, Damietta University, Damietta, Egypt

M. A. Farag Faculty of Engineering, Menoufia University, Shebin El-Kom, Egypt

Fethi Filali Qatar Mobility Innovations Center (QMIC), Qatar University, Doha, Qatar

Thippa Reddy Gadekallu Vellore Institute of Technology, Vellore, India

Puneet Goswami SRM University, Sonepat, India

Deepa Gupta Department of Computer Science and Engineering, Amrita School of Engineering, Amrita Vishwa Vidyapeetham, Bengaluru, India

Jayesh Gupta Maharaja Agrasen Institute of Technology, Delhi, India

Shubham Gupta Department of Computer Science and Engineering, Amrita School of Engineering, Amrita Vishwa Vidyapeetham, Bengaluru, India

Sonam Gupta Ajay Kumar Garg Engineering College, Ghaziabad, India

Yagna Gurjala Department of Computer Science, Kennesaw State University, Marietta, GA, USA

H. Hakan Kilinc Orion Innovation Turkey, Istanbul, Turkey

Zeeshan Hameed Mir Faculty of Computer Information Science, Higher Colleges of Technology (HCT), Fujairah, United Arab Emirates

Samer Sami Hasan Computer Science Department, Faculty of Science, University of Baghdad, Baghdad, Iraq

Aboul Ella Hassanien Faculty of Computers and Information, Cairo University, Giza, Egypt;
Scientific Research Group in Egypt, Giza, Egypt

Medromi Hicham Research Foundation for Development and Innovation in Science and Engineering, Casablanca, Morocco;
System Architecture Team (EAS), Engineering Research Laboratory (LRI), National High School of Electricity and Mechanic (ENSEM), Hassan II University, Casablanca, Morocco

Linh Duong Vu Hoang Viettel Solutions, Hanoi, Vietnam

Huynh Thai Hoc Faculty of Management and Economics, Tomas Bata University in Zlin, Zlin, Czech Republic

Vo Viet Hung Faculty of Management and Economics, Tomas Bata University in Zlin, Zlin, Czech Republic

Hamzaoui Ikhlasse Research Foundation for Development and Innovation in Science and Engineering, Casablanca, Morocco;
System Architecture Team (EAS), Engineering Research Laboratory (LRI), National High School of Electricity and Mechanic (ENSEM), Hassan II University, Casablanca, Morocco;
EIGSI, La Rochelle, France

Shayla Islam Institute of Computer Science and Digital Innovation (ICSDI), UCSI University, Kuala Lumpur (South Wing), Malaysia

Şahin Işık Department of Computer Engineering, Eskisehir Osmangazi University, Eskisehir, Turkey

Dipika Jain Department of Computer Science and Engineering, Delhi Technological University, New Delhi, India

C. Jyotsna Department of Computer Science and Engineering, Amrita School of Engineering, Amrita Vishwa Vidyapeetham, Bengaluru, India

Kannan Kaliyan Department of CSE, Sree Vidyanikethan Engineering College, Tirupati, India

Amandeep Kaur Guru Tegh Bahadur Institute of Technology, New Delhi, India

Zeynep Kaya Department of Electrical and Electronics Engineering, Eskisehir Osmangazi University, Eskisehir, Turkey

Arjun S Kedlaya Department of Computer Science and Engineering, Amrita School of Engineering, Amrita Vishwa Vidyapeetham, Bengaluru, India

Farzil Kidwai Maharaja Agrasen Institute of Technology, Delhi, India

Nizameddin Koca Department of Internal Medicine, University of Health Sciences, Bursa, Turkey

Sajitha Krishnan Department of Computer Science and Engineering, Amrita School of Engineering, Amrita Vishwa Vidyapeetham, Bengaluru, India

Akshi Kumar Department of Information Technology, Netaji Subhas University of Technology, New Delhi, India

Zuhal Kurt Department of Computer Engineering, Atilim University, Ankara, Turkey

Ali Azawii Abdul Lateef Human Resources Department, University of Anbar, Anbar, Iraq

Komal Lawand Vidyalankar Institute of Technology, Mumbai, India

Dengxu Li School of Software, Dalian University of Technology, Dalian, China

MingChu Li School of Software, Dalian University of Technology, Dalian, China

Chuan Lin School of Software, Northeastern University, Shengyang, China

Suman Madan Jagan Institute of Management Studies, Delhi, India

P. S. S. Madhulika Department of Computer Science and Engineering, Amrita School of Engineering, Amrita Vishwa Vidyapeetham, Bengaluru, India

Cosmena Mahapatra University School of Information Communication and Technology, Guru Gobind Singh Indraprastha University, Delhi, India;
Vivekananda Institute of Professional Studies, Guru Gobind Singh Indraprastha University, Delhi, India

Ho Nguyen Manh Viettel Solutions, Hanoi, Vietnam

Ning Mao School of Software, Dalian University of Technology, Dalian, China

Jason S. Metcalfe US Devcom Army Research Laboratory, Human Research and Engineering Directorate, Aberdeen Proving Ground, MD, USA

Auxilia Michael Department of Artificial Intelligence and Data Science, Sri Manakula Vinayagar Engineering College, Puducherry, India

Faouzi Moussa LCOMS, Université de Lorraine, Metz, France

Ch. A. S. Murty Centre for Development of Advanced Computing (C-DAC), Hyderabad, India

V. Muthumanikandan Vellore Institute of Technology, Chennai, Tamil Nadu, India

Ramanathan Nachiappan Department of Computer Science and Engineering, SRM Institute of Science and Technology, Ramapuram, Chennai, India

Manoj Nandakumar Department of Computer Science and Engineering, SRM Institute of Science and Technology, Ramapuram, Chennai, India

Liqaa Nawaf Cardiff Metropolitan University, Cardiff, UK

Hemalata Nawale Sandip University, Nashik, India

João C. Neves IT: Instituto de Telecomunicações, University of Beira Interior, Covilhã, Portugal

Thanh Nguyen Ngoc Viettel Solutions, Hanoi, Vietnam

Ahmed Assim Nsaif Informatics Institute for Postgraduate Student, Baghdad, Iraq

Mahesh Parihar College of Engineering Pune, Pune, Maharashtra, India

Roshan Pathak Centre for Development of Advanced Computing (C-DAC), Hyderabad, India

Shital Patil Ramrao Adik Institute of Technology, Mumbai, India

Ashish Payal University School of Information Communication and Technology, Guru Gobind Singh Indraprastha University, Delhi, India

Quynh Giao Ngoc Pham Faculty of Multimedia Communications, Tomas Bata University in Zlin, Zlin, Czech Republic

Hugo Pedro Proença IT: Instituto de Telecomunicações, University of Beira Interior, Covilhã, Portugal

Pushp Jawaharlal Nehru University, New Delhi, India

Sang Nguyen Quang Viettel Solutions, Hanoi, Vietnam

R. Radhika SRM Institute of Science and Technology, Kattankulathur, Chennai, Tamil Nadu, India

K Raja Department of CSE, SRM Institute of Science and Technology, Ramapuram, Chennai, India

Harmesh Rana Centre for Development of Advanced Computing (C-DAC), Hyderabad, India

Shalli Rani Chitkara University Institute of Engineering and Technology, Chitkara University, Rajpura, Punjab, India

Meriem Riahi High National School of Engineers of Tunis (ENSIT), University of Tunis, Tunis, Tunisia

Rizk M. Rizk-Allah Faculty of Engineering, Menoufia University, Shebin El-Kom, Egypt

Parag H. Rughani National Forensic Sciences University (NFSU), Gandhinagar, India

Nalini Sampath Department of Computer Science and Engineering, Amrita School of Engineering, Amrita Vishwa Vidyapeetham, Bengaluru, India

Mithileysh Sathiyanarayanan MIT Square Services Private Limited, London, UK

Jahnavi Seth Maharaja Agrasen Institute of Technology, Delhi, India

Kshitij Sharma Paralaxiom Pvt. Ltd., Bangalore, India

Moolchand Sharma Maharaja Agrasen Institute of Technology, Delhi, India

Tariq Hussain Sheikh Department of Computer Science, Government Degree College Poonch, Poonch, J&K, India

Anupriya Shrivastava Department of Computer Science and Engineering, Amrita School of Engineering, Amrita Vishwa Vidyapeetham, Bengaluru, India

Bharti Shukla Ajay Kumar Garg Engineering College, Ghaziabad, India

Abhijeet Singh Guru Tegh Bahadur Institute of Technology, New Delhi, India

Himanshu Kumar Singh National Institute of Technology Patna, Patna, India

Jyoti Prakash Singh National Institute of Technology Patna, Patna, India

Arushi Sondhi Maharaja Agrasen Institute of Technology, Delhi, India

Satyendra Kumar Srivastav Delhi Technological University, Delhi, India

Jane Jaleel Stephan University of Information Technology and Communications, Baghdad, Iraq

Akhil Krishnan Sunil Department of Computer Science and Engineering, SRM Institute of Science and Technology, Ramapuram, Chennai, India

Senem Tanberk Orion Innovation Turkey, Istanbul, Turkey

Sushma Tejwani Narayana Nethralaya, Bommasandra, Bengaluru, India

Anand Shanker Tewari National Institute of Technology Patna, Patna, India

Manmohan Singh Thakur Department of Computer Science and Engineering, Amrita School of Engineering, Amrita Vishwa Vidyapeetham, Bengaluru, India

Cong Le Thanh Viettel Solutions, Hanoi, Vietnam

Chee Ling Thong Institute of Computer Science and Digital Innovation (ICSDI), UCSI University, Kuala Lumpur (South Wing), Malaysia

Yuan Tian School of Economics and Management, Dalian University of Technology, Dalian, China

M. M. Tripathi Delhi Technological University, Delhi, India

Anh Dang Trung Viettel Solutions, Hanoi, Vietnam

Tsutomu Tsuboi New Business Creative Division, Nagoya Electric Works Co. Ltd, Ama-shi, Japan

Neha Tuniya MmM Ltd, Navi Mumbai, Maharashtra, India

Dilek Bilgin Tükel Doğuş University, Istanbul, Turkey

Sreedevi Uday Department of Computer Science and Engineering, Amrita School of Engineering, Amrita Vishwa Vidyapeetham, Bengaluru, India

M. Ugur Seker Orion Innovation Turkey, Istanbul, Turkey

Huong Nguyen Van Viettel Solutions, Hanoi, Vietnam

Linh Nguyen Van Viettel Solutions, Hanoi, Vietnam

Quang Than Van Viettel Solutions, Hanoi, Vietnam

Emrecan Varyok Department of Automotive Engineering, Atılım University, Ankara, Turkey

Praveena Vasudevan Department of ECE, SRM Institute of Science and Technology, Deemed to be University, Kattankulathur, Chennnai, India

Rachit Verma Centre for Development of Advanced Computing (C-DAC), Hyderabad, India

Courboulay Vincent IT, Image, Interaction Laboratory (L3I), University of La Rochelle, La Rochelle, France

Sonali Vyas University of Petroleum and Energy Studies, Dehradun, India

Arun Kumar Yadav National Institute of Technology, Hamirpur, HP, India

Divakar Yadav National Institute of Technology, Hamirpur, HP, India

Qifan Yang School of Software, Dalian University of Technology, Dalian, China

Xulu Yao Manchester Metropolitan University, Manchester, UK

Moi Hoon Yap Manchester Metropolitan University, Manchester, UK

Tan Shy Yu Institute of Computer Science and Digital Innovation (ICSDI), UCSI University, Kuala Lumpur (South Wing), Malaysia

Atef Zaguia Department of Computer Science, College of Computers and Information Technology, Taif University, Taif, Saudi Arabia

Yanlong Zhang Manchester Metropolitan University, Manchester, UK

Xiao Zheng School of Computer Science and Technology, Shandong University of Technology, Zibo, China

Intelligent System for Acquiring Knowledge by Converting Arabic Speech to Text

Amany S. Elsharawy, Atta E. Alalfy, Elsaeed E. AbdElrazek, and Ahmed A. Ewees

Abstract The purpose of this study is to acquire the required knowledge resulting from converting Arabic speech into text. This system passes through some stages. The initial stage is "system training." This stage starts with identifying the lecturer's sound in different frequencies and then saving it upon an audio file and then converting it from analogue waves into digital ones; this is called "the preprocessing path." The properties of this audio file are then extracted and stored within a template. The second stage is "system testing." This stage compares the new sound file with the one previously stored on the system directory by comparing certain features. When the sound characteristics are matched using a similarity scale, the selected sound file has been identified. The third stage is printing the words into a text document. The next stage revises the resulting words according to specific rules identified by the proposed system itself. The final stage extracts the knowledge from the revised text. The proposed system is evaluated using Arabic dataset; it contains Arabic speech collected from different person types and ages. The experiment tests the system under three different recording modes, namely normal, quick, and loud. The results showed that the proposed system is effective in acquiring knowledge from converted Arabic speech to text especially with normal recording mode in Arabic speech recognition and classifying the fields.

Keywords Arabic speech recognition · Features extraction · Speech to text · Extracting knowledge

---

A. S. Elsharawy · E. E. AbdElrazek · A. A. Ewees (✉)
Computer Department, Damietta University, Damietta, Egypt
e-mail: a.ewees@hotmail.com; ewees@du.edu.eg

E. E. AbdElrazek
e-mail: elsaeed2004@du.edu.eg

A. E. Alalfy
Computer Department, Mansoura University, Mansoura, Egypt

A. K. Bashir et al. (eds.), *Proceedings of International Conference on Computing and Communication Networks*, Lecture Notes in Networks and Systems 394,
https://doi.org/10.1007/978-981-19-0604-6_1

1 Introduction

Despite the progress accomplished in the area of speech communication and Arabic speech recognition (ASR), it is still considered one of the most important fields calling for extensive research [1]. The conversion of speech into text is among the most important fields of language processing. Converting Arabic speech into text is a challenge to most researchers dealing with Arabic language. Arabic is known to be one of the most challenging morphological complex languages [2–8]. As a result, the development of ASR or automatic speech recognition is a significantly difficult task. Arabic language is known to have a high number of dialects, which we use in our daily communications. There is a considerable difference between these dialects and the research dealing with them [9]. Splitting speech targets the split of speech input signals into a number of basic units or words, and defining the beginnings and ends of these units (words, syllables, or phonemes) [10]. Another approach to identify the pronunciation variation is by using the phonological rules of this language that considers the phonetic variation through ASR pronunciation dictionaries [11]. The influence of phonological rules in Arabic speech recognition shows that the performance of ASR can be improved by eliminating the mismatch between the speech and the text used in training the audio model [12]. Employing phonological rules for the ASR dictionary adaptation is classified as a knowledge-based (kb) approach; however, data-driven is another option for the pronunciation variation. Hence, both approaches introduce some variants to generate phonetically rich dictionary pronunciation that might alleviate the acoustic changes on the performance. Modeling pronunciation variation includes two types, the within-word and the crossword pronunciation variation [13]. Although extracting knowledge is creating knowledge from structured (like relational databases and XML) and unstructured sources (like images, text, and documents), the obtained knowledge must be in a specific format that is readable and interpreted by machines, and must express knowledge in a facilitated manner. Even though natural language processing (NLP) and extract transform load (ETL) extraction of information are methodologically similar, the most important factor is that the extraction results go beyond creating structured information or transforming into a relational scheme. It needs the reuse of the existing information or generating a new scheme based on the source data [13].

Many studies were involved in the ASR field as the authors of [14] proposed a system for converting Arabic speech to text using recurrent neural networks (RNNs). It used a character-based decoder to avoid using a word lexicon. It also applied connectionist temporal classification as an objective function. The corpus contained 1200 h of broadcast programs. The error rate reached of 12.03% for non-overlapped speech. Furthermore, another ASR system was presented by [15] to recognize 260 h of recorded files, and its experiment provided a good result than the compared method and indicated that providing a less information in training phase about the data was better than providing inaccurate information.

In this regard, this paper proposes an intelligent system to obtain knowledge from a text resulting from converting Arabic speech into this text. Therefore, the main contribution of this paper is to propose a system to convert Arabic speech to text besides classifying the domains of this text automatically.

The proposed system includes many stages, which are: (1) the training stage, (2) the testing stage, (3) the extracting knowledge stages. In the next sections, we will review the details of the previously mentioned stages, whereas Sect. 2 illustrates the used method. Section 3 includes the proposed framework. Section 4 includes building-specific sound dictionary. Section 5 includes converting sound to text. Section 6 includes extracting knowledge phase. Section 7 shows the experiment. Section 8 concludes the paper.

2 Methods and Material

This section describes different methods used in this paper.

2.1 *Speech Duration*

It is very important to determine the time durations of each emotion statement. Both voiced and non-voiced parts are included in these times, which contribute to emotion. The ratios of voiced and unvoiced durations of emotion statements are considered as one of the parameters of feature recognition [14]. In order to separate the voiced and unvoiced parts of a signal, zero crossing rate (ZCR) and short time energy (STE) methods are used. The STE method is used in this paper. The process of detecting the signal voiced and unvoiced parts using STE method is illustrated as follows [16]: (1) Calculate the average energy (Eav) for (Ev) and use it as a threshold level. (2) Compare elements of (Ev) with (Eav). (3) If En > Eav, then Framen is "voiced frame": Add n to (FNv), else Framen is "unvoiced frame," where FNv vector contains frame numbers for the voiced parts. (4) Reconstruct signal for the voiced parts using (FNv). These durations are calculated from original signal and signal for the voiced parts as follows:

$$\mathrm{Ds} = \mathrm{Len(x)/fs}; \quad \mathrm{Dv} = \mathrm{Len(xv)/fs}; \quad \mathrm{Du} = \mathrm{Ds} - \mathrm{Dv} \tag{1}$$

where Ds is duration of the original signal (voiced + unvoiced) parts, Dv is duration of voiced parts, bDu is duration of unvoiced parts, Len(x) is length of original signal, and Len(xv) is length of signal for the voiced parts. Consequently, five proposed features are extracted from signal durations.

2.2 Speech Rate

There is a significant connection between speech rate and different modes. Humans tend to speak faster when they get excited than in cool mood. So, anger, fear, or high-frequency content emotions might have a higher speech rate than neutral or sad or low-frequency voices. For an utterance, the average speaking rate (S_r) can be estimated as follows [17]: $S_r = \frac{n_w}{\mathrm{Ds}}$, where nw is the number of words in utterance. Ds is the duration of the original signal (voiced + unvoiced) parts. Calculating number of words in a signal is explained as follows [16]: (1) Detect difference between FNv elements and put differences in DFv vector. (2) If Difn > 1, then replace Difn by "1," else replace Difn by "0." (3) Count the number of DFv elements that is equal to "1" and put the output in C. (4) Calculate nw as follows: ($nw = C + 1$). (5) Compute the average of speaking rate through nw divided by Ds.

2.3 Energy

The intensity of the spoken signal generated by energy is highly associated with different modes. Speech signals of happy and angry emotions have much higher energy than sad emotions [18, 19]. The function of short-term energy is to extract the value of energy in each speech frame. The energy of each frame is calculated by [20] $E = \sum_{n=0}^{N-1} |Si(n)|^2$, where E is energy of each frame, $si(n)$ denotes the ith frame of the speech signal $s(n)$, and N is the total number of samples in a frame or a window (frame length). After applying previous equation for all frames in a signal, a vector of energies (Ev) is obtained. It consists of number of energy values for all frames in the signal. Consequently, in this paper, five proposed features are extracted from Ev. These features are: mean (Em), max (Emx), min (Emn), median (Emd), and standard division (Esd).

2.4 Formant

Formants, or vocal tract resonances, have a significant role in studying both speech production and perception, particularly with using vowels [21]. Formant is one of the most important parameters to reflect sound track features. First, linear prediction is applied in order to calculate the 14 order prediction coefficients. Those coefficients are then used to estimate the sound track frequency response curve. Finally, the peak picking method is adopted to calculate the frequency of every formant [22]. Emotional states alter the shape of the vocal tract. Formant is an acoustic resonance of the human vocal tract. It is measured as an amplitude peak in the frequency spectrum of the sound [23]. The first three formants F1, F2, and F3 are extracted by using linear predictive coding (LPC) filter as shown in [24]. Three vectors (F1v, F2v, and F3v) are acquired for all frames in a signal. Then, the average for each vector is calculated to obtain total F1, F2, F3 for all frames. Finally, five proposed features are extracted from signal formant frequencies as: FSet3 = {F1, F2, F3, F2/ F1, F3/ F1}.

2.5 *Pitch*

Pitch is fundamental frequency of speech signal. It is the periodic time of a wave pulse produced by compressing air through the glottis from the lungs. When the emotions of a person change, his/her biological characteristics like blood pressure and flow of air from the lungs change as well. So, extracting this feature helps to identify the emotions of a person [25]. By using the autocorrelation method, the pitch of all the speech can be estimated, as explained in [26]. After applying it for all frames in the signal, a vector of frequency pitch (FPv) is obtained. Consequently, five proposed features are extracted from FPv. These features are: mean (FPm), max (FPmx), min (FPmn), median (FPmd), and standard division (FPsd).

2.6 *MFCC*

MFCC is considered as among the best distinctive features of recognition problems. MFCC is based on the characteristics of the human ear's hearing and perception that utilizes a nonlinear frequency unit to simulate the human auditory system [27]. MFCC includes a few steps. These steps are explained in [28]. After applying MFCC for input utterance, it is transformed into a sequence of acoustic vector. The statistical features mean (Mm), standard division (Msd), median (Mmd), skewness (Msk), and kurtosis (Mku) are calculated from the acoustic vector. As a result, five proposed features are extracted from signal MFCC.

3 The Proposed Framework

The proposed framework contains three stages as follows: The first stage (the training stage) is considered to be the basic structure of the sound system. It is based and trained on the input sounds that the system identifies and stores. This stage consists of some many substages, the first of which is obtaining the required sound by recording a person's sound in various tone frequencies. The input file is then converted from analogue waves to digital ones. This is known as the "preprocessing path." The next step removes noise and pollution, to obtain pure sound that can be processed. The sound words are then decoded in a specific sound dictionary template, a text name is added, and an equivalent domain code is used. Each word has been recorded a number of times, depending on the recording mode. The next stage is to extract some sound features to be stored within a template for later in the stages for processing the sound to be converted. The next stage comes to extract some of sound features stored in forms within the template to dependence later in the stages of processing the sound to be converted. The second stage is the testing stage. In this stage, the system is tested by converting the selected sound file into a text file, passing through several substages starting with reading the sound file and generating an initial processing for it. Another substage of processing starts with the process of splitting the sound file into words by using a certain set of parameters that will be listed later. These words are organized in a queue, so that it will be easy to handle their comparison and testing according to the system of the initial input word. Each word in the queue is compared to the sound dictionary stored in the databases, by comparing extract features for each case. When the characteristics are matched by a similarity scale, this means that the desired result has been reached, which is identifying the sound file. The next step is printing the word in a text document. The third stage) the extracting knowledge stage) includes a number of substages. This stage can be explained as how to extract kb from the text. This text was obtained from recognition, processing, and printing of the converted spoken sound to a text file. Figure 1 shows the steps of the proposed method.

4 Building Spoken Word Dictionary

The main step in the conversion of spoken words into written text is generating a sound word dictionary. This dictionary is used as a reference source in the matching of words. The detailed steps of this flowchart are explained as follows:

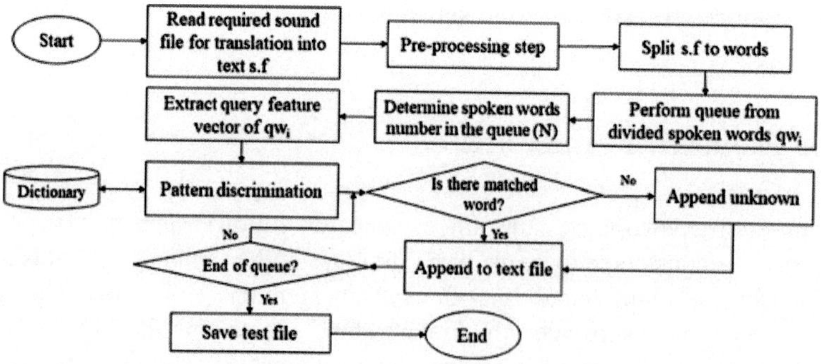

Fig. 1 Steps of the proposed method

4.1 *Perform Word Template*

The dictionary is made by using a word template. This template was designed by the authors. Each word is recorded by the user in different modes (normal, quick, and loud). The words are related to different domains.

4.1.1 Word Class Number

The first field in the template represents the word class number (WID). It consists of three digits beginning with the number (000) and ending with the number (999). This number of digits may be increased or decreased according to number of word classes.

4.1.2 Spoken File Name

The second field in the template represents the spoken file name. This field includes two blocks. The first block is the sound word name (*W*) in which the spoken word is recorded. The second block is called the recorded word mode. The selected recording modes are normal, quick, and loud.

4.1.3 Equivalent Text Word(s)

The third field in the template is the equivalent text word(s) (EW). Different words can be expressed for each recorded word, according to their classification within the used field. A single Arabic word may have various equivalent meanings according to the given domain. Each word is recorded by the user in different modes. For example, the word "علم" has more than one meaning that varies according to the domain such as "رايه" and "مشهور". This part is useful in case of refining the translation.

4.1.4 Domain Code

The fourth field in the template is the domain code (DC). For eight domains, DC may include three digits to express a specific domain. These fields are categorized to make it easier to determine where the words are stored for the keyword. For example, the domain name "Artificial intelligence" is coded with [0 0 0], whereas the "databases" is coded [0 0 1] and so on.

4.1.5 Feature Extract of Spoken Words

The fifth field is called the extracted features (EF). It is the vector that consists of the sound features. These features will be demonstrated later.

4.2 Record Sound Word

In this stage, the words in the spoken word dictionary, which have been compiled from specialized journals and research in the selected field, are recorded.

4.3 Retrieve Sound Word

In this stage, the previously recorded sounds in the audio files are retrieved one by one.

4.4 Extract Features of Spoken Word

In this stage, the features of the stored file are extracted. A set of features are drawn, as in Sect. 2, in order to make each audio file distinct from its peers. In the step of signal recognition, certain features are used to identify each class of signal from the other. The signal space has a high dimensionality. The purpose of feature extraction is to help the system to identify the inputs [29, 30].

5 Converting Sound to Text

The converting of sound into text passes through different stages as in the following procedural steps: Read required sound files (s.f). (2) Preprocess (s.f). (3) Split (s.f) into words. (4) Perform queue from split spoken words. (5) Determine number of spoken words in the queue. (6) Extract query features vector of each word in the queue. (7) Pattern discrimination. (8) Open a text file and add recognized words or add (unknown). (9) Consult kb for text refinement. (10) Save text file.

5.1 *Sound Acquisition*

This stage includes the process of recording different sounds using the microphone in specific environmental conditions to avoid high rates of noise distractions and disturbances that influence the quality of the entered sounds. Each voice speaks for a specified period of time and specific number of words, sentences, or paragraphs related to the pronunciation method to clearly distinguish the entered words.

5.2 *Sound Preprocessing*

Before feature extraction, the speech signals are normally preprocessed. This enhances the accuracy and efficiency of the process of feature extraction. The preprocessing stages are filtering, framing, and windowing.

5.2.1 Filtering

In order to decrease the noise effect, filter operation is performed. That can be done using the high-pass filter. Special flattening is performed by the process of pre-emphasis using a first-order finite impulse response (FIR) filter. Compensating the highfrequency part suppressed during the sound production mechanism of humans—is the main goal of pre-emphasis [31]. The following formula shows how pre-emphasis is done $s2(n) = s(n) - -a * s(n-1)$, where $s(n)$ is the input spoken signal, $s_2(n)$ is the output signal of filter, and A is the pre-emphasis parameter.

5.2.2 Framing

There has always been a finite length to the pre-emphasized speech signal. Its quasi-stationary nature is the cause of its not being usually processed. Through blocking the speech signal into short frames of N samples, it can be regarded as stationary [20].

5.2.3 Windowing

To minimize the disruptions at both the starting and the end of each frame, windowing is done. The ultimate goal here is reducing the spectral distortion through using the window to reduce the signal to zero at the beginning and end of each frame [32]. The Hamming window is used in this system, which has the form below [33]:

$$w(n) = 0.54 - 0.46\cos\left(\frac{2\pi n}{N-1}\right), 0 \leq n \leq N-1 \quad (2)$$

5.2.4 Framing and Windowing

It is a fact that speech signal is non-stationary, but this signal normally has stationary at a specific time range (20_40 ms), basically short windows or frame. Speech signal will be split up into various frames and then processed in the framing process [34].

5.3 Splitting Sound

Recently, great efforts have been exerted to find a solution to the problem of classifying speech into: silence/voiced/unvoiced parts of splitting audio frames and classifying it [35]. In this procedure, the formerly attained audio file is divided into words depending on certain technique using the MFCC that are useful to increase both the performance and the accuracy of recognition systems.

5.4 The Queue of Words

Queue is defined as a sequence or a line of some people or even vehicles waiting for their turn in order to be served or to proceed. Here, they are represented in a stored spoken word list in order to be recoverable in a clear order, normally the in-insertion order [36]. The first word waits till the speech end fills the queue and this is recognized as enqueue. Various operations are performed with special audio processing. On extracting the special qualities, a process to compare them with the words stored in the dictionary is performed to attain the corresponding word and insert them into a text file before writing them. We call this process dequeue. A linear data structure or a type of abstract data is also definitions of queue, exactly like stack data structure, where we insert the first element from the end called the REAR (tail) removing an already existing element occurring from the other end which is called FRONT (head). This makes queue what is called first-in, first-out (FIFO) structure of data, which explicit that first inserted element will be cleared away first. The process of adding an element into queue is known as enqueue while, dequeue refers to the process of clearing away an element from queue.

5.4.1 Perform Queue for Spoken Words

This is an expression that refers to the audio files already extracted and arranged in the queue. Every single spoken word is finished in its due order in this line, and then every word in order is reached to the last one. Then, any set of words that are pronounced can be stored in the queue so as to fill the first word from the speaker until the conversation end.

5.4.2 Methodology of Queue

Life stacks is an abstract data structure. A queue is open at both ends. Enqueue is one end that is used for data insertion, while "Dequeue" is the other end used for data removal. First-in, first-out methodology is adopted by queue. FIFO order of data items—maintained by queue—is useful in producer–consumer situations, where a code portion is creating data that can be used by some other portion [37].

5.4.3 The Operations of Queue

Initializing the queue may be involved by queue operations and then can be utilized and after that removing it completely off the memory. The following are the main operations related to queues [38]—(1) Enqueue: Add (store) an item to the queue; (2) Dequeue: Remove (access) an item from the queue.

5.5 Extract Query Feature Vector of Queue

Here, in this stage, the features of every single word in the queue are extracted in the order of receipt. These features are stored in the queue word feature vector divided by the number of these words.

5.6 Pattern Matching

In this step, every single word in the queue word feature vector that is extracted will be compared with the (Swd). When there is a matching with a feature of a word, then it can be decoded and appended to the file of the text.

5.7 The Result of Transforming Sound to Text

Determining similar words are included in this stage, even (one word, binary and also three-words). Next phase, the matching words will be appended to the file of the text.

6 Extracting Knowledge

It is important to extract modern knowledge from speech; such knowledge can be used in the process of education. This stage explains how to extract kb from the text. The text resulted from recognition, processing, and printing of the converted spoken sound to a text file as follows: 1—Build a knowledge base containing a set of general and special rules. 2—Refine translation by comparing the text written in the rule set of knowledge rules. 3—Determine the number of terms (singular—binary—trilateral). 4—Terminology in its own domain. 5—Determine the number of words for each domain. 6—Determine the domain type of the research paper.

Some general rules: If word contain possessive pronouns such as: {كتابي – كتابه – كتابك – كتابهم –كتابكم – كتابكما – كتابهم – كتابهما}, then strip off word to ˋˋكتاب'. If one word is member of the set: {كتب – كتبت – كتبتما – كتبا – كتبوا – يكتب – تكتبين – تكتبي – اكتب – اكتبي – اكتبا – اكتبوا – اكتبن}, then strip off it to the verb ˋˋكتب'. If the WS_i or WD_j or WT_n includes the possessive pronouns such as: {له – لها – لهما – لهن – لي – لك – لكما – لنا – لكم - لكن}, then the repetition (p) = 0.

7 Experiment

The main difference between voice and speech recognition is how they are being used. Both of them are completely different from each other. Speech recognition aims to reach the spoken words; thus, programs of speech recognition remove all the individual idiosyncrasies like accents to identify words, whereas the purpose of voice recognition is to identify the individual speaking words instead of the words themselves. Voice recognition is also known as speaker recognition, as it ignores the words and helps to identify the speaker.

This study illustrates how the speech is recognized and converted into text, rather than recognizing the individual's voice and interpreting it. This research has completed its practical part using the Google Speech (API) application programming interface. As it is considered one of the first studies that use this interface in applications related to sound recognition, as mentioned in Abdel Hamid et al. [39], it supports modern standard Arabic language. Google developed cloud speech-to-text service application used to convert Arabic speech or audio file to text using a deep learning neural network algorithm. Cloud speech-to-text service allows its translator system to directly accept the spoken word to be converted to text and then translated. The service offers an API for developers with multiple recognition features. This application is composed of 2 steps:

The first step is about speech recognition, it works get a previous recorded sound in a wav file format, and this file should contain a spoken paragraph which will be converted to a written paragraph. This audio file can be acquired by choosing a file from pc or recording the audio file using the application. After that, the application reads the audio file and normalize it to reduce the variety in the frequency. When we are talking about the speech recognition, it is easier to recognize small files than big files. In order to do so, we need to split the big sound file into small files, each file can contain a word or two, and with this way the recognition process will be more accurate. The application splits the sound file by detecting the small frequencies in its wave; these small frequencies represent the silence between words. After splitting the sound file, we can recognize it using Google Speech API. The system GUI deals with the speech, by either calling it if it was previously stored or direct recording of the speaker to start the recognition task.

The second step is to classify the recognized text into its field, in order to do that we need a pre-trained model that can classify text. We used SANAD dataset to perform this step. To train the model we used random forest as a classifier. After training the model, we saved it and load it in the application so the recognized text can be classified easily. The application was conducted on a sample of 10 individuals, five females and five males, so that each individual recorded the audio file of the selected piece three times; each time, the recording was done in a different mood. Both correct and confusing words were extracted which have not been identified definitively and collecting data, and the achieved result is shown in Table 1a.

Table 1a shows the results of the "Mann–Whitney" test for the significance between the averages of the words number for both the males and females' groups in *word case* for *normal mod*, where the values of "Z" for all the word case have proven insignificant, which indicates that there is no difference between the two groups: males and females. The average value of the words number for the case *correct* for the males' group has reached (140.60) with percentage (96.30%), whereas the average of the words number for the females' group has reached (142.80) with percentage (97.81%). And the average of the words number for the case *confused* for the males' group has reached (3.40) with percentage (2.33%), whereas the average of the words number for the females' group has reached (2.60) with percentage (1.78%). And the average of the words number for the case *unknown* for the males' group has reached (2.0) with percentage (1.37%), whereas the average for the words number for the females' group has reached (0.60) with percentage (0.41%). Table 1b shows the results of the "Mann–Whitney" test for the significance between the averages of the words number for both the males and females' groups in *word case* for *quick mod*, where the values of "Z" for all the word case have proven insignificant, which indicates that there is no difference between the two groups: males and females. The average value of the words number for the case correct for the males' group has reached (134.60) with

Table 1 Significance of the differences between the averages of words for both the males and females' groups in all word cases

Word case	Sex	N	(A) Normal					(B) Quick					(C) Loud				
			Mean	SD	%	Mann–Whitney		Mean	SD	%	Mann–Whitney		Mean	SD	%	Mann–Whitney	
						Z	P-value				Z	P-value				Z	P-value
Correct	Male	5	140.6	3.65	**96.30**	1.17	0.242	134.6	3.44	92.19	0.95	0.343	138	4.85	94.52	0.53	0.595
	Female	5	142.8	2.28	**97.81**			136.8	1.79	93.70			140.2	1.48	96.03		
Confused	Male	5	3.4	1.67	**2.33**	0.64	0.519	6.4	1.95	4.38	0.22	0.827	4.2	1.64	2.88	0.56	0.572
	Female	5	2.6	1.14	**1.78**			6	0.71	4.11			3.6	0.89	2.47		
Unknown	Male	5	2	2	**1.37**	1.57	0.116	5	1.58	3.42	1.61	0.108	3.8	4.71	2.60	0.43	0.67
	Female	5	0.6	1.34	**0.41**			3.2	1.64	2.19			2.2	1.79	1.51		
Total	Male	5	146		100			146		100			146		100		
	Female	5	146		100			146		100			146		100		

percentage (92.19%), whereas the average of the words number for the females group has reached (136.80) with percentage (93.70%). And the average of the words number for the case *confused* for the males' group has reached (6.40) with percentage (4.38%), whereas the average of the words number for the females' group has reached (6.0) with percentage (4.11%). And the average of the words number for the case *unknown* for the males' group has reached (5.0) with percentage (3.42%), whereas the average for the words number for the females' group has reached (3.20) with percentage (2.19%). Table 1c shows the results of the "Mann–Whitney" test for the significance between the averages of the words number for both the males and females' groups in *word case* for *loud mod*, where the values of "Z" for all the word case have proven insignificant, which indicates that there is no difference between the two groups: males and females.

The average value of the words number for the case *correct* for the males' group has reached (138.0) with percentage (94.52%), whereas the average of the words number for the females' group has reached (142.29) with percentage (96.03%). And the average of the words number for the case *confused* for the males' group has reached (4.20) with percentage (2.88%), whereas the average of the words number for the females' group has reached (3.60) with percentage (2.47%). And the average of the words number for the case *unknown* for the males' group has reached (3.80) with percentage (2.60%), whereas the average for the words number for the females' group has reached (2.20) with percentage (1.51%). The proposed system showed acceptable performance in recognizing the words in different modes, as it correctly recognized 95.09%, 92.95%, and 95.27% of words in normal, quick, and loud mode, respectively. It also recognizes the words as "confused" with 2.05%, 4.25%, and 2.67% of words in normal, quick, and loud mode, respectively. In addition, it recognizes the words as unknown words with 0.89%, 2.81%, and 2.05% of words in normal, quick, and loud mode, respectively, as shown in Table 2.

The text classification part is performed in two steps as follows. First step is the preprocessing phase to generate features to help the classifier in the classification task. Second step is the evaluating phase of the results using a machine learning classifier. In the preprocessing, we first used count vectorization. Count vectorization is used to transform a given text into a vector on the basis of the frequency (count) of each word that occurs in the entire text. Then, we extracted TFIDF features. Whereas in the classification step we used the random forest classifier, we split the dataset into training and testing (70%: 30%); then, we manually compute

Table 2 Ratios of the word cases during the different modes

Word case	Speech mod			
	Normal (%)	Quick (%)	Loud (%)	Average (%)
Correct	**97.05**	92.95	95.27	95.09%
Confused	**2.05**	4.25	2.67	2.99
Unknown	**0.89**	2.81	2.05	1.92
Total	100	100	100	100

the results as in the following: accuracy: 96.1, sensitivity: 88.6, specificity: 99.5, and f-measure: 90.3. These results prove that the proposed system effectively converts Arabic speech to text and successfully classify them in different fields.

8 Conclusions and Future Works

Nowadays, Arabic speech recognition is considered as an important trend among the researchers. It is a useful way to activate electronic and technological to facilitate the educational burden. This is in line with what is also recommended by various scientific papers in activating the use of artificial intelligence in our various fields of life and education. This paper proposed an intelligent system for acquiring knowledge by converting Arabic speech into text, where the research presented to deal with the recognition of the readable voice in the Arabic language, activating the use of Google ABI application interface, and extracting knowledge by specifying the field to which the converted text belongs according to the specific sample. The paper's experiment relied on a dataset collected from different participants in different ages. The results showed good performances for the proposed system and promising future for Arabic speech recognition. In the future work, the proposed system may be applied and developed on different platforms such as mobile phones and tablets.

References

1. G. Hemakumar, P. Punitha, Speech recognition technology: a survey on Indian languages. Int. J. Inf. Sci. Intell. Syst. **2**(4), 1–38 (2013)
2. A. Ali, Y. Zhang, P. Cardinal, N. Dahak, S. Vogel, J. Glass, A complete kalbi recipe for building Arabic speech recognition systems, in *Spoken Language Technology Workshop (SLT)* (IEEE, 2014), pp. 525–529
3. A. Farghaly, K. Shaalan, Arabic natural language processing: challenges and solutions. ACM Trans. Asian Lang. Inf. Proces. (TALIP) **8**(4), 14 (2009)
4. E. Othman, K. Shaalan, A. Rafea, Towards resolving ambiguity in understanding Arabic sentence, in *International Conference on Arabic Language Resources and Tools*, (NEMLAR, 2004), pp. 118–122.
5. M.M. Gaheen, R.M. ElEraky, A.A. Ewees, Automated students Arabic essay scoring using trained neural network by e-jaya optimization to support personalized system of instruction. Educ. Inf. Technol. **26**(1), 1165–1181 (2021)
6. A.A. Bialy, M.A. Gaheen, R.M. ElEraky, A.F. ElGamal, A.A. Ewees, Single Arabic document summarization using natural language processing technique, in *Recent Advances in NLP: The Case of Arabic Language* (Springer, Cham, 2020), pp. 17–37
7. M.N. Arafa, R. Elbarougy, A.A. Ewees, G.M. Behery, A dataset for speech recognition to support Arabic phoneme pronunciation. Int. J. Image Graph. Signal Process. **11**(4), 31 (2018)
8. S.M. Elatawy, D.M. Hawa, A.A. Ewees, A.M. Saad, Recognition system for alphabet Arabic sign language using neutrosophic and fuzzy c-means. Educ. Inf. Technol. **25**, 5601–5616 (2020)

9. M. Menacer, O. Mella, D. Fohr, D. Jouvet, D. Langlois, K. Smaili, An enhanced automatic speech recognition system for Arabic, in *The third Arabic Natural Language Processing Workshop-EACL 2017* (2017)
10. H. Frihia, H. Bahi, Embedded learning segmentation approach for Arabic speech recognition, in *International Conference on Text, Speech, and Dialogue*. (Springer International Publishing, 2016), pp. 383–390
11. F.S. Al-Anzi, D. AbuZeina, The impact of phonological rules on Arabic speech recognition. Int. J. Speech Technol. **20**(3), 715–723 (2017)
12. A. Ramsay, I. Alsharhan, H. Ahmed, Generation of a phonetic transcription for modern standard Arabic: A knowledge-based model. Comput. Speech Lang. **28**(4), 959–978 (2014)
13. Wikipedia (Accessed date 15–5–2021). Knowledge extraction. https://en.wikipedia.org/wiki/Knowledge_extraction
14. J. Zhengbiao, Z. Feng, Z. Ming, An algorithm study for speech emotion recognition based speech feature analysis. Int. J. Multimedia Ubiquitous Eng. **10**(11), 33–42 (2015)
15. A. Ahmed, Y. Hifny, K. Shaalan, S. Toral, *End-to-End lexicon free Arabic speech recognition using recurrent neural networks*, in *Computational Linguistics, Speech and Image Processing for Arabic Language* (2019), pp. 231–248
16. E. Alsharhan, A. Ramsay, H. Ahmed, Evaluating the effect of using different transcription schemes in building a speech recognition system for Arabic. Int. J. Speech Technol. 1–14
17. S.A. Shaban et al., A novel advisory system for the psychological guidance of university students. Int. J. Comput. Sci. Trends Technol. (IJCST) **6**(3) (2018)
18. H.K. Palo et al., Emotion analysis from speech of different age groups, in *Proceedings of the Second International Conference on Research in Intelligent and Computing in Engineering*, vol. 10, 283–287 (2017)
19. S. Pathak, V. Kolhe, Emotion recognition from speech signals using deep learning methods. Imperial J. Interdiscip. Res. **2**(9), 19–24 (2016)
20. A. Shaw et al., Emotion recognition and classification in speech using artificial neural networks. Int. J. Comput. Appl. (0975 – 8887), **145**(8), 5–9 (2016)
21. J. Raj, S. Kumar, Gender based affection recognition of speech signals using spectral and prosodic feature extraction. Int. J. Eng. Res. General Sci. **3**(2), 898–905 (2015)
22. M. Kiefte, Formants in speech perception (2016). Retrieve from https://asa.scitation.org/doi/10.1121/1.4969927
23. J. Clark, C. Yallop, J. Fletcher, An introduction to phonetics and phonology, 3rd Ed. (Blackwell Publishers, Malden, Ma, USA, 2007)
24. A.A. Khulage, B.V. Pathak, Analysis of speech under stress using Linear techniques and Non-Linear techniques for emotion recognition system, in *International Conference of Advanced Computer Science & Information Technology*, (2012), pp.1–10
25. A. Firoz Shah, Study and analysis of speech emotion recognition, Unpublished thesis (Department of Information Technology, Kannur University, 2016)
26. M. Ahsan, M. Kumari, Physical features based speech emotion recognition using predictive classification. Int. J. Comput. Sci. Inf. Technol. (IJCSIT) **8**(2), 63–74 (2016)
27. S.S. Kumar, T. RangaBabu, Emotion and gender recognition of speech signals using SVM, Int. J. Eng. Sci. Innovative Technol. **4**(3), 128–137 (2015)
28. C. Prakash et al., Analysis of emotion recognition system through speech signal using KNN & GMM classifier. J. Electr. Commun. Eng. **10**(2), 55–61 (2015)
29. R.M. Sneha, K.L. Hemalatha, Implementation of MFCC extraction architecture and DTW technique in speech recognition system. Int. J. Emerg. Trends Sci. Technol. **3**(5), 753–757 (2016)
30. E. Ibrahim, A.A. Ewees, M. Eisa, Proposed method for segmenting skin lesions images, in: *Emerging Trends in Electrical, Communications, and Information Technologies. Lecture Notes in Electrical Engineering*, vol. 569. (Springer, 2020), pp. 13–23
31. G. Khairy, A.A. Ewees, M. Eisa, A proposed approach for Arabic semantic annotation, in *International Conference on Advanced Machine Learning Technologies and Applications*. (Springer, Cham, 2020), pp. 556–565

32. I. Trabelsi, D. Ben Ayed., On the use of different feature extraction methods for linear and non-Linear kernels, in *6th International Conference on Sciences of Electronics, Technologies of Information and Telecommunications* (Sousse, Tunisia, 2012), pp. 1–8 (21–24 Mar 2012)
33. Nitisha, A. Bansal, Speaker recognition using MFCC front end analysis and VQ modeling technique for Hindi words using MATLAB. Int. J. Comput. Appl. (0975–8887) **45**(24), 48–52 (2012)
34. N.M. Krishna et al., Emotion recognition using dynamic time warping technique for isolated words. Int. J. Comput. Sci. **8**(5, 1), 306–309 (2011)
35. E.S. Wahyuni, Arabic speech recognition using MFCC feature extraction and ANN classification, in *International Conferences on Information Technology, Information Systems and Electrical Engineering (ICITISEE)* (2017), p. 23
36. P. Sharma, A.K. Rajpoot, Automatic identification of silence, unvoiced and voiced chunks in speech. J. Comput. Sci. Inf. Technol. (CS & IT) **3**(5), 87–96 (2013)
37. T. point, Data structure and algorithms-queue, what is a queue data structure? Retrieved from https://www.tutorialspoint.com/data_structures_algorithms/dsa_queue.htm. Accessed date: 16 Apr 2021 (2018)
38. National Instruments, (Accessed date: 1 June 2021), https://knowledge.ni.com/KnowledgeArticleDetails?id=kA00Z000000P7OfSAK
39. Tutorials Point, Basic Operations (2018). Retrieved from https://www.tutorialspoint.com/data_structures_algorithms/dsa_queue.htm
40. A.A. Abdelhamid, H.A. Alsayadi, I. Hegazy, Z.T. Fayed, End-to-end Arabic speech recognition: a review, in *Conference: The 19th Conference of Language Engineering (ESOLEC'19)* (Alexandria, Egypt, At, 2020)

Towards a Deep Learning Approach for Automatic GUI Layout Generation

Xulu Yao, Moi Hoon Yap, and Yanlong Zhang

Abstract Various studies have been carried out over the past few years to solve the problem of automatically converting image models into source code. In 2018, pix2code has inspired and promoted the work in this domain. This paper presents a new model architecture to improve the framework of pix2code. We design a framework that can automatically generate a specific platform code for a given graphic user interface screenshot as an input. Although bilingual evaluation understudy (BLEU) is natural language processing metric, it has been adopted for source code evaluation. To overcome the limitation of BLEU in domain-specific language (DSL) tokens evaluation, we introduce a modified BLEU score (MBLEU). Our results show our proposed frameworks outperform the state-of-the-art methods in BLEU and MBLEU. We show MBLEU is suitable for DSL similarity evaluation, but further work is necessary to establish this new metric.

Keywords GUI · Deep learning · BLEU · DSL

1 Introduction

When people communicate with machines, the user interface (UI) is an indispensable tool [1]. Most user-oriented software applications rely on an attractive graphical user interface (GUI) to attract customers to facilitate the effective completion of a computing task [1]. When developing any GUI-based application, an important step is to draft and prototype design models, which help UI instantiate to evaluate or prove abstract design concepts. In large-scale industrial environments, this process

X. Yao (✉) · M. H. Yap · Y. Zhang
Manchester Metropolitan University, Manchester M1 5GD, UK
e-mail: xulu.yao@stu.mmu.ac.uk

M. H. Yap
e-mail: m.yap@mmu.ac.uk

Y. Zhang
e-mail: y.zhang@mmu.ac.uk

A. K. Bashir et al. (eds.), *Proceedings of International Conference on Computing and Communication Networks*, Lecture Notes in Networks and Systems 394,
https://doi.org/10.1007/978-981-19-0604-6_2

is usually accomplished by professional designers who have expertise in this field and can use image-editing software such as photoshop [2] or sketch [3] to generate attractive, intuitive GUIs. After these initial design drafts are created, it is important to faithfully translate them into code so that the end-user experiences the design and expected form of the user interface can be achieved (Fig. 1).

Previous works have shown that this process (usually involving multiple iterations) is challenging, time-consuming, and error-prone [4], especially if the design and implementation are performed by different teams (industries usually have all these settings [5]). In addition, UI teams usually adopt an iterative design process to collect feedback on GUI effectiveness at an early stage. It is best to use prototypes because more detailed feedback can be collected; however, using current practices and tools is often too expensive [6]. In addition, past work on detecting GUI design violations in mobile applications has emphasized the importance of this issue from an industrial perspective [5]. Instead of spending scarce time and resources on iterative design and user interface coding, it is better to choose an accurate automation method. This will enable smaller companies to focus more on features and values rather than turning design into operational application code. Given the setbacks faced by front-end developers and designers in building accurate GUIs, automation support is clearly needed.

To help ease this process, some modern IDEs, such as Xcode [7], Visual Studio [8], and Android Studio [9], provide built-in GUI editors. However, recent studies have shown that using these editors to create complex, high-performance GUIs are cumbersome and difficult because users are prone to introduce errors and demonstration failures, even with simple tasks [10]. Other business solutions include collaborative GUI design and interactive preview design on target devices or browsers (with limited functionality using custom frameworks), but none provides an end-to-end solution that automatically converts mockups. Obviously, a tool that can partially automate this process can significantly reduce the burden of the design and development process.

The aims of this paper are twofold: first is to design a framework that can automatically generate a specific platform code for a given GUI screenshot as an input; and second is to investigate into the performance metrics used in domain-specific language evaluation. We infer that the extended version of this method may potentially reduce the time for manual GUIs coding process.

Fig. 1 A web UI design workflow

2 Related Work

Models and prototypes are used to collect feedback at the beginning of the design process. They help improve visual design and are meant to be used by design teams as communication tools to focus on the user's appearance and solve layout problems for websites or applications [11]. The problem of automatically generating computer programs from a specification has been studied since the early days of artificial intelligence (AI) [12]; however, recent research has focused on the possibility of generating source code from design models to save developers from labour-intensive and repetitive parts of the design process. As a result, deep learning applications are being explored as a potential solution to this problem. Because computing power is the biggest obstacle to front-end development automation, the application of deep learning in the design model field has been initiated. User interface code development for applications is a cumbersome and expensive practice, and users expect mobile and computer user interfaces to be highly customized and optimized for specific tasks at hand [13]. A gap has been observed during production, and the conversion of user interface concepts to a working user interface code is done manually by programmers in a cumbersome, error-prone, and expensive manner [4].

Many of the approaches discussed so far have relied on domain-specific languages (DSL) (languages for specialized domains that are more restrictive than full-featured computer languages). The use of domain-specific languages limits the complexity of programming languages that need to be modelled and reduces the size of the space to be searched [14].

Beltramelli [14] describes an important development in this area, explaining how deep learning can transform screenshots of the GUI created by designers into computer code. Beltramelli achieves 77% accuracy on three different platforms (iOS, Android, and Web-based technologies) by using deep learning to train the model's code and automatically generate a single image end to end. The authors believe that this is the first attempt to solve the problem of generating GUI code from visual input by using machine learning to understand potential variables rather than complex problem-solving engineering [14]. The paper further states that UI components are synthetically generated, but the author does not offer a way to generate DSL code.

Another related work is a project developed by Wallner [15], which is another Keras-based implementation of pix2code, using the same dataset. It differentiates itself from pix2code by replacing the pre-trained image features with a light CNN. Instead of using max-pooling to increase information density, it increases the strides. Lee [16] also made improvements on the basis of pix2code. Unlike the single end-to-end pix2code model, their system follows an image-captioning model previously created for PyTorch, with an encoder CNN and a decoder RNN. As a whole, the system takes in a screenshot as input and outputs a sequence of indices (based on DSL language's vocabulary), which are then converted into valid HTML.

3 Approach

3.1 Datasets

Beltramelli's pix2code dataset [14] contains 1750 screenshots of synthetically generated websites and their associated source code, which are used as the features for training. Because each site generated in the dataset comprises only a few simple bootstrap elements (such as buttons, text boxes, and divs), the "vocabulary" of the model is limited to these features. However, this approach could be generalized to a larger vocabulary by simply increasing the number of elements. The DSL file is compiled with reference to the code in the JSON format. The source code for each example comprises tokens in a DSL file, with each token corresponding to a piece of HTML code, as illustrated in Fig. 2. The compiler is used to convert the DSL file into working HTML code.

Since pix2code is focused on GUI layouts, graphical components, and their relationships, the actual text value of the tag is ignored, and the resulting text portion is replaced with a specified number of random letters. In this case, we used DSL tokens from the pix2code dataset as input features to improve the training efficiency of the pix2code model and the accuracy of the output results.

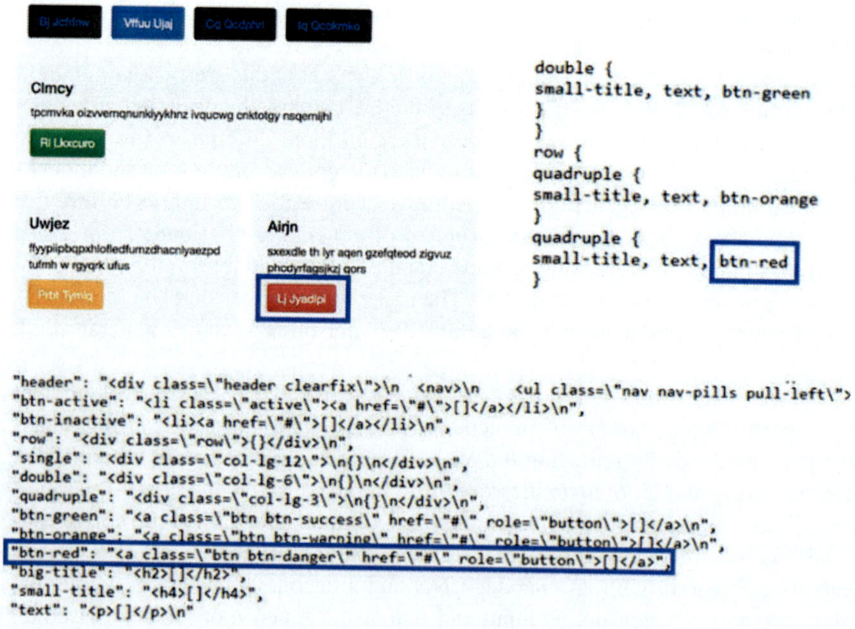

Fig. 2 The web DSL token mapping from pix2code

3.2 Model Architecture

As the framework of pix2code is based on Vinyals's image captioning model [17], the first input of LSTM comes from the feature vector extracted by CNN. The feature vector of the image is the first input of the LSTM, and its information is captured in the hidden state of the LSTM. This can cause some of the information in the feature vector to be discarded as the length of the caption increases, thereby affecting the overall performance of the model [18].

To solve this issue, we redesigned the framework of pix2code's model (Fig. 3). The LSTMs is replaced by gated recurrent unit (GRU) [19] to improve the training rate, and the CNN's feature vector connects to the GRU's input as an embedded input to the GRU. Thus, in theory, the model can capture all the information of the feature vector during the DSL token generation process.

Referring to the pix2code dataset, we preprocess the data by resizing the input image to 256 × 256 pixels without retaining its aspect ratio and then normalized the pixel values. We used VGG16, VGG19 [20], and ResNet34 [21] as the encoders for our experiment. We adjust the embedded size of GRU to 50, with three layers and 256 hidden units as a decoder.

3.3 Training and Sampling

When training the model, we divided an input into an image and its DSL token sequence, the label of which is the next token in the DSL file. The model compares the next token prediction of the model with the actual next token prediction using a cross-entropy loss function.

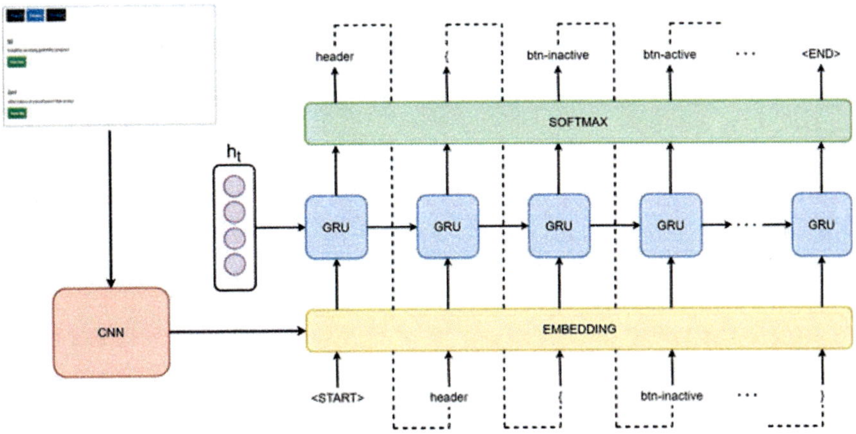

Fig. 3 Our proposed framework based on pix2code [14]

During sampling, the image is still processed through the CNN network, but text processing is only a seed of the starting sequence. In each step, the model's prediction of the next token in the sequence is appended to the current input sequence and entered into the model as a new input sequence. This process is repeated, beginning with the <START> token, until the model predicts an <END> token or the process reaches a predefined limit on the number of tokens in each DSL file. After the model generates the predicted token, the compiler converts the DSL token into HTML code that can be rendered in a web browser.

4 Experimental Results

In the automatic translation evaluation, BLEU [22] is an algorithm that must be mentioned. The impact of BLEU is epoch-making, although there have been similar views before. The basic assumption of BLEU is that if there are more N-grams to be co-produced with the reference translation, the more similar the description, the higher the quality of the translation. By counting the number of co-occurring N-grams and adding a penalty factor to short sentences, the translation of the same topic can be evaluated through a reference translation. Since the field of code metrics has emerged recently, there is no corresponding method to measure the accuracy of DSL at this stage. Therefore, we designed a modified BLEU score (MBLEU) to evaluate the model. The formula to calculate the MBLEU score is following:

$$\mathrm{MBLEU} = \mathrm{BP} \cdot \exp\left(\sum_{n=1}^{N}(w_n \log P_n)\right) \tag{1}$$

where BP is a penalty factor for translations whose length is less than the reference value:

$$\mathrm{BP} = \begin{cases} \mathrm{e}^{1-c/r}, & \text{if } c > r \\ \mathrm{e}^{1-r/c}, & \text{if } c \le r \end{cases} \tag{2}$$

and P_n is the n-gram matching rate.

Take a sequence of tokens in a DSL file as an example (Fig. 4). MBLEU divides the sentence from one to four token sequences into four n-grams. In the prediction below, “btn-orange” is a false prediction, and the actual correct token should be “btn-green”. When $n = 4$ (4-gram), w_n will be 1/4 or 0.25. Then the score of MBLEU will be $(11/12) * 0.25 + (9/10) * 0.25 + (4/6) * 0.25 + (4/5) * 0.25 = 0.22 + 0.22 + 0.16 + 0.2 = 0.8$.

The sum also needs to be multiplied by the penalty BP of a sentence length. In the example of 3-gram, the token length is outside the measurement range, so the BP at this time equals 1, and the result of the above becomes our final score. If MBLEU gets a score of 1.0, there will be correct elements in the correct position of

```
1-gram: <START> @ header @ { @ btn-inactive @ } @ row @ { @ double @ { @ btn-orange @ } @ <END> (11/12)
2-gram: <START> header @ { btn-inactive @ } row @ { double @ { btn-orange @ } double @ { text @ } double @ { } @ } <END> (9/10)
3-gram: <START> header { @ btn-inactive } row @ { double { @ btn-orange } double @ { btn-orange } @ }  <END> (4/6)
4-gram: <START> header { btn-inactive @ } row { double @ { btn-orange } double @ { text } double @ { } } <END> (4/5)
```

Fig. 4 An example of MBLEU score in DSL tokens

Table 1 Evaluation results from the metrics of pix2code dataset

Methods	CNN types	BLEU-1	BLEU-2	BLEU-3	BLEU-4	MBLEU
Pix2code model	VGG16	0.527	0.452	0.397	0.322	0.79
Model-A	VGG16	0.535	0.463	0.404	0.337	0.82
Model-B	VGG19	0.556	0.485	0.417	0.346	0.88
Model-C	ResNet34	**0.577**	**0.502**	**0.452**	**0.376**	**0.93**

Bold font indicates that Model-C with ResNet34 has the best score on BLEU and MBLEU

the given source image. The lower the score, the greater the difference between the generated DSL token sequence and the real result, and the decoded HTML code will be different from the input sketch.

We compared the experimental scores of our model on MBLEU and BLEU-N [22] with the model from pix2code [14] in three different CNN types (Model-A: VGG16, Model-B: VGG19, Model-C: ResNet34). Each score on methods ranges from 0 to 1, and a higher value gains a more accurate result of GUI code generation (Table 1). The results of evaluation metrics may change with further experiments in future.

5 Discussion

Through the experimental results for the pix2code dataset, we obtained a higher score when compared with pix2code model in the same CNN type (Model-A in Table 1). It can also be seen that our models delivered the best performances on ResNet34 when compared with the other two CNN types (VGG16, VGG19).

At present, the model still has some limitations, which illustrate the possible follow-up steps:

- Due to the limitations of the existing pix2code dataset, the model only trains a vocabulary of 16 elements, so it can only predict the DSL token specified in the data.
- Because CSS lacks style changes when making sketch parts of existing datasets, there is still a big difference compared to human hand-drawn sketches.
- There are some shortcomings in the NLP evaluation metrics. The existing methods lack the ability to judge the dependency relationship between different DSL tokens and its importance to the whole web page. It is necessary to improve the penalty factor or find a more suitable alternative.

6 Conclusion

Existing work in the area of automatic GUI generation is still in the early stages of development; models like Beltramelli [14] have so far contained only a few parameters and have been trained on small datasets. There is a further scope to focus on the more limited areas of a Web-based GUI that do not require data synthesis. Because a large number of websites are already available online, and because new websites are created every day, this situation provides almost unlimited training data to extend deep learning methods and transform Web-based design models into HTML/CSS code [14]. In this paper, we proposed a modified framework for solving the problem of feature vector losses in pix2code framework to a certain extent. The preliminary experimental results outperformed the state-of-the-art methods based on BLEU and MBLEU. We demonstrate MBLEU is suitable for DSL evaluation but it is inconclusive due to lack of datasets for further evaluation.

The next step could be trying to create more elements to generate additional web examples such as actual texts, images, drop-down menus, forms, and bootstrap components. With the increasing performance of computer hardware, it is better to create a dataset that can be directly trained by HTML/CSS code than a DSL token sequence in future. A good way to generate more variants in hand-drawn sketch data might be to create a realistic hand-drawn website image using a generative adversarial network (GAN).

References

1. J.A.T. Hackos, J. Redish, *User and Task Analysis for Interface Design* (Wiley, New York, 1998)
2. Adobe Inc., Adobe photoshop. https://www.adobe.com/Adobe/Photoshop/ (2019)
3. Bohemian Coding, Sketch. https://www.sketch.com/ (2019)
4. T.A. Nguyen, C. Csallner, Reverse engineering mobile application user interfaces with remaui (t), in *2015 30th IEEE/ACM International Conference on Automated Software Engineering (ASE)* (IEEE, 2015), pp. 248–259
5. K. Moran, B. Li, C. Bernal-Cárdenas, D. Jelf, D. Poshyvanyk, Automated reporting of GUI design violations for mobile apps, in *Proceedings of the 40th International Conference on Software Engineering* (ACM, 2018), pp. 165–175
6. B. Myers, S.Y. Park, Y. Nakano, G. Mueller, A. Ko, How designers design and program interactive behaviors, in *2008 IEEE Symposium on Visual Languages and Human-Centric Computing* (IEEE, 2008), pp. 177–184
7. Apple Inc., Xcode. https://developer.apple.com/xcode/ (2019)
8. Microsoft, Microsoft visual studio. https://visualstudio.microsoft.com/ (2019)
9. Google, Android studio. https://developer.android.com/studio/index.html/ (2019)
10. C. Zeidler, C. Lutteroth, W. Stuerzlinger, G. Weber, Evaluating direct manipulation operations for constraint-based layout, in *IFIP Conference on Human-Computer Interaction* (Springer, 2013), pp. 513–529
11. T. Brinck, D. Gergle, S.D. Wood, *Designing Web Sites That Work: Usability for the Web* (Morgan Kaufmann Publishers, 2002)
12. J. Devlin, J. Uesato, S. Bhupatiraju, R. Singh, A.-R. Mohamed, P. Kohli, Robustfill: neural program learning under noisy i/o. arXiv preprint arXiv:1703.07469 (2017)

13. G. Nudelman, *Android Design Patterns: Interaction Design Solutions for Developers* (Wiley, 2013)
14. T. Beltramelli, pix2code: generating code from a graphical user interface screenshot, in *Proceedings of the ACM SIGCHI Symposium on Engineering Interactive Computing Systems* (ACM, 2018), p. 3
15. Screenshot to code, https://github.com/emilwallner/Screenshot-to-code/blob/master/README.md. Accessed 22 Jan 2019
16. A. Lee, *Generating Webpages from Screenshots* (Stanford University, 2018)
17. O. Vinyals, A. Toshev, S. Bengio, D. Erhan, Show and tell: a neural image caption generator, in *Proceedings of the IEEE Conference on Computer Vision and Pattern Recognition* (2015), pp. 3156–3164
18. R. Jozefowicz, O. Vinyals, M. Schuster, N. Shazeer, Y. Wu, Exploring the limits of language modeling. arXiv preprint arXiv:1602.02410 (2016)
19. K. Cho, B. Van Merriënboer, C. Gulcehre, D. Bahdanau, F. Bougares, H. Schwenk, Y. Bengio, Learning phrase representations using RNN encoder-decoder for statistical machine translation. arXiv preprint arXiv:1406.1078 (2014)
20. K. Simonyan, A. Zisserman, Very deep convolutional networks for large-scale image recognition. arXiv preprint arXiv:1409.1556 (2014)
21. K. He, X. Zhang, S. Ren, J. Sun, Deep residual learning for image recognition, in *Proceedings of the IEEE Conference on Computer Vision and Pattern Recognition* (2016), pp. 770–778
22. K. Papineni, S. Roukos, T. Ward, W.-J. Zhu, Bleu: a method for automatic evaluation of machine translation, in *Proceedings of the 40th Annual Meeting on Association for Computational Linguistics* (Association for Computational Linguistics, 2002), pp. 311–318

Insurance Sales Forecast Using Machine Learning Algorithms

Zuhal Kurt, Emrecan Varyok, Ege Baran Ayhan, Mehmet Turhan Bilgin, and Duygu Duru

Abstract Car accidents and the possible resulting loss of assets or life are issues for every car owner that must contend with some point in their driving life. Driving is an inherently dangerous act, even if it does not seem so at first, resulting in greater than 33,000 fatal vehi le crashes in USA in 2019 alone. However, the loss of life and possible damages can be reduced with the help of insurances. Insurance is an arrangement under which a person or agency receives financial security or reimbursement from an insurance provider in the form of a policy. Insurances help limit the losses of the customers when an undesirable event occurs, such as a car crash or a heart attack. Vehicle insurance provides customers monetary compensation after unfortunate accidents, provided they annually pay premium fees to the companies first. Our goal is to develop a machine learning algorithm that predicts customers who are interested in getting or renewing their vehicle insurance with the help of personal, vehicle, contact, and previous insurance data. The insurance sales forecast is helpful to companies, since they can then accordingly plan its communication strategy to reach out to those customers and optimize its business model and revenue, while also being beneficial to customers, who can go through the process and the aftermath of car accidents easier thanks to their monetary compensation. In this paper, the Health Insurance Cross-Sell Prediction dataset is used. The proposed model tries getting the value by training itself on a train and test dataset and will result in a categorical response feature based on the aforementioned data with the aid of well-known machine learning algorithms: k-nearest neighbors, random forest, support vector machines, Naive Bayes, and logistic regression.

Keywords Insurance prediction · Data analysis · Machine learning algorithm

Z. Kurt (✉) · M. T. Bilgin
Department of Computer Engineering, Atılım University, Ankara, Turkey
e-mail: zuhal.kurt@atilim.edu.tr

E. Varyok · E. B. Ayhan
Department of Automotive Engineering, Atılım University, Ankara, Turkey

D. Duru
Department of Chemical Engineering, Atılım University, Ankara, Turkey

A. K. Bashir et al. (eds.), *Proceedings of International Conference on Computing and Communication Networks*, Lecture Notes in Networks and Systems 394,
https://doi.org/10.1007/978-981-19-0604-6_3

1 Introduction

The objective of this paper is to classify customers based on their probability and desire to buy insurance based on their personal information, personal preferences, and the data of their owned cars. The machine learning (ML) algorithm tried predicting the likelihood of getting a positive or negative response of the customers getting an insurance by learning from labeled or tagged data and ending up with a classified response feature. This task needs multiple definitions to be fully understood. The first definition of this task comes with the options of classification or regression. Since the insurance prediction algorithm should classify the response and should categorize classes, the output class of this algorithm response is not considered continuous, so the task needs to be defined as a classification problem [1, 2].

The second decision of the insurance prediction algorithm is the selection of a supervised or unsupervised algorithm. All the data in this paper are labeled, and the data are separated as a train and test dataset, with the training dataset providing the 'response' feature as a variable. The model then makes predictions on the test class by assigning the categorical 'response' feature to the given dataset. Since supervised tasks predict classes, unsupervised tasks predict groups, and our task is a classification problem, all of these senses push our proposed algorithm to be a supervised learning algorithm [1, 2]. Finally, our model response feature is outputted as either a '0' or a '1'. So, all in all, with the previously discussed topics, our task can be defined as a binary classification problem with supervised learning.

The dataset used for this paper is, namely Health Insurance Cross-Sell Prediction, and gathered from the Kaggle website [3, 4]. The dataset contains three different *.csv files named as sample_submission, test, and train. The train data have 12 features and 381,109 records, the test data have 11 features and 127,038 records, and finally, the sample submission data have the same number of records with the same data, except it includes only the I.D. and 'response feature.' This final dataset is used to figure out if the final predictions are correct since 'response' is the target feature of this proposed model. There are no missing or mismatched values in this dataset. The training dataset is used and modified for this paper since during this study an uneven match of response features has been observed, so the train data were reduced to 165,582 records in order to even up the response rates and end up with a more efficient algorithm, which has succeeded.

With this goal, this paper presents an insurance prediction algorithm. Initially, we give a simple overview of the proposed machine learning algorithm, and then we conducted this model by using the Health Insurance Cross-Sell Prediction dataset to explain how this model can be used in practice. The remainder of the paper is coordinated as follows: The detailed representation of the proposed prediction model is explained in Sect. 2. The evaluation measurements that are used in this study are given in Sect. 3. The application of experiments on a real-world dataset and discussion of the experimental results are included in Sect. 4. Finally, the conclusion of this study is summarized.

2 Motivation—ML Algorithms

The main aim while working with the Health Insurance Cross-Sell Prediction dataset is to make the necessary classifications according to gathered information about customers and their vehicles to predict their response. Firstly, the proposed model is trained, and then it is tested based on a test dataset that has not been 'seen' from the model. Therefore, our task can be considered a supervised binary classification, the following ML algorithms are deemed the most appropriate, and they are k-nearest neighbors (k-NN), Naive Bayes, random forest (RF), logistic regression, and support vector machines (SVM) [5, 6].

2.1 Insurance Prediction Model

This paper predicted the likelihood of customers' interest in getting or renewing car insurance based on personal and vehicle data. To make this proposed model as efficient, a binary classification ML algorithm is considered as the best, the original data are preprocessed, appropriate features are selected, and the sample size is reduced to get the best prediction result. The best ML algorithms for this task can be considered such as k-NN, Naive Bayes, logistic regression, random forest, and SVM.

After the evaluations, the random forest algorithm is deemed the most successful, with 97% accuracy on the training dataset, and 91% results on the previously untested test dataset. This result shows a slight overfit, which the proposed model cannot solve completely, yet it can be able to reduce significantly compared to the original results thanks to preprocessing. For the most basic explanation, a random forest is ensembles of a group of decision trees. Random forest is a supervised learning method that outputs good results with various regression and classification problems.

The random forest algorithm also compensates for the common drawback of decision trees. The main problem of decision trees is that they tend to overfit on test on training data. In the random forest method, since every tree is randomized, their overfit condition also gets randomized. This randomization reduces overfit by averaging the results of decision trees in the forest.

Like some algorithms before, random forest works well with large and low-dimensional datasets. The random forest also fixes the overfitting problem that plagues decision trees and does not require the rescaling of data, unlike some other algorithms. The disadvantage of this algorithm is that it is a slow worker, needing much time to predict and train the data; however, our personal computers have high-end RAMs and CPUs, we have decided to use them anyway. The random forest test scheme inside the Orange3 environment is shown in Fig. 1.

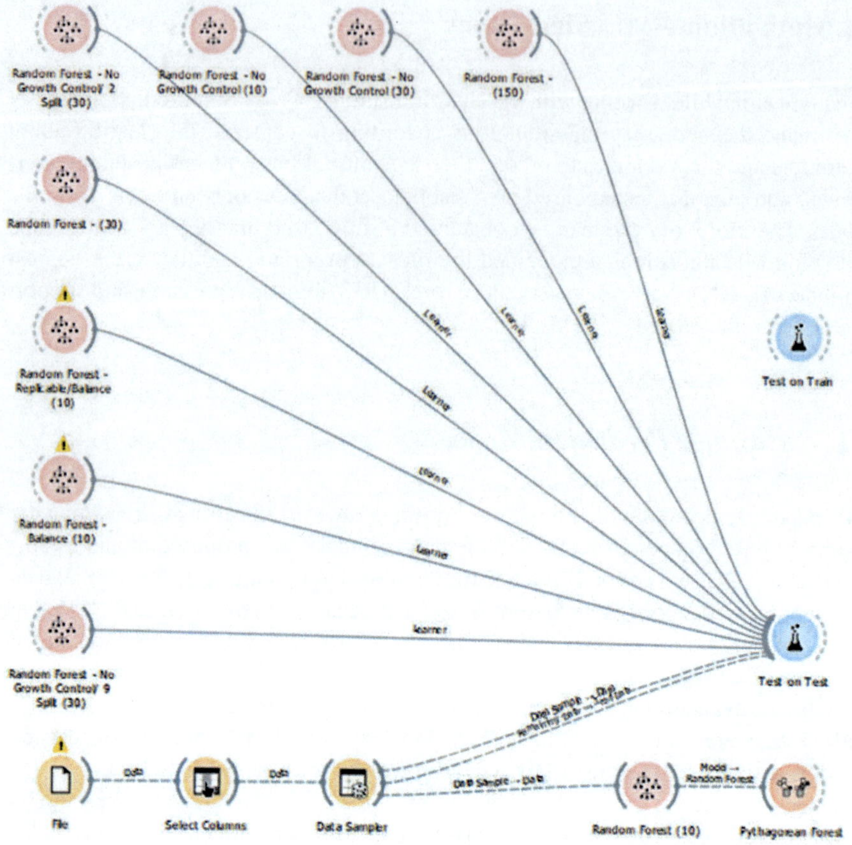

Fig. 1 Random forest test scheme inside Orange3 environment

2.2 *Random Forest Parameters*

In the Orange3, widget for the random forest algorithm has few options to modify (See in Fig. 2). Besides main parameters like the number of trees and the number of attributes, controlling the growth of every tree is possible.

The main parameter that changes results dramatically is the number of trees. The higher number of trees is tended to have better accuracies. However, the random forest is quite a source-hungry learning method. In about 150 trees, results were converged in a stable accuracy. Hence, the number of trees is set up to 150. Then, we changed other parameters around it to get different outputs. Initial results with different parameters show a ~10% difference between test and train accuracies. We tried to reduce overfitting by changing parameters, but it did not close the gap too much. Initial results were not good as assumptions; even 85.6% is not a terrible accuracy. The main problem is overfitting. Therefore, a simple preprocessing method is applied to the dataset, overfitting reduced as 5%. With such a little manipulation on

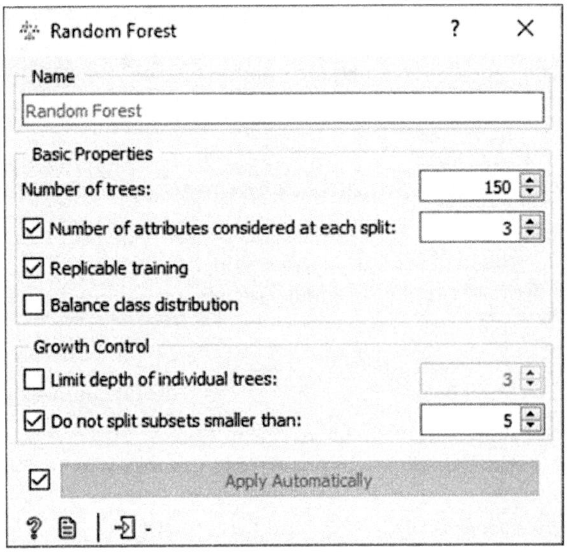

Fig. 2 Parameter screen of random forest in Orange3 environment

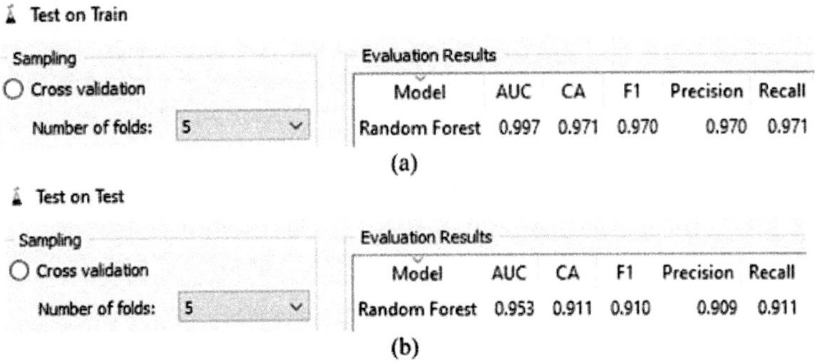

Fig. 3 **a** Test on train data results and **b** Test on test data results

the dataset, getting such improvement is satisfying for this study. After the preprocessing method is applied, the test on train data and test on test data results are shown in Fig. 3a, b.

3 Evaluation Measurements

3.1 Area Under Curve (AUC)

The area under the curve (AUC) is a performance metric for binary classifiers. By comparing the ROC curves with the AUC, it captures the extent to which the curve is up in the northwest corner. A higher AUC is expected. A score of 0.5 is no better than random guessing, or a score of 0.9 can be considered a very good result, but a score of 0.9999 can be too good to be true and will indicate overfitting.

3.2 *Classification Accuracy (CA)*

Classification accuracy is represented as the ratio of the number of correct predictions to the total number of input samples.

$$\text{Classification Accuracy} = \frac{(\text{Number of Correct predictions})}{(\text{Total number of predictions made})} \tag{1}$$

3.3 *Precision*

Precision can be defined as the number of correct positive results divided by the number of positive results predicted by the classifier.

$$\text{Precision} = [\text{True Positives}/(\text{True Positives} + \text{False Positives})] \tag{3}$$

3.4 *Recall*

It can be defined as the ratio of the number of correct positive results to the number of relevant samples (all samples that should have been described as positive).

$$\text{Recall} = \left[\text{True Positives}/(\text{True Positives} + \text{False Negatives})\right] \tag{4}$$

3.5 *F1-Score*

F1-score can be referred as the harmonic mean between precision and recall. The range for F1-score is [0, 1]. This metric informs you how precise your classifier is (how many instances it classifies correctly), as well as how robust it is (it does not miss a significant number of instances).

High precision but lower recall gives you an extremely accurate, but it then misses many instances that are difficult to classify. Hence, the better performance of any model can be expressed by the greater value of the F1-score. It can be mathematically represented as follows:

$$F1 = 2 * \left[1/(1/\text{precision}) + (1/\text{recall})\right] \quad (2)$$

4 Experimental Results and Datasets

Because of the research on related works, multiple similar types of research on car insurance prediction were found, two of which will be included here. For comparison, the results for the efficiency of our algorithms can be seen in Table 1. Effects of important features for insurance auto-renewal with classification ML algorithms are represented in [7]. In this research, the most successful models are random forest, gradient-lifting tree (GBDT), and lifting machine algorithm (LightGBM), with LightGBM producing the best result at 0.8045 AUC. However, our proposed method instead opted to use different methods, and due to the quality of our dataset, no missing data, independent features, and no necessary feature generation helped the proposed model to come out with better results with logistic regression. Our proposed model is resulting in a 0.924 AUC and is shown in Fig. 4. The data features of the dataset are mainly independent; hence, logistic regression works best with these types of data (See in Fig. 4). The experimental results of the proposed Health Insurance Cross-Sell Prediction model are given in Table 1.

Another related work, which used binary classification algorithms on a dataset gathered from a Brazilian car insurance company, tries to enhance the sales of a car insurance customer service [8]. Many methods similar to our approach used in this research, the eight common ML models are used, showed the best results using the random forest algorithms. The results of this research can be seen in Table 2. This research was also plagued with similar problems we have encountered during our study, with the binary classification algorithm incorrectly classifying 1 values as 0 due to noise and imbalanced data, with their final accuracy of the 1 values being 71%. The auto insurance model experimental results are given in Table 2. We have fixed this problem by removing 0-leaning data and reducing some unnecessary samples.

Table 1 Proposed Health Insurance Cross-Sell Prediction model performance

Measures (%)	Recall	Accuracy	Precision	F1-score	AUC
Logistic regression	0.919	0.919	0.916	0.917	0.958
Random forest	**0.920**	**0.920**	**0.917**	**0.918**	**0.961**
SVM	0.801	0.801	0.665	0.715	0.500
k-NN	0.919	0.919	0.916	0.917	0.955
Naïve Bayes	0.875	0.875	0.901	0.822	0.944

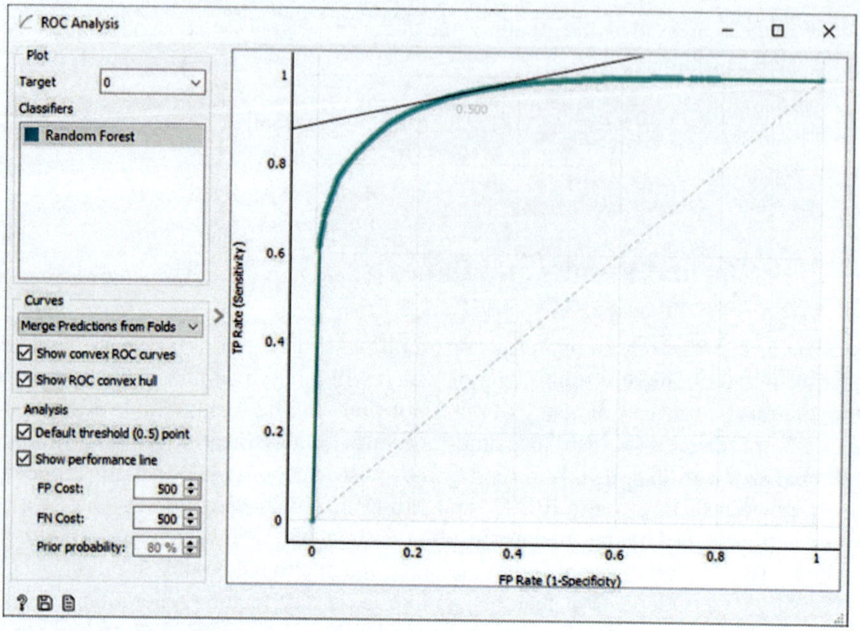

Fig. 4 Area under curve graph of random forest

Table 2 Auto insurance model performances [8]

Model	Accuracy	Precision	Recall	F1-score	AUC
RF	0.8677	0.9429	0.71	0.8101	0.84
C50	0.7913	0.7717	0.6743	0.7197	0.769
XGBoost	0.7067	0.6777	0.4994	0.575	0.671
J48	0.6994	0.6174	0.6399	0.6284	0.689
k-NN	0.6629	0.6167	0.4003	0.4855	0.628
LR	0.6192	0.55	0.2296	0.3239	0.615
Caret	0.6148	0.5601	0.1422	0.2268	0.534
Naïve Bayes	0.6056	0.6558	0.7273	0.6897	0.574

The performance is successful, with one of our best-proposed algorithms correctly classifying 85% of our 1 valued data, thanks to feature and sample reduction.

To obtain the general result of our proposed model, we also used tenfold cross-validation technique. The experimental results of the proposed model with tenfold cross-validation are shown in Fig. 5.

Fig. 5 Experimental results of the proposed model with ten-fold cross-validation

Evaluation Results

Model	AUC	CA	F1	Precision	Recall
kNN-25	0.955	0.919	0.917	0.916	0.919
SVM c=10 g=10	0.500	0.801	0.715	0.665	0.801
Random forest-150	0.961	0.920	0.918	0.917	0.920
Naive Bayes	0.944	0.875	0.882	0.901	0.875
Logistic Regression 11	0.958	0.919	0.917	0.916	0.919

5 Conclusion

This paper aims to predict the likelihood of customers' interest in getting or renewing car insurance based on personal and vehicle data. It is a binary classification task, and we try finding a ML algorithm to solve this task. Furthermore, the original data are preprocessed, the important features are selected, and the sample size is reduced to get the best prediction result, in this paper. The most useful evaluation metrics are thought to be the confusion matrices on each result, as well the classification results on each model, and so these metrics are used. Overall, the algorithms worked significantly well on guessing the negative responses with almost all algorithms' final 0 response prediction rates being above 90%, yet the main shortcomings of the selected methods and algorithms were guessing the 1 response feature since the original data are heavily skewed toward the 0 responses. Even though that problem is eventually fixed with sample size reductions, the algorithms still have trouble with guessing 1 response rates. This problem can be solved with possible feature additions; however, any additional features can be created during this study, which is another shortcoming in and of itself.

The experimental result shows that the proposed model is solving this task efficiently. Furthermore, the experimental results demonstrate the superior performance of our proposed model by using five quality measurements: CA, recall, precision, AUC, and F1-score on the Health Insurance Cross-Sell Prediction dataset, as well as its flexibility to incorporate different information sources. To start with, from the companies' perspective, this study can lead to optimal insurance pricing, increasing company profits, and insurance customers. From the customer's perspective, since they are now more likely to be insured, they can have a monetary compensation in cases of vehicle accidents, resulting in better time and money management since insurance companies can help with the aftermath of the accident process. Finally, the drivers around the world can be safer around the roads since higher insurance having driver percentages can lead to safer roads and less reckless driving thanks to the drivers now being more careful because of their insurance driving guidelines, making the affected driver-used roads of this research safer in the long term.

References

1. S. Rawat, A. Rawat, D. Kumar, A.S. Sabitha, Application of machine learning and data visualization techniques for decision support in the insurance sector. Int. J. Inf. Manage. Data Insights **1**(2) (2021)
2. D.R. Gopagoni, P.V. Lakshmi, P. Siripurapu, Predicting the sales conversion rate of car insurance promotional calls, in *Rising Threats in Expert Applications and Solutions. Advances in Intelligent Systems and Computing*, vol. 1187, eds by V.S. Rathore, N. Dey, V. Piuri, R. Babo, Z. Polkowski, J.M.R.S. Tavares (Springer, Singapore, 2021)
3. IIHS HLDI, Fatality facts 2019 state by state: https://www.iihs.org/topics/fatality-statistics/detail/state-by-state#:~:text=There%20were%2033%2C244%20fatal%20motor,Columbia%20to%2025.4%20in%20Wyoming, Last accessed 31 June 2021 (2021)
4. K. Anmol, Health ınsurance cross sell prediction: https://www.kaggle.com/anmolkumar/health-insurance-cross-sell-prediction, Last accessed 31 June 2021 (2020)
5. B.D. Sommers, A.A. Gawande, K. Baicker, Health insurance coverage and health—what the recent evidence tells us. N. Engl. J. Med. **377**(6), 586–593 (2017)
6. Tutorials Point, KNN Algorithm—Finding nearest neighbors: https://www.tutorialspoint.com/machine_learning_with_python/machine_learning_with_python_knn_algorithm_finding_nearest_neighbors.htm, Last accessed 31 June 2021
7. W.H. Dong, Research on the features of car ınsurance data based on machine learning. Procedia Comput. Sci. **166** (2020)
8. M. Hanafy, R. Ming, Machine learning approaches for auto insurance big data. Risks **9**(2), 42 (2021)

Ensemble Learning with CNN–LSTM Combination for Speech Emotion Recognition

Senem Tanberk and Dilek Bilgin Tükel

Abstract Speech plays the most significant role in communication between people. The voice enables a speaker's unique characteristics to be mapped with biometric properties as well as carrying emotions. Emotion contains many non-linguistic signals to express ourselves as humans. Emotion recognition in human speech is a challenging task in different applications in fields such as healthcare, services, telecommunications, video conferencing, and human–computer interaction (HCI). Deep learning techniques are becoming a significant focus in recent research in the speech emotion recognition (SER) domain. In this paper, we present an ensemble learning approach based on various combinations of CNN and LSTM networks to address the limitations of the existing SER models. The proposed system is evaluated using the RAVDESS dataset. More specifically, the LSTM, CNN, and CNN and LSTM models achieved an accuracy rate of 0.64, 0.73, and 0.71, respectively. The simulation outcomes confirm that ensemble learning of the three deep model combinations contributes to the effectiveness of SER.

Keywords Speech emotion recognition · Deep learning · Convolutional neural network · Long short-term memory · Ensemble learning

1 Introduction

Communication is the act of transferring information, thoughts, intentions, or meanings from one mind or one machine to another. There are different categories of communication between one person to another, including verbal, nonverbal, and written forms. In other words, "wanting to tell" a certain thing also includes the act of conveying a propositional attitude (sadness, fear, etc.) to a listener by linguistic

S. Tanberk (✉)
Orion Innovation Turkey, Istanbul, Turkey

D. B. Tükel
Doğuş University, Istanbul, Turkey

A. K. Bashir et al. (eds.), *Proceedings of International Conference on Computing and Communication Networks*, Lecture Notes in Networks and Systems 394,
https://doi.org/10.1007/978-981-19-0604-6_5

or other means. To achieve natural human–machine interfaces the machine should have sufficient intelligence to recognize human voices and emotions.

Human psychology is the combination of cognition, emotion, and behavior. Emotion is a process that consists of many physiological changes and our evolutionary and personal past. During the 1970s, psychologist Paul Eckman identified six universal basic emotions that were experienced in all human cultures. The emotions he identified were happiness, sadness, disgust, fear, surprise, and anger. He later expanded his list of basic emotions to include such things as pride, shame, embarrassment, and excitement. Each emotion has unique signals, the most identifiable being in the face and the voice.

Since the mid-1970s, automatic speech recognition has improved considerably [1], and there exist a quickly growing number of commercial applications. Speech emotion recognition (SER), however, is very challenging. Understanding the emotions of the other person is in some cases complicated even for a human being.

Words are not emotions; they are simple representations. Speech features such as acoustic variability, speaking speed, and styles determine the emotion content. These features can be extracted from pitch and energy content. Continuous hidden Markov models and Gaussian mixture models [2] are used to detect speech pitch and energy contours. This method classified human emotions in speech with good recognition rates. Parlak and Diri [3] also used signal processing techniques to determine the fundamental frequency of a speech signal, F0, and classified emotions using the Emol and BerlinDB datasets.

Recent advances in the era of deep neural networks have helped researchers to improve the performances of SER algorithms. Zhang et al. [4] studied speech and song emotion recognition. They proposed a shared emotion recognition classifier using visual features and spoken or only sung data. Yoon et al. [5] used a dual recurrent neural networks model to encode the information from audio and text. Deep learning methods have been used in many SER studies [6–11].

Mel-frequency cepstral coefficients and convolutional neural networks were used to detect emotions in the RAVDESS dataset [12]. The visual model was trained with feature extraction gradient descent. The other model was trained using voice samples. Nagula [13] applied deep learning to build two convolutional neural network models to perform multiclass classification of emotion based on the face (video) and voice (audio) from a real-time video. Shau used an LSTM-based classifier model [14] and tested their algorithm on the IEMOCAP dataset. Huang and Bao [15] used conventional vocal feature extraction with deep learning and improved emotion-level classification. Iqbal [16] also used the RAVDESS dataset and implemented a real-time detection algorithm using a support vector machine (SVM) and K-nearest neighbor (KNN). Speech SimCLR [17] applies augmentation to raw speech and its spectrogram to maximize agreement between differently augmented samples in the latent space and reconstruction loss of input representation. A new self-supervised speech representation learning was presented. Zisad [18] proposed CNN and data augmentation to recognize emotion in the speech of a neurologically disordered person.

Hetterscheid [19] proposed a model consisting of recurrent neural network (RNN), long short-term memory (LSTM), and bidirectional LSTM layers. The datasets RAVDESS, TESS, and ElderReact are used to train and test.

Ensemble learning techniques combine multiple models to improve the solution for classification or prediction problems. Stacking, bagging, and boosting are commonly used classes of ensemble learning that are expected to produce better and generalizable results compared to a single model. The success of these methods depends on the learning success of the base learner and their differences from each other. Deng [20] applied linear and log-linear stacking methods for ensemble learning to convolutional, recurrent, and fully connected deep neural networks. Zvarevashe [21] experimented with different ensemble learning algorithms and concluded that random decision forest ensemble learning with hybrid acoustic features is effective.

In our project, we proposed a new ensemble learning architecture that combines CNN and LSTM networks. The major contribution of the present study is the design of a new heterogeneous ensemble architecture based on three deep models to take advantage of each model for SER. The paper is organized as follows: the proposed ensemble architecture is given in Sect. 2, the experimental results are displayed in Sect. 3, and conclusions are drawn in Sect. 4.

2 Proposed Ensemble Architecture

In this section, we present the proposed SER architecture and its main components including emotion recognition. The proposed system model consists of the three main blocks as seen in Fig. 1. The first block uses the CNN-1 model for SER. In the second main block, LSTM-1 is used. The last block consists of a two-stream deep model including CNN-2 and LSTM-2. After successful training, two types of max approach are evaluated as a heterogeneous ensemble. In the first approach, the maximum value of predicted probability values for the three models is considered for classification. In the second approach, simple max voting is used.

We used Keras functional API and TensorFlow on GPU for SER implementation. It is possible to use multiple inputs in a deep network via Keras functional API. We used the dropout technique to avoid overfitting.

3 Experiments

In this section, we describe experiments to evaluate the effectiveness of our combined SER model using the RAVDESS dataset. The dataset is described in the subsections below. Results for several experiments are discussed in the subsequent section.

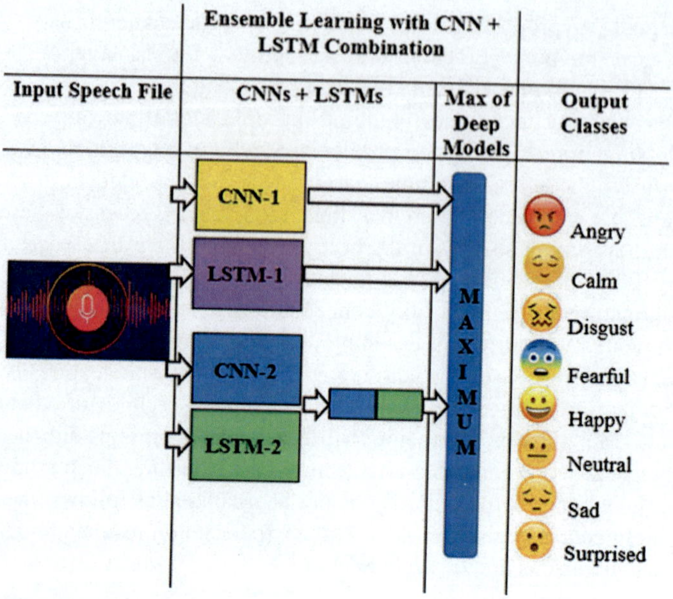

Fig. 1 The proposed SER architecture using ensemble learning with various CNN and LSTM network combinations

3.1 RAVDESS Dataset

The Ryerson Audio-Visual Database of Emotional Speech and Song (RAVDESS) [12] is a dataset that contains the emotions of 24 professional actors (12 female, 12 male). It is recorded in English. It is a multiclass dataset consisting of eight classes labeled as calm, happy, sad, angry, fearful, surprised, neutral, and disgust emotions. It has three modality formats: audio-only (16 bit, 48 kHz,.wav), audio–video (720p H.264, AAC 48 kHz,.mp4), and video-only (no sound). We used audio-only files in the present study, because we focused on speech emotion recognition. A total of 1440 audio files were used. Table 1 shows the total count of audio files for the calm, happy, sad, angry, fearful, surprised, disgust, and emotion classes.

Table 1 Explanation of the distribution of emotions for each class of RAVDESS

Class	Angry	Calm	Disgust	Fearful	Happy	Neutral	Sad	Surprised	Total
Total files	192	192	192	192	192	96	192	192	**1440**

3.2 Experimental Evaluations and Discussion

The main focus of the present study is to compare the performance of speech emotion recognition among three deep learning models, namely CNN, LSTM, and a hybrid CNN and LSTM model. Then, extensive experiments are conducted via two types of ensemble learning with the highest value of predicted probabilities and simple max voting to increase the effectiveness of classification. Conventional machine learning algorithms have been used for decades for SER. However, in recent years, deep learning methods have become popular because of their high performance without handcrafted features. The limitation is that the recognition rate for the RAVDESS dataset still needs further improvement. Therefore, we set up various experiments using the RAVDESS dataset with three deep models and an ensemble model for SER.

Table 2 contains the test accuracy results of three deep models for SER. In this experiment, the test accuracy of the CNN model was higher than that of the others.

The confusion matrix allows us to visualize the actual and predicted values of emotion classes and a result of confusion for each class. The classification performance of the three deep models and the suggested ensemble system for SER are shown in Figs. 2 and 3, respectively. In the experiment by the LSTM model in Fig. 2, we obtained 85% accuracy for surprised, 78% for angry, and 73% for calm. The recognition rate was low for fearful and happy in this experiment. In the experiment by the CNN model in Fig. 2, we obtained 88% accuracy for surprised, 78% for happy, and 73% for calm. The recognition rate was low for angry and disgust in this experiment. In the experiment by the CNN and LSTM model in Fig. 3, we obtained 79% accuracy for neutral, 77% for calm, and 76% for surprised. The recognition rate was low for fearful and angry in this experiment too.

Table 2 Test accuracy results for the three models using the RAVDESS dataset

Model	LSTM	CNN	CNN and LSTM
Test accuracy %	0.6594	0.7292	0.7118

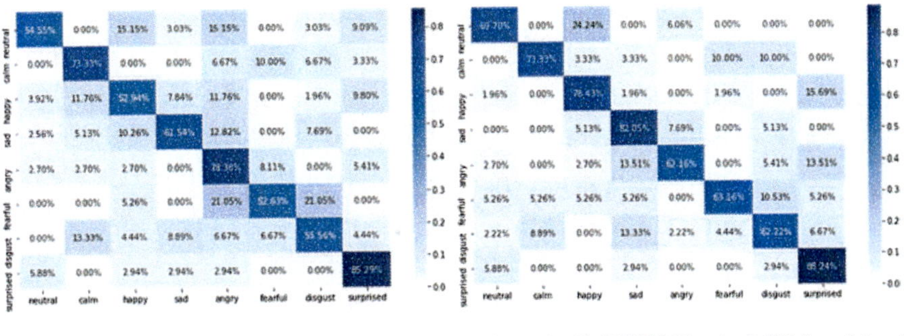

Fig. 2 Confusion matrices of speech emotion prediction using RAVDESS by the LSTM model and CNN model. In the corresponding row, the accuracy value and confusion among emotion classes of "actual" and "predicted" are shown

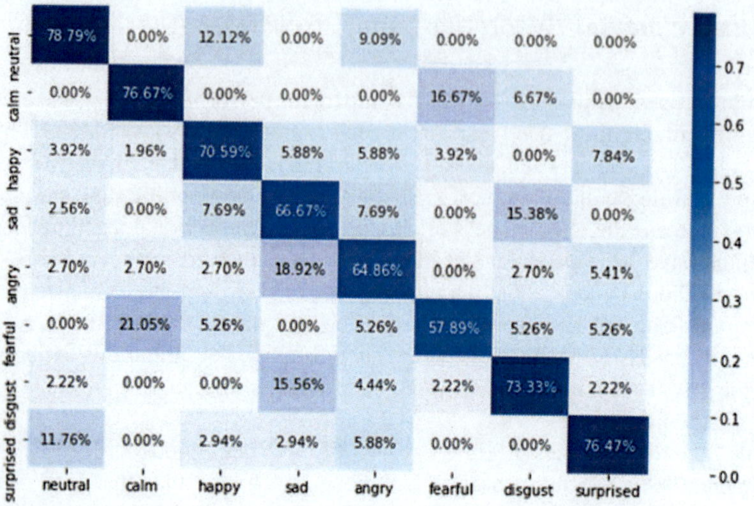

Fig. 3 Confusion matrix of SER using RAVDESS by the CNN and LSTM model

The overall model performance of all deep models is shown in terms of emotion class level precision, recall, and F1 score in Table 3. In addition, accuracy, macro avg., and weighted avg. are presented. Overall, LSTM predicted angry better than the others, while CNN predicted surprised and sad better and CNN and LSTM predicted neutral and calm better. In the experiment by the ensemble learning model in Table 4, ensemble learning benefited from the advantages of these three models in specific cases.

Table 3 Percentage precision, recall, F1-score, and accuracy with RAVDESS by the LSTM, CNN, and CNN and LSTM models

Model	LSTM			CNN			CNN and LSTM		
Emotion class	Precision	Recall	F1 score	Precision	Recall	F1 score	Precision	Recall	F1 score
Neutral	0.75	0.55	0.63	0.79	0.70	0.74	0.74	0.79	0.76
Calm	0.59	0.73	0.66	0.81	0.73	0.77	0.79	0.77	0.78
Happy	0.66	0.53	0.59	0.75	0.78	0.77	0.78	0.71	0.74
Sad	0.71	0.62	0.66	0.68	0.82	0.74	0.59	0.67	0.63
Angry	0.53	0.78	0.63	0.79	0.62	0.70	0.63	0.65	0.64
Fearful	0.53	0.53	0.53	0.67	0.63	0.65	0.58	0.58	0.58
Disgust	0.69	0.56	0.62	0.74	0.62	0.67	0.77	0.73	0.75
Surprised	0.69	0.85	0.76	0.64	0.88	0.74	0.76	0.76	0.76
Accuracy			0.64			0.73			0.71
Macro avg.	0.64	0.64	0.63	0.73	0.72	0.72	0.71	0.71	0.71
Weighted avg.	0.65	0.64	0.64	0.74	0.73	0.73	0.72	0.71	0.71

Table 4 Sample prediction results of the ensemble model for emotion classes

	Emotion	Emotion index	LSTM	LSTM prob (%)	CNN	CNN prob (%)	CNN&LSTM	CNN&STM prob (%)	Ensemble result	Max prob value/Max voting (%)
Ensemble learning with max value of predicted probabilities by deep models	Neutral	0	2	70	**0**	**74**	4	32	0	74
	Calm	1	1	95	**1**	**98**	1	72	1	98
	Happy	2	3	56	4	91	**2**	**97**	2	97
	Sad	3	**3**	**93**	1	62	0	35	3	93
	Angry	4	**4**	**99**	6	63	4	85	4	99
	Fearful	5	**5**	**99**	2	92	7	44	5	99
	Disgust	6	2	84	**6**	**98**	6	82	6	98
	Surprised	7	**7**	**95**	6	53	7	52	7	95
Ensemble learning with simple max voting	Neutral	0	1	99	**0**	53	**0**	54	0	54
	Calm	1	**1**	99	**1**	96	**1**	73	1	99
	Happy	2	3	98	**2**	75	**2**	69	2	75
	Sad	3	5	99	**3**	89	**3**	93	3	93
	Angry	4	7	55	**4**	85	**4**	81	4	85
	Fearful	5	2	85	**5**	58	**5**	85	5	85
	Disgust	6	2	84	**6**	98	**6**	82	6	98
	Surprised	7	4	99	**7**	66	**7**	67	7	67

4 Conclusion

The literature on SER includes many challenges, such as reducing the computational complexity and improving recognition accuracy. In the present study, we planned a novel ensemble approach for SER to overcome these limitations. We proposed a new ensemble learning architecture that consists of combinations of CNN and LSTM networks. The effectiveness of the proposed SER model was evaluated using the RAVDESS dataset. The experimental results for the proposed model are convincing for correctly recognizing the speaker's emotional state. The proposed ensemble method achieves a more accurate prediction by taking advantage of each deep model's ability to better learn specific emotions. These results proved the significance and effectiveness for SER. The proposed architecture can be further enriched in the future by adding GRU as well and used as an alternative model to the existing CNNs/LSTMs system of SER. In addition, the heterogeneous ensemble strategy can be implemented by adding an advanced software function as a next step. The proposed model can be used in many use cases such as speaker identification.

References

1. J.M. Baker, L. Deng, J. Glass, S. Khudanpur, C.H. Lee, N. Morgan, D. O'Shaughnessy, Developments and directions in speech recognition and understanding, Part 1 [DSP Education]. IEEE Signal Process. Mag. **26**(3), 75–80 (2009)
2. B. Schuller, G. Rigoll, M. Lang, Hidden Markov model-based speech emotion recognition, in *2003 IEEE International Conference on Acoustics, Speech, and Signal Processing*, vol. 2 (IEEE, 2003), pp. II-1
3. C. Parlak, B. Diri, Emotion recognition from the human voice, in *2013 21st Signal Processing and Communications Applications Conference (SIU)* (IEEE, 2013), pp. 1–4
4. B. Zhang, G. Essl, E.M. Provost, Recognizing emotion from singing and speaking using shared models, in *2015 International Conference on Affective Computing and Intelligent Interaction (acii)* (IEEE, 2015), pp. 139–145
5. S. Yoon, S. Byun, K. Jung, Multimodal speech emotion recognition using audio and text, in *2018 IEEE Spoken Language Technology Workshop (SLT)* (IEEE, 2018), pp. 112–118
6. B.J. Abbaschian, D. Sierra-Sosa, A. Elmaghraby, Deep learning techniques for speech emotion recognition, from databases to models. Sensors **21**(4), 1249 (2021)
7. X. Ai, V.S. Sheng, W. Fang, C.X. Ling, C. Li, Ensemble learning with attention-integrated convolutional recurrent neural network for imbalanced speech emotion recognition. IEEE Access **8**, 199909–199919 (2020)
8. S. Kwon, MLT-DNet: Speech emotion recognition using 1D dilated CNN based on multi-learning trick approach. Expert Syst. Appl. **167**, 114177 (2021)
9. D. Issa, M.F. Demirci, A. Yazici, Speech emotion recognition with deep convolutional neural networks. Biomed. Signal Process. Control **59**, 101894 (2020)
10. M. Sajjad, S. Kwon, Clustering-based speech emotion recognition by incorporating learned features and deep BiLSTM. IEEE Access **8**, 79861–79875 (2020)
11. S. Kwon, A CNN-assisted enhanced audio signal processing for speech emotion recognition. Sensors **20**(1), 183 (2020)
12. S.R. Livingstone, F.A. Russo, The Ryerson audio-visual database of emotional speech and song (RAVDESS): a dynamic, multimodal set of facial and vocal expressions in North American English. PloS one **13**(5), e0196391 (2018)

13. K. Nagula, K. Priya, G.Y. Kavya, R.S. Sunitha, Analytical comparison of emotion using real time video and audio. Int. J. Innovations Eng. Technol. (IJIET) (2019, June)
14. G. Sahu, Multimodal speech emotion recognition and ambiguity resolution. arXiv preprint arXiv:1904.06022 (2019)
15. A. Huang, P. Bao, Human vocal sentiment analysis. arXiv preprint arXiv:1905.08632 (2019)
16. A. Iqbal, K. Barua, A real-time emotion recognition from speech using gradient boosting, in *2019 International Conference on Electrical, Computer and Communication Engineering (ECCE)* (IEEE, 2019), pp. 1–5
17. D. Jiang, W. Li, M. Cao, R. Zhang, W. Zou, K. Han, X. Li, Speech SIMCLR: combining contrastive and reconstruction objective for self-supervised speech representation learning. arXiv preprint arXiv:2010.13991 (2020)
18. S.N. Zisad, M.S. Hossain, K. Andersson, Speech emotion recognition in neurological disorders using convolutional neural network, in *International Conference on Brain Informatics* (Springer, Cham, 2020), pp. 287–296
19. K.J.T. Hetterscheid, Detecting agitated speech: a neural network approach (Bachelor's thesis, University of Twente) (2020)
20. L. Deng, J.C. Platt, Ensemble deep learning for speech recognition, in *Fifteenth Annual Conference of the International Speech Communication Association* (2014)
21. K. Zvarevashe, O. Olugbara, Ensemble learning of hybrid acoustic features for speech emotion recognition. Algorithms **13**(3), 70 (2020)

Novel WSN Localization Optimization Algorithm Using MVCRSA

Cosmena Mahapatra, Ashish Payal, and Meenu Chopra

Abstract The application of wireless sensor networks (WSNs) has grown widely in today's world. Even with the coming of IoT, WSN has not diminished but in fact it has hugely benefited by this new technology. Nowadays, IT giants like Microsoft have come up with WSN and IoT applications. The scope of WSN has grown from merely being a research topic, to being the basis of several missions and life-critical systems such as health monitoring system, security system for border control, smart city surveillance system, and industrial drill temperature monitoring system (oil wells). For mission-critical systems where the WSN nodes are mobile, there is an additional need for localizing the nodes accurately without overheads. Over the decades, much work has been done to crack the issue; however, scope still remains. Recent COVID-19 has thrown unprecedented challenges on human survival. Primary education givers have also been rattled, as parents are apprehensive of sending their small children to schools, as young children are unable to themselves tell their givers if they feel unwell. In order to provide reassurance to the parents, we propose to build mobile WSN system, which would involve the children wearing WSN-based bands, which shall capture their health-related data, while they are inside the same premises of their schools and send it for further real-time processing to IoT-based cloud environment, so that the health of a student can be monitored live, and the data sent to parents for their comfort. Such a system would require mobile WSN-based localization algorithm, which would in minimum time period fulfill the goal. This paper proposes a data mobile WSN which uses nature-inspired algorithm to localize the nodes correctly using MVCRSA (Multi-verse Crow search algorithm for localization, which has through simulation provided successful results).

C. Mahapatra (✉) · A. Payal
University School of Information Communication and Technology, Guru Gobind Singh Indraprastha University, Delhi, India
e-mail: cosmenamahapatra1@gmail.com

A. Payal
e-mail: ashish@ipu.ac.in

C. Mahapatra · M. Chopra
Vivekananda Institute of Professional Studies, Guru Gobind Singh Indraprastha University, Delhi, India

A. K. Bashir et al. (eds.), *Proceedings of International Conference on Computing and Communication Networks*, Lecture Notes in Networks and Systems 394,
https://doi.org/10.1007/978-981-19-0604-6_6

Keywords WSN · Optimization · MVCRSA · GWCSA · Localization · Querying · Processing · Real time

1 Introduction to Mobile WSN

In mobile WSN otherwise called as mobile ad hoc networks (MANET), sensors are portable, they not only transmit data from areas which are isolated, but they also distribute the energy consumption of sensors to successfully send data. Mobility in wireless sensor networks (MWSNs) has always been a challenging task to manage, yet this has made WSN are popular and have been named as per their characteristics and functionality. Also, much research has been done in identifying the types of movement associated with a mobile WSN. In brief, they may be summarized as follows [1]:

Controlled Movements: Mobile WSN nodes whose movement can be planned to a particular route and are controlled come under this category. For example, of such a movement is movement of mobile base stations to distributed sensor nodes for collecting data.

Expected Movements: The movements, which cannot be controlled, however a planned route for the sensors can be designed come under this category.

Unexpected Movements: Moments which can neither be controlled nor planned, which are unpredictable come under unexpected movement category.

For example, children were moving inside a primary wing while playing or in play period [2].

The MANETS/mobile WSNs may be summarized as they being self- reliant in nature. This means the networks have self-forming capabilities, if a mobile network goes out, then network heals the breach itself without manual inputs by forming a constant peer-to-peer connectivity with the remaining nodes. Thus, in short the characteristics of these may be defined as follows: [3] **Self-preserving and self-driving behavior**: These WSNs require minimum human interventions and are able to function both as a cluster head or a cluster member whenever needed.

Robust and on demand topology: The mobile WSNs are able to create directional and bidirectional associations with each other, thus making the basis of a topology changing nature.

IoT strengthening Security: Though in themselves WSNs have always been in the midst of security-related debates, when clubbed with IoT infrastructural platform, the security aspects have been easily tackled (Fig. 1).

Energy Constraining Characteristic: WSN's one major area of research is energy saving. Due to the limited sources of electricity (batteries), often nowadays researchers are resorting to innovative solutions (solar, movement generation, etc.) and optimizing algorithms to recharge and save electricity, respectively [4].

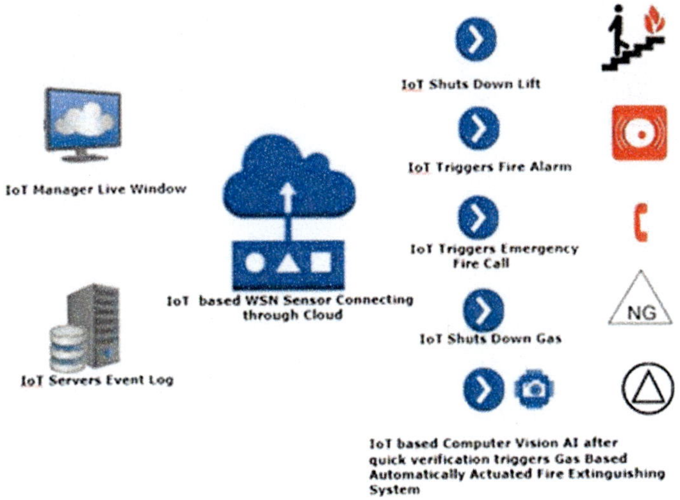

Fig. 1 Example of how WSN may be integrated with IoT

Limited and variable bandwidth and connectivity: Often the nodes in a mobile WSN have varying connectivity parameters, and on the top of that, the available bandwidth plays a major spoil spot [4].

The overall design of a WSN can be classified into two categories, flat and hierarchical. Flat wireless sensor networks are generally comprised of 20–30 sensors spread over a limited area; however as the number of sensors increases, the need for a more hierarchical architecture arises in which the data communication is controlled by multiple cluster heads each in control of their own set of sensors, thus making data communication complex. WSN designers have over the decade faced and conquered to an extent a lot of challenges such as limited battery life, memory, and bandwidth availability, apart from difficult terrain, where communication among the deployed sensors becomes a challenge. Success in the above parameters was seen by the designers with the turn of this decade, and now WSN has been recognized as an essential network for monitoring and control situations. Like every communication protocol, WSN to make uses of a set of network protocols to implement its architectural functions such as localization, clustering, routing, and synchronization, and last but not the least security.

Although the previously mentioned functions of WSN have to an extent been resolved, however much scope is left in refining quality of service (QoS) parameters of wireless sensor networks, these parameters are—scalability, deployment of nodes, coverage, and connectivity. The detailed analysis of these parameters is given as follows [5]:

Scalability—The scalability of a WSN may be defined as the increase and decrease in the number of nodes (sensors), present in the network depending upon the availability, need, and other application-based factors. Scalability (Fig. 2).

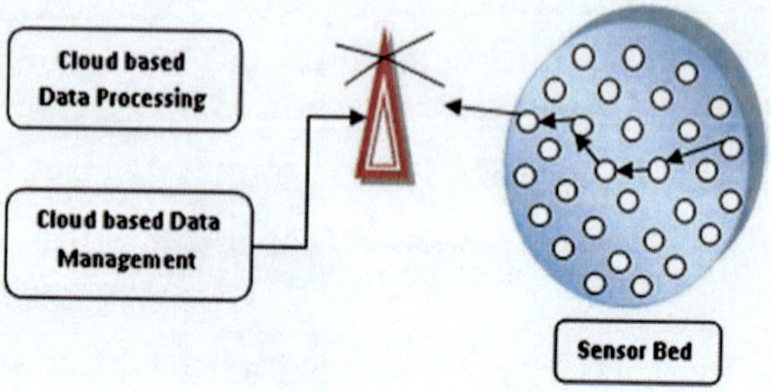

Fig. 2 Wireless sensor network and cloud of the network decides the efficiency of routing protocol which has been implemented in the WSN

Deployment of Nodes—Positioning of the sensors in WSN is a major factor upon which QoS of the network is defined. Deployment of nodes decides the architecture of the WSN in general. It also affects the efficiency of the network on parameters such as energy consumption, data aggregation, data communication (delay in transfer of messages) [6].

Coverage and Connectivity—WSN coverage is classified into two types, full and partial. The idea of full coverage is theoretical, as to get full coverage of an area, we must plant sensors in such a way that they would cover all the points in that area. A more realistic coverage technique would be to plant a sensor in such a way that it covers a particular set of points having same or almost same data generation, this however means that the sensing region of the sensor increases which means that it would need more energy, thus here arises a need to balance the coverage with sensing area and its energy consumption. There are various ways of deciding coverage: K-coverage, square grid coverage, THT cell, etc. Connectivity is an output of coverage, the sensors communicate with their base stations or cluster heads to send and receive data [7].

Though WSNs have become popular in many areas of research, they are still are various issues which are plaguing mobile sensor nodes; these are broadly divided into communication and data management capabilities. In this paper, we shall be focusing on data communication issue, namely localization issue. In order for us to locate the exact coordinates of a sensor, there have been many types of algorithms and described research; these can be clubbed into this following three broad categories: range-based WSN, range-free WSN, mobility-based WSN. These issues of localization may be solved through the use of following types of methods: distance/angle estimation, position computation last but not the least localization algorithm. In this paper, we have focused ourselves in designing a mobile WSN localization algorithm which is optimized by nature-inspired algorithm (multivariate crow search optimization algorithm) (Table 1).

Table 1 Understanding the basis for physics chemistry-based nature-inspired algorithm

S. No	Physics chemistry-based nature-inspired algorithms	Year
1	Simulated annealing	1983
2	Stochastic diffusion search	1989
3	Self-propelled particles	1995
4	Harmony search	2001
5	River formulation dynamics	2007
6	Central force optimization	2007
7	Gravitational search	2009
8	Intelligent water droplet	2009
9	Charged system search	2010
10	Spiral optimization	2010
11	Galaxy-based search algorithm	2011
12	Water cycle algorithm	2012
13	Big bang big crunch	2012
14	Black hole	2013
15	Electromagnetic like optimization	2014
16	Multi-verse optimizer	2016

1.1 Brief Review

Various studies have been done to model and implement physics and chemistry-based nature-inspired algorithms for optimization purposes.

Authors [8] **proposed, probably for the first time, a novel attempt to the fairly** accurate solution of complex combinatorial optimization problems using stimulated annealing algorithm which in turn was based on analogy of physical annealing process. They proved a profound relationship between the statistical mechanics and multivariate or combinatorial optimization. Using the classical traveling salesman model, the usefulness of simulated annealing in solving multiple optimization problems was validated. The research thus illustrated a computer's ability to arrive at an optimal solution through purely stochastic processes without intervention of human intelligence. Carrying forward the work on stochastic cells, Bishop [9] described stochastic search network to arrive at high-performance solutions and deal with stimulus equivalence in pattern recognition problems.

Authors of this paper [10] came up with another physics–chemistry-based optimization algorithm and investigated the emergence of a self-ordered movement in particle systems through a newly introduced particle dynamics model to examine clustering, transport, and phase transition. The particles were moved with a constant velocity, and it was found that at each time step a random perturbation was added and the direction of the particles assumed the average value of that of its neighboring particles at a radius r. The results demonstrated a transition of the kinetic phase from zero velocity to finite through spontaneous breaking of the rotational symmetry.

Paper [11] devised an algorithm called harmony search based on the analogy of the music performance process. Authors reviewed a number of traditional optimization techniques and existing heuristic methods before coming up with it. Through traveling salesman problem and least-cost pipe network design, the advantageous features of HS were illustrated that were, namely its ability to create new vector by considering all vectors rather than just parent vectors, and omitting the need to set up initial values of decision variables. Similarly, Rabanal et al. [12] presented another type of heuristic algorithm RFD, based on natural dynamics of river formation to get acceptable solutions to N-P hard problems in a relatively better way than other efficient heuristic methods like ant colony optimization algorithm. RFD was applied to the traveling salesman problem to prove its efficacy above other algorithms in achieving optimal solutions.

Study [13] introduced a metaheuristic optimization search algorithm called central force optimization. CFO's effectiveness over other optimization methods was established through preliminary analysis in which an equalizer for the canonical Fano load was designed as well as through the synthesis of a 32—element linear array. The results were equally optimal or better than other optimization algorithms. Rashedi et al. [14] created a gravitational search algorithm on the basis of Newtonian law of gravity and concept of mass interactions. The mass systems existed in isolation and transfer of information between each mass, in relation to other, occurred using the laws of motion and gravitational force. The authors evaluated the results of the algorithm on a set of standard functions and found it to be superior to renowned heuristic search methods like PSO, RGA, and CFO.

Likewise, research [15] used a nature-inspired optimization algorithm based on swarm of water drops to solve three different problems, the n-queen puzzle, the TSP, and the MKP. The algorithm was modified to solve the TSP problem (MIWD-TSP), and it leads to better comparative results with respect to standard MKP-TSP algorithm. The authors also tested some new MKPs and found optimal or near-optimal solutions to the given problems. It was concluded that the IWD algorithm was efficient in finding optimal solutions to optimization problems, though the authors admitted the scope for modifications in the algorithm and utilizing other natural dynamics of rivers along with add-on heuristic search systems to make IWD more resourceful.

Adding novelty to optimization techniques, [16] used Coulomb law of electrostatics and Newtonian laws of mechanics to present a new optimization algorithm called charged system search (CSS). The agents of the CSS were termed as charged particle having a charged sphere of radius r and uniform density with an electric force imposed on other charged particles. The CPs could exert an attractive force on each other based on their separation distances. Authors applied the self-adaptation, cooperation, and competition step in the algorithm and did the comparative analysis on standard functions to prove CSS's superiority over other optimization algorithms. Its application can be widespread for either non-smooth or non-convex design problems.

Study [17] proposed an innovative two-dimensional metaheuristics, spiral optimization, based on natural spiral phenomenon such as nautilus shell and spiral galaxies. The optimization algorithm was found to be effective and has potential for further research through simulation results. However, following domains were still considered as worth improving, adding randomness, extension to n dimension and establishing center x^*, αi and βi. However, Shah-Hosseini [18] improvised the spiral optimization technique and formulated a metaheuristic called galaxy-based search algorithm, GbSA-PCA, for principal component analysis (PCA). The algorithm utilizes the spiral mechanics of galaxies to search the PCA space and is aided by chaotic sequence for enhanced results. Author also used local search algorithm to adjust the solution generated by spiral motion of GbSA-PCA. Initial experimental results showed promising use of this metaheuristic in PCA estimation as well as in real applications having PCA involved in problem-solving.

A new population-based, global optimization technique named as water cycle algorithm (WCA) was presented by [19], deriving inspiration from natural water flow in rivers and streams to the sea. The algorithm was applied to various constrained optimization and engineering design problems. Comparative analysis of achieved results proved its efficacy over other optimizers as well as its increased efficiency in terms of computational cost, depending upon the type and difficulty level of the given problem. Author proposed the application of WCA for real-world optimization problems with scope of further study on its efficiency in solving wide-ranging optimization problems.

Based on the two big bang and big crunch theories, study [20] constructed an optimization algorithm called big bang big crunch (BBBC) that employed generation of random points in the first phase and subsequent shrinkage of those points to a single representative point through center of mass or minimal cost approach in the second phase. The algorithm was developed for the improvement of voltage stability and authors employed the technique to solve ORPF problems. Simulations were done on IEEE 30 and IEEE 118—bus test systems, and their results were compared with those obtained by PSO. Comparative analysis proved the efficacy of BB-BC method in achieving improved quality solutions and rapid convergence rate of the algorithm.

Authors [21] invented a population-based algorithm inspired by black hole phenomenon for cluster analysis. Among initial pool of candidate solutions, the best one was chosen as black hole which pulled other candidates around it and eliminated anyone that gets too close. It gives rise to a new candidate solution to be placed in the search zone for next iteration. The algorithm demonstrated superior performance in comparison with traditional heuristic methods on the six benchmark datasets. Further, the technique is uncomplicated and free from problems of parameter adjustment. Author proposed it for varied optimization problems independently or in combination with other search algorithms.

Taking a step further, [22] demonstrated the use of electromagnetism-like optimization in circle detection process using a combination of three points on edges as parameters to identify circle images in the given scene. The identified circle by the objective function was used by EMO to find the optimum circle candidates which better related with it. The ultimate algorithm had the capacity to rapidly detect a circle with accuracy up to sub-pixel level even under complex and noisy image conditions. To evaluate its performance, several experimental tests were employed such as multiple circle detection and circular approximation.

Author [23] came up with a novel nature-inspired algorithm which they have named as multi-verse optimizer. The authors have based the algorithm on physics principals of white hole, black hole and wormhole. The authors have successfully validated there finding with gray wolf optimization algorithm, particle swarm optimization algorithm, genetic algorithm, and gravitational search algorithm. By the results of the comparisons, they have been able to successfully prove the high potential of MVO in major mission-critical applications.

2 Simulation and Result

Our proposal for solving localization problem is hybrid form of multi-verse optimization by integrating it with crow search optimization algorithm [24], which has been proven to be a better algorithm [25] then cuckoo search.

The basic integration criteria for the algorithms are as follows:

1. Using multi-verse theory to establish the existence of white holes, black holes, and wormholes.
2. Black holes would lead the data to a universe where the speed of enhancement is very less.
3. If speed of growth of a universe is too fast, it will result in big bang and creation of multi-verses.
4. A wormhole may be used to transmit data (s) the best-fit solution to a multi-verse universe where it may be utilized to find a better best fit. The data (s) to be transmitted to find global optimum solution would be identified throw crow search algorithm.
5. Crows always live in a flock, they share food sources, and the best group is able to defend its food source from other birds/animals. Thus, the best group of data is identified and fired through the wormhole to the best-fit multi-verse for finding global optimum solutions.

Figure 3 depicts the success of MVCRSA, and it has been able to map the mobile WSN node to its actual target with almost 93% success rate.

3 Conclusion

Education specially the primary education system has been severely hit by COVID-19 pandemic threat. Parents are apprehensive of sending their wards to schools, least they get infected. In these times, it is only logical to have a mobile WSN be set up for tracking the children's health parameters while they are inside their primary wing

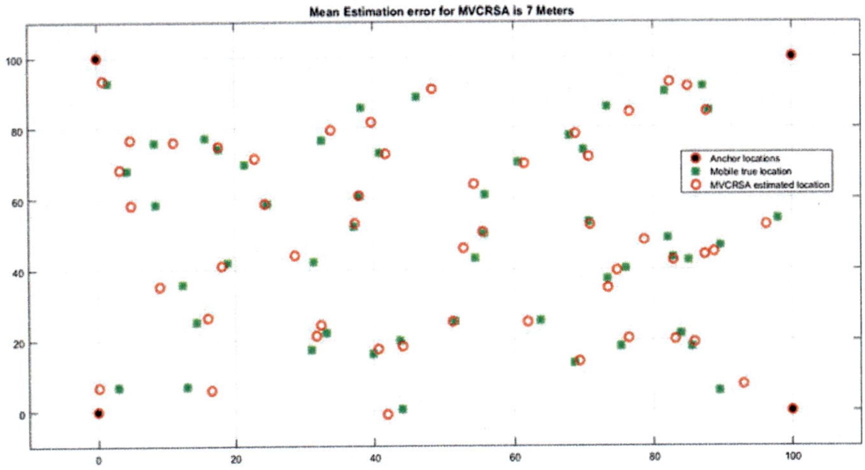

Fig. 3 WSN localization result

schools which would be monitored through live feed least a student's health takes a bad turn. For this use, in this paper we propose a hybrid localization algorithm—multi-verse crow search optimization algorithm, which may be used for locating and monitoring critical wireless sensor networks nodes (MVCRSA). The algorithm has shown considerable success in MATLAB environment.

References

1. A. Adeel, M. Gogate, S. Farooq, C. Ieracitano, K. Dashtipour, H. Larijani, A. Hussain et al., A survey on the role of wireless sensor networks and IoT in disaster management, in *Geological Disaster Monitoring Based on Sensor Networks* (Springer, Singapore, 2019)
2. A. Katti, Target coverage in random wireless sensor networks using cover sets. J. King Saud Univ-Comput Inf Sci (2019)
3. H. Zou, Clustering algorithm and its application in data mining. Wireless Pers Commun **110**, 21–30 (2020)
4. P. Mukherjee, A. Das, Nature-inspired algorithms for reliable, low- latency communication in wireless sensor networks for pervasive healthcare applications, in Springer Tracts in Nature-Inspired Computing eds by D. De, A. Mukherjee, S.K. Das, N. Dey (Springer, Berlin, 2020)
5. **K. Xu, Z. Zhao, Y. Luo, G. Hui, L. Hu et al., An energy-efficient clustering routing protocol** based on a high-QoS node deployment with an inter-cluster routing mechanism in WSNs. Sensors **19**(12), 2752–2752 (2019)
6. B. Rashid, M.H. Rehmani et al.: Applications of wireless sensor networks for urban areas: a survey. J. Network Comput. Appl. (2016)
7. A. Boukerche, P. Sun, (2018). https://doi.org/10.1016/j.adhoc.2018.07.003
8. S. Kirkpatrick, C.D. Gelatt, M.P. Vecchi, (1983)
9. J. Bishop, Stochastic Searching Networks. in *Proceedings of the 1st IEE International Conference Artificial Neural Networks*, (1989), London, UK, 16–18 October, Volume **313** (1989)

10. T. Vicsek, A. Czirók, E. Ben-Jacob, I. Cohen, O. Shochet, Novel type of phase transition in a system of self-driven particles. Phys. Rev. Lett. **75**(6), 1226–1226 (1995)
11. Z.W. Geem, J.H. Kim, G.V. Loganathan (2001)
12. P. Rabanal, I. Rodríguez, F. Rubio, Using river formation dynamics to design heuristic algorithms. in *Unconventional computation*, UC'07, LNCS 4618. Springer, pp 163–177 (2007)
13. R.A. Formato, Central force optimization. Prog Electromagn Res **77,** 425–491 (2007)
14. E. Rashedi, H. Nezamabadipour, S. Saryazdi, GSA: A Gravitational Search Algorithm. Inf. Sci. **179**(13), 2232–2248, ISSN 0020-0255, (2009), https://doi.org/10.1016/j.ins.2009.03.004
15. H. Shah-Hosseini, The intelligent water drops algorithm: a nature-inspired swarm-based optimization algorithm. Int. J. Bio-inspired Comput. **1**(1–2), 71–79 (2009)
16. A. Kaveh, S. Talatahari, A novel heuristic optimization method: charged system search. Acta Mech. **213**(3), 267–289 (2010)
17. M. Tamura, M. Miki, A. Okamoto, K. Kusunose, A. Uchikoshi, D. Igarashi, T. Hamajima, vol. 664. (U.S, Washington, DC, 2010)
18. H.S. Hosseini (2011) Principal components analysis by the galaxy-based Search algorithm: a novel metaheuristic for continuous optimisation. Int. J. Comput. Sci. Eng. **6**(132), (2011) https://doi.org/10.1504/ijcse.2011.041221
19. H. Eskandar, A. Sadollah, A. Bahreininejad, M. Hamdi, Water cycle algorithm—a novel metaheuristic optimization method for solving constrained engineering optimization problems. Comput. Struct. 151–166 (2012)
20. Z. Zandi, E. Afjei, M. Sedighizadeh, Reactive power dispatch using big bang-big crunch optimization algorithm for voltage stability enhancement, in *2012 IEEE International Conference on Power and Energy (PECon)*, pp. 239–244 (2012)
21. A. Hatamlou, Black hole: a new heuristic optimization approach for data clustering. Inf. Sci. **222**, 175–184 (2013)
22. D. Oliva, E. Cuevas, G. Pajares, D. Zaldivar, V. Osuna, A multilevel thresholding algorithm using electromagnetism optimization. Neurocomputing **139**, 357–381 (2014)
23. S. Mirjalili, S.M. Mirjalili, A. Hatamlou, Multi-verse optimizer: a nature-inspired algorithm for global optimization. Neural Comput Applic **27**, 495–513 (2016)
24. A. Askarzadeh, A novel metaheuristic method for solving constrained engineering optimization problems: crow search Algorithm. Comput. Struct. **169** (2016)
25. A.S. Joshi, O. Kulkarni, G.M. Kakandikar, V.M. Nandedkar et al.: Cuckoo search optimization—a review, in *Materials Today: Proceedings* (2017), pp. 7262–7269. (https://www.sciencedirect.com/science/article/pii/S2214785317313433)

An Enhanced QR Code-Based Smart Parking System for Mobile Environment

Tan Shy Yu, Shayla Islam, and Chee Ling Thong

Abstract This paper presents a mobile application system that allows users to reserve a parking lot by using their smartphones. The aim of this paper is to support the efficient parking at university level by developing a QR code-based mobile smart parking system. When the smartphone is connected to the Internet, the students are only required to start the parking time through scan QR code. A parking lot will start the parking time for the students. Before the users exits the university, the students are only required to pay the balance through their wallet of the application. This reduces the hassle of queuing up for in front of the auto pay machine. Moreover, after payment, students can directly exit the university.

Keywords Mobile smart parking system · QR code · Mobile phone · University parking slot

1 Introduction

Urban parking is an area for people to parking in the city. It is also the place with the highest frequency of traffic jams. The exhaust gas from people's daily find for parking in urban parking has a huge impact on the environment. Especially in large cities, the air pollution caused by motor vehicle emissions accounts for about 70%, more than twice the air pollution caused by heat source 4 [1]. Therefore, people began to try to make smart parking to solve these related problems and reduce emissions.

Nowadays in urban areas car parking is a time-consuming activity, which has a significant impact on people's daily life. Parking needs to pay the parking fees are everyone to know. However, in the process of paying parking fees we all know that sometimes it is very troublesome. Usually, finding the right parking lot needs to

T. S. Yu · S. Islam (✉) · C. L. Thong
Institute of Computer Science and Digital Innovation (ICSDI), UCSI University, 56000 Kuala Lumpur (South Wing), Malaysia
e-mail: shayla@ucsiuniversity.edu.my

C. L. Thong
e-mail: chloethong@ucsiuniversity.edu.my

A. K. Bashir et al. (eds.), *Proceedings of International Conference on Computing and Communication Networks*, Lecture Notes in Networks and Systems 394,
https://doi.org/10.1007/978-981-19-0604-6_7

wander around in unpredictable time, which wastes time, fuel, pollution, and traffic jams. In this regard, efficient urban parking is very important for solving traffic jams and inconveniences.

The main purpose of smart parking lot is to use big data to collect and disseminate real-time information, cooperate with the application of relevant technologies and equipment, and take the interests of car owners as the ultimate goal to improve parking efficiency and reduce the consumption of customer time and resource costs. In the USA, there are more than 40,000 parking lots. Most of the business models here are dependent and cooperative management models. Through cooperation with third-party intelligent parking operators (software suppliers) and equipment suppliers, we provide optimized solutions to reduce the investment cost of owners and maximize the profit income.

An example is IoT-based smart parking system. The concept of the Internet of things (IoT) began with the Internet of things with identity communication devices. A remote computer connected via the Internet can be used to track, control, or monitor the device. IoT expands the use of the Internet to provide communication, thereby enabling the interconnection of devices with physical objects or "things" (Abhirup Khanna 2016). Two words in the Internet of things are “Internet” and “things.” The Internet refers to a vast global network of connected servers, computers, tablets, and mobile phones using protocols and connection systems used internationally. The Internet makes it possible to send, receive, or communicate information. The dictionary meaning of “things” is a term used to refer to physical objects, actions or ideas, situations, or activities that we do not want to be precise. The Internet of things usually consists of a network of devices and physical objects, and many objects can collect data at remote locations and communicate with units that manage, acquire, organize, and analyze data in processes and services. It provides a vision to make things (wearables, watches, alarm clocks, home devices, and surrounding objects) smart and sense, compute through embedded small devices (interacting with remote objects or people through connections), and communication becomes lively. The scalability and robustness of cloud computing allow developers to create and host their applications on it. The cloud is an ideal partner for the Internet of things because it acts as a platform that can store and access all sensor data from remote locations. These reasons led to the fusion of two technologies, and a new technology called the Internet of things (IoT) was formed. In IoT, things (nodes) can be accessed, monitored, and controlled through the cloud from any remote location. Due to the high scalability in the cloud, any number of nodes can be added or removed from the IoT system in real time. In simple terms, the Internet of things can be explained with equations:

Physical objects + controllers, sensors, and actuators + Internet = Internet of things based on the advent of the Internet of things, the ideal of creating a smart city is now possible. In today’s cities, finding available parking slot is always difficult for drivers, and as the number of private car users increases, it becomes more and more difficult. This situation can be seen as an opportunity for smart cities to take action to improve the efficiency of their parking resources, thereby reducing search time, traffic congestion, and road accidents. If drivers can be informed in advance of the availability of their intended destination and the surrounding parking slot, issues related to parking and traffic congestion can be addressed. Recent advances in creating low-cost, low-power embedded systems are helping developers build new applications for the Internet of things. With the development of sensor technology, many modern cities choose to deploy various IoT-based systems in-and-around the city for monitoring.

Sensor technology is considered to be an effective means of effective parking management. Although these solutions have feasibility and advantages, the effect of sensor arrangement hinders their practical application. On the contrary, the widespread use of mobile devices has opened up new prospects, far beyond simple support for parking fees. Parking fees and mobile devices can be combined, which will greatly solve many people's parking problems.

In this paper, we will show a mobile application that supports wide area parking service, from search to parking start/end, with reservation considered. The identification of any single parking point can rely on QR code and/or GPS positioning, thus reducing system costs and speeding up automatic parking procedures, including fees charging. Guidance provided by dedicated mobile applications can improve the efficiency of parking, have a positive impact on the quality of life, and reduce fuel consumption. Besides that, other city monitoring applications can effectively utilize real-time data on occupancy of specific parking lots to better plan and manage traffic and public city services. The proposed application can help users to find the most convenient parking slot and can complete the payment of parking fees directly in the application. Search for the nearest parking location in real time. Given destination, for planned travel, browse, and book available locations in advance. A typical parking procedure begins with a parking search action. After that, after arriving at the parking slot, the actual occupation is performed by identifying the place where the car is placed as illustrated in Fig. 1.

Fig. 1 QR codes parking slot

2 Literature Review

This literature review will focus on Android operating system and application projects related to mobile smart parking system. The development of smart parking systems requires very high costs. Besides that, the maintenance fee is not a small amount [2]. Despite the high cost, an intelligent parking system is still required. It is because it can greatly reduce the air pollution of people in big cities and save people time in finding parking spaces [3].

In order to perform smart parking, people nowadays use some methods such as RFID to record and keep track of their parking [2]. RFID is an acronym for "radio-frequency identification" and refers to a technology whereby digital data encoded in RFID tags or smart labels (defined below) are captured by a reader via radio waves [4]. However, some more advanced smart parking systems are still under study and some are in trial such as QR code smart parking system.

There are some limitations listed in the previous study [5] such as cost of the system being more, complexity of design increases, no multilevel parking inside an infrastructure, and system only allows one by one parking. It is because the system needs to build base on the terrain, not all place can be built. Besides that, the system is open for 24 h and will cause easy damage, it needs regular maintenance.

There is a wide variety of smart parking system such as intelligent parking lot application using wireless sensor networks, smart routing technology, IoT smart parking system, and others. Most of the projects are aimed to develop a system or application to make people finding the parking become easy and limitation of air pollution of parking. Most of the parking systems developed are ultrasonic sensor, the sensor can let the parking or car receive a message and perform a series of actions [6]. For example, when the driver gets the parking ticket in the outside, the parking gate can detect the car and open the gate to let the driver enter the parking lot. In the parking slot, when the car is parking in the slot already, the parking slot will be detecting the car and the light above will change from green to red. It can reduce the inconvenience of drivers looking for a parking space while driving a car. They only need to look at the lights on the parking space to know whether there is a parking space.

Other technologies are in mobile application, and these technologies use street maps to let drivers know where to park [7]. The technologies are more used for outdoor parking and garden parking lot. It is very useful for both because finding parking in outside area easy to cause traffic jams. Therefore, if drivers can know where the parking space in advance through the application are, it can reduce air pollution and traffic jams as existing architecture is illustrated in Fig. 2.

- Issue 1:

  With more and more university students, university vehicles have gradually increased. For example, in UCSI University, when after school, students pay parking fees and drive home. At this moment, the problem is coming, and the flow of people would be blocked at the elevator entrances of Block G and doorway of

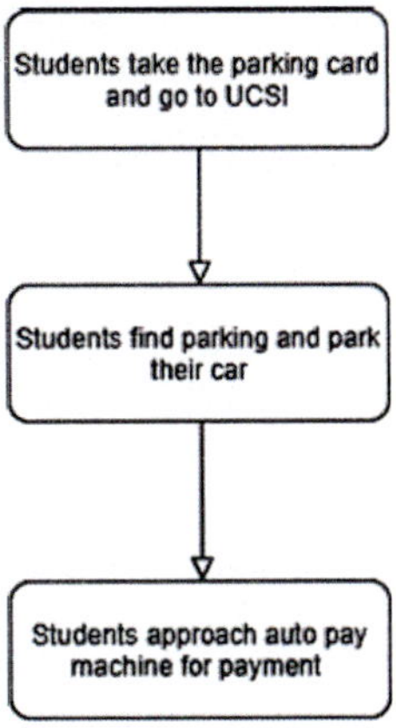

Fig. 2 Illustration of existing architecture flow

Block C. It is because there is an auto pay machine there, and most students pay parking fees in both places. Thence, once it is time for after school, there will be a lot of people in both places as shown in Figure 3.

- Issue 2:

  The auto pay machine and parking kiosk machine should be repaired and services if it down or failure (shown in Figure 4). Thence, repaired and services need some money to do, regular maintenance increases university monthly expenses. For the reason, students may be at risk of increasing tuition fees. University tuition

Fig. 3 People line up to pay parking fees

Fig. 4 System down of UCSI auto pay machine

fees are not cheap, suppose that even higher tuition fees may affect the reputation and interests of university, then students will choose another university as well as university ranking will go down in the ranking list.

- Issue 3:

  If the auto pay machine system is down, other auto pay machine will be increasing the burden as depicted in Figure 5. For example, UCSI has four auto pay machines,

Fig. 5 Traffic jam because students pay the parking fees and insert card to leave

but some auto pay machine has problems and causes the machine system down. System down of an auto pay machine will increase the risk of other auto pay machine. For the reason, UCSI often has system down with the auto pay machine, and students and teachers are adapted to it. Moreover, the parking kiosk machine in the guardhouse also have more problems. It is because, those machines were active 24 h for every day, so sometimes those machines cannot exit the card normally. Sometimes the guard will let those machines out of the card based on manually. But in the worst-case scenario, students and lecturers or another visitor need to go around the guardhouse at the front or back door to get into UCSI. It is a waste of time, and when they encounter traffic jam, they will be late to the class or meeting. It will have a huge negative impact.

3 Proposed Architecture

The proposed architecture as shown in Fig. 6 intends to allow users who UCSI students only. They can login to the system using their student ID and password same with their CN and IIS. Each student will have their own account. However, if they forgot the password or unable to login, they need to find the related department to solve this problem, it may need some processing fees. Moreover, students need to install the application before parking in UCSI. Besides that, they must ensure that their accounts have sufficient amounts before payment, otherwise they cannot make payments. They can top up the wallet on the wallet interface, and the interface will transfer to the bank select interface to allow students to complete their processing of top-up. In addition with that, each student parking records will be carefully recorded in the relevant UCSI department. If the students have a violation record and will be penalty. In severe cases, their account will be blocked and the student will not be able to park at the UCSI. The penalty fees may be RM 50 between RM 100, but the unlocked fees are more than RM 500. Therefore, students should be careful about the rules and do not violate the rules.

The proposed system consists of two sub-systems—management control panel and smart parking application as described in Figs. 7 and 8.

The management control panel allows students to process some action on the application. The management control panel is a webpage using HTML.PHP framework which allows students to view their home page that can be displayed in smart parking applications. In addition, students can also access to the wallet page to view their balance in the application (Fig. 9).

The students are the users who use applications installed in their devices provided by UCSI as shown in Table 1. It is part of the mobile smart parking system.

Security guard are the UCSI employee who manage all situation about the parking lot (shown in Table 2). They control the students parking times and manage the parking lot.

Admin is the system administrators who are responsible for providing security guard access to the system as described in Table 3.

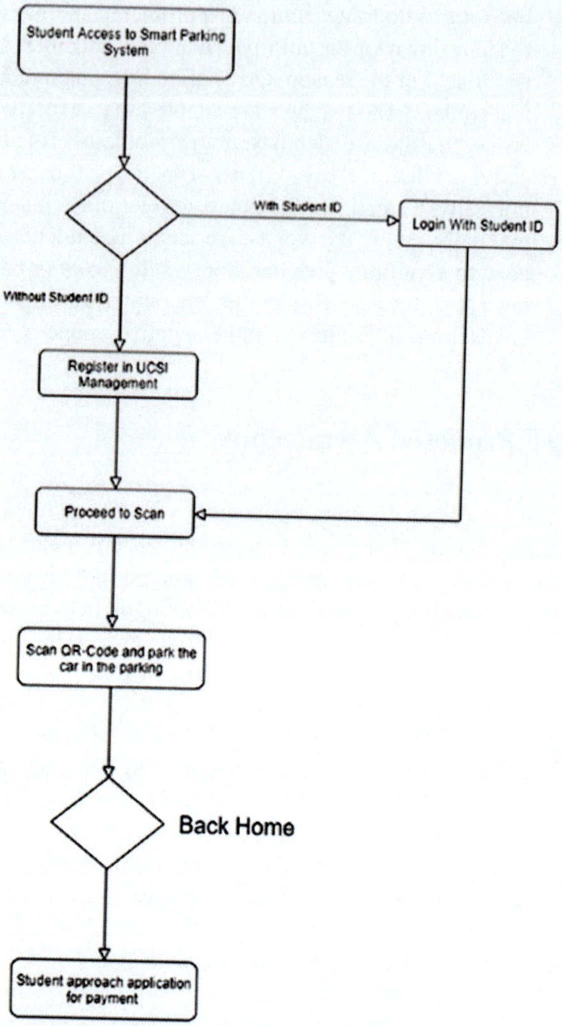

Fig. 6 Illustration of proposed architecture flow

The entity–relationship diagram below is a representation of the process of the system as shown in Fig. 10.

A first page in Fig. 11 is let the students to login their account for the application.

Students can top up their wallet and check the amount through the wallet page, and it can use to pay the parking fees in the UCSI as shown in Fig. 12.

Each QR code represents each parking slot, which has information on the parking slot (shown in Fig. 13). Students can view their personal information in profile page, but cannot edit the information such as IIS system as depicted in Fig. 14.

The students will login their account first and then can start to use the application as shown in Figs. 15 and 16. If students scan the code but cannot transition to the

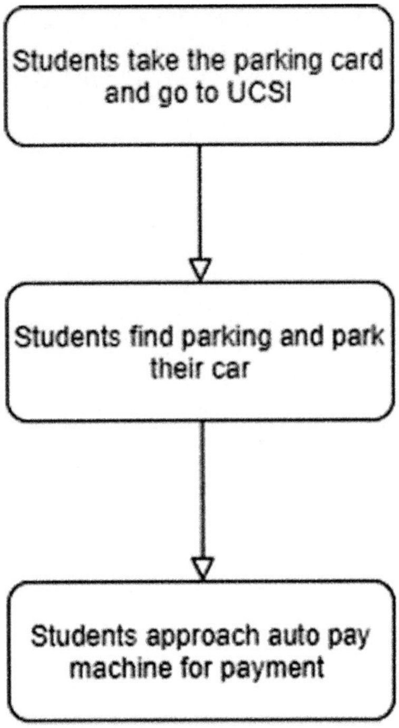

Fig. 7 Illustration of current architecture flow

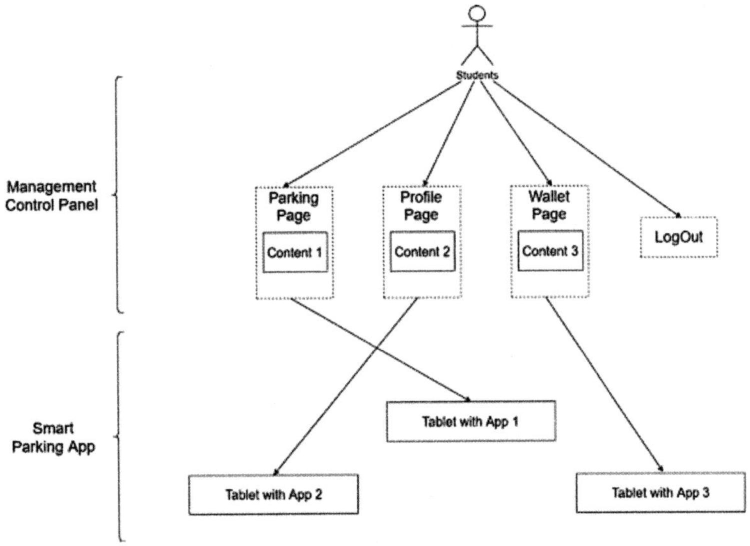

Fig. 8 Sub-system dependency and vendor ownership illustration

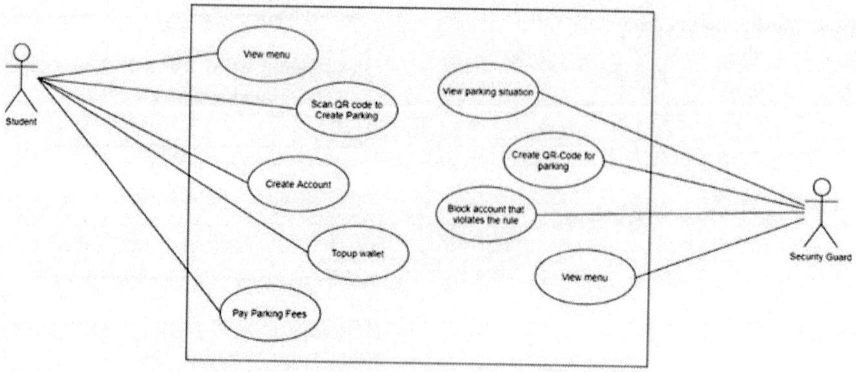

Fig. 9 Use case for mobile smart parking system

Table 1 Users who use applications installed in their devices provided by UCSI

Case	Description
View menu	Menu items are listed on the application interface
Scan QR code to create parking	Students can proceed to park their car by scanning the QR code in the parking
Create account	Students can sign up for a new account in UCSI management to be used for the service
Top-up wallet	Students can top up money in their wallet to make payment
Pay parking fees	Students can proceed to payment by button "Pay" shown on the application interface

Table 2 UCSI employee who manage all situation about the parking lot

Case	Description
View menu	Menu items are listed on the application interface
View parking situation	Parking situation will be listed on the home page
Create QR code for parking	If QR code of parking invalid or broken, guard will be created new one for it
Block account that violates the rule	If some account violates the rule, they have the privilege to block the account

Table 3 System administrators who are responsible for providing security guard access to the system

Case	Description
Add security guard	Admins can assign an account as security guard by entering the account holder's email address and phone number
Remove security guard	Admins also can remove the security guard privilege from an account

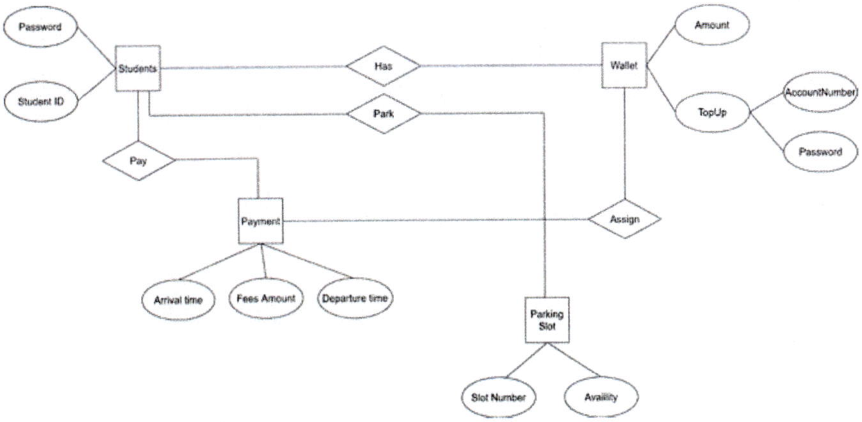

Fig. 10 ER diagram for mobile smart parking system

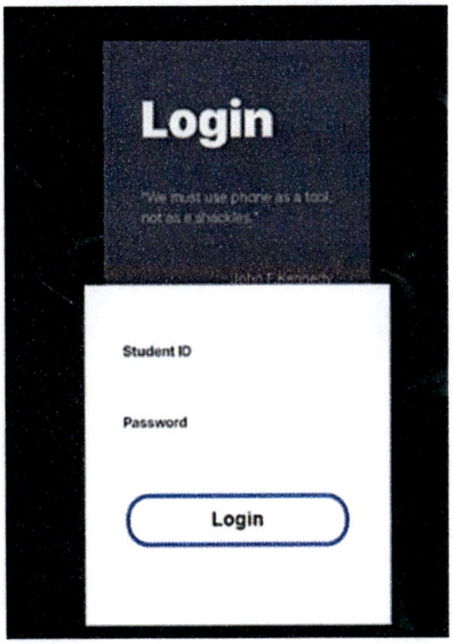

Fig. 11 Login page for mobile smart parking application

parking slot page, it maybe is QR code unclear or network unstable, students can scan the code again to complete their parking.

Before students drive away from the parking slot need to complete the payment as illustrated in Fig. 17. Students can click the button "Pay" in the home page. If cannot complete payment maybe is insufficient balance in the wallet or network problem. Students should try again to avoid the tire will block by security guard. Moreover, students need to ensure that their wallets have sufficient balance to make payment; otherwise, they will not be able to complete the payment. By the way if students have balance but also cannot complete the payment, it maybe network problems or other technical problems.

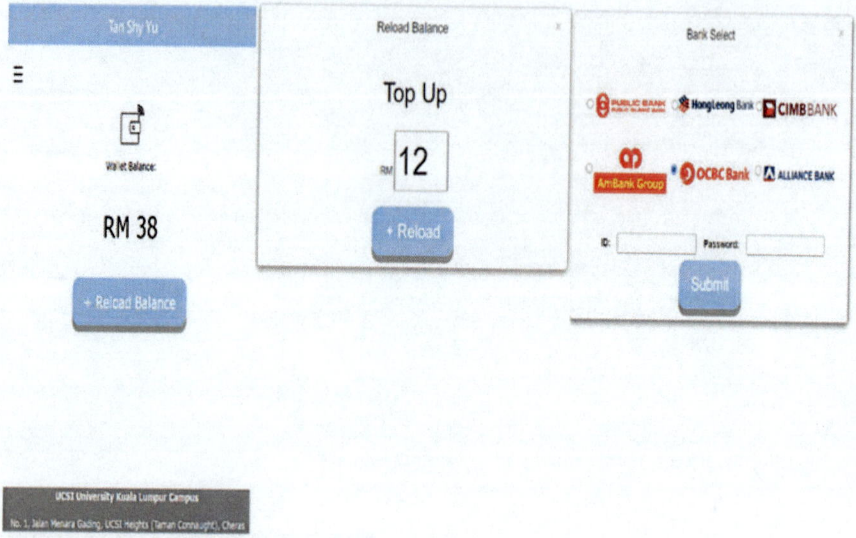

Fig. 12 Wallet page for mobile smart parking application

Fig. 13 Parking slot information for mobile smart parking application

4 Evaluation

The section will discuss the procedures used in assuring the project is delivering the required products and some of the techniques used in testing the system. Throughout the system development, unit testing is frequently conducted before every major implementation. This is especially important when two subsystems are dependent on one another. Since the web services of the system are first deployed before the development of the app, the API components must go through rigorous testing before it should be used on the application to prevent further issues. After the development

Fig. 14 Profile for mobile smart parking application

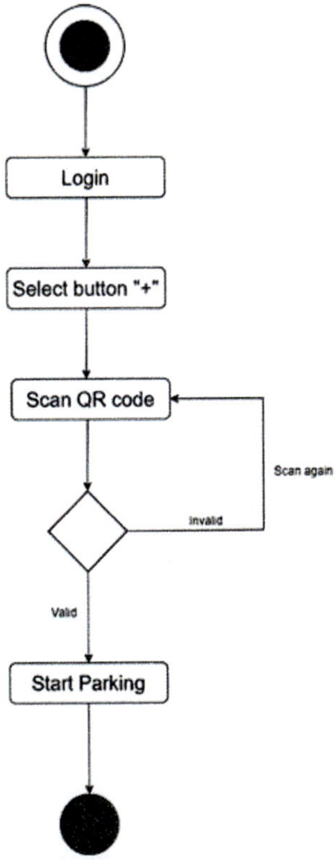

Fig. 15 Activity diagram to illustrate the parking process

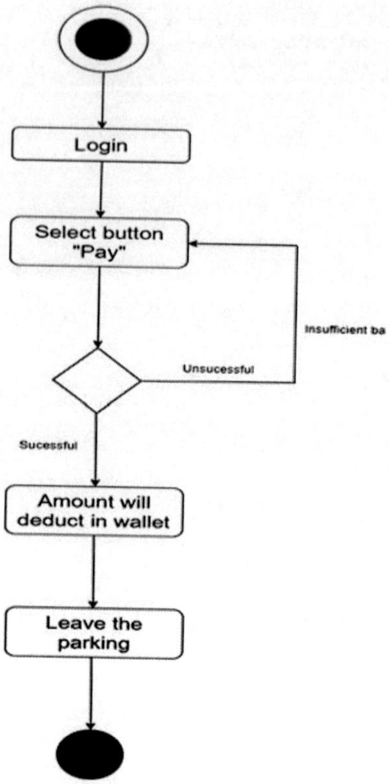

Fig. 16 Activity diagram to illustrate the payment process

of the application is completed, integration testing is a level of software testing where individual units are combined and tested as a group. The application must go through rigorous testing including the core functionally, function process, and bug of application. It needs to ensure subsystems work together each other. After the core functionality of the system is developed, subsequent improvements and addition implementations are accompanied by smoke testing to ensure the new implementations do not affect the system stability. Smoke testing is conducted every day.

User acceptance testing is conducted once the system development has arrived at the stage where all core functionality of the system becomes usable. In the system, 8 students have been volunteered to participate in the UAT as illustrated in Table 4. They have been briefed on the working of the system and its purpose and are given a list of system deliverable tasks before the testing begins.

5 Conclusion

With QR code technology, this paper has demonstrated the ability for smart parking system to integrate payment system to further minimize crowded and time-saving.

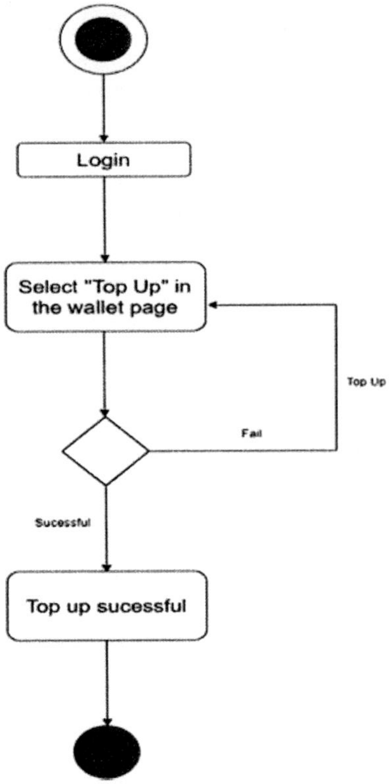

Fig. 17 Activity diagram to illustrate the top-up process

With the use of only auto pay machine, people need to spend the time to line up to pay their parking fees. By the way if the machine system down, it will cause more problem in the process of pay parking fees. However, mobile smart parking system solve all, users can make QR payment easily directly without having to install additional apps. The application top-up function is linked with the FPX. The application will be developed and used in UCSI, and students can directly login through their student ID and password, which is the same as IIS and CS, allowing the system to provide a more secure parking payment approach. It is highly likely that users will not pay parking fees in the proposed system, it is because these managements depend on security guards. Therefore, only when they are on patrol can they know which user did not pay the parking fee. However, most users do not pay parking fees when the security guard is not patrolling. Besides that, the application is easily affected by the network, it is because sometimes the UCSI network will be unstable, and some devices of students do not have Internet data to support, which will cause students cannot start parking through this application, or even cannot complete the payment. This will waste the time of students in waiting. By implementing offline version into the system, it can provide students with a mobile smart parking system that is completely unaffected by the Internet, especially in raining. Students do not have to worry about sometimes being unable to pay or scan codes. Moreover, social

Table 4 System deliverable tasks

No	Area		Item	Status
1	Management control panel (App)	Login page	Login panel	Passed
2	Management control panel (App)	Menu page	Multiple users operation (for multiple students)	Passed
3	Management control panel (App)	Menu page	Singular user operation	Passed
4	Management control panel (App)	Menu page	Listing of menu	Passed
5	Management control panel (App)	Menu page	Removing menu	Passed
6	Management control panel (App)	Home page	Multiple users operation (for multiple students)	Passed
7	Management control panel (App)	Home page	Singular user operation	Passed
8	Management control panel (App)	Home page	Feature of scanner	Passed
9	Management control panel (App)	Home page	Display of parking	Passed
10	Management control panel (App)	Profile page	Information march with users	Passed
11	Management control panel (App)	Payment page	Amount is march with parking hour	Passed
12	Management control panel (App)	Payment page	Payment successful	Passed
13	Management control panel (App)	Wallet page	Display of amount	Passed
14	Management control panel (App)	Wallet page	Feature of top-up	Passed
15	Management control panel (App)	Logout	Logout successful	Passed

media like Facebook provides API to bind external accounts with their accounts. This allows students account in UCSI linking to social media accounts. Users who link their accounts with social media account can easily access to the system through any devices. Students can scan QR code by the Facebook, not just in the mobile smart parking application. However, if students in scan in other social media application, the payment method will be converted to online transfer, not the wallet provided.

References

1. S.K.A.A.I.H. Topacoglu, Effect of exhaust emissions on carbon monoxide levels in employees working at indoor car wash facilities. [Online]. Available at: https://www.ncbi.nlm.nih.gov/pmc/articles/PMC4103039/. [Accessed 23 Jan 2020]
2. L.P.L.P.M.L.S.R.V.L. Mainetti, Integration of RFID and WSN technologies in a smart parking system. IEEE Xplore -(-) 22 (2017)
3. W.Y.D.B.R.S.O. Gongjun Yan, SmartParking: a secure and intelligent parking system. IEEE Xplore -(-) 19 (2018)
4. R. Weinstein, RFID: a technical overview and its application to the enterprise. IEEE Xplore -(-) 23 (2017)
5. K. Hassoune, W. Dachry, F. Moutaouakkil, H. Medromi, Smart parking systems: a survey. IEEE Xplore -(-) 24 (2016)
6. W.-J. Park et al., Parking space detection using ultrasonic sensor in parking assistance system. IEEE Xplore -(-) 28 (2019)
7. S.R. Dario Di Nocera, A social-aware smart parking application. RearchGate -(-) 19 (2015)

Traffic Congestion Analysis in India and Fluid Flow Reduction Loss Theory

Tsutomu Tsuboi

Abstract This study is a series of traffic congestion analysis in India for more than one-year observation of joint government project between Japan and India since April 2017. The uniqueness of this study is a method to identify traffic congestion by bottleneck theory of fluid flow reduction loss theory. In general, it is well-known that traffic flow analysis is used fluid flow analysis, and several famous traffic flow formulas are born such as Greenshields, Greenburg, and Underwood, and shockwave analysis theory is also used for traffic congestion condition. In this study, it is the first time to use shockwave theory of fluid flow for finding out traffic congestion timing in daily basis traffic flow data. Based on this study, fluid flow analysis about flow reduction theory and shock wave is valid for Indian traffic congestion analysis.

Keywords Traffic flow · Traffic congestion · Fluid flow analysis · Shock wave

1 Introduction

This study is one of series of traffic flow analysis in India, which is joint government-founded project between Japan and India (i.e., Program ID JPMJSA1606 of the International Science and Technology Cooperation Program (SATREPS) for global challenges in 2016) [1]. In general, traffic flow analysis in developing countries such as India is always challenging because there is not enough actual traffic data under chaotic traffic condition. Several studies have been performed on traffic flow research [2, 3]. Studies have also been carried out specifically on Indian traffic. For example, A. Salim et al. used traffic density and space headway parameters to analyze traffic congestion [4], but measurement data were only obtained over four days in Chennai, India. M. Goutham and B. Chanda reported vehicle probe data in terms of the traffic volume and speed in Hyderabad, India, based on the Indian Road Standard IRC-106-1990 [5].

T. Tsuboi (✉)
New Business Creative Division, Nagoya Electric Works Co. Ltd, Ama-shi 29-1, Mentoku Shinoda, 490-1294, Japan
e-mail: t_tsuboi@nagoya-denki.co.jp

A. K. Bashir et al. (eds.), *Proceedings of International Conference on Computing and Communication Networks*, Lecture Notes in Networks and Systems 394,
https://doi.org/10.1007/978-981-19-0604-6_8

From related work, we have already had some of typical conclusion as follows. All traffic data has been collected at Ahmedabad City in India and moreover a year measurement.

- The characteristics of traffic flow are not followed with traffic flow theory. But by using envelope observation of traffic flow characteristics, it is capable to estimate traffic flow parameter such as traffic jam density, traffic-free speed, critical traffic volume, and so on [6, 7].
- Traffic congestion happens in the evening time rather than in the morning in Ahmedabad City which has been observed during 2019 whole years. The reason of this evening traffic congestion is not clear [8, 9].
- Traffic congestion condition is represented by traffic occupancy parameter than traffic volume, which means traffic congestion is not occurred at maximum number of traffic on roads [10].
- There is some time lag between traffic volume and occupancy in traffic congestion in Ahmedabad [11].

Based on the above work, the traffic condition becomes a little bit clear and it is recognized that "fluid mechanics" is useful for traffic flow analysis. In terms of traffic flow theory, it is known that basic traffic flow equations such as Greenshields [12], Greenburg [13], and Underwood [14] equations are born from "fluid mechanics." In this study, it focuses on traffic congestion mechanism in more detail with "fluid mechanics" and reaches evening traffic congestion mechanism.

The next Sect. 2 introduces test field information and measurement of traffic flow, and Sect. 3 is theoretical analysis for traffic congestion. In Sect. 4, discussion of traffic congestion is introduced. In Sect. 5, it summaries conclusion.

2 Test Field and Measurement

2.1 Test Field

In this study, we choose one of the major cities in India which is Ahmedabad City where locates in the west part of India. Its population is about 6 million at 2011 shown in Ahmedabad profile Table 1 [15]. The current population is over 8 million [16] in 2018 from 5 million in 2011, and the number of vehicles is about 4 million in 2017 [17]. More than 70% vehicle is two wheelers, which is typical percentage in developing countries. Based on city government or Ahmedabad Municipal Corporation (AMC) agreement, the case study field was chosen in west side of the city called "New City" where there will have more commercial business and new building constructions.

There are 31 traffic monitoring cameras in the city, and its traffic data is collected at every minutes 24 h per day. The camera's installation points are shown in Fig. 1. The number shows each traffic monitoring camera location. Two digit number cameras are

Table 1 Ahmedabad profile

Coordinates	23.03° N 72.58° E
Area	466 km^2 (year 2006)
Population	5,577,940 (year 2011 Census)
Density	11,948/km^2
Literacy rate	89.60%
Average annual rainfall	782 mm
Popularly known as	Amdavad
STD code	079

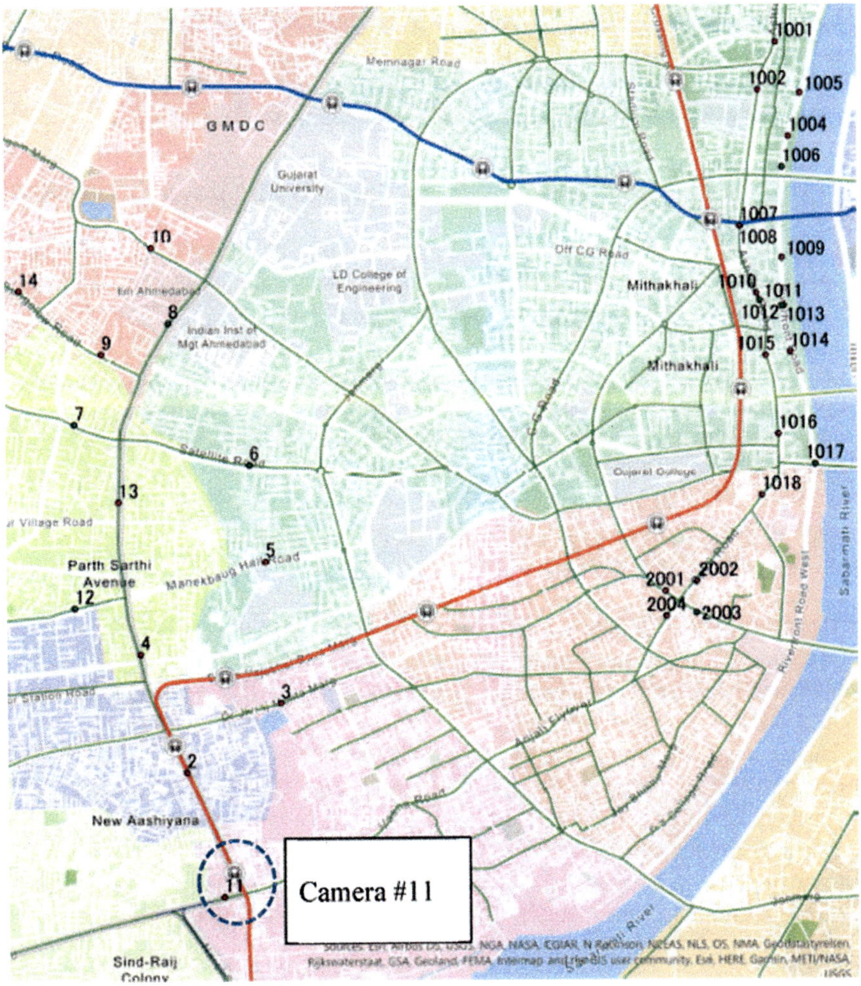

Fig. 1 Traffic monitoring camera location (each number is indicated each camera ID)

Fig. 2 Traffic monitoring camera example in Ahmedabad City

located in west side of block along with "132 Feet Ring Road" where there are many office, shops, and market. The four digit number cameras are located near the river "Sabarmti River" so-called RiverFront and new development area. The RiverFront is control area by city government, and it is not allowed to enter by auto-rickshaw and city bus. This road is currently used for a way to airport from city center. The main study location is on these both area because it is highly developing for business and the traffic grows heavily year by year. And there is another aspect which is metro development plan. The blue and red lines in the map are metro lines which are under development by 2023. This project is handled by Japanese International Corporation Agency or JICA and this metro is the first line in Ahmedabad followed by other major city such as Delhi metro.

The traffic monitoring camera and FLIR's company "TrafiCam™" are shown in Fig. 2. The cameras are relatively cost-effective and easy installation with reasonable performance. This camera is able to measures the number of vehicles, speed of vehicles, and other traffic flow parameters such as traffic density, traffic volume (Q), headway (HD) (length between the end of front parameter), vehicle length (L), and so on. The actual measurement data example is shown in Fig. 3. The camera measures traffic data at each one minute, 24 h, one month, number is measurement point, day is measured day of each month, and hour is from 0:00 to 23:59, cam_id is camera number which is same as in Fig. 1.

Figure 4 shows typical Ahmedabad traffic condition scene, and there are heavy traffic congestion these days.

No.	day	hour	cam_id	zone_id	density	idx	vehicles	speed	Q	OC	HD	L
1	1	0	11	161	6	0	3	33	198	4	15	6
2	1	0	11	162	9	0	5	39	351	5	6	6
3	1	0	11	161	7	0	3	27	189	4	14	6
4	1	0	11	162	8	0	4	35	280	5	23	6
5	1	0	11	161	11	0	5	29	319	7	16	6
6	1	0	11	162	7	0	3	30	210	4	11	6
7	1	0	11	161	11	0	4	24	264	7	10	6
8	1	0	11	162	7	0	6	52	364	4	13	6
9	1	0	11	161	2	0	1	26	52	1	20	6
10	1	0	11	161	7	0	3	30	210	4	20	6
11	1	0	11	162	9	0	5	37	333	5	15	6
12	1	0	11	161	10	0	6	36	360	6	9	6
13	1	0	11	162	7	0	5	42	294	4	9	6
14	1	0	11	161	10	0	5	34	340	6	12	6
15	1	0	11	162	10	0	5	37	370	8	12	8
16	1	0	11	161	3	0	2	35	105	2	14	6
					~							
82920	31	23	11	162	11	0	6	39	429	8	8	7
82921	31	23	11	161	10	0	5	36	360	6	10	6
82922	31	23	11	162	9	0	6	42	378	5	11	6
82923	31	23	11	161	6	0	3	31	186	4	18	6
82924	31	23	11	162	12	0	7	37	444	7	10	6
82925	31	23	11	161	28	0	10	28	784	21	4	7
82926	31	23	11	162	26	0	10	30	780	18	4	7
82927	31	23	11	161	9	0	3	20	180	5	7	6
82928	31	23	11	162	27	0	10	30	810	19	3	8
82929	31	23	11	161	15	0	7	30	450	9	9	6
82930	31	23	11	162	28	0	15	34	952	16	3	6

Fig.3 Traffic flow measurement data example

Fig. 4 Traffic condition example in Ahmedabad City

2.2 *Measurement Data*

As for traffic flow analysis, the location at camera #11 (doted black circle in Fig. 1) is chosen in this study where is most crowded area in our previous research. In Fig. 5, it shows three types of time zone-based traffic characters. The measurement data is in October 2020 one month. The traffic volume is defined as the number of vehicles passing a point on road or given lane or direction of a road in specific time and usually expressed as vehicles/hour. The average speed is defined as the average vehicle speed in a road. The occupancy is defined proportion of time that a detector is occupied by

Fig. 5 Basic traffic flow characteristics at camera #11 in October 2020

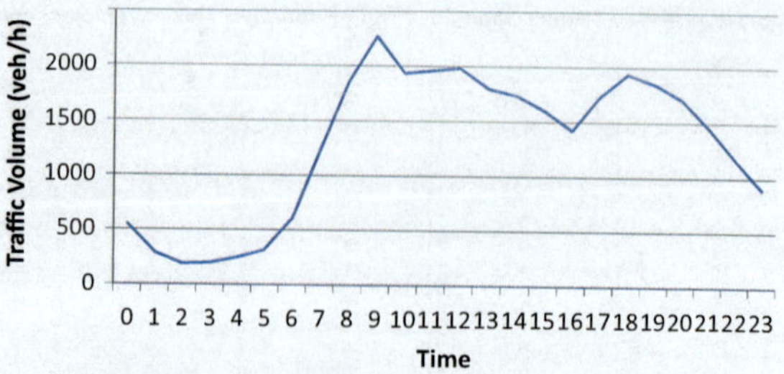

(a) Average Traffic volume time zone (Time is 24 hours' time)

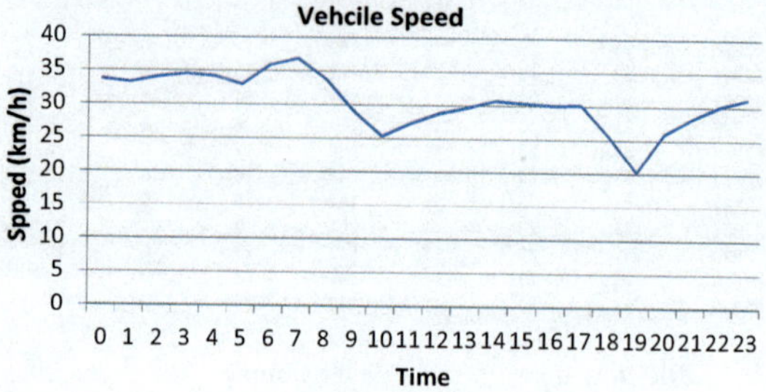

(b) Average Vehicle speed

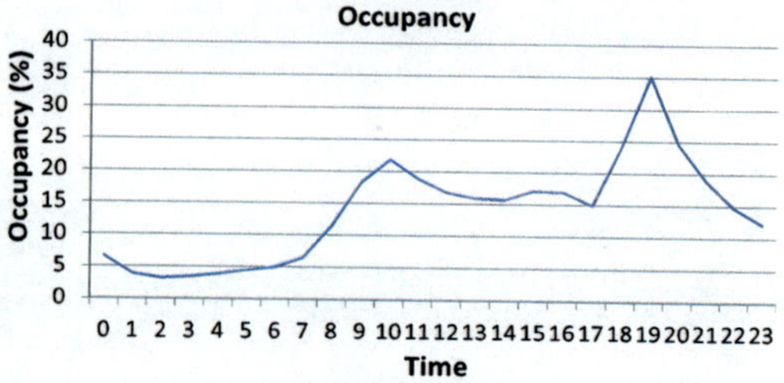

(c) Occupancy

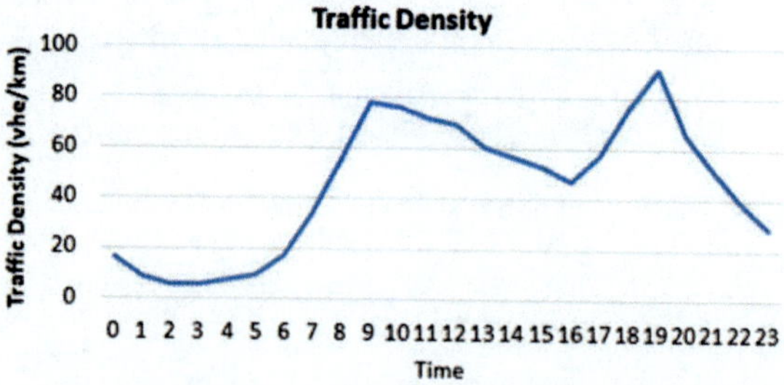

(d) Traffic Density

a vehicle on defined time period and is expressed percentage. It is used as parameter for traffic congestion condition in general. The traffic density is defined as direct measurement of traffic demand in certain length of a road and expressed normally vehicles/km.

From Fig. 5a–d, there are two peaks—one is at 10:00 and the other is at 19:00. From Fig. 5b, the average vehicle speed at 10:00 and 19:00 is 25 km/h and 20 km/h, so vehicle speed is slow down compared with other time zone. From Fig. 5c, there are also two peaks in traffic occupancy characteristics and each peak timing is at 10:00 and at 19:00. As previous mentioned, occupancy represents traffic congestion; therefore, traffic congestion is heavy at 19:00 compared with 10:00 in the morning. This has been already known in our previous study mentioned in. Introduction section. And there is also peak gap between traffic volume and occupancy. In Fig. 5a, c, traffic volume peak in the evening is at 18:00 and occupancy peak is at 19:00. The peak gap is about one hour. From correlation coefficient analysis between occupancy and volume, Fig. 6 shows its mutual correlation coefficient characteristics. From Fig. 6, it is clear that the lag is one hour delay from traffic volume to occupancy. The ACF is autocorrelation function which is correlation between different points on the time series. In this case, there is one hour time lag between traffic volume and occupancy at Camera #11. In our previous study [10], two hours lag at Paldi junction at Camera #2001, #2002, #2003, and #2004 in Fig. 1.

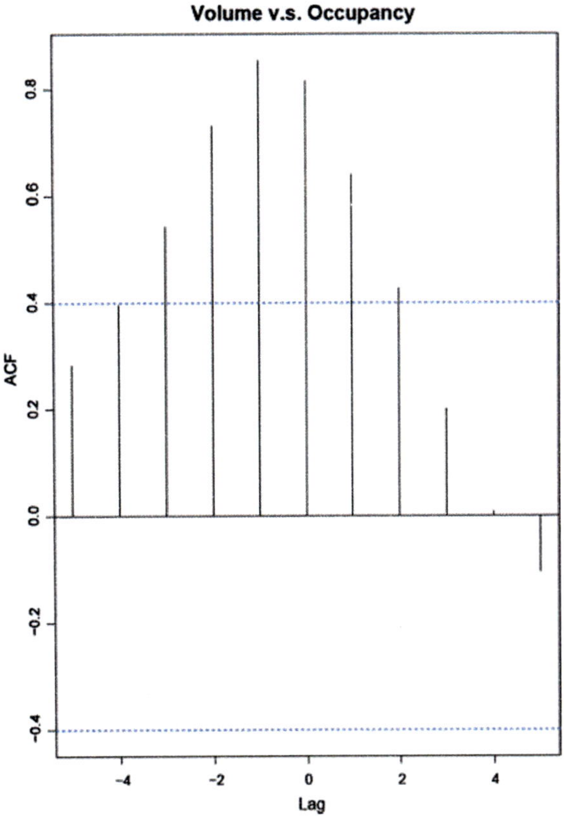

Fig. 6 Mutual correlation coefficient characteristics

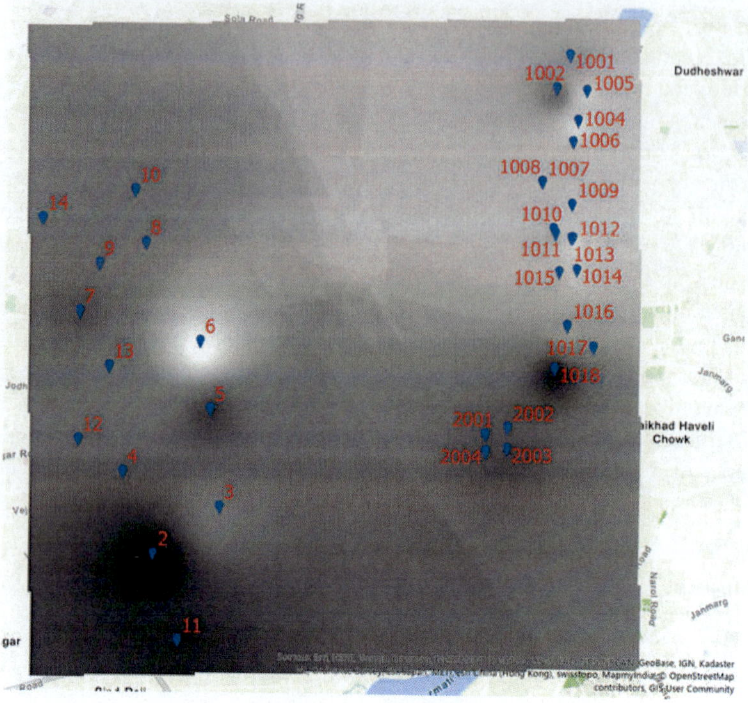

Fig. 7 Traffic density condition in Ahmedabad on October 2, 2020, at 18:00

So far, the characteristics shown in this section are based on Camera #11. When we look at all other location, Fig. 7 shows total traffic density in measurement area as an example of traffic flow characteristics at 18:00 on October 2, 2020, as an example. From Fig. 7, we see seral hot traffic density locations such as Camera #2, #15, and #1018. From our previous study, the location Camera #2 is always congested are and we know the reason from new flyover bridge construction between Camera #2 to Camera #2003 area. In the evening, majority vehicle direction is from Camera #2 to RiverFront area where more people lives. In order to traffic congestion analysis, we take camera #11 which is the second traffic congested area followed by Camera #2.

3 Traffic Congestion Theory

3.1 Fluid Flow Reduction Model

There are several traffic congestion theories. The major theoretical congestion analysis is used from the fundamental traffic flow characteristics such as traffic density to average vehicle speed such as Greenshields, Greenberg, and Underwood. This

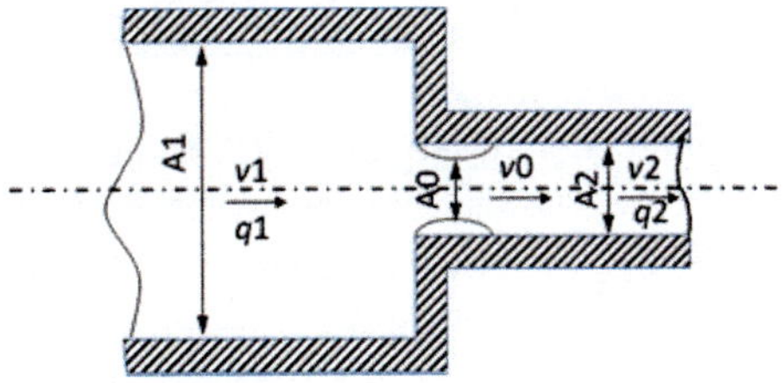

Fig. 8 Fluid flow reduction model with two different cross-sectional area

theoretical analysis has been done in previous study. Therefore, it is introduced fluid flow reduction loss theory at this time.

As introduced in introduction section, traffic flow theory comes from fluid flow mechanism. In order to flow congestion analysis, we take fluid flow reduction model in this study. In Fig. 8, it shows fluid pipe which has two different cross-sectional area, one is small diameter followed by large diameter. This model is good example to change free fluid flow to congested flow. In large diameter A_1 area, fluid flow runs freely at speed v_1. When fluid runs into the narrow diameter A_0 area, fluid flow reduction loss occurs with speed v_0. After through the entrance of the narrow diameter, the fluid flow runs with speed v_2 at diameter A_2.

In Fig. 8 model, the flowing Eq. (1) is given from fluid flow theory [18].

$$h_s = \left(\frac{A_2}{A_0} - 1\right)^2 \frac{v_2}{2g} \tag{1}$$

where h_s is head fluid loss and g is gravity constant. When fluid loss factor is ζ, Eq. (2) is established.

$$\zeta = \left(\frac{A_2}{A_0} - 1\right)^2 \tag{2}$$

When fluid volume is defined each diameter area q_1 at A_1 diameter and q_2 at A_2 diameter area, Eq. (3) is established.

$$q_2 = (1 - \zeta) \times q_1 \tag{3}$$

Then A_0/A_2 is given as Eq. (4) from Eq. (1)–(3).

$$\frac{A_0}{A_2} = \frac{1}{1 + \sqrt{1 - \frac{q_2}{q_1}}} \tag{4}$$

where under the condition of $q_2 \leqq q_1$.

3.2 Traffic Flow Shock Wave

In this section, it introduces the traffic flow shockwave theory. When there are two phases condition of a road which is illustrated in Fig. 9, the number of vehicle passing boundary S is same between cross section A_1 and cross section A_2. When its number of vehicle at boundary S is N, Eq. (5) is established.

$$N = (v_1 - c)k_1 t = (v_2 - c)k_2 t \tag{5}$$

where c is moving speed at boundary S, k is traffic density at each areas 1 and 2, and t is time.

From Eq. (5), Eq. (6) is established.

$$(v_1 - c)k_1 = (v_2 - c)k_2 \tag{6}$$

Therefore, Eq. (7) is established.

$$c = \frac{v_2 k_2 - v_1 k_1}{k_2 - k_1} = \frac{q_2 - q_1}{k_2 - k_1} \tag{7}$$

When $q_2 - q_1 = \Delta q$ and $k_2 - k_1 = \Delta k$, Eq. (7) is shown as Eq. (8) by small changes of each parameters.

$$\mathrm{c} = \frac{\Delta q}{\Delta k} = \frac{\mathrm{d}q}{\mathrm{d}k} \tag{8}$$

When it takes the differentiation by (k) of both side, Eq. (9) is established because of $q = kv$.

$$c = \frac{\mathrm{d}q}{\mathrm{d}k} = v + k\frac{\mathrm{d}v}{\mathrm{d}k} \tag{9}$$

From Greenshields equation, $\mathrm{d}v/\mathrm{d}k < 0$ is established. Then Eq. (10) is established.

$$c < v \tag{10}$$

Fig. 9 Traffic flow shockwave model

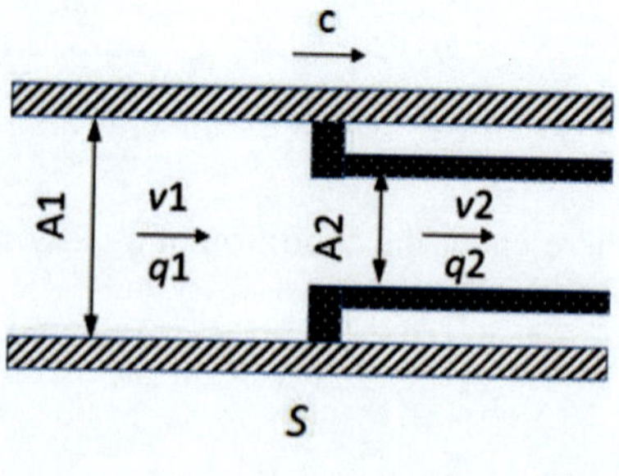

Therefore, shock wave is transferred from the preceding vehicle in traffic flow to the subsequent vehicle. From traffic flow theory, critical traffic density is k_c and jam traffic density is k_{j}, Eq. (11) is established and shock wave is transferred forward direction and backward direction.

$$\begin{gathered} c > 0 \quad \text{where} \quad 0 \leq k < k_c \\ c = 0 \quad \text{where} \quad k = k_c \\ c < 0 \quad \text{where} \quad k_c \leq \mathrm{k} < k_j \end{gathered} \tag{11}$$

In terms of shockwave analysis, there are several study before [19, 20]. From analysis experience, shock wave is negative value which means the wave is transferred backwards under congestion condition.

4 Discussion

4.1 Traffic Congestion from Data

From Sect. 2.2 measurement data at Camera #11, the traffic congestion occurs during the time frame from 17:00 to 21:00. Let us pick up measurement data from traffic volume, traffic density, and occupancy in Fig. 10.

From Fig. 10, the traffic congestion occurs at 19:00 from occupancy. Traffic volume declines from 18:00 to 19:00 but traffic density grows from 18:00 to 19:00. Accordingly Sect. 3.1, when traffic flow is at area A1 as stage 1 and area 2 as stage 2, the result of Eq. (4) is summarized in Table 2 by using Eq. (4).

From Table 2, each area size relation of Fig. 6 is following Eq. (12),

$$A_0 (= 0.82 A_2) < A_2 < A_1 \tag{12}$$

Equation (12) means that congested point area becomes narrow about 80% from A_2 area and a road width becomes relatively narrow from normal condition. From our actual traffic observation in Ahmedabad, there are many vehicle parking along the road, especially in the evening. Therefore, its road width becomes narrow physically. This reason is based on our actual observation during this study and the measurement time was during pandemic of COVID-19 in India in October 2020. Therefore, all works had to return home before 20:00. This may cause many vehicles moving at the same time frame. But it is necessary to have more detail analysis in the future.

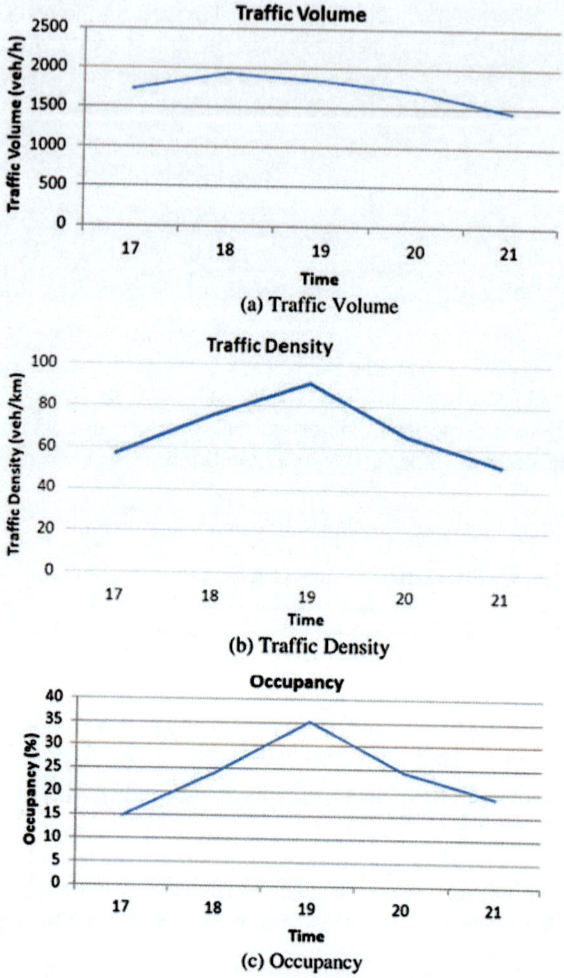

Fig. 10 Traffic congested time frame characteristics

Table 2 Parameters from measurement value

Stage	1	2
Time	18:00	19:00
q	1937	1849
K	76	91
A_0/A_2	–	0.82

4.2 *Traffic Shock Wave from* Data

Based on Sect. 3.2, it is able to get traffic shockwave data from measurement data of Fig. 10. The summary is shown in Table 3 by using Eq. (7).

Table 3 Traffic shockwave parameter from measurement value

Stage	1	2
Time	18:00	19:00
q	1937	1849
k	76	91
q_2/q_1	0.95	
c	−5.69	

From this Table 3 result, it is able to identify traffic congestion time frame using total traffic measurement value in Table 4. In Table 4, it is highlighted value in q_{i+1}/q_i, and c column, where q_{i+1}/q_i, is less than 1.0, and c is negative. Under both condition matchings with q_{i+1}/q_i are less than 1.0, and c is negative is at 19:00 o'clock. This

Table 4 Identify traffic congestion time frame from measurement value at Camera #11

Time	Speed	Density	Volume	$q_i + 1/q_i$	c	Occupancy
0	33.67745	16.25957	547.5806	NA	NA	6.680339
1	33.25795	8.758516	291.2903	**0.531959**	34.16727	3.931052
2	34.06058	5.578296	190	**0.65227**	31.85009	3.199501
3	34.4822	5.799159	199.9677	1.052462	45.13075	3.455209
4	34.16009	7.341145	250.7742	1.254073	32.94871	3.922187
5	32.98177	9.651469	318.3226	1.269359	29.23763	4.508371
6	35.87269	16.93894	607.6452	1.908897	39.70139	5.027115
7	36.87139	33.63041	1240	2.040665	37.88491	6.551053
8	33.84547	54.88227	1857.516	1.497997	29.05704	11.54952
9	29.01919	78.30526	2272.355	1.22333	17.71075	18.25217
10	25.38498	76.22744	1935.032	**0.851554**	162.3446	21.8481
11	27.2762	71.89417	1961	1.01342	**-5.99265**	18.84659
12	28.75329	69.02399	1984.667	1.012069	**−8.2457**	16.78016
13	29.69614	60.55445	1798.233	**0.906063**	22.01221	15.94807
14	30.72042	56.54762	1737.167	**0.966041**	15.24066	15.75688
15	30.37589	52.76225	1602.7	**0.922594**	35.52264	17.11845
16	30.11052	47.28359	1423.733	**0.888334**	32.66617	17.0206
17	30.17894	57.40316	1732.367	1.216777	30.49867	14.98445
18	25.54571	75.83791	1937.333	1.118316	11.11849	24.12406
19	20.23354	91.38132	1848.968	*0.954388*	*-5.68508*	**35.18498**
20	26.03964	65.71124	1711.097	**0.925434**	5.370881	24.53213
21	28.28883	51.62417	1460.387	**0.85348**	17.79715	18.88877
22	30.05933	38.78674	1165.903	**0.798352**	22.93947	14.74714
23	31.20345	28.81298	899.0645	**0.771131**	26.75409	12.27578

result is matching with the result of Fig. 5 basic traffic flow characteristics. In general, traffic congestion is judged by occupancy value and its value is greater than 25%. The occupancy at 19:00 from Table 4 is 35%, and it can be said it is very congested condition.

As for comparison with other non-congestion location such as Camera#10, Table 5 shows each parameters of the same category of Table 4. From Table 5, there is no negative value of shock wave (*c*). And there is higher value of occupancy than 20%, which means there is no traffic congestion at Camera#10. As the result, the traffic congestion analysis by using shockwave method is valid.

Table 5 Identify traffic congestion time frame from measurement value at Camera #10

Time	Speed	Density	Volume	q_{i+1}/q_i	c	Occupancy
0	35.3654	4.6397	131.0333	NA	NA	1.5539
1	33.6712	2.8605	52.1667	**0.3981**	44.3261	0.7413
2	33.3390	2.4658	28.5667	**0.5476**	59.8000	0.5106
3	33.6704	2.5435	39.7000	1.3897	143.3767	0.4892
4	34.5366	2.7328	63.2667	1.5936	124.4989	0.5772
5	33.7823	3.2661	88.5333	1.3994	47.3787	0.8337
6	36.1014	4.3909	138.4667	1.5640	44.3913	1.7049
7	34.2269	7.8134	311.3333	2.2484	50.5089	2.6550
8	34.1840	13.0212	529.1000	1.6995	41.8154	4.8803
9	34.5464	24.7220	1074.3333	2.0305	46.5981	8.5922
10	34.3634	30.7293	1390.2414	1.2941	52.5873	10.4057
11	34.2160	29.0135	1281.0357	**0.9214**	63.6475	9.6122
12	34.4769	26.4472	1170.5172	**0.9137**	43.0659	8.5939
13	34.9151	24.6933	1072.9310	**0.9166**	55.6368	8.2199
14	35.1698	22.6529	925.2414	**0.8623**	72.3827	7.8970
15	34.3944	22.4425	827.2759	**0.8941**	465.6512	8.4542
16	33.2330	23.9959	841.5862	1.0173	9.2121	9.2993
17	33.7304	25.3821	1063.7931	1.2640	160.2975	8.5608
18	33.3436	30.1549	1342.1034	1.2616	58.3126	11.3874
19	32.8251	29.2365	1245.4000	**0.9279**	105.2957	12.9033
20	33.9966	21.6439	949.7241	**0.7626**	38.9431	9.1390
21	33.9530	18.4296	781.5172	**0.8229**	52.3294	7.6630
22	34.3300	15.2071	562.4828	**0.7197**	67.9718	6.4684
23	34.4075	9.8858	299.3667	**0.5322**	49.4458	4.0643

5 Conclusion

In this study, we bring fluid flow theory for identify traffic congestion. There are two unique congestion analysis methods. One is fluid flow reduction model, and it enables to explain traffic congestion by pseudo-narrowing road width which means shoulder of road is occupied by parking vehicles. The other is traffic shock wave, and its negative value is identified as one of traffic congestion condition. And one more parameter for traffic congestion is traffic volume decrease. Both conditions are able to identify traffic congestion time frame from traffic measurement data.

The above this method is valid for Indian traffic analysis. It is necessary to continue to collect data from the current test field and check with any other monthly differentiation condition.

In the future work, there are more than one-year measurement data in all locations in Fig. 1. Therefore, it is necessary to location-based traffic congestion analysis in other locations and finds any reason for traffic congestion condition among each month on whole year. And in 2020, there is pandemic condition by COVID-19. Therefore it is also necessary to consider about this pandemic situation for traffic congestion.

Acknowledgements This research is part of SATREPS program 2017 (ID: JPMJSA160 between India and Japan

References

1. T. Tsuboi, Y. Tatsugami, C. Kikuchi, B. Mallesh, A smart mobility of visual traffic information by India ITS project using PPP business model. SMART CITIES 2015 INDIA Forum, 20–22 May 2015
2. N. Gartner, C.J. Messer, A.K. Rathi, Traffic Flow Theory A State-of-the-Art Report: Committee on Traffic Flow Theory and Characteristics (AHB45) (2001)
3. C. Millikarjuna, K.R. Rao, Area occupancy characteristics of heterogeneous traffic. J. Transportmetrica **2**(3) (2006)
4. A. Salim, L. Vanajakshi, C. Subramanian, Estimation of average space headway under heterogeneous traffic conditions. Int. Recent Trends Eng. Technol. **3**(5) (2010)
5. M. Goutham, B. Chanda, Introduction to the selection of corridor and requirement, implementation of IHVS (Intelligent Vehicle Highway System). In Hyderabad, Int. J. Mod. Eng. Res. **4**(7), 49–54 (2014)
6. T. Tsuboi, Quantitative traffic congestion analysis approach in Ahmedabad. Adv. Sci. Technol. Eng. Syst. J. **4**(3), 183–189 (2019)
7. T. Tsuboi, N. Yoshikawa, Traffic flow analysis in emerging country (India). CODATU XVII (2017)
8. T. Tsuboi, Time zone impact for traffic flow analysis of Ahmedabad city in India. In *4th International Conference on Vehicle Technology and Intelligent Transport Systems 2019 (VEHITS)*
9. T. Tsuboi, Traffic congestion visualization by traffic parameters in India. In *The 2nd International Conference on Intelligent Control and Computing (ICICC 2019)*

10. T. Tsuboi, Traffic congestion "Gap" analysis in India. In *6th International Conference on Vehicle Technology and Intelligent Transport Systems 2021 (VEHITS)*
11. T. Tsuboi, Traffic congestion triangle" based on more than one-month real traffic big data analysis in India. In *International Virtual Conference on Multidisciplinary Research (IVCMR 2020)*
12. B.D. Greenshields, A study of traffic Capaci. In *Proceedings of H.R.B*, vol. 14 (1935), pp. 448–477
13. H. Greenberg, An analysis of traffic flow. Oper. Res. **7**, 79–85 (1959)
14. R.T. Underwood, Speed, volume and density relationships, in *quality and theory of traffic flow*. (Bureau of Highway Traffic, Yale University, 1961), pp. 141–188
15. Ahmedabad Municipal Corporation Available From: https://ahmedabadcity.gov.in/portal/jsp/Static_pages/introduction_of_amdavad.jsp. Accessed: 2020–08–21
16. Population of India [Internet] 2020. Available From: https://indiapopulation2019.com/population-of-ahmedabad-2019.html. Accessed: 2020–08–21
17. Registered number of vehicles Ahmedabad India FY 2006–2017. Available From: https://www.statista.com/statistics/665754/total-number-of-vehicles-in-ahmedabad-india/. Accessed: 2020-08-21
18. S. Mastuoka, T. Aoyama, et al., *Fluid Mechanics—Fundamentals and Exercise*. Corona Publishing Co. Ltd. (2001), pp. 152–163
19. M. Fukui, Y. Sugiyama, M. Schreckenberg, D. E. Wolf (eds.) Traffic and Granular Flow '01. Springer (2003) to appear. http://traffic.eken.phys.nagoya-u.ac.jp/tgf01/
20. M. Suzuki, T. Tokuda, H Shigeno, Shockwave control method considering CACC stability. In *The 75th National Convention of Information Processing Society of Japan* (2013), pp. 3-84–85

Building an AI/ML Based Classification Framework for Dark Web Text Data

Ch. A. S. Murty, Harmesh Rana, Rachit Verma, Roshan Pathak, and Parag H. Rughani

Abstract The dark web is that integral part of WWW that provides freedom of Content Hosting. The dark web is accessible with specially designed browsers and tools of having peer-to-peer network technology such as TOR, IP2, and FREENET etc. These tools help users exchange information on the dark web while remaining anonymous; We used TOR (The Onion Router) browser in our research for understanding the dark web. It provides excellent anonymity to its users. The users mainly utilized the dark web for illegal activities; although accessing it is legal in most countries, its usage can arouse suspicion with the law. Categories like Adult, Counterfeits, illicit markets, and weapons are prevalent. This research provides an analytical framework for automating the classification of web pages with scraping and analysis of its hosted content on the dark web. The method we used can easily crawl data, classify the hosted content by machine learning model, and categorize that the hosted content is illegal and legal. The proposed framework contains Machine Learning Classifier Algorithms that are Naïve Bayes with an accuracy of 0.87%, Random Forest with an accuracy of 0.91%, Linear SVM with an accuracy of 0.91%, and Logistic Regression with an accuracy of 0.94%. This study created a machine learning framework that can classify hosted text content on dark web websites—models trained to classify the content and evaluated them based on accuracy. The results from our study validate the effectiveness of the proposed classification framework for analyzing the text

Ch. A. S. Murty (✉) · H. Rana · R. Verma · R. Pathak
Centre for Development of Advanced Computing (C-DAC), Hyderabad, India
e-mail: chasmurty@cdac.in

H. Rana
e-mail: harmeshr@cdac.in

R. Verma
e-mail: rachitv@cdac.in

R. Pathak
e-mail: roshanp@cdac.in

P. H. Rughani
National Forensic Sciences University (NFSU), Gandhinagar, India
e-mail: parag.rughani@nfsu.ac.in

A. K. Bashir et al. (eds.), *Proceedings of International Conference on Computing and Communication Networks*, Lecture Notes in Networks and Systems 394,
https://doi.org/10.1007/978-981-19-0604-6_9

data, which has more relevance with smaller datasets. Also, it is encouraging further studying this growing phenomenon and for investigators examining illegal activities on the Dark Web.

1 Introduction

WWW (Surface Web) is relatively easy to access and traverse, using traditional web browsers like Internet Explorer, Google Chrome, Apple's Safari etc. The surface web extended to the deep web, where the data in the deep web is not accessible and not indexed. For example, traditional search engines such as Google, Bing cannot search the data inside an organization rather than the meta-data of an organization. As part of the surface web, the content exchanged is with the indexed data. This unindexed portion of WWW that intentionally hidden and inaccessible by Standard Browsers referred to as the Dark Web [1]. Some Studies state that the Surface Web is only 4 (Percentage) of Web Content indexed by the publicly available search engines like Google, Bing, etc. According to TOR Metrics Project, more than 200 K registered. Onion-based addresses as of May 2020, and on average, daily 2 million users use TOR Browser [2] to access the deep or dark web.

The World Wide Web is like an iceberg where we see the small portion above the surface during the significant part of the data with internal or hidden as the large iceberg under the sea. For accessing the web, the Internet Protocol (IP) address is required and, the same is provides by the Internet Service Provider (ISP). This IP Address identifies a system as a unique identifier as part of communication over the Internet. Also, it helps in finding the geolocation of any device connected over the Internet. However, in the dark web, end-to-end device communication is encrypted by providing anonymity of end devices, routers, communication over the dark web. These Routers communicate through a network of relays run by volunteers supporting around the globe Alnabulsi et al. [3].

The deep and dark web provides a venue for malicious actors to coordinate cyber-attacks [4], illicit activities such as human trafficking, terrorist activities, etc. The main reason for using the dark web is its features such as encrypted protocols, anonymity, hidden in nature. While accessing the dark web, the dark web users encouraged neither the risk of getting caught by law enforcement nor being censored by a website Rosenbach et al. [5]. The US military researchers created a method of trading information over the TOR network anonymously in the 1990s by hiding all exchanged data's origin and destination. The nature of anonymous implemented in services over the TOR network and hence called "Hidden Services". The anonymity afforded by TOR has led to the dark web garnering a reputation as a gateway for illegal activity. In the TOR network, we may find famous marketplaces like Silk Road, Black market, and tons of legal and illegal services such as hacking, selling, illicit drugs, and purchasing products banned by the land law. In October 2013, D. P. Roberts was arrested by the FBI. The estimation was around 1.2 billion dollars for the Silk Road market and, approximately 4000 vendors and 150,000 anonymous customer accounts on the Silk Road database shows the popularity of the dark web.

Internet is divide into three main parts, but there are also levels [6] in the web such as Level 0—Common Web where everything is accessible, Level 1—The Surface Web where All information is publicly available, Level 2—Bergie Web: At this level, PROXY required for accessing the bergie web, as it loaded with FTP Servers, Web Servers, Google Locked Results, Jailbait, etc., Level 3—Deep Web loaded with only subscription-based content, Banned books/videos Forgery Documents, illicit material, Financial Records Censorship, Private Corporates database, Scientific Experiments, etc., Level 4—Charter Web divided into two parts as the TOR network and Special level Access, where the TOR browser cannot reach. The computers and servers are insufficient to reach this level. The Level-5 web is Mariana Web, where most of the portion is yet to discover.

The Law Enforcement Agencies (LEA) show [7] interest to understand the type of data hosted and their activities for any illegal activities in TOR-based network due to its services for any unlawful activities. The complex URL structure of the dark web text data is hard to compare with conventional web to map the website content with the website name, e.g. google.com is a search engine, facebook.com is a social media platform. Because of this, the end-users, including Law Enforcement Agencies (LEAs), need to rely on dark web search engines, hidden wikis, forums, etc., to know about a type of URL to access the data in the dark web. Further, Darknet is an overlay network within the Internet and accessed by specific software like TOR, I2P, Freenet, etc., part of the dark web to provide encrypted, decentralized, and anonymous communication to the stakeholders. Due to the anonymity and encrypted transmission nature of the dark web, scraping and crawling data from the dark web for such classification is more challenging when compared with the surface web.

The classification of data from onion domain URLs based on linguistic information has been an approach followed by many researchers. Also, some researchers worked on text data from the URL itself by dividing it into words, characters in URL as tokens for classification. Adopting characters in URLs as tokens for text classification is considered a faster method as it does not require any content in the respective website. Text Classification is an Automatic text classification (which also known as text categorization or topic spotting) is the task of assigning a label to a document based on several predefined categories Nalini et al. [8]. The Goal of Classification is to Categorize the text into a particular category. A data mining classification Model has x number of Documents as input of documents that are labeled with a class L, and determine a classification model.

$$X = (X_1, X_2, \dots X_n) \tag{1}$$

$$F : X \to L \text{ i.e where } (X) = L \tag{2}$$

Therefore, after training F as a model given X as a Document, the model will predict L as its label/category, assigning the correct class to a new document "X" of the domain Mehler et al. [9]. Data mining has various techniques such as classification of data, clustering data, building decision trees, neural networks, and data mining applications are text mining and web mining. Data mining is also called Knowledge Discovery from data (KDD), as it discovers and refers to knowledge concerning that data. Researchers may use different approaches to fetch their required data from the Internet. Data mining mainly deals with structured data organised in a database (DB) [10], while text mining deals with unstructured data/text. The mining of text-based keywords for the classification based on transactions, data hosted in web pages of the dark web is a challenge as the hidden nature of the dark web. Web mining [11] lies in between and copes with semi-structured data and unstructured data. *Web mining* is an application that uses data mining to analyse and discover interesting data patterns such as user's transactions or data usage on the web. Web content mining is the process of understanding information from hosted web page content that may consist of text, image, audio, or video data on the web. This framework also consists of a Machine Learning Model that predicts the category, legal/illegal, based on hosted content. A text analyser, which analysis the text and produces the result for respective analysis. The framework is comprised of three main steps and is reported in Fig. 1. The outcome of our study is relevant to both academicians and investigators seeking to examine the content and related nefarious activates on the dark web. The difference between surface web, deep web, dark web, and Darknet has been presented in [12].

A brief description of this research paper comes along with an abstract and an introduction to data and web mining, text classification, dark web, and Tor Network. The 2nd section discusses an overview of the literary works from earlier days to recent years. The System Architecture is briefly described in Sect. 3 and provides more information on how authors collected the data that portrays how we implemented this research scenario and framework. Dataset and data attributes are explained in Sect. 4. Section 5 shows the result achieved, and finally, in the 6th section author concludes the paper.

2 Literary Work

Nowadays, the research community has raised its interest in recognizing TOR and its Hidden Services, and machine learning is one of the emerging fields for classifying data and prediction. Website Classification is one of the processes for organizing the data of a website based on its hosted content such as text, images, etc. The classification includes the scraping of data, i.e. text, images from web pages, text classification required weighing techniques like BOW (Bag of words) and Term Frequency-Inverse Document Frequency (TF-IDF) [13]. Cleaning text is another step that ultimately re-moves the burden from the model to process irrelevant text. The Classification algorithms such as Gaussian Naive Bayes, Random Forest, Linear SVM, and Logistic Regression, etc., are used to build a classifier to predict. A very few members in the field of the Dark Web classification due to challenges like anonymity,

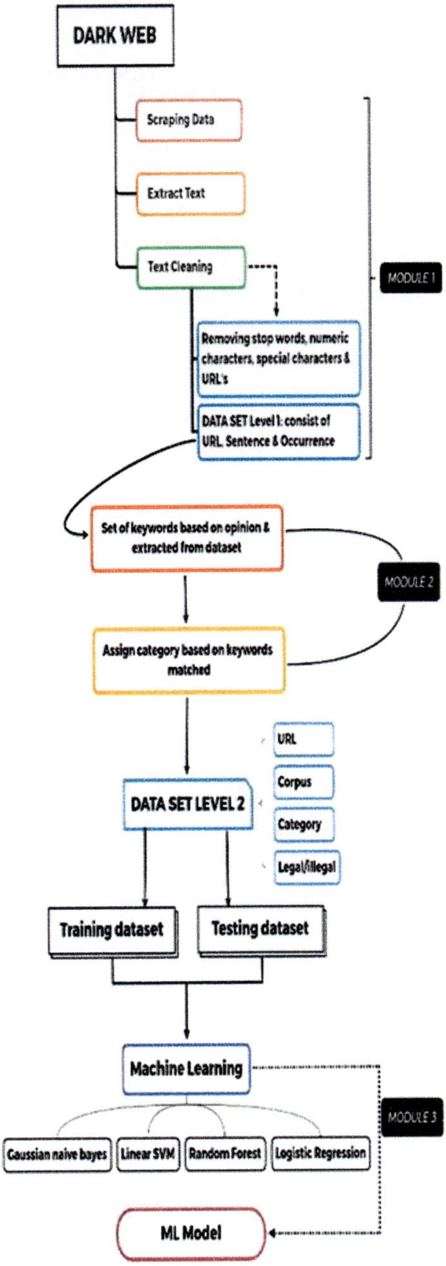

Fig. 1 Flow diagram of the research methodology

hidden in nature, difficulty in text data mining, low bandwidth speed, and need for more computational power.

A document of RAND Corporation stated that the expert panel of law enforcement practitioners, academic researchers, and civil rights advocates generally agreed that the research initiatives targeted new problems posed by the dark web. Improving capacities and information sharing are likely to significantly impact these problems posed by criminal activity on the dark web. Lack of knowledge about the dark web shows and how criminals have begun to leverage it is a key problem. Investigating officers often overlook physical artefacts indicative of the dark web activities when collecting evidence during a criminal investigation. These artefacts might include cryptocurrency wallets, encryption keys, or dark web addresses. The anonymity and encryption associated with the dark web activities make it much more difficult for investigators to assemble the evidence puzzle and prove that a crime has been committed [14]. In research on improvised explosive device webpages, Chen [15], presented a classifying framework about the extremists, how they can spread information about improvised explosive devices (IEDs) on the Internet. The author used a complex feature extraction, Extended Feature Representation, and Support Vector Machine (SVM) learning algorithm. Chen. H collected the web pages containing information about IED's data from the deep web. The accuracy of the approach was around 88%. The Genre classification has been implemented through the testbed on over two thousand and five hundred web pages. Manual categorization using a domain expert has identified material pages with approximately two thousand and five hundred discussions.

Robert [16], in his book "Archives for the Dark Web: A Field Guide for Study (2018)" has done some great work of several years of study about Dark web markets, search engines, and social networking sites. Weaving the Dark Web provides a history of Freenet, Tor, and I2P and details the politics of Dark Web markets, search engines, and social networking sites. In the book, the writer draws on three main streams of data: participant observation, digital archives, and the experience of running the routing soft-ware required to access the Dark Web.

Ozkaya and Islam [17], in their book "Inside the Dark Web", help to understand/learn the core concept of the dark web, emerging cyber- crime threats, the forms of cybercriminal activity through the dark web, and the technological and "social engineering" methods used to undertake such crimes. The dark web ecosystem with cutting edge areas like IoT, forensics, and threat intelligence, and so on. The dark web-related research and applications and up-to-date on the latest technologies and research findings in this area. The following Table 1 shows the differences among the three webs.

A survey performed by Michael [18] found that there is 400–550 times more information on the Deep Web than is available as public information on the Surface Web. The deep web contains 7500 terabytes of data compared to nineteen tera-bytes of information in the surface Web.

Researchers in the early 2000s have also worked on the surface web category due to the classification of web-based content. Dumais and Chen [19] performed for hierarchical classification of web content. Sun et al. [20] for Web classification

Table 1 Data summary

Attributes	Summary
URL	Address of the onion website
CORPUS	Extracted text data from the onion website
CATEGORY	Based on keywords matching in which class does website falls such as Adult, Crypto, etc
LEGAL/ILLEGAL	Content is legal/illegal

using support vector machines. Kan et al. [21] proposed to segment the URL into meaningful chunks with add-on components such as sequential and orthographic features to the model as silent features using the maximum entropy model and achieving more success. Prabhjot [22] surveyed several algorithms to make structural data after mining web content data. Further, the survey made on feature selection, analysis, and enhancement techniques as part of the data mining process. Su et al. [23] have proposed combining the Support Vector Machines (SVM) algorithm that classifies structured deep web data. Barbosa et al. [24] address organizing hidden databases by proposing clustering approaches for web forms and metadata. Noor et al. [25] discussed supervised learning algorithms and related standard extraction techniques for structured, unstructured deep web. Also, Xian et al. [26] proposed" Visible Form Features" for classification and, Khelghati et al. [27] also contributed to the monitoring methods for efficient web harvesting strategies.

Concerning the dark web classification, [28] proposed TOR hidden service classification for an Automatic product categorization for anonymous market places. Initially, they experimented with 5000 TOR onion-based web- sites as a sample and segregated them into 12 categories using an SVM classifier. Michał and Kevin [29] proposed pipelines to classify the black market named Agora on Dark-net. They classified into12 categories using TF-IDF for feature selection and PCA for feature selection in their pipeline architecture of SVM classification with an accuracy of 0.79.

Al Nabki et al. [30] discussed various supervised algorithms, especially on the Logistic Regression-based classifier for multiple activities of the TOR net-work and their illegality based on the dark web content. With 10-Fold cross-validation, achieved an accuracy of 96.6 and 93.7% of *F*-Score values with the Darknet Usage Text Addresses (DUTA) Dataset. In research of Classification of Illegal Activities on Dark Web, Al Nabki proposed a classification method that uses 'Federal Code of United States of America' as training data to their model, which gave them the accuracy of 0.935. Al-Nabki et al. presented a web-text-content-based classification pipe-line containing TOR darknet illegal activities. They have used two well-known text representation techniques (Frequency Inverse Document Frequency and Bag-of-Words) together with three different supervised classifiers (Logistic Regression, SVM, and Naive Bayes).

Takoma [31] proposed a system based on unique keywords by extracting from any website and segregated into a specific class with automation, and classified tourism websites into different categories. The paper discussed the dynamic tourism-based Internet directory. The method proposed by them classifies websites into different categories with a higher degree of precision and sets a threshold evaluation for other experiments. Further, it also detects unrelated websites not classified in-to any type.

Text-based classification techniques mainly rely on content attributes of the meta-data and tags of the web pages. Shibu et al. [32] proposed an approach where malicious attempts to get a particular page listed can be overcome and achieved ac-curate results with comparatively minimal time constraints. Alnabulsi and Islam [33] identified illegal activities from forums within the dark web and used those discussions for analysis. They also trained their model to classify those posts into different activities upon testing a new URL set. Siyu et al. [34], in "Classification of Illegal Activities on Dark web", proposed a classification method that uses the 'Federal Code of United States of America' as training data for their model, which gave them an accuracy of 0.935.

We proposed this automated framework that mines dark web URLs and scrapes the hosted text data. An Analysis of scraped data provides the basis for a subsequent investigation of legal and illegal Content, Categorizing the hosted content into Categories like Adult, Counterfeits, Market, Cryptocurrency, Drug, Services, and Weapons.

3 System Architecture

The system architecture is proposed and divided into three modules:

In the first module, we enabled data mining techniques such as scraping, extraction of keywords from hosted pages of the dark web. Two approaches are integrated as one is an automated collection of text-based keywords directly from web pages themselves and, another is dumping the website using tools and extracting keywords.

In the second module, we used techniques like feature selection, feature analysis to categories the keywords that correspond to a particular category based on an opinion of experts and also by analysing their occurrence and density percentage to form a dataset. The dataset made the independent (classes) and dependent features (Legal/illegal) for the classification problems for the dark web. Based on the percentage of keywords matched for legal/illegal against to categories of keywords assigned.

In the third module, we used supervised machine learning algorithms for selecting the best model to fit and, the module was enhanced for predictions based on classification work. The supervised learning classifier algorithms, i.e., Gaussian naïve Bayes, Linear SVM, Random Forest, Logistic Regression algorithms considered for comparing accuracy and efficiency for proposed classification.

Our System Architecture consists of all required elements for classifying and predicting the legality of content hosted on a website hosted on the dark web. Every aspect is taken into consideration while developing the tool based on the proposed framework. During the development phase, there are several challenges like the continuous availability of hosted URLs, security implementations, anti-scaping methods, and scraping multiple onion websites simultaneously, etc. We solved those problems by integrating and developing a customised script in our first module.

4 Data Description and Evaluation

This section discusses the methodology adopted to select the significance of guaranteeing precise and suitable data collection, text keyword extraction, Data Preprocessing, Categorization of Keywords, and respective processes as part of module 1 of system architecture.

4.1 Data Collection

Data Gathering is the path regarding the access for a piece of information (text-based keywords) to prepare a dataset for variables of independence through a developed action plan that enables classifying or predicting illegal content of a website on the dark web. The possible data from a website in the dark web is defined in Table 2 in general.

It is also important to note that while the degree of impact from broken data gathering may move by discipline and the possibility of assessment, it is likely to cause unbalanced naughtiness when these investigation results are used to help open-plan proposition. Here, Data Collection utilising a lab environment consists of Tor Browser, Python Programming Language, and Beautiful Soup, which is utilised parallelly to gather the information. Initially, the authors mined more than 20 thousand URLs by developing a customised script in Python that visits several dark web search engines and mines URLs based on a particular query.

When the required number of URLs in the data set is accomplished, authors further extract the hosted text data and classify it into a category to build a model based on machine learning concerning efficiency and prediction. For any machine learning algorithm, we need some training set and test dataset for training the model and testing the accuracy. Hence to create a dataset for the model, we already have the text from different websites for classifying them according to the keywords, and

Table 2 Accuracy over the URLs

Model—Url	1000	2000	3000	4000	5000
Random Forest	0.66	0.8	0.84	0.9	0.91
Gaussian NB	0.58	0.76	0.81	0.83	0.87
LR	0.73	0.84	0.88	0.92	0.94
Linear SVM	0.71	0.84	0.87	0.9	0.91

then apply the results in the next module. The approach here is to have keywords belonging to the particular category and match those keywords with the text and the class with the maximum Matching value.

Data collection was the foremost step taken by the authors, which is an essential and challenging part of analysing any data. It is crucial because analysis and results are based on the collected data, and challenging to extract any text-based data. The continuous availability of that URL is no guarantee as most URLs are temporarily down or inactive. It is also a task to tackle such problems, that author writes an indigenous script that gives an output of the website's status. The script returns the HTTP response code, the title of a website, and it could also follow the redirection.

4.2 Data Extraction

Data Extraction is a process that includes the recovery of information from different sources—data extraction achieved by utilising the beautiful soup library. In our case, we created a customised script that mines the text data and cleans the unwanted text.

4.2.1 Sample Dataset

The dataset has Three Columns URL, Category, and legal/illegal. URL is the address of the onion web- site from where we will extract hosted text data. Corpus is the text what we scrape from dark web websites. The category is a type of content of the text, i.e., Adult, Drugs, etc. (Refer to Fig. 2).

Fig. 2 Sample dataset

4.3 *Data Pre-processing*

Data pre-processing is a strategy utilised to change raw data helpfully and efficiently, i.e., the process would convert raw data into text-based keywords to classify efficiently. The data can have irrelevant and unnecessary keywords. Data cleaning is a process as part of data pre-processing—the data transformation technique used to transform the data in appropriate forms for a suitable mining process. Initially, we eliminated all the HTML tags, URLs, special characters, numeric characters and converted all characters into the Lower case for nor-mailing the data into a format. Then we left with the text for classification.

4.3.1 Selection of Dataset Size

Data mining is a technique that used to handle a massive amount of data; while working with the vast volume of data, analysis became harder in such cases. To get rid of this, we use the data reduction technique. It aims to increase storage efficiency and reduce data storage and analysis costs. From the entire dataset, we collected around 5000 URLs. Before selection, we cleaned the text and removed a lot of redundancy in the dataset.

4.3.2 Extraction of Data

The feature extraction implemented in two ways —the first part with an automated script, which collected keywords and respective sentences. In this method, the sentences from the index page of the URLs were organized and stored in a database. The keywords extracted from the dark web pages and their enhancement need to work out for handling high dimensional vector space in the dataset. It carefully needs to do select the features of high value from vector space.

Also, removing the highfrequency words, auxiliary verbs, etc., using conventional techniques is essential for classification—one of the word removal techniques to provide high weightage to keywords for a category of textual content. The feature selection in segregating key-words required a subset of input variables to eliminate features with little or predictive information. The cleaning method of elements used in this model is first by removing stop words (stop words = ['now','also','often','at''is',] and also using a combination of keywords and making sentences into one of the main classes.

4.3.3 Output

The output after pre-processing the data is put away as an excel file for supervised machine earning algorithms. The File contains the separated data in the correct way.

5 Methodology Used

Our goal was to developed and validate the empirical of an analytical framework for Scraping Dark Web of interesting category; we did not worry about any violations since most of the website are unauthorized, through Our Research, we determined that most websites have not implemented any methodology for preventing scraping. Thus, We Did not need any bypass mechanisms and did not face any tough challenges to mine text data. This classification pipeline comprises seven main stages: System Setup, URLs Mining, Data Scraping, Keyword Set, Text Pre-processing, Feature extraction, and performing classification on the dataset using a supervised machine learning Classifier Algorithms. We used famous text representation techniques across four different supervised classifiers. We examined every pipeline to figure out the best combination with the best parameters to achieve high performance.

5.1 System Setup to Access Dark Web

We began with the installation of the Tor Browser from torproject.org on our hardware systems. Before browsing, we need lots of. Onion URLs. We Began our research initially with and the URLs.

5.2 URL Mining

Dark Web is an unindexed web. The Format of the Dark web URL is very complex and uses a more robust, more recent crypto algorithm V3 URLs are 56 bytes long. Therefore, remembering the URLs is very difficult. The properties of the dark web make the mining process difficult, so we created a script that mines URLs by visiting different search engines. It takes keywords as input and websites related to that keyword by visiting several dark web search engines and scrapes the URLs from their result set.

5.3 Data Scraping

Web Scraping, also known as harvesting, means data extraction from the websites. The framework has a Classification Module that will predict the category based on the hosted text of the dark web websites. Thus, writes a script that is the solution to all problems which is capable of scraping text data as well as cleaning the data.

5.4 *Keyword Set*

After preparing the dataset of dark web text data from different dark web web-sites, to test the accuracy concerning a machine learning model, we need some training data and test data for training the model and testing the accuracy. Hence, we will use those machine learning models for classifying the data using collected text data of keywords. The approach here is that we will have specific keywords belonging to the particular category, and we will match those keywords with the text and the class with the maximum matching value.

$$\begin{aligned}\text{Matching value} = {} & \text{(Number of keywords matched with one Category)}\\ & /\text{(Total number of keywords matched)}\end{aligned}$$

5.5 *Feature Extraction*

After pre-processing the text, we used famous text representation/ features extraction techniques. Term Frequency—Inverse Document Frequency model (TF-IDF) is a statistical model that assigns weights for the vocabularies where it emphasizes the words that frequently occur in a given document while at the same time de- emphasizes words that frequently occur in many forms. However, even though the BOW and TF-IDF do not consider the order of the terms, they are simple, computationally e client, and compatible with medium dataset sizes.

5.6 *Classifier Selection*

We examined Four different supervised machine learning algorithms for each feature representation method: Support Vector Machine (SVM) [35], Logistic Regression (LR) [36], Gaussian Naive Bayes Hand and Yu [37], and Random Forest [38].

6 Result Analysis

Since we are working on a Multiclass problem, we collected around 20,000 URLs. After that, we can Scrape the data of 8000 URLs due to the availability of websites. Duplicity is another problem in the dataset, and then we removed the duplicity based on the Con-tent/Corpus and URL. We have chosen a threshold value for the removal of duplicity. Finally, we got is our Dataset for Training Machine Learning Model.

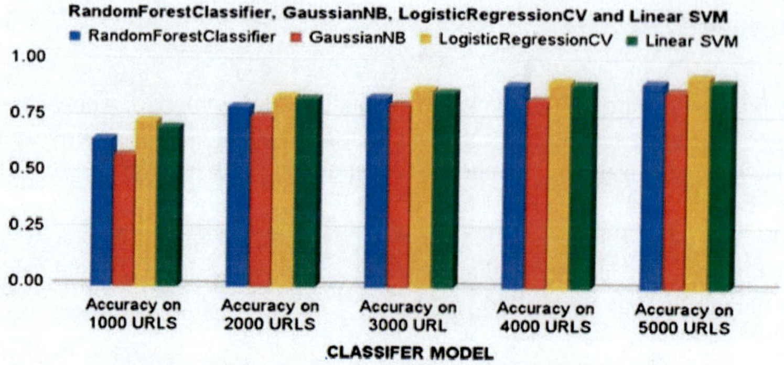

Fig. 3 Model accuracy comparison

6.1 Accuracy Comparison of Models

We have Chosen Four Classification algorithms, i.e., Random Forest, Gaussian NB, Logistic Regression, Linear SVM, to classify the data. Each Classifier Model has its pros/cons in any manner. We Scarp the data from the Websites in batches of URLs; in each batch, the script has given 500–1000 URLs. Initially, we have a huge dataset. Based on Random Sampling, we Divided the Dataset into frames consisting of 0–1000, 0–2000, 0–3000, 0–4000, 0–5000 data of the website for analysis about the model's accuracy over the N number of entries. The training and Testing ratio are 70–30. We recorded around 79,107 features with the TF-IDF Feature extraction algorithm on the 5000 URL dataset.

Table 2 shows the accuracy of the respective number of URLs.

We Performed Accuracy Comparison: Our Goal of the Study is to develop a framework and a Product that could be useful for academic research purposes. We believe this product can be helpful for LEA's as well. We expected our tool to perform in the best way; we compare models because it has to be in that product.

Table 2 and Fig. 3 state that Data is Directly Proportional to Accuracy and model performance. As we increased data accuracy across the models also increased 0.0.94, the highest accuracy we achieved over 5000 URLs on Logistic regression. Linear SVM and Random Forest Scored 0.91 Accuracy.

6.2 The Framework

We developed this framework to integrate all three modules into a web application framework, which is easy to use and supports end-to-end classification of the dark web. This framework has a variety of features listed below:

1. Exception Handling with Requests.
2. Dashboard Look for easy understanding.

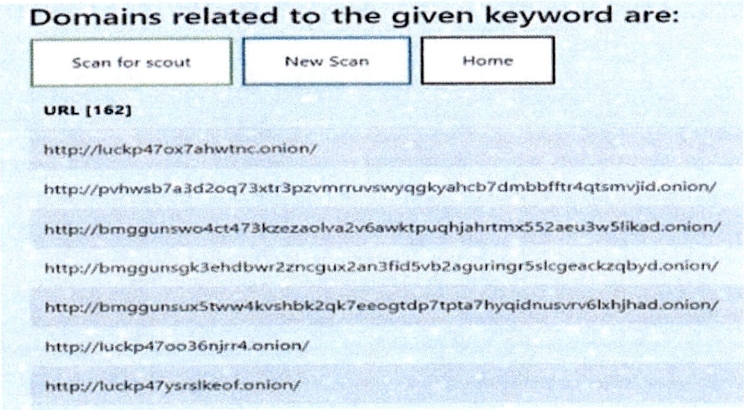

Fig. 4 Web application output of url mining on given a keyword

3. Download Data in CSV Format.
4. Add and retrieve from Database.
5. Multiple Search Engine.
6. Integration with Tor Browser.

There is a wide range of information on the dark web. Covering all categories is difficult, therefore we've compiled a list of the most popular ones from our database. In the dark web environment, regional languages such as Arabic, Russian etc., are also extensively utilized. Language Barrier is also a limitation because the framework can only classify a text in one of the categories if the text is written in English (Fig. 4).

6.3 Deep Learning Model

KERAs is a deep learning/neural network library written in Python running on top of the machine learning platform Tensor ow. We also performed training on a Neural Network with a dataset having 5000 URLs scraped data. The network Performs best with the epochs value of 10, which has a minimum loss value compared to other epochs value. Figure 5 shows the accuracy and loss on the same. So, the result we achieved using neural network is 0.91 is the accuracy similar to random forest or Linear SVM classifier models. Figure 6 shows the compassion of test loss and testing accuracy on different epoch values. The below image concludes the result from the neural network model that we built. We can achieve an accuracy of 0.92, which is not the best but has given good results. We can see increasing epoch value does not change the accuracy; thus, we get 30% of loss when the epochs stated that increasing the number of epochs does not improve accuracy (Fig. 7).

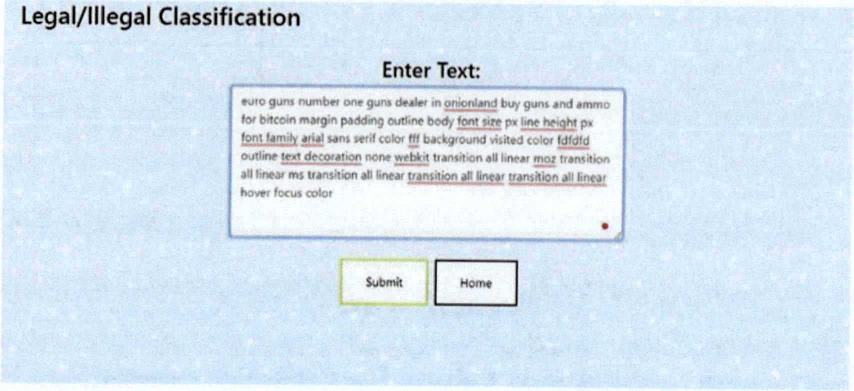

Fig. 5 Legal/illegal prediction

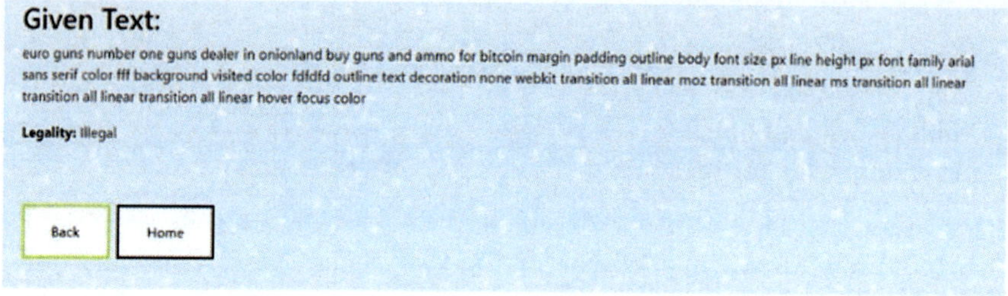

Fig. 6 Result from prediction

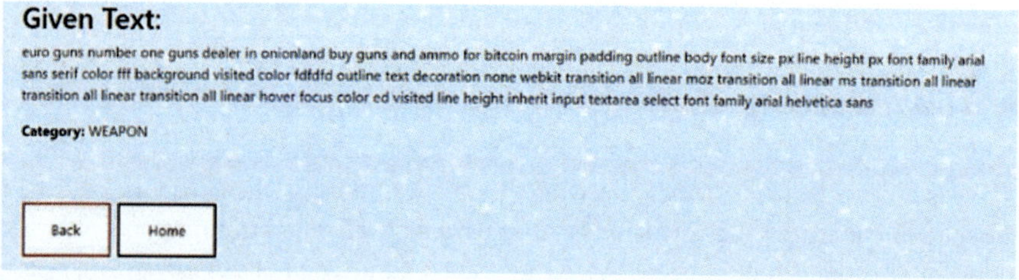

Fig. 7 Text classification and its result

7 Conclusion

The dark web is an emerging area for research purposes, and machine learning adds an extra edge to this domain. Classification of web pages is a subset of web mining at the early stage of this framework; we are performing mining. Further, we are performing classification based on mining. In this paper, we have examined four classification algorithms on the Dataset of text which is mines from the thousands

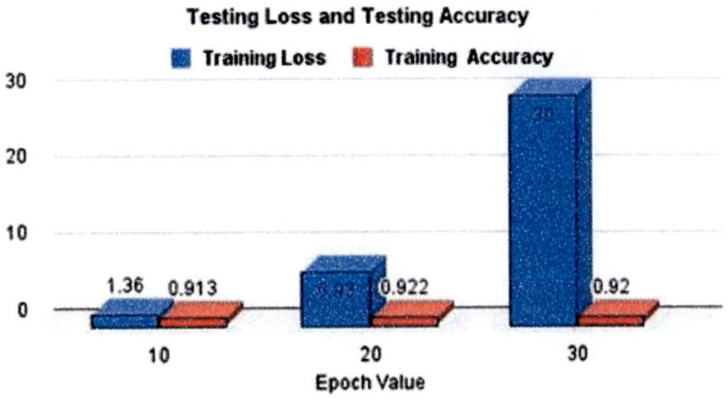

Fig. 8 Accuracy on the model built on Keras

of dark web websites and get classified into Seven classes [‘Adult’, ‘Drug’, ‘Counterfeits’, ‘Crypto’, ‘Market’, ‘Service’, ‘Weapon’]. We compared accuracy at every thousand entries till we have the data of around 5000 websites. We can achieve 94% in logistic regression, whereas 91% in Linear SVM and Random Forest. Gaussian NB performs poorly compared to other Models, which states that the nature of this problem is not suitable for it (Fig. 8).

We also observed that improvements are in the classification showing only when adding more data into the model and whenever we improve our keyword set for each class. At the initial stage, the accuracy score is low, but as soon we improve the key-words set, the model can now classify the text more efficiently.

We also compared the Neural Network model built over Keras with Supervised Machine Learning Classification Algorithms. We observed that the best parameter model could perform the same as the two classifier algorithms, Random Forest and Linear SVM. Although Logistic Regression performed best out of all.

In this framework, we developed a content analyser, that tells what percentage of the text belongs to which category. As soon we give the text to the analyser, it gives out-put as the graph that tells how much text falls in which class and how much text is legal/illegal.

The Goal of Classification is to generate more precise and accurate results. Also, this framework will help LEA, Academic institutes in various activities like mining URLs scouting of URLs that show whether an onion website is alive or dead at that moment. A Classification model predicts that content belongs to which class, legal/illegal.

References

1. D. Hayes, F. Cappa, J. Cardon, A framework for more effective Dark Web market-place investigations. Information **9**(8), 186 (2018). https://doi.org/10.3390/info9080186
2. The Onion Router project metrics (2021). URL at https:// metrics.torproject.org/
3. H. Alnabulsi, R. Islam, Identification of Illegal Forum Activities Inside the Dark Net. In *2018 International conference on machine learning and data engineering (iCMLDE)* (2018). https://doi.org/10.1109/icmlde.2018.00015 (2018)

4. N. Tavabi, N. Bartley, A. Abeliuk, S. Soni, E. Ferrara, K. Lerman, Charac-terizing activity on the deep and Dark Web. In: Companion proceedings of the 2019 world wide web conference (2019). https://doi.org/10.1145/3308560.3316502
5. A. Kumar, E. Rosenbach, The truth about the dark web (2019). At https://www.imf.org/external/pubs/ft/fandd/2019/09/the-truth-about-the-dark-web-kumar.htm
6. Deep web -the hidden side of Internet. URL at https://tharjournal.com/deep-web/
7. M. Mirea, V. Wang, J. Jung, The not so dark side of the darknet: a qualitative study. Secur. J. **32**(2), 102–118 (2018). https://doi.org/10.1057/s41284-018-0150-5
8. K. Nalini, L.J. Sheela, Survey on text classification. Int. J. Innov. Res. Adv. Eng. **1**(6), 412–417 (2014)
9. A. Mehler, C. Wolff, *Text Mining*. Themenheft des LDV-Forum (2005)
10. S. Brindha, K. Prabha, S. Sukumaran, A survey on classification techniques for text mining. In *2016 3rd International Conference on Advanced Computing and Communication Systems (ICACCS)* (2016). https://doi.org/10.1109/icaccs.2016.7586371
11. Z. Xu, D. Zhao, Research on mobile learning system based on web mining. In *2012 Third International Conference on Intelligent Control and Information Processing* (2012). https://doi.org/10.1109/icicip.2012.6391484
12. Cybersecurity Spotlight—The Surface Web, Dark Web, and Deep Web at URL https://www.cisecurity.org/spotlight/cybersecurity-spotlight-the-surface-web-dark-web-and-deep-web/
13. S. Sarlis, I. Maglogiannis, On the Reusability of sentiment analysis datasets in applications with dissimilar contexts. In *IFIP Advances in Information and Communication Technology*, vol. 34 (2020), pp. 409–418. https://doi.org/10.1007/978-3-030-49161-1
14. F. Thomaz, C. Salge, E. Karahanna, J. Hulland, Learning from the Dark Web: Leveraging conversational agents in the era of hyper-privacy to enhance marketing. J. Acad. Mark. Sci. **48**(1), 43–63 (2019). https://doi.org/10.1007/s11747-019-00704-3
15. H. Chen, IEDs in the Dark Web: Genre classification of improvised explosive device web pages. In *2008 IEEE International Conference on Intelligence and Security Informatics* (2008). https://doi.org/10.1109/isi.2008.4565036
16. R. W. Gehl, Archives for the Dark Web: A field guide for study. In *Research methods for the digital humanities* (2018), pp. 31–51. https://doi.org/10.1007/978-3-319-96713-43
17. R. Islam, E. Ozkaya, *Inside the Dark Web* (CRC Press, 2019)
18. M. K. Bergman, White paper: The Deep Web: surfacing hidden value. J. Electron. Publish. **7**(1) (2001). https://doi.org/10.3998/3336451.0007.104
19. S. Dumais, H. Chen, Hierarchical classification of web content. In *Proceedings of the 23rd Annual International ACM SIGIR Conference on Research and Development in Information Retrieval—SIGIR '00* (2000). https://doi.org/10.1145/345508.345593
20. A. Sun, E. Lim, W. Ng, Web classification using support vector machine. In *Proceedings of the Fourth International Workshop on Web Information and Data Management—WIDM '02* (2002). https://doi.org/10.1145/584931.584952
21. M. Kan, H.O. Thi, Fast webpage classification using URL features. In *Proceedings of the 14th ACM International Conference on Information and Knowledge Management—CIKM '05* (2005). https://doi.org/10.1145/1099554.1099649
22. P. Kaur, Web content classification: A survey. Int. J. Comput. Trends Technol. **10**(2), 97–101 (2014). https://doi.org/10.14445/22312803/ijctt-v10p117
23. W. Su, J. Wang, F. Lochovsky, Automatic hierarchical classification of structured deep web databases. In *International Conference on Web Information Systems Engineering* (pp. 210–221). Springer (2006)
24. L. Barbosa, J. Freire, A. Silva, Organizing hidden-web databases by clustering Visible Web documents. In *2007 IEEE 23rd International Conference on Data Engineering* (2007). https://doi.org/10.1109/icde.2007.367878
25. U. Noor, Z. Rashid, A. Rauf, A survey of automatic Deep Web classification techniques. Int. J. Comput. Appl. **19**(6), 43–50 (2011). https://doi.org/10.5120/2362-3099
26. X. Xian, P. Zhao, W. Fang, J. Xin, Z. Cui, Automatic classification of Deep Web databases with simple query interface. In *2009 International Conference on Industrial Mechatronics and Automation* (2009). https://doi.org/10.1109/icima.2009.5156566

27. M. Khelghati, D. Hiemstra, M. Van Keulen, Efficient web harvesting strategies for monitoring Deep Web content. In *Proceedings of the 18th International Conference on Information Integration and Web-based Applications and Services* (2016). https://doi.org/10.1145/3011141.3011198
28. D.R. Moore, Thomas, Cryptopolitik and the Darknet. Survival **58**, 7–38 (2016). 1080/00396338.2016.1142085
29. K. Kinningham, M. Graczyk, Automatic product categorization for anonymous marketplaces Kevin Kinningham project overview (2015)
30. M.W. Al Nabki, E. Fidalgo, E. Alegre, I. De Paz, Classifying illegal ac-tivities on TOR network based on web textual contents. In *Proceedings of the 15th Conference of the European Chapter of the Association for Computational Linguistics*, vol. 1, Long Papers (2017). https://doi.org/10.18653/v1/e17-1004
31. T. Honda, M. Yamamoto, A. Ohuchi, Automatic classification of web-sites based on keyword extraction of nouns. Inf. Commun. Technol. Tourism **2006**, 263–272 (2006). https://doi.org/10.1007/3-211-32710-x38
32. S. Shibu, A. Vishwakarma, N. Bhargava, A combination approach for web page classificationusing page rank and feature selection technique. Int. J. Comput. Theory Eng. 897–900 (2010). https://doi.org/10.7763/ijcte.2010.v2.259
33. Alnabulsi, H., Islam, R. (2018). Identification of illegal forum activities inside the dark net. In: *2018 International Conference on Machine Learning and Data Engineering (iCMLDE).* https://doi.org/10.1109/icmlde.2018.00015
34. S. He, Y. He, M. Li, Classification of illegal activities on the Dark Web. In *Proceedings of the 2019 2nd International Conference on Information Science and Systems—ICISS 2019* (2019). https://doi.org/10.1145/3322645.3322691
35. C. Cortes, W Support-vector network. Mach. Learn. **20**, 1–25 (1995)
36. D.R. Cox, The regression analysis of binary sequences. J. Roy. Stat. Soc.: Ser. B (Methodol.) **20**(2), 215–232 (1958)
37. D.J. Hand, K. Yu, Idiot's Bayes: Not so stupid after all? Int. Statist. Rev./Revue Internationale de Statistique **69**(3), 385 (2001). https://doi.org/10.2307/1403452
38. Breiman, L. (2001). Mach. Learn. **45**(1), 5–32. https://doi.org/10.1023/a:1010933404324

Speech Gender Recognition Using a Multilayer Feature Extraction Method

Husam Ali Abdulmohsin, Belal Al-Khateeb, and Samer Sami Hasan

Abstract Human speech contains paralinguistic properties used in automatic speech recognition (ASR) systems. These properties are used in many ASR applications such as gender recognition, which is the main goal of this paper. Gender recognition has been the target of many researchers since recognizing the human gender (female or male) is essential in many applications especially in security applications. Through this work, an ASR has been proposed and implemented. The main goal of any ASR system is to determine the best features that can address the required recognition. The features deployed in this work are smoothness, pitch, the first two formants and spectral centroid variability (SCV). The new approach proposed in this work was using the analysis of variance (ANOVA) as a feature selector to choose the best combination of features that can lead to the best classification accuracy, and then apply the decision tree feature selection algorithm to choose the best group of features. Then use backpropagation neural network (NN), Gaussian mixture models (GMM) and SVM as separate classifiers. The common voice dataset was used as benchmark dataset through all experiments of this work. The best result gained with respect to the three genders was 74.87% using the pitch and the first two formant features and classified by NN. The best result gained with respect to the two genders (female and male) was 97.71% using the pitch, and the first two formant features are classified by NN.

Keywords Automatic speech recognition · Speech gender recognition · Backpropagation NN · GMM · Common voice dataset

---

H. A. Abdulmohsin (✉) · S. S. Hasan
Computer Science Department, Faculty of Science, University of Baghdad, Baghdad, Iraq
e-mail: husam.a@sc.uobaghdad.edu.iq

S. S. Hasan
e-mail: ssami@uob.edu.iq

B. Al-Khateeb
Computer Science Department, College of Computer Science and Information Technology, University of Anbar, Anbar, Iraq
e-mail: belal-alhkhateeb@uoanbar.edu.iq

A. K. Bashir et al. (eds.), *Proceedings of International Conference on Computing and Communication Networks*, Lecture Notes in Networks and Systems 394,
https://doi.org/10.1007/978-981-19-0604-6_10

1 Introduction

When understanding the speech production process in female and male, it can be noticed that formants of females are higher in frequency than those of their male counterparts and the spectrum of female voiced sounds are lower than in male sounds, since the spectrum usually decreases in amplitude with increasing the frequency. All these acoustic effects are caused by the production of speech. Therefore, it is possible to find gender-specific features represented in acoustic speech signals [1].

Due to the transgender phenomena that appeared in the last years, and the surgeries applied to the vocal cords to adapt with the new gender, it is essentially required to determine the gender of the human being, regardless of its new gender for security identification reasons.

Acoustic voiced sounds are generated through the vibration of the vocal cords that in turn generates the periodic behavior of voice. This oscillation in frequency is called pitch. The pitch feature and other frequency-related features used in this work have a clear relationship with gender, and especially the low frequencies that contain the most important speech properties are useful in ASR. It is important to mention that male speech has a low-frequency behavior compared to female speech. The pitch in particular depends on the glottis physical characteristics, which are mass and elasticity [2, 3]. Speech is always represented as a discrete signal $x(n)$; therefore, the pitch represents the fundamental frequency (f_0) where the signal is repeated. The inverse is the fundamental period (T_0). Statistically, there are certain frequency intervals for men regarding each language, for example, the pitch of the Spanish men lay in the frequency interval, 50–300 Hz, and the Spanish women and children can reach to 500 Hz frequency [4].

Many researches have been published in the field of gender recognition, but none has studied the third gender's psychological effect on voice. Through this study, we tried to analyze the effect of the third gender on the human voice, and this study is considered a new field of study, since there are no researches published under this problem statement. Through our work, we were able to recognize the third gender voice, but as can be seen from the results and discussion section, that we did not achieve high classification accuracies. The limitation of our work lays in the features extracted. More features need to be extracted and more related to the third gender. The other limitation is the need for a dataset that shows third gender voice samples that have been under vocal tract transformation surgery, to change the voice from one gender to the other, in order to study the original voice features in sound that can differ the third gender from the other two genders.

2 Related Work

Many researches have confirmed the existence of unique acoustic and physiologic features in each of the female and male voices that can be used to recognize male from female voice, but till now, they have not reached the required classification accuracy [5, 6]. In the past few years, gender recognition has gained the interest of many researchers.

In 2015, Faek [7] used the first four formants and twelve MFCC's as features and used the SVM as a classifier. A special feature selection is used based on the frequency range that the feature represents. A total of 114 speech samples uttered in Kurdish language were used in this work. The model of two classes (adult males and adult females) of gender recognition reached 96% recognition accuracy.

In 2019, Shaqra et al. [8] designed four emotion recognition models to present the relationship between gender/ age and emotion. The results showed that when using the same classifier for all four models on each gender and on different age limits, the results were higher compared to the performance of the system on all samples without categorization. This proves that gender and age affect the emotion recognition accuracy for its direct effect on voice.

In 2019, Alkhawaldeh et al. [9] provided an analysis about the gender-related features in speech and experimented three feature selection algorithms to find the best features. Then studied different machine learning (ML) models with different theoretical background, to find the best gender-related ML models. The best result gained was 99.7% classification accuracy.

In 2019, Abdulsatar [10] proposed a two-part system. The first part was called pre-processing and feature extraction to select the best feature, which are the first four formant frequencies and twelve MFCCs used to extract relevant features to recognize the gender and K-NN for classification. The highest accuracy classification obtained was 66%.

In 2021, Kwasny [11] applied d-vector and x-vector as deep neural network (DNN)-based embedder architectures to gender classification experiments. Then applied a transfer learning-based training scheme with pre-training the embedder network. The best overall performance achieved gender recognition was 99.60%.

Many challenges were faced; one of the challenges was the psychological situation of the human being and its effect on voice. When the human being is psychologically unstable, his vowels will be differently announced when he is psychologically stable [8, 12]. The other challenge was dealing with similarity in children voices of age 3 to 7 years; therefore, children were avoided in the experiments applied to the method proposed in this work.

The limitation of the state of art works is avoiding working on the third gender (not female neither male). Through this work, many experiments were conducted on the third gender, which is considered the first paper that studies the difference between the three genders (female, male and others).

3 Materials and Methodology

- Dataset
  The dataset utilized in this work is the common voice dataset version en_2637h_2021-07-21. The common voice dataset contains 60 different languages recorded in different percentages as 23% US English, 8% England English, 7% India and South Asia (India, Pakistan, Sri Lanka), 3% Australian English, 3% Canadian English, 2% Scottish English, 1% Irish English, 1% Southern African (South Africa, Zimbabwe, Namibia), 1% New Zealand English and many other languages.
  The common voice dataset was recorded by 75,879 different individual voices with different age limits. 6% of the individuals were younger than 19 years old that means 4552 individuals of age less than 18 have participated in recording this online public dataset and 94% of the dataset is recorded by individuals older than 18 years. All voices related to under legal age (<18) were neglected from our experiments, and only adult voices were included.
  The common voice dataset contains adult voice for three genders, 45% male, 15% female and 40% third gender [13]. The number of samples tested through the experiments of this work was 71,327 samples. A total of 32,098 (45%) of those samples were male voices, 10,699 (15%) of the samples were female voices and 28,530 (40%) of the samples were for other genders.

4 Framework

The block diagram of the proposed method is shown in Fig. 1. In the following three sections, the main steps of the proposed method will be illustrated and discussed in details.

- Pre-processing
  All samples were segmented according to the ratio (0.05%) of the original signal, and the overlap ratio deployed in this work was (0.025%), which provides (50%) overlap, and the total number of segments generated will be $(n - 1)$, where n is the number of original segments.
- Feature Extraction
  A feature is a measurable property established from the material being observed [14]. The most important aspect in feature extraction is extracting the most relevant features to the problem statement. In the case of this work, the features extracted were smoothness, pitch, the first two formants and spectral centroid variability (SCV) features, which are strongly related to human being gender, and were selected according to the experiments conducted.
  Smoothness is defined as the transaction of speech through air, as much as the speech was smooth as much as its transaction was slower, and when speech is rougher, the transaction of speech is faster [6]. The smoothness was calculated

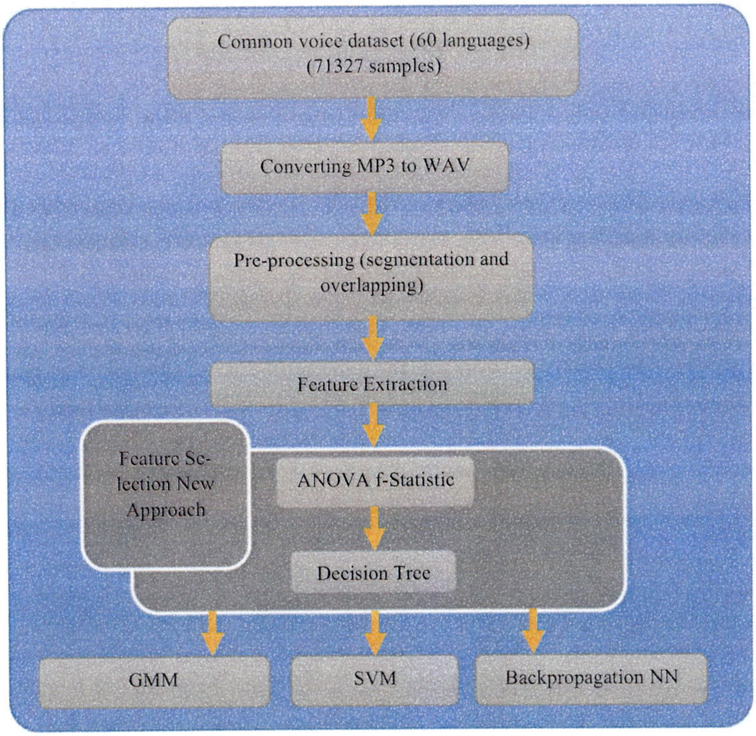

Fig. 1 Block diagram of the method proposed in this work

through two domains, first the time domain, the second, the spectral domain. Smoothness is calculated through Eq. (1 and 2) [15].

$$GV_t = \frac{1}{P}\sqrt{\sum_{j=1}^{P} (\text{var}_t(j))^2} \quad (1)$$

$$GV_s = \frac{1}{N}\sqrt{\sum_{i=1}^{N} (\text{var}_s(i))^2} \quad (2)$$

where var_t and var_s represent the variances in time and spectral domain of the spectral feature, P is the dimension of the feature, N is the length in the time domain of the feature

- Feature Selection
  In this work, the ANOVA feature selection algorithm was used to filter features ahead of constructing the decision tree, to remove all irrelevant features, then pass the selected features to the decision tree.

  Decision tree is used for classification purposes or used as a feature selection algorithm of type embedded, and also used in data mining and machine learning [16, 17].

Most of the decision tree implementations in the previous state-of-the-art works such as ID3 [18], C4.5 [19] and CART [20] did not measure the importance of each feature regarding other classes and the final classification results. Therefore, in this work, we will determine the importance of each feature regarding each class. In this work, each feature will be weighted and the weight will be used in feature selection and finally in the decision tree construction. Feature weight will be calculated respectively, the feature with the highest weight will be selected as the root feature of the next layer and so on, the decision tree will be constructed.

A filter ranking method was used through this work for three reasons. First, to filter the less relevant variables. Second, to benefit from the criteria of variable selection by order of the variable ranking techniques. Third and finally, their simplicity and good success are reported from online applications. A ranking criterion is used to score each variable, then a threshold is fixed through experiment, and used to remove variables below that threshold [14].

Feature selection methods that are applied before classification are considered filter feature selection methods, that's why ranking methods are considered filter methods. The main principle of feature selection methods is to select unique features that contain useful information of different classes in the dataset through using a basic property of that feature. This property is called feature relevance that measures the power of that feature classifying different classes [14, 21, 22]. The chi square and analysis of variance (ANOVA) statistical feature selection methods were used in this work to measure the independence of two selected features.

The chi square feature selection method was calculated through Eq. (3) [23].

$$x_c^2 = \sum \frac{(O_i - E_i)^2}{E_i} \tag{3}$$

where c is the degree of freedom, $O(s)$ are the observed values, which are s number of values, and $E(s)$ are the expected values.

ANOVA that is developed by the statistician Ronald Fisher [24] is a set of statistical models and their related estimation procedures, like the variation among and between groups of features that are used to analyze differences between means. ANOVA is based on total variance law, where the variance observed in a specific variable is partitioned into attribute components to other sources of variation. ANOVA in the simplest form provides a statistical experiment of whether two or more feature groups means are equal, not as the t-test that involves two means only.

The proportion of variance in ANOVA represented by a feature or groups of features can be found through Eq. (4).

$$\text{Variance} = \frac{\text{SST}}{\text{TotalSS}} \tag{4}$$

where the SST is the treatment sum of squares and the Total SS is the total sum of squares. As much as the higher ratio, the more groups of features can represent the data. In other words, the groups of features with higher proportion must be selected [25].

- Feature Classification
  The backpropagation NN and the GMM classifiers were selected in this work, to classify the three genders.

5 Results and Discussion

The aim of the experiments deployed in this work was to test which of the features (smoothness, pitch, the first two formants and spectral centroid variability (SCV)) or group of features can act better in gender recognition with respect to each of the two classification algorithms which are backpropagation NN and the GMM classifiers. The experiments also aim to evaluate the performance of the new proposed feature selection algorithm with respect to feature and classifier.

- Gender recognition results with respect to each feature and classifier

The four types of features will be tested individually with each of the two classifiers, after applying the feature selection algorithm that is proposed in this work.

Table 1 shows the female gender classification performance according to feature and classifier. The results show the outperformance of backpropagation NN on GMM through all types of features and also show that the first two formant features gained the highest classification accuracy in female gender speech recognition, more than the other three types of features.

Table 2 shows the male gender classification performance according to feature

Table 1 Female recognition results

Feature type	Backpropagation NN (%)	GMM (%)
Smoothness	53.22	48.39
Pitch	46.34	44.4
First two formants	57.45	52.12
Spectral centroid variability (SCV)	50.89	45.23

Table 2 Male recognition results

Feature type	Backpropagation NN (%)	GMM (%)
Smoothness	33.09	38.5
Pitch	52.41	48.36
First two formants	38.79	40.21
Spectral centroid variability (SCV)	47.34	41.56

Table 3 Third gender recognition results

Feature type	Backpropagation NN (%)	GMM (%)
Smoothness	21.02	12.08
Pitch	32.57	30.88
First two formants	25.55	22.04
Spectral centroid variability (SCV)	16.04	19.89

and classifier. The results show the outperformance of backpropagation NN on GMM through all types of features and also show that the pitch feature gained the highest classification accuracy in male gender speech recognition, more than the other three types of features.

Table 3 shows the third gender classification performance according to feature and classifier. The results show the outperformance of backpropagation NN on GMM through all types of features and also show that the pitch features gained the highest classification accuracy in third gender speech recognition, more than the other three types of features. But as a overall conclusion, the third gender was misclassified, mostly to the male gender, that is why the pitch feature outperformed other features like in male gender speech recognition, that is explained by the similar properties in the speech signal of the male and third gender. Because of the misclassification of the third gender, Table 3 shows the low classification accuracy gained with respect to the other two genders classification accuracy mentioned in Tables 1 and 2.

- Gender recognition results with respect to best feature group and classifier

After going through all six combination possibilities of the four feature types with best features selected, it was found that the highest accuracy results gained were through deploying the pitch and the first two formants in speech, with respect to the backpropagation NN classifier as shown in Table 4. The highest accuracy gained was 74.87% with respect to all three genders. If the third gender was excluded from the experiments, the highest classification accuracy achieved to classify female and male genders (without the third gender) is 97.71% with respect to the backpropagation NN, and 91.03% with respect to the GMM classifier using the pitch and first two formants' features.

6 Conclusion

To design a system that can recognize age through the same setting was challenging, because age is related to language, and each language has a different range of frequencies for the male and female and children.

Table 4 All genders' recognition results

Feature type	Backpropagation NN				GMM			
	Female (%)	Male (%)	Third gender (%)	Total (%)	Female (%)	Male (%)	Third gender (%)	Total (%)
Smoothness, pitch	83.23	77.98	25	62.07	79.05	79.6	23.4	60.68
Smoothness, first two formants	88.34	72.45	24.34	61.71	85.03	75.54	19.21	59.89
Smoothness, spectral centroid variability (SCV)	83.06	79.02	20.26	60.78	81.6	82.45	17.08	60.37
Pitch, first two formants	98.32	97.11	29.2	74.87	92.06	90	24	68.68
Pitch, spectral centroid variability (SCV)	84.3	80.34	22.24	62.29	83.2	81.09	18.2	60.83
First two formants, spectral centroid variability (SCV)	75.55	74.44	21.22	57.07	70	79.33	20.5	56.61

The similar frequency behavior between the male voices and third gender caused a lot of ambiguity to the system designed in this work, regardless of the strong gender related features extracted and the new designed feature selection method.

It is clearly noticed that the third gender recognition classification accuracies achieved are very low with respect to the other two genders, which proves two things, first there are some special properties in the transgender's speech that differs them from other genders, but those properties are weak or were not extracted perfectly. Second, a lot of the third gender speeches were misclassified as male voices and vice versa that led to the low accuracy classification in both genders, the male and the third gender, which either is explained that the original gender of the transgenders was a male gender, and the original speech properties retain in the speech of the transgenders regardless of the new gender selected willingly.

Acknowledgements Many thanks to Mozilla (voice.mozilla.org) for creating such a global speech dataset, with many languages and many specifications with huge details.

References

1. J. Harrington, S. Cassidy, The acoustic theory of speech production, in *Techniques in Speech Acoustics*. (Springer, 1999), pp. 29–56
2. L. Rabiner, R. W. Schafer, *Digital Processing of Speech Signal* (1978)
3. J.R. Deller Jr, J.G. Proakis, J.H. Hansen, *Discrete Time Processing of Speech Signals* (Prentice Hall PTR) (1993)
4. Rabiner, L.R., B.J.E.C.P.-H. Gold, *Theory and application of digital signal processing* (1975)
5. Titze, I.R., *Physiologic and Acoustic Differences Between Male and Female Voices*. J. Acoust. Soc. Am. **85**(4), 1699–1707 (1989)
6. G. Fant, *Acoustic Theory of Speech Production* (Walter de Gruyter, 1970)
7. K.F. Fatima, *Objective gender and age recognition from speech sentences*. ARO Sci. J. Koya Univ. **3**(2), 24–29 (2015)
8. F.A. Shaqra, R. Duwairi, M.J.P.C.S. Al-Ayyoub, *Recognizing emotion from speech based on age and gender using hierarchical models*. Procedia Comput. Sci. **151**, 37–44 (2019)
9. R.S.J.S.P. Alkhawaldeh, DGR: gender recognition of human speech using one-dimensional conventional neural network. Sci Program **2019** (2019)
10. A.A. Abdulsatar, et al., *Age and gender recognition from speech signals*. J. Phys. Conf. Ser. (2019)
11. D. Kwasny, D.J.S. Hemmerling, Gender and age estimation methods based on speech using deep neural networks. Sensors **21**(14), 4785 (2021)
12. H.A. Husam Ali Abdulmohsin, A.M.J.A. Hossen, J. Mech. Continua Math. Sci. Speech Emot. Recogn. Survey **15**(9), 24 (2020)
13. R. Ardila, M. Branson, K. Davis, M. Henretty, M. Kohler, J. Meyer, R. Morais, L. Saunders, F.M. Tyers, G. Weber, Common voice: a massively-multilingual speech corpus, in *Proceedings of the 12th Conference on Language Resources and Evaluation (LREC 2020)* (2020), pp. 4211—4215
14. G. Chandrashekar, F.J.C. Sahin, A survey on feature selection methods. Comput. Electr. Eng. **40**(1), 16–28 (2014)
15. P.T. Nghia, et al., A measure of smoothness in synthesized speech. Electron Commun **6**(1–2) (2016)
16. H. Zhou, et al., A feature selection algorithm of decision tree based on feature weight. Exp. Syst. Appl. **164**, 113842 (2021)
17. H. Sun, X.J.C. Hu, I.L. Systems, Attribute selection for decision tree learning with class constraint. Chemometr. Intell. Lab. Syst. **163**, 16–23 (2017)
18. , J.R. Quinlan, Induction of decision trees. Mach. Learn. **1**(1), 81–106 (1986)
19. J.R. Quinlan, *C4. 5: Programs for Machine Learning* (Elsevier, 2014)
20. C.-H. Yeh, *Classification and Regression Trees (CART)* (Elsevier, 1991)
21. H.A. Abdulmohsin et al., A new hybrid feature selection method using T-test and fitness function. CMC-Comput. Mater. Continua **68**(3), 3997–4016 (2021)
22. R. Kohavi, G.H John, Wrappers for feature subset selection. Artif. Intell. **97**(1–2), 273–324 (1997)
23. S. Gajawada, Chi-square test for feature selection in machine learning (2019). Retrieved from Towards Data Science: https://towardsdatascience.com/chi….
24. P. Moran, C.J.T.o.t.R.S.o.E. Smith, The correlation between relatives on the supposition of mendelian inheritance. Earth Environ. Sci. Trans. R. Soc. Edinburgh **52**, 899–438 (1918)
25. A. Anderson, *Business Statistics for Dummies* (Wiley, 2013)

Speech Age Estimation Using a Ranking Convolutional Neural Network

Husam Ali Abdulmohsin, Jane Jaleel Stephan, Belal Al-Khateeb, and Samer Sami Hasan

Abstract The age of 18 has been chosen as the legal age to enter many sites, receive any service or to get some license. Since age has a huge effect on the human being voice, many researchers have worked on automatic age estimation (AAE) in speech analysis. Through this work, a new approach has been designed to estimate the age of the human being depending on his speech. This work has regarded common voice dataset in its experiments with 60 different languages and seven age limits. The features depended were the smoothness and pitch features for their strong capability in recognizing the human voice frequency properties that have a strong relationship with human age. The chi square feature selection was utilized in this work. A ranking convolutional neural network (CNN) was used to calculate the performance of the designed approach. The results gained through this work outperformed the results gained through the state of the art in the field of age recognition. The highest accuracy age estimation was 87.97%, gained through the common voice dataset, testing both genders and all age limits.

Keywords Automatic age estimation · Common voice dataset · Smoothness features · Convolutional neural network

H. A. Abdulmohsin (✉) · S. S. Hasan
Computer Science Department, Faculty of Science, University of Baghdad, Baghdad, Iraq
e-mail: husam.a@sc.uobaghdad.edu.iq

S. S. Hasan
e-mail: ssami@uob.edu.iq

J. J. Stephan
University of Information Technology and Communications, Baghdad, Iraq
e-mail: janejaleel@uoitc.wedu.iq

B. Al-Khateeb
Computer Science Department, College of Computer Science and Information Technology, University of Anbar, Anbar, Iraq
e-mail: belal-alhkhateeb@uoanbar.edu.iq

A. K. Bashir et al. (eds.), *Proceedings of International Conference on Computing and Communication Networks*, Lecture Notes in Networks and Systems 394,
https://doi.org/10.1007/978-981-19-0604-6_11

1 Introduction

Controlling the access of children and underage youth to the pornography websites has been the interest of many governments and child protection organizations. The UK parliament, in 2017, passed a law script requiring legal websites of pornography to implement restricted age verification checks, regardless of the slight age verification that has been implemented for many years from now. The action was built according to studies published in 2021 [1] that showed many children are accessing pornography materials through legal websites. Therefore, speech age verification is a targeted research field in the state-of-the-art research. Through the past few years, many applications consider age as the most important category to decide whether to authorize the user to register or not, to use the service the application provides or not, to authorize the user to enter their website or not, if the user is above 18 or not. All authorization system relies on questioning the user about his age, and depending totally on his answer, without any further checking. This is considered an extremely weak point in all online authorization systems. This research is suggesting adding a new age detection approach using speech that will add more reliability to authorization systems online.

The most important phase in automatic speech recognition in general is extracting features that are most related to the problem. In speech age estimation, many of the researchers stated that Mel frequency cepstral coefficient (MFCC) is the strongest feature in age estimation, but the opposite of this statement was proved by our results and by Abdulsattar et al. [2], where they stated that MFCC is not efficient in age estimation when working with speech that has been recorded in different places using different recording devices, and to improve the efficiency of the results, fundamental frequency acoustic features have to be combined with the MFCC features. According to the simple literature survey conducted through this work, it was noticed that fundamental frequency acoustic features are extremely related to age estimation, but still there are many other variables that affect age estimation, such as the preprocessing, feature selection and the classification method deployed.

Many challenges were faced; one of the challenges was the psychological situation of the human being and its effect on voice. When the human being is psychologically unstable, his vowels will be differently announced when he is psychologically stable [3, 4]. The other challenge was dealing with similarity in children voices of age 3 to 7 years.

The limitation of the state of art works is finding the features that are strongly related to human being age. Through the extensive survey on speech features and depending on the huge number of experiments implemented through this work, it was noticed that smoothness, loudness and pitch are strongly related to age.

2 Related Work

According to the state-of-the-art research, many approaches were followed in age estimation. In 2011 [5], a group of researchers proposed an AAE system that estimated age into 6 limits, deploying a Persian dataset. Two types of features were used, perceptual linear predictive (PLP) and Mel frequency cepstral coefficients (MFCCs). SVM was utilized for classification. In 2014, a group of researchers conducted experiments on six age categories, including children aged between 7 and 12 years, and the experiments revealed that the fuzzy fusion classifier's output was 53.33% in age estimation accuracy [6]. In 2016, [7] used the analog-to-digital converter (ADC) to find the frequency ranges, and estimate age regarding the frequency and pitch of the signal. In 2019, Abdulsattar et al. [2] used K-NN and the first two formants in speech to estimate age, stating that the best information gain in speech lays in the low frequencies of the speech signal. In 2020, Bugdol et al. [8] used the random forest (RF) for regression and was tested using a tenfold cross-validation. The five vowels were recorded and examined for every child, to extend the phonation. Six voice features of were extracted from each recording and used in classification. In 2021, Damian et al. [9] applied different architectures of deep neural network-based embedder, such as d-vector and x-vector in age estimation, and then applied a training scheme of transfer learning with pre-training the embedded networks used in speaker recognition tasks that used vox-celeb1 dataset, to tune it in age estimation.

3 Framework

The block diagram of the proposed method is shown in Fig. 1. In the following three sections, the main steps of the proposed method will be illustrated and discussed in detail.

- Dataset

The dataset utilized in this work is the common voice dataset version en_2637h_2021-07-21. The common voice dataset contains 60 different languages recorded in different percentages as 23% US English, 8% England English, 7% India and South Asia (India, Pakistan, Sri Lanka), 3% Australian English, 3% Canadian English, 2% Scottish English, 1% Irish English, 1% Southern African (South Africa, Zimbabwe, Namibia), 1% New Zealand English. The common voice dataset was recorded by 75,879 different individual voices with different age limits. 6% of the individuals were younger than 19 years old that means 4552 individuals of age less than 19 have participated in recording this online public dataset. And 94% of the dataset is recorded by individuals older than 18 years such as 24% (19–29), 13% (30–39), 10% (40–49), 4% (50–59), 4% (60–69), and finally 1% of the individuals are in the age limit (70–79).

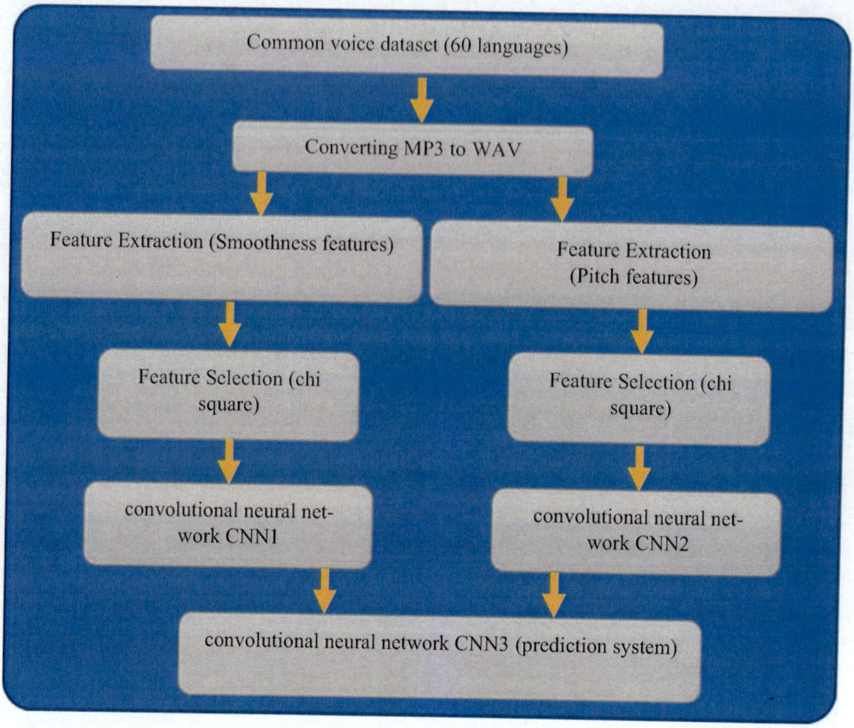

Fig. 1 Block diagram of the method proposed in this work

- Feature extraction

A feature is a measurable property established from the material being observed [10]. The most important aspect in feature extraction is extracting the most relevant features to the problem statement. In the case of this work, smoothness and pitch features are strongly related to human being age and were selected accordingly.

Smoothness if defined as the transaction of speech through air, as much as the speech was smooth as much as its transaction was slower, and when speech is rougher, the transaction of speech is faster [11]. The smoothness was calculated through two domains, first the time domain, and second, the spectral domain. Smoothness is calculated through Eqs. (1 and 2) [12].

$$GV_t = \frac{1}{P}\sqrt{\sum_{j=1}^{P}(\mathrm{var}_t(j))^2} \quad (1)$$

$$GV_s = \frac{1}{N}\sqrt{\sum_{i=1}^{N}(\mathrm{var}_s(i))^2} \quad (2)$$

where var_t and var_s represent the variances in time and spectral domain of the spectral feature, P is the dimension of the feature, N is the length in the time domain of the feature.

The frequency and pitch features were used for its clear relationship with age, since if the frequency ranges are high, age might be limited between 1 and 16 years. It is important to mention that men speeches have low frequency as compared to women speeches. The pitch feature has a close relationship to gender, since men have lower pitch as compared to women. Through experiment, it was noticed that a person with a moderate pitch and moderate frequency, being men or women, can be in the age range of 20–39 years, and the relationship between age and pitch and frequency is opposite, as humans get older their speech pitch and frequency get lower. If the pitch and frequency are low, the age can be limited to 40–60 years, compared to the threshold fixed in the proposed system. If there is any trembling in voice with unclear pronunciation of words, then age can be 60 years or above.

The vibration of vocal cords generates a periodic behavior for the acoustic voiced sounds; this oscillation in frequency is called pitch. The pitch depends on the glottis physical characteristics, which are mass and elasticity [13, 14]. Speech is always represented as a discrete signal $x(n)$; therefore, the pitch represents the fundamental frequency f_o where the signal is repeated. The inverse is the fundamental period T_0. Statistically, there are certain frequency intervals for men regarding each language, for example, the pitch of the Spanish men lays in the frequency interval, 50–300 Hz, and the Spanish women and children can reach to 500 Hz frequency. A methodology used to determining the pitch in this is autocorrelation, which is considered one of the cross-correlation cases [15].

- Feature Selection

A filter ranking method was used through this work for three reasons—first, to filter the less relevant variables, second, to benefit from the criteria of variable selection by order of the variable ranking techniques and third and finally, their simplicity and good success as reported from online applications. A ranking criterion is used to score each variable, then a threshold is fixed through experiment, and used to remove variables below that threshold [10].

Feature selection methods that are applied before classification are considered filter feature selection methods, that's why ranking methods are considered filter methods. The main principle of feature selection methods is to select unique features that contains useful information of different classes in the dataset through using a basic property of that feature. This property is called feature relevance that measures the power of that feature classifying different classes [10, 16, 17]. The chi square statistical feature selection method was used in this work to measure the independence of two selected features.

The chi square feature selection method was calculated through Eq. (3) [18].

$$x_c^2 = \sum \frac{(O_i - E_i)^2}{E_i} \tag{3}$$

- Feature Classification

A ranking convolutional neural network was designed and implemented through this work. Each convolutional neural network was trained on a specific group of features. The first convolutional neural network (CNN1) was trained on the smoothness features, and the second convolutional neural network (CNN2) was trained on the pitch features. Both results were sent to a third convolutional neural network (CNN3) that was used to predict the best results gained from CNN1 and CNN2.

4 Results and Discussion

The results gained in this work will be discussed in detail—two kinds of experiments were applied in this work. First, the samples were first divided into two groups, female and male, and the age estimation was conducted on each gender individually. Second, all samples were tested together regardless of gender, checking if the voice belongs to a human being that is under 18 or higher than 18 years old or equal.

- Age Estimation Results with Each Genders Tested Separately

The results shown in Table 1 show that the age estimation of the male gender is higher in all cases, when estimating age less than 18, when estimating age older or equal to 18 and when considering both genders in the same experiment. This can be justified for the common behavior of children and women voices, which are both specified with high frequencies, which can confuse the system in estimating between woman and children voices. The classification accuracy achieved through testing the male voices was 94.4%, which is achieved by 90.3% of the male samples younger than 18 were correctly classified, and 98.5% of the male samples older than 17 were correctly classified. Whereas the classification accuracy achieved through testing the female voices was 81.55%, which is achieved by 80.4% of the female samples younger than 18 were correctly classified, and 82.7% of the female samples older than 17 were correctly classified, which shows that the age estimation with males are higher than females.

Table 1 Results gained through this work

	6% underage (<18) (4552 samples) (%)	94% legal age (≥18) (71,327 samples) (%)	Total accuracy (75,879 samples) (%)
Both genders	85.35	90.6	87.97
Female	80.4	82.7	81.55
Male	90.3	98.5	94.4

- Age Estimation Results with Both Genders

Through this experiment, all samples were tested together, to estimate age represented in each voice in the dataset. The results gained from this experiment are shown in Table 1. The highest classification accuracy achieved through our experiments was 87.97%, which is a combination of 85.35% of both gender samples representing voices younger than 18, were correctly classified and 90.6% of both gender samples older than 17 were correctly classified. It is obviously clear that the female samples affected the overall accuracy classification.

5 Conclusion

To design a system that can recognize age through the same setting was challenging, because age is related to language, and each language has a different range of frequencies for the male and female and children.

The similar frequency behavior between the female voices and children caused a lot of ambiguity to the system designed in this work, regardless of the strong age-related features deployed.

Acknowledgements Many thanks to Mozilla (voice.mozilla.org) for creating such a global speech dataset, with many languages and many specifications with huge details.

References

1. N. Thurman, F. Obster, The regulation of internet pornography: What a survey of under-18s tells us about the necessity for and potential efficacy of emerging legislative approaches. Policy & Internet (2021)
2. A.A. Abdulsatar, et al., Age and gender recognition from speech signals. In J. Phys. Conf. Ser. (2019)
3. F.A. Shaqra, R. Duwairi, M. Al-Ayyoub, Recognizing emotion from speech based on age and gender using hierarchical models. Procedia Comput. Sci. **151**, 37–44 (2019)
4. H.A. Abdulmohsin, H.A.W. Abdul Mohssen Jaber Abdul hossen, Speech emotion recognition survey. J. Mech. Continua Math. Sci. **15**(9), 24 (2020)
5. D. Mahmoodi, et al., Age estimation based on speech features and support vector machine, in *2011 3rd Computer Science and Electronic Engineering Conference (CEEC)*. IEEE (2011)
6. S.M. Mirhassani, A. Zourmand, H.-N. Ting, Age estimation based on children's voice: a fuzzy-based decision fusion strategy. Sci. World J. **2014** (2014)
7. Gayathri, S., B. Mugundhan, Identification of age using voice recognition. Int. J. Adv. Electron. Comput. Sci. **4**(7), 3 (2016)
8. M.D. Bugdol et al., Adolescent age estimation using voice features. Biomed. Tech. **65**(4), 429–434 (2020)
9. D. Kwasny, D. Hemmerling, Gender and age estimation methods based on speech using deep neural networks. Sensors **21**(14), 4785 (2021)
10. G. Chandrashekar, F.J.C. Sahin, E. Engineering, A survey on feature selection methods. Comput. Electr. Eng. **40**(1), 16–28 (2014)

11. G. Fant, *Acoustic Theory of Speech Production* (Walter de Gruyter, 1970)
12. P.T. Nghia, et al., A Measure of Smoothness in Synthesized Speech. Electron. Commun. **6**(1–2) (2016)
13. L. Rabiner, R. W. Schafer, *Digital Processing of Speech Signal* (1978)
14. J.R. Deller Jr, J.G. Proakis, J.H. Hansen, *Discrete Time Processing of Speech Signals* (Prentice Hall PTR, 1993)
15. L. R. Rabiner, B. Gold, C. K. Yuen, *Theory and Application of Digital Signal Processing* (1975)
16. H.A. Abdulmohsin et al., A new hybrid feature selection method using T-test and fitness function. CMC-Comput. Mater. Continua **68**(3), 3997–4016 (2021)
17. R. Kohavi, G.H. John, Wrappers for feature subset selection. Artif. Intell. **97**(1–2), 273–324 (1997)
18. S. Gajawada, *Chi-Square Test for Feature Selection in Machine learning* (2019). Retrieved from Towards Data Science: https://towardsdatascience.com/chi

Efficient Data Aggregation Strategy in Wireless Sensor Networks: Challenges and Significant Applications

Ahmed Subhi Abdalkafor and Salah A. Aliesawi

Abstract In recent years, WSNs have been considered one of the most sensitive fields of research due to their widespread use in many important and critical applications. However, utilizing the full potential of this network is very hard due to numerous challenges. WSNs face a hurdle in collecting raw data since most of the transmitted data from the sensor nodes to the base station are redundant and lead in many cases to reduce the overall performance of the network. Therefore, a key research challenge has been the study of data aggregation strategy for reducing the number of data transmissions by eliminating redundant data and hence improving the overall network performance. This paper is divided into third parts. Firstly, we discussed the various challenges associated with data aggregation for WSNs. Secondly, we have classified the applications into basic classes and provided a broad overview of significant applications. Finally, we tried to summarize the required features that sensor networks should have for each type of application like range communication, reliability, security, robust node, and network tolerance.

Keywords Wireless sensor networks (WSNs) · Data aggregation · Challenges · Classification applications

1 Introduction

In the past few years, WSNs have attracted great interest from the research community. The main reason behind recent research efforts and the rapid development of WSN is their applications in a wide range of environments including surveillance

A. S. Abdalkafor (✉)
Career Development Center, University Of Anbar, Anbar, Iraq
e-mail: ahmed.abdalkafor@uoanbar.edu.iq

S. A. Aliesawi
College of Computer Science and Information Technology, University Of Anbar, Anbar, Iraq
e-mail: salah_eng1996@uoanbar.edu.iq

A. K. Bashir et al. (eds.), *Proceedings of International Conference on Computing and Communication Networks*, Lecture Notes in Networks and Systems 394,
https://doi.org/10.1007/978-981-19-0604-6_12

systems, environmental monitoring, health care, military operations, and other important applications [1]. WSNs consist of small computing devices that collect data from an area of interest and transmit it to the central station via multiple hops.

These nodes have several advantages, including small size, computing, communication and sensing capabilities, low cost, low energy consumption, and lightweight [2]. These nodes communicate via short-range radio signals and then cooperate with others to accomplish common tasks in sensing phenomena in the target environment such as light, temperature, pressure, and other phenomena, and then deliver this data to the sunken node. The data sent by the adjacent sensor nodes are redundant and results in an amount of data that is too large to be processed in the sump [3, 4]. Therefore, it is imperative to use data aggregation techniques to produce high-resolution information in the sensor nodes that can limit the number and quantity of packets sent to the sink. Actually, the technique of data aggregation is the preeminent way to prevent duplicate and redundant data from being sent to the sink node, which leads to increased network performance [5]. The main contributions of this paper are stated as follows:

1. Clarifying the importance of the data aggregation strategy in WSNs and explain potential differences after implementing this strategy.
2. Provide an insight into the most challenges facing WSNs that affect the overall network performance.
3. Classify the applications of WSNs into basic classes and a comprehensive survey of the significant applications that used these networks.
4. Summarizing the required features that sensor networks should have for each type of application.

This paper is organized as follows. In the next section, we define the data aggregation strategy in WSNs and state the potential differences when applying this strategy. Then, we cover the discussion of the most challenges in WSNs that affect the performance of this network in Sect. 3. Section 4 describes the classification of WSN applications, followed by the most essential apps employed in this environment. Finally, concludes this paper in Sect. 5.

2 Data Aggregation Strategy in WSNs

Rapid technical progress in the world, especially in WSNs, led to the expansion of data and became more complicated with the inputs and outputs for each process. A typical sensor network scenarios include the deployment of a very large number of sensor nodes in an area from which data are periodically obtained. Sensor nodes will transfer the data to a sink node after sensing the target area and can be made accessible to the user. Using routing operations, each sensor node sends its data independently so that it generates numerous redundant traffic in case the data are geographically linked. To clarify this, if two nodes are neighboring, it is expected that the measured values will be very close. So, the definition of data aggregation is

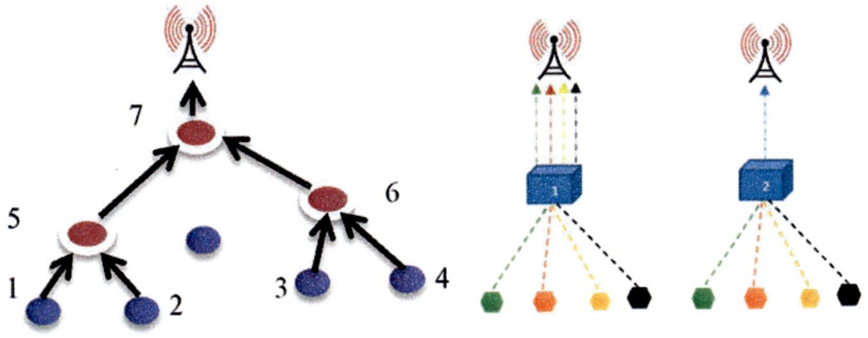

Fig. 1 Idea of data aggregation in WSNs

to avoid redundant data if it is handled in the sensors by applying some functions such as maximum, minimum, sum, average, and count, which can be implemented either periodically or on-demand [6–8]. For getting a clear idea of data aggregation, let us consider seven sensor nodes, namely '1', '2', '3', '4', '5', '6', '7', with a base station as shown in Fig. 1.

Initially, sensor node '5' aggregates the data from both '1' and '2' sensor nodes. In the same process, node '6' aggregates data from '3' and '4' sensor nodes, and the sensor node '7' gathers data from '5' and '6' and finally sends the data to the base station. Also, in the same figure (right side), the (No.1, 2) are two nodes that perform sensing and aggregation data. The four data packet is transmitted into the network in the first type (node No.1) and forwarded it all of them to the base station, while in (node No. 2), the four data packets moved within a network with the type of aggregation, but only, one data packet was transmitted to the upper layer. So, data aggregation is defined as the set of automated ways of merging the data into a set of meaningful data that comes from several sensor nodes. Data aggregation is an encouraging solution for decreasing the communication overhead through merging redundant data, and it reduces the traffic load and conserves the energy of the sensors. Also, it is enhancing the accuracy of the information obtained by the entire network. Figure 2 illustrates the construction of the data aggregation process [9].

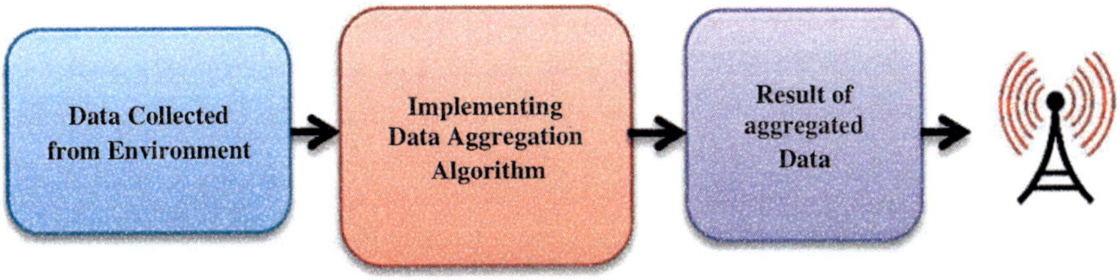

Fig. 2 Process of data aggregation strategy

3 Challenges of WSNs with Data Aggregation Strategy

Despite WSNs are useful for many applications after applying data aggregation strategy through deploying the sensor nodes in the target environment and obtaining sensed data, they suffer from many challenges that affect the design, efficiency, and performance of these networks. In the following subsections, details of the main challenges are presented as shown in Fig. 3.

- Conservation of Energy

The challenge associated with designing WSNs is that sensor nodes operate with limited energy, which is the principal source of their operation and must be replacing or recharging (e.g., via solar power) when depleted. But, neither of the two options is suitable when deploying these nodes in different environments that may be harsh, hostile, or remote so that they cannot be accessed by humans, such as military applications or volcanic monitoring [10]. That is, once it runs out of energy, it will be disposed of, and here lies the challenge. For sensor nodes that have batteries that cannot be charged, they must be able to operate until the end of the task time so that the life of the WSN must be long enough to meet the requirements of the application to be performed [11].

- Security

Another challenge facing WSNs is security, where these networks collect sensitive information when used in critical applications such as military applications, as well as in monitoring places such as airports and hospitals [11]. Operating unattended sensor nodes deployed in remote areas increases their vulnerability to malicious attacks and breaches. their vulnerability to malicious attacks and breaches. Likewise, it is easy for opponents to listen to sensor transmissions when using wireless communications [12]. Accordingly, it is necessary to build a complete system that takes into account the physical security for the sensors as well as the data security when used in such applications.

- Coverage

Another challenge in WSNs is to maintain the maximum possible coverage of the sensor nodes in the target environment and obtain the required data. It must be ensured that this coverage is satisfactory during the network's lifetime. Some critical and dangerous applications require full coverage of the field of interest [12].

Fig. 3 Challenges of WSNs

- Routing

A routing can be defined as to find out a path between the source node and the destination node. So, another challenge that can be exposed to WSNs is how to find a path between the base station and the sensor nodes and maintain these paths [10]. The routing protocols face some exigent factors when designing some of them focus on the capabilities of the sensor nodes such as the transmission range is restricted, the capacity of energy, processing, and storage is limited, and others focus on the inherent features of the network such as topological changes, sensor locations, self-configurable, and finally error tolerance. Therefore, some routing strategies must be used by these protocols to reduce this challenge [10].

4 Classification of WSNs Applications

WSNs have achieved wide popularity in society because they provide an encouraging infrastructure for many monitoring and control applications. These networks are simple and cost-effective and allow monitoring operations to be carried out remotely as well as in real-time without human intervention as little as possible [13, 14]. WSNs consist of a large number of sensors, containing magnetic, optical, seismic, acoustic, thermal, infrared, and radar sensors. These devices are characterized by their ability to monitor various environmental conditions, including noise levels, temperature, pressure, movement, direction, speed, and soil composition [15, 16]. The applications of WSNs can be classified into four classes as shown in Fig. 4.

- Event Detection

The purpose of WSNs in this class is to detect rare events and inform the sink of these incidents immediately, such as intrusions or forest fires. This network should immediately send event reports to the sink node, and this report should contain some information to describe the nature of the event [16].

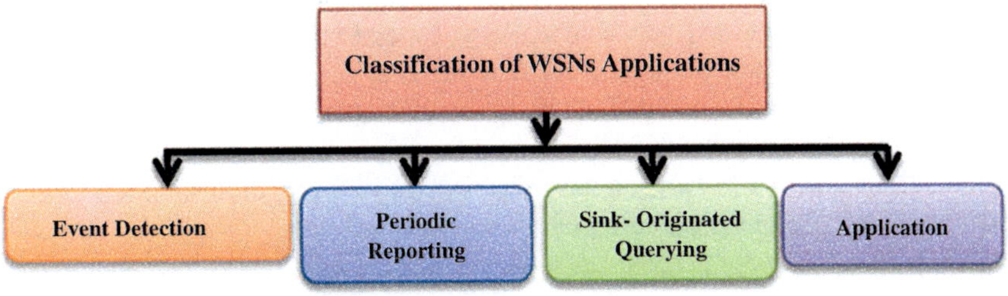

Fig. 4 Classification of applications in WSNs

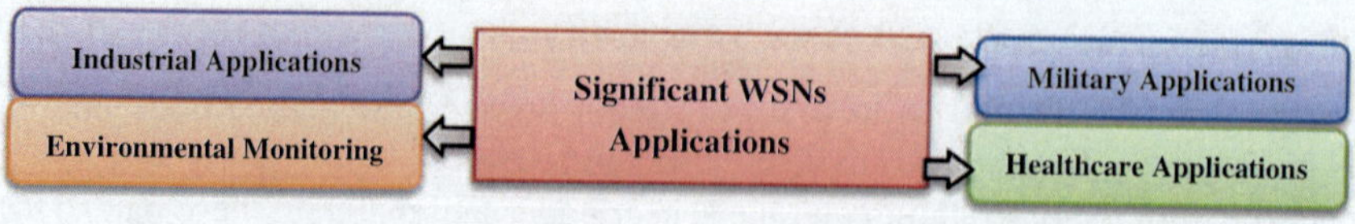

Fig. 5 Popular applications of WSNs

- Periodic Reporting

In this type of application, regular periodic updates are sent to the sink node, so there is a constancy in the data collection phase as well as a stationary flow of data that is sent by sensor nodes to the sink node [17]. However, the measures of neighboring sensor nodes are possible to be related. Therefore, data aggregation of sensor nodes is useful for these class applications because it reduces energy consumption on communication between sensor nodes and reduces data redundancy, thus extending network lifetime [6, 18].

- Sink-Originated Querying

The sink node can query a group from sensors for its measures instead of the sensor nodes periodically reporting their measurements as in the previous application class. This work allows the sink to extract data granularity from various areas in the target environment [16].

- Application Tracking

This application class involves tracking an intruder or the movements of an object after it has been detected and located, such as in military or border surveillance, and then the transfer of relevant information to the sink node while continuing to inquire from appropriate sensors. This class combines some of the features of the previous application classes [16]. Therefore, WSNs opened the way for the innovation of a new generation of applications in a variety of fields as shown in Fig. 5.

- Industrial Applications

WSNs can be used to solve many problems related to this application. This network provides health monitoring of machines, logistics, and robots in terms of analysis and control processes and helps to determine the best operating performance [18]. Many applications can be specified in this framework such as monitoring the manufacturing process or equipment condition, monitoring production performance. Also, this network can manage a set of complex machines and production system conditions for factories in terms of temperature, humidity, vibration, etc.

- Healthcare Application

Developments in smart sensors and implantable biomedical devices have enabled sensor networks to apply in biomedical applications, where the sensor network has played a vital role in health care in terms of its effectiveness and its low cost [15]. Due to the increasing number of patients, continuous monitoring has become necessary to provide health care to patients, whether in hospitals or at homes. One of the examples of sensor networks is the wireless body area network (WBAN), which includes many sensors that monitor physiological conditions. The data that are monitored is periodically sent to a remote location without human intervention [17] where this data are transferred to the specialist doctor for interpreting the patient's condition then provide appropriate treatment.

- Environmental Monitoring

Due to the high impact of human society's development on the environment, environmental monitoring is one of the significant efforts made to improve environmental protection. Environmental monitoring is one of the necessary applications in WSNs to control or reduce the process of environmental degradation also can be described as observing and analyzing natural environments and phenomena to gain a good understanding of these environments [19]. WSN-based environmental monitoring applications include monitoring meteorology and natural catastrophic events and pollution studies, tracing the behavior of insects, birds, small animals, observing environmental conditions that affect livestock and crops, detecting forest fires as well as floods [15].

- Military Applications

The WSNs in this application are essential because it is characterized by rapid propagation and self-organization. It considers a better method for battlefields because they depend on the dense deployment of sensor nodes. Also, the destruction of several nodes by a hostile action does not affect military operations as much as it impacts the destruction of the central node. Some WSN-based military applications include battlefield monitoring, equipment, ammunition and opposition force reconnaissance, battle damage assessment, and others [15].

Table 1 shows the required features that sensor networks should have for each type of application mentioned in this manuscript.

Table 1 Required features of WSNs for applications

Application type	Range	Reliability	Security	Robustness	Network tolerance
Health applications	Small	Highly	High	High	High
Industrial applications	Depend on application	Highly	High	Highly	Highly
Environmental monitoring	Wide	High	High	Highly	Highly
Military applications	Wide	Highly	Highly	Highly	Highly

5 Conclusion

WSNs are becoming more popular due to their increasing application in remote sensing in many fields. This paper provides inspect of the main challenges associated with data aggregation strategy that facing WSNs. In addition, it discusses significant applications that use this network after classifying it into basic classes. Finally, the requirements that the sensor networks should have for each type of application are summarized. This paper is expected to contribute further understanding to overcome or reduce raw data collecting problems and assist sensor network users in using this network for their applications, thus make effective use of WSNs.

References

1. R. Elhabyan, W. Shi, M. St-Hilaire, Coverage protocols for wireless sensor networks: Review and future directions. J. Commun. Networks **21**(1), 45–60 (2019)
2. S.P. Ardakani, Data aggregation routing protocols in wireless sensor networks: a taxonomy. arXiv Prepr. arXiv1704.04588 (2017)
3. K. Al-Ani, A. Abdalkafor, A. Nassar, An overview of wireless sensor network and its applications. Indones. J. Electr. Eng. Comput. Sci. **17**(3), 1480–1486 (2020)
4. A. Tripathi, H.P. Gupta, T. Dutta, R. Mishra, K.K. Shukla, S. Jit, Coverage and connectivity in WSNs: a survey, research issues and challenges. IEEE Access **6**, 26971–26992 (2018)
5. I. Mosavvar, A. Ghaffari, Data aggregation in wireless sensor networks using firefly algorithm. Wirel. Pers. Commun. **104**(1), 307–324 (2019)
6. H. Ramezanifar, M. Ghazvini, M. Shojaei, A new data aggregation approach for WSNs based on open pits mining. Wirel. Netw. 1–13 (2020)
7. A. Abdalkafor, S. Aliesawi, Data aggregation techniques in wireless sensors networks (WSNs): taxonomy and an accurate literature survey, in *The First Virtual International Conference on Sciences (VICS2021)*, Accepted (2021)
8. A.S. Abdalkafor, S. Aliesawi, The impact of data aggregation strategy on a performance of wireless sensor networks (WSNs), in *The 14th International Conference on Developments in eSystems Engineering (DeSE)*, Accepted (2021)
9. D. Vinodha, E.A.M. Anita, D.M. Geetha, A novel multi functional multi parameter concealed cluster based data aggregation scheme for wireless sensor networks (NMFMP-CDA). Wirel. Netw. 1–18 (2020)
10. S.P. Singh, S.C. Sharma, A survey on cluster based routing protocols in wireless sensor networks. Procedia Comput. Sci. **45**, 687–695 (2015)
11. W. Dargie, C. Poellabauer, *Fundamentals of wireless sensor networks: theory and practice* (Wiley, 2010)
12. X. Deng, Y. Jiang, L.T. Yang, M. Lin, L. Yi, M. Wang, Data fusion based coverage optimization in heterogeneous sensor networks: a survey. Inf. Fusion **52**, 90–105 (2019)
13. M. Pule, A. Yahya, J. Chuma, Wireless sensor networks: a survey on monitoring water quality. J. Appl. Res. Technol. **15**(6), 562–570 (2017)
14. K.W. Al-Ani, A.S. Abdalkafor, A.M. Nassar, Smart city applications: a survey, in *Proceedings of the 9th International Conference on Information Systems and Technologies* (2019), pp. 1–4.
15. I.F. Akyildiz, M.C. Vuran, *Wireless Sensor Networks*, vol. 4 (Wiley, 2010)
16. A. Iyer, S.S. Kulkarni, V. Mhatre, C.P. Rosenberg, A taxonomy-based approach to design of large-scale sensor networks, in *Wireless Sensor Networks and Applications* (Springer, 2008), pp. 3–33

17. M. Ghamari, B. Janko, R.S. Sherratt, W. Harwin, R. Piechockic, C. Soltanpur, A survey on wireless body area networks for ehealthcare systems in residential environments. Sensors **16**(6), 831 (2016)
18. D. Kandris, C. Nakas, D. Vomvas, G. Koulouras, Applications of wireless sensor networks: an up-to-date survey. Appl. Syst. Innov. **3**(1), 14 (2020)
19. S. Majumder, M.J. Deen, Smartphone sensors for health monitoring and diagnosis. Sensors **19**(9), 2164 (2019)

An Internet of Things-Empowered Disaster Management Framework

Adil Chekati, Meriem Riahi, and Faouzi Moussa

Abstract Each year, natural disaster (floods, earthquakes, forest fires, tsunamis are only a few examples) and man-made disasters (for instance, industrial explosion, leakage in an oil pipeline or gas productions) cause damage to infrastructure, distress, financial losses, injuries, and a large death toll. However, with the help of recent technological advances, it is now vital to change from traditional monitoring systems to smart prediction systems that incorporate governments and people affected by these disasters equally. Motivated by this, in this paper we present an IoT-empowered disaster management framework applied on natural flood disaster. The main aim of this framework is to monitor direct measures of climate such as rainfall, humidity, temperature, pressure, and water levels, as well as to determine their temporal correlations for flood prediction. For data collection via sensors, an IoT technique is employed, an efficient data classification phase is established, and an agent-based approach is used for decision making in flood prediction and management.

Keywords Internet of Things · Disaster management · Smart objects · Data classification · Decision making

1 Introduction

According to the World Health Organization (WHO): "A disaster is an occurrence disrupting the normal conditions of existence and causing a level of suffering that exceeds the capacity of adjustment of the affected community" [1]. Disasters often strike in the vicinity of human livelihood. Natural disasters impacted 217 million

A. Chekati (✉)
Faculty of Sciences of Tunis (FST), University of Tunis ElManar, LIPAH-LR11ES14 Tunis, Tunisia
e-mail: adil.chekati@fst.utm.tn

M. Riahi
High National School of Engineers of Tunis (ENSIT), University of Tunis, Tunis, Tunisia

F. Moussa
LCOMS, Université de Lorraine, 57070 Metz, France

A. K. Bashir et al. (eds.), *Proceedings of International Conference on Computing and Communication Networks*, Lecture Notes in Networks and Systems 394,
https://doi.org/10.1007/978-981-19-0604-6_13

people annually between 1994 and 2013, according to estimates [2], and around 97.6 million people have directly or indirectly got affected during the year of 2019 only [3]. Natural disasters are also said to have caused $15.3 billion in damage and a million new cases of cancer between 2000 and 2011. In 2016, this resulted in over 65 million immigrants and refugees fleeing around the globe [4].

Therefore, there is a need to identify and exploit the data needed for effective disaster management operations in case of natural disasters. Early detection of these disasters will keep everyone safe from such calamities.

In recent years, the Internet of Things (IoT) has become an integral part of the human's daily life. IoT is a technology that integrates embedded system components with a wireless communication channel to transfer data from sensor nodes to a computing device for real-time analysis. In fact, this technology has the potential to connect the world and allow people to communicate with each other and with their environment, what gives it the potential to save lives. For example in disaster management, IoT's role is critical and ubiquitous, and it has the potential to reduce the risk [5]. As a consequence, IoT devices may be used to capture data and identify risks during disasters, as well as to locate wounded person [6].

Fortunately, IoT is a promising technology used in several applications including disaster management. The Internet of Things has proven to be fundamentally capable of providing more significant, scalable, adaptable, and energy-efficient solutions to many disaster management difficulties. As a result of these concerns, a comprehensive understanding how IoT is presently monitoring and managing catastrophes becomes critical.

Motivated by this, in this paper an IoT-empowered framework with an agent-based approach for decision making has been performed in anticipation for the establishment of a flood forecasting and management system. The system is made up of sensors that detect their surroundings, and a local sensor server analyzes the collected information; infrastructure for communication is based on Lora WAN protocol, and a decision tree classifier is employed in order to forecast the current condition of rainfall flood level, while a multi-agent approach would finally work to elaborate the flood disaster solutions.

The remainder of the paper is structured as follows: Sect. 2 explores the extant literature on disaster management and the existing IoT solutions; Sect. 3 outlines the disaster management scenario using the proposed framework. Furthermore, Sect. 4 provides the details of the different phases of the framework; Sect. 5 discusses some of the key challenges for IoT-based disaster management. Finally, Sect. 6 presents the concluding remarks and future works.

2 Related Work

Several studies on flood disaster management and flood forecasting systems have already been conducted, and this section provides a summary of the present literature relevant to the proposed solution described in this paper.

Authors in [7] presented an IoT-based method to deliver real-time information on disaster-affected regions in order to allow quick and efficient recommendations about rescue activities. The task-technology fit technique is used to verify the proposed solution.

In [8], ideas and methods for the detection of flood disaster based on IoT, big data, and convolutional deep neural network (CDNN) have been proposed to overcome disaster difficulties.

An IoT-based flood monitoring and artificial neural network (ANN)-based flood prediction are presented in [9] with the aim of enhancing the scalability and reliability of flood management system.

Sood et al. [10] gift the case of the big data and high performance computing (HPC) Integration IoT flood management method and demonstrate IoT's usability and efficiency, with the aim to classify geographical areas into a web of hexagonal for effective installation of energy efficient IoT devices.

With the aim to forecast possibility of future disasters by using data mining algorithm, authors in [11] proposed a real-time flash-flood monitoring system, where river water level and diverse meteorological variables such as temperature, humidity, and vibration are measured and forecasted using data mining and wireless sensor networks.

In [12], it is proposed to provide interactive and real-time information on the current water level and rain intensity, as well as alarm alerts. Through a smartphone application, Internet of Things might aid flood victims in monitoring floods, groundwater level, and precipitation intensities.

These relevant contributions with the aim of developing an IoT solution for flood disaster management have been compared regarding various design and application perspectives. This comparison based on features offered by previously presented IoT-enabled solutions is described in Table 1. A few of the main pieces as (i) IoT-based architecture, (ii) enabled Machine Learning for data classification, (iii) Real-Time interaction, (iv) enabled early warning notification, and (v) provided forecasting for future floods.

Table 1 Comparison between Internet of Things disaster management systems

Article	IoT architecture	ML classification	Real time	Early warning	Forecasting
[7]	Yes	No	Yes	Yes	No
[8]	Yes	Yes	No	Yes	No
[9]	Yes	No	Yes	Yes	Yes
[10]	Yes	No	Yes	Yes	Yes
[11]	Yes	No	Yes	Yes	Yes
[12]	Yes	No	Yes	Yes	No

3 Disaster Management Scenario

Flooding is a natural catastrophes that continues to occur in many places around the globe as a direct result of the planet's climate change. This natural calamity has struck numerous nations in recent years (i.e., Oman, Germany, North Africa, etc.). Human will be less likely to be affected by such disasters if they are detected early.

Flooded regions are frequently coated with silt and dirt when the floodwaters recede. Hazardous pollutants such as chemicals, sharp objects, gasoline, and untreated sewage can contaminate the water and landscape. Mold blooms that are potentially hazardous can swiftly overtake water-soaked buildings. Flooded communities may lose electricity and access to safe drinking water, resulting in epidemics of fatal waterborne illnesses such as cholera, hepatitis, and typhoid [13]. Disaster management is extremely sensitive during a natural catastrophe. To preserve people's lives and reduce destruction, various services and systems, as well as the presence of many governmental authorities, must be engaged. A smart city structure will undoubtedly help all those parties in handling the disaster.

The IoT technology available today is quite advanced, and it has the potential to be extremely valuable in disaster scenarios. Disaster management is significantly influenced by the area's climatic conditions, habitat, and other factors, as well as the available resources. In disaster management, IoT technologies offer advantages in terms of monitoring, tracking, controlling, and sensing the environment using real-time data [14].

As a reason, we have decided to implement the proposed IoT-empowered framework on flood disaster management as a case of study. In the case of flood disaster, exploiting the newest technologies is vital, as it can help to take the best decision in order to perform some action like evacuation, contingency planning, and so forth. In such a circumstance, all ministries of the environment, health, defense, and local authorities must collaborate to better identify issues, save people and animals' lives, and limit damage. As a result, an IoT platform might be very beneficial. Water bodies are equipped with a variety of sensors and actuators that are controlled by the processing systems of the water bodies department. To guarantee comprehensive collaboration between multiple institutions, all of these equipment are in real-time contact with other equipment at the weather forecast services and authorities concerned. A machine learning (ML) classifier will handle data processing and classification, while a multi-agent system (MAS) approach helps for decision making.

4 Flood Disaster Management Based on IoT, ML Classification, and MAS: SADM-SmartObject

Flood is one of the most common natural catastrophes that can devastate any city [15]. Since this disaster poses a risk to humanity existence, an effective countermeasure or system of warning should be established for notifying people as soon as possible

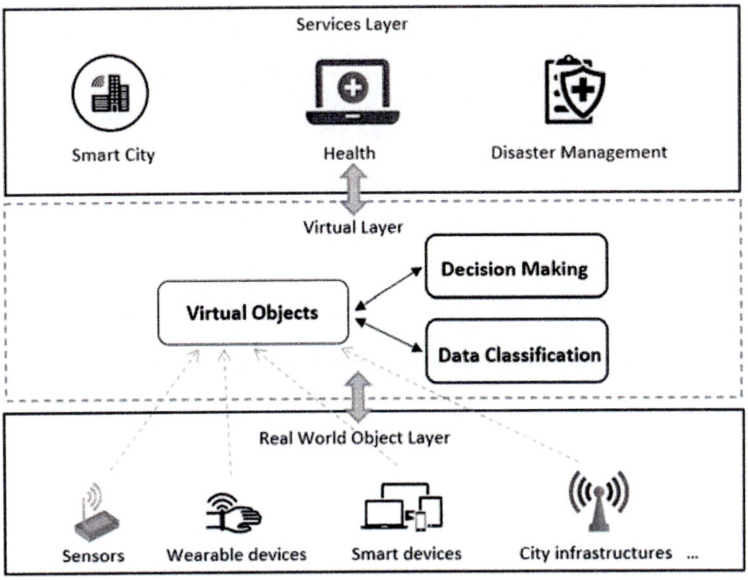

Fig. 1 SADM-SmartObject framework architecture

so that security measures can be taken to avoid any tragedy [16]. Hence, Internet of Things is useful for a better understanding of the situation through real-time analysis of data collected by ground-based sensors, vehicles, etc. These mounted sensors are able to monitor a high-risk area and transmit this information for further processing.

SADM-SmartObject is a self-adaptation and decision making for smart object framework, designed as a multi-layer paradigm (Fig. 1). Its intention is to lay forth a framework for developing Internet of Things applications and proposing an optimal solution for autonomous decision making [17]. It is a promising system that can be used in several applications including disaster management.

Internet of Things has the potential to become one of the enabling technologies in disaster management. The proposed SADM-SmartObject framework is aimed to provide the following functionalities:

1. Disaster risk minimization and prevention: Keeping track on potential disasters via relevant measurements gathered by multiple types of mounted sensors, analysis of predefined high risks and vulnerabilities used for forecasting to generate accurate and timely warnings.
2. Emergency response: Immediate rescue and emergency actions require real-time communication. Spread risks information and warnings to high-risk individuals and institutions around the crisis area.
3. Disaster recovery: The process of intervention based on rescue plans that exploit local resources and preparations to react in the case of natural disaster.

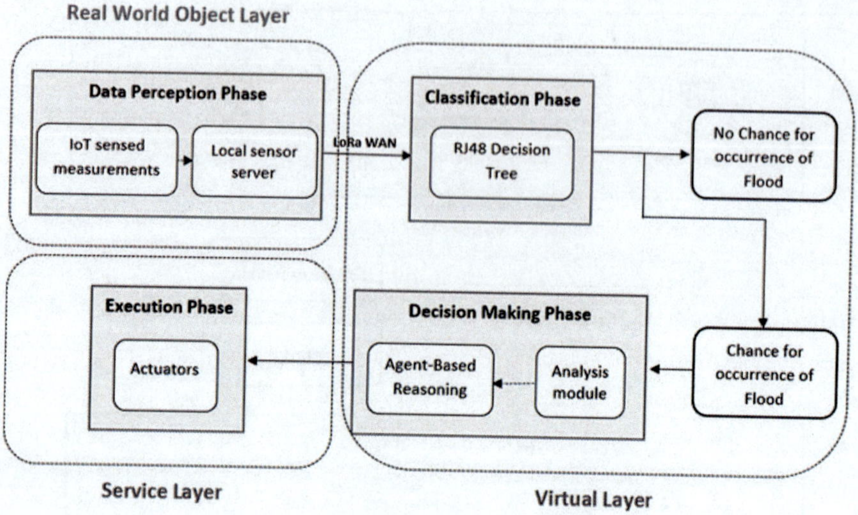

Fig. 2 Block diagram for the proposed SADM-SmartObject framework

Figure 2 depicts a detailed block diagram of the proposed framework flood disaster framework.

4.1 Data Perception Phase

The lowest layer, or the physical layer, is represented by the real-world object layer (RWO layer). It collects all of the IoT system's real-world objects.

In the data perception phase, the input data is taken from mounted sensors, namely temperature, pressure, humidity, wind, rain, and water level. Next, the data is collected and normalized at a local sensor server. Next, the preprocessed data is given as input to the classifier at the virtual layer which classifies (a) the chances of occurrence of flood disaster and (b) no chances of occurrence of the flood disaster. The classifier is explained in Sect. 4.2.

4.2 Classification Phase

The intermediate level is represented by the virtual layer. It guarantees that every physical object in the system is virtualized. It permits the most efficient use of data provided from the RWO layer by analyzing, processing, and enhancing decision making in real time. The classifier is the first module of this layer.

Table 2 Accuracy of the tested classification techniques [19]

Algorithms	Correctly classified	Incorrectly classified
J48 decision tree	92.81	7.19
Random forest	91.50	8.50
KNN	88.89	11.11

An experimental study was conducted on three effective data classification algorithms (J48 decision tree, random forest, and K-nearest nighbor) in order to figure out which classifier is the most effective for our smart objects framework "SADM-SmartObject" according to a range of performance evaluation metrics. We took use of real-world sensor datasets from the Kaggle data repository [18] for the experiments. Daily weather reports from a large number of Australian weather stations are included in the dataset. It is a big dataset comprising data from a lot of different weather sensors and measuring devices. Temperature, rainfall, evaporation, sunlight, wind direction and strength, humidity, pressure, and clouds are among the variables measured. These sensors were used to gather weather data throughout the day. The experiment has shown that the best approach is J48 decision tree. As seen in previous work [19], the J48 decision tree technique outperformed its competitors in terms of IoT data classification, demonstrating that it is a potentially convenient and powerful classification algorithm for IoT data. The experiment findings indicate that decision tree has the highest precision and the lowest execution time, which are two of the most significant evaluation criteria for an IoT application. Table 2 summarizes the precision of each classification approach obtained after the experiment on IoT data.

Next, the classification result values are forwarded to decision making, which is explained in Sect. 4.3.

4.3 Decision-Making Phase

Also a part of virtual layer, this phase aims to present an IoT solution using collected data to make decisions and take actions in real time depending on that data.

A multi-agent system (MAS)-based approach is used for decision making in SADM-SmartObject framework. This aims to take advantage of MAS capabilities, to provide the aspect of self-adaptation and decision making for existing connected objects [17]. The BDI paradigm, which is based on a theory of how people operate [20], is used in the SADM-SmartObject framework for decision making. Objects have a set of beliefs about the world. They also have a desire list that they intend to fulfill. Finally, they have a set of intents, which are the present activities they are taking to attain their goals [21].

In another word, it empowers IoT applications to advance from basic data gathering to autonomous behavior and decisions in real time. The MAS paradigm, in fact, facilitates the generation of (i) self-configuring, self-protecting, and self-optimizing Iot devices and (ii) IoT systems with context-aware and adaptive smart objects.

IoT permits interconnections of different devices, and it establishes seamless communication, monitoring, and management of smart embedded objects. In this context, the integration of MAS capabilities with the power of IoT that offer more efficient methods to analyze, monitor, predict, and manage flood disaster.

A simulation of the decision-making process is implemented by means of JADE platform to demonstrate the efficiency of this agent-based approach:

- Different types of meteorological sensors (such as temperature, water level, humidity, wind, and rain) are used to gather weather data throughout the day; this sensor reading is transmitted in real time, and many actions must be implemented immediately to avert a disaster.
- A reactive agent is generated as a virtual representation for each physical sensor. We define: AgentHumid, AgentWaterLevel, AgentRain, AgentTemp, and AgentWind.
- Reasoning agents are a group deliberative agent, representing the central smart aspect of the decision-making process. Reasoning agents are of the type deliberative agent.
- Coordination agent is used to ensure coherency between reasoning agents, in the aim of making the accurate decision. Coordination agent is of the type deliberative agent.

Figure 3 shows the application's deployment on the JADE platform. The interaction between the agents is also illustrated, and the values from the sensors are continuously communicated to reasoning agents via various sensing agents. While this main processing module receives real-time updates on weather conditions, it also delivers personalized recommendations to coordination agents on how to safeguard the area from any potential flood threat.

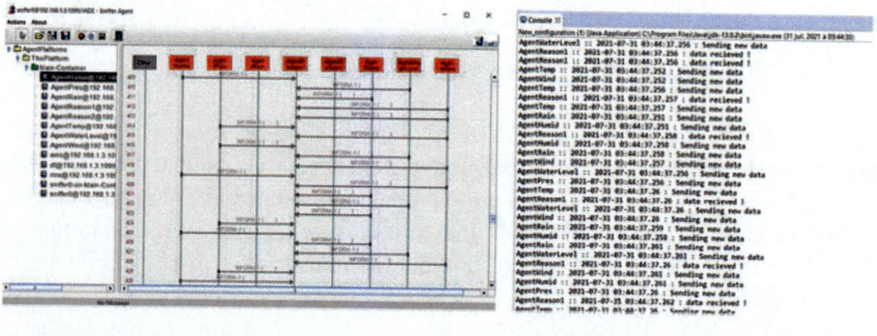

Fig. 3 JADE platform log and sniffer interface

4.4 *Execution Phase*

The highest layer is represented by the service layer, which might be considered a front-end layer owing to its direct connection with users. In this phase, virtual actuators are evoked to execute the solution obtained from the decision-making phase: To effectively handle and deal with the issue, a flood prevention and management mechanism is triggered. At the same time as executing the solution, a backup request for the established solution is sent back to the previous phase, and notifications are instantly dispatched to civil protection services, national security, and local authorities.

5 Challenges for IoT-Empowered Disaster Management System

In this section, we go through some, and not all, of the key challenges that must be surmounted in order to put the pillars of IoT-based disaster management systems.

5.1 *Security*

The first critical challenge concerns with security and data privacy. As personal and private data is being collected continuously before, during, and after disasters, security concerns are particularly relevant. Malicious actors cannot interfere with the data collected from damaged areas or incidents. This security issues are very critical throughout the disaster management process. As a result, security mechanisms must be added to disaster management system in order to ensure data integrity, data, and network security in all the phases of a disaster management operations.

5.2 *Standardization*

We admit that our proposed framework, as well as other current works, does not conform to the standardized format in terms of data presentation or processing. This presents another challenge which is standardization, as it is usually recognized that different types of disasters require different responses and have distinct standardization. Disaster management is tough to implement, but the three key issues are security requirements, connectivity standards, and identification standards, all of which must be implemented in conjunction with the deployment of IoT technologies' meaning.

5.3 Real-Time Analytic

To manage the dynamic nature of the disaster, real-time processing is one of the most important requirements in IoT-enabled technologies for disaster management systems. Although, as mentioned in Sect. 2. Not all disaster management proposed approaches consider this requirement. In our proposed framework, a real-time approach is used in the decision-making phase.

5.4 Cost Effectiveness

Due to the numerous devices needed to deploy an IoT-based disaster management system, cost effectiveness is becoming a key challenge in which the presence of the vast number of sensors is a primordial task to aggregate data and also establish communication infrastructures to ensure data transmission. Multiple researches are being conducted with the aim to optimize the device's performance while lowering the hardware cost.

6 Conclusion

Flood is one of the most common disastrous events that take place in different countries every year around the globe. Hence, adoption of new technologies could minimize the death toll and large-scale facilities damaged as a result of natural and man-made disasters. IoT has been able to be applied to save the livelihood among flood-affected areas in recent times. This paper has proposed ideas and methods for the flood disaster management based on and IoT-empowered framework to overcome such difficulties. The first phase starts with the collecting sensed data from different mounted sensors like water level, rain sensor, humidity, etc., followed by the machine learning classification phase which divides the flood prediction into chances and no chances. Proceeding to the decision-making step, an agent-based reasoning process is conducted in order to elaborate the proposed solutions.

References

1. WHO Disasters & Emergencies Definitions, https://apps.who.int/disasters/repo/7656.pdf. Accessed 29 June 2021
2. M.T. Basu et al., IoT based forest fire detection system. Int. J. Eng. Technol. **7**(2.7), 124–126 (2018)
3. World Disasters Report 2020, https://media.ifrc.org/ifrc/world-disaster-report-2020/. Accessed 29 June 2021

4. K. Dorofeev et al., Device adapter concept towards enabling plug&produce production environments, in *22nd IEEE International Conference on Emerging Technologies and Factory Automation* (IEEE, 2017)
5. K. Sharma et al., A disaster management framework using internet of things-based interconnected devices. Math. Probl. Eng. **2021**, 1–21 (2021)
6. A. Boukerche, R.W. Coutinho, Smart disaster detection and response system for smart cities, in *2018 IEEE Symposium on Computers and Communications (ISCC)* (IEEE, 2018), pp. 1102–1107
7. A. Sinha et al., Impact of internet of things (IoT) in disaster management: a task-technology fit perspective. Ann. Oper. Res. **283**(1–2), 759–794 (2019)
8. M. Anbarasan et al., Detection of flood disaster system based on IoT, big data and convolutional deep neural network. Comput. Commun. **150**, 150–155 (2020)
9. S. Bande, V.V. Shete, Smart flood disaster prediction system using IoT & neural networks, in *Proceedings of the 2017 International Conference on Smart Technology for Smart Nation* (2018)
10. S.K. Sood et al., IoT big data and HPC based smart flood management framework. Sustain. Comput. Inform. Syst. **20**, 102–117 (2018)
11. S.S. Mane, M.K. Mokashi, Real-time flash-flood monitoring alerting and forecasting system using data mining and wireless sensor network, in *2015 International Conference on Communication and Signal Processing* (2015)
12. M. Sabre, S. Shahrum et al., Flood warning and monitoring system utilizing internet of things technology. Kinetik **4**(3), 287–296 (2019)
13. Floods Explained, https://www.nationalgeographic.com/environment/naturaldisasters/floods/. Accessed 8 July 2019
14. A. Ghapar et al., Internet of things (IoT) architecture for flood data management. Int. J. Future Gener. Commun. Netw. **11**(1), 55–62 (2018)
15. A. Yusoff, I.S. Mustafa et al., Green cloud platform for flood early detection warning system in smart city, in *5th National Symposium on Information Technology Towards New Smart World (NSITNSW)* (IEEE, 2015)
16. E. Shalini, S. Subbulakshmi, P. Surya, R. Thirumurugan, Cooperative flood detection using sms through IoT. Int. J. Adv. Res. Electr. Electron. Instrum. Eng. **5**(3), 1–7 (2016)
17. A. Chekati, M. Riahi, F. Moussa, Agent-based modelling approach for decision making in an IoT framework, in *Advanced Information Networking and Applications. AINA 2021*, ed. by L. Barolli, I. Woungang, T. Enokido. Lecture Notes in Networks and Systems, vol. 226 (Springer, 2021). https://doi.org/10.1007/978-3-030-75075-6_21
18. J. Young, Rain in Australia. Kaggle, 11 Dec 2020. [Online]. Available: https://www.kaggle.com/jsphyg/weather-dataset-rattle-package. Accessed 04 Apr 2020
19. A. Chekati, M. Riahi, F. Moussa, Data classification in internet of things for smart objects framework, in *International Conference on Software Telecommunications and Computer Networks SoftCOM* (2020). https://doi.org/10.23919/SoftCOM50211.2020.9238186
20. H. Fouad, I. Moskowitz, Meta-agents using multi-agent networks to manage dynamic changes in the internet of things, in *Artificial Intelligence for the Internet of Everything* (Elsevier Inc, 2019), pp. 271–281
21. P. Stone, Learning and multiagent reasoning for autonomous agents—a goal of AI, in *IJCAI 2007* (2007), pp. 13–30
22. Y. Hirabayashi, R. Mahendran, S. Koirala et al., Global flood risk under climate change. Nat. Clim. Change **3**, 816–821 (2013)

Priority-Based Buffer Management Technique Against Dropping Attack in Opportunistic Networks

Satyendra Kumar Srivastav and M. M. Tripathi

Abstract The opportunistic network is intermittent. It is based on the principle to carry, store, and forward the messages for routing from source to destination. It is very prominent in recent years as it provides an alternative to conventional networks. But, it has low-delivery probability and high latency. In the proposed model, the performance of the system can be increased by the use of the priority-based buffer management technique (PBMT) with a threshold. It not only forwards the message according to priority but also controlled the random dropping of the message. Various simulations are analyzed by comparing the proposed mechanism with the other two benchmark buffer mechanisms such as first in first out (FIFO) and random by using real-time mobility data-trace. PBMT with threshold provides higher delivery probability 7.85%, 16.29% with the TTL, 2.98%, 4.35% with the buffer size, and 6.10%, 14.41% with the message generation interval as compared to random and FIFO, respectively. It also provides lower average latency 4.05%, 16.81% with the TTL, 4.54%, 2.34% with the buffer size as compared to random and FIFO, respectively. Results of these simulations confirm that the PBMT with threshold shows a significant improvement in the performance of opportunistic networks (OppNets).

Keywords Buffer management · Opportunistic network · FIFO · Priority · Random

1 Introduction

In a crisis, it is difficult to maintain the connectivity through the conventional network, so the OppNets are very promising but they incur very high-delay, low-delivery, and high-intermittent connection among intermediate nodes. This results less assurance of message to reach from source to destination. The performance of OppNets can be improved by the use of efficient buffer management. Conventional wireless

S. K. Srivastav (✉) · M. M. Tripathi
Delhi Technological University, Delhi, India
e-mail: dtu.satyendra@gmail.com

A. K. Bashir et al. (eds.), *Proceedings of International Conference on Computing and Communication Networks*, Lecture Notes in Networks and Systems 394,
https://doi.org/10.1007/978-981-19-0604-6_14

networks such as ad hoc networks are exposed in such an environment where end-to-end connections have existed. The conventional networks use a carry and forward mechanism, whereas OppNets use carries, store, and forward mechanisms to transmit message from source to destination. In OppNets, each node carries a amount of buffer space. This buffer space is a very critical resource to fight against dropping attacks. The inappropriate use of buffer will lead to high-delay and low-delivery rate when nodes are exposed to challenging environmental conditions.

This paper proposes a particular scenario of crisis when conventional networks are not working properly. An opportunistic network is used to handle such conditions. It provide an alternative way of connectivity but delivery probability is very low due to dropping attack. Dropping attack can be handled by increasing buffer space but it increases the delay which is not the favorable condition for the real-time message. Buffer management plays a major role in reducing dropping attacks. This paper proposed a priority with threshold-based efficient buffer mechanism against dropping attacks which minimizes the dropping attack as well as delay of the message simultaneously.

In this mechanism, the priority of the message (Pmsg) is defined according to type of message. The message can classified as a service data message, management data message, control data message, normal data message, and emergency data message and priority assigned at time of creation to each type 1, 2, 3, 4, and 5. Service data messages are relayed between network devices and other end devices, management data messages are generated by the management station to manage the operating network, and control data messages are used to create or operate the network itself. Normal data messages are user-created messages relayed in the network according tó the destination IP address. Emergency data messages are used for the real-time exchange of data such as life-saving information against disaster situations.

2 Related Work

Buffer management policy plays a key role in scheduling and dropping decisions. In scheduling policy, it decides the sequences of messages to be transferred at the time of contact with the neighboring node. In dropping the policy, it selects the message to be dropped at the time of overflow of the buffer. The four basic buffer management policies such as random drop, drop least recently received, drop the oldest drop, and least encounter were evaluated. The performance of drop least encounter is better than drop oldest because dropping decision can be taken according to meeting probability [1]. In MaxProp, path likelihood and hop counts are used to decide the route and dropping priorities. It provides more opportunities for newer messages [2]. In probabilistic routing protocol using history of encounters and transitivity (PRoPHET), nodes move with a certain meeting pattern according to past the meetings of nodes and delivery probability between two nodes [3, 4]. In epidemic routing protocol using drop head, drop tail, and source-prioritized drop head, where the best performance is given by source-prioritized drop head among all [5]. N-drop policy message dropping decision is taken when a message is forwarded N times or greater than N as well as the buffer is filled and it is needed to drop any message [6]. Dropping decisions have been also evaluated according to drop head dropping the policy, remaining lifetime dropping, replication count of messages [7]. Bjurefors et al. evaluated different dropping decision policies like Least interested (LI), most interested (MI), max copies (MAX), most forwarded (MF), least forwarded (LF), random dropping, and compare all such different dropping policies against infinite buffer and no buffer [8, 9]. In drop largest (DLA), the dropping decision is taken according to the size of the message from the buffer in case of buffer overflows and drops the largest size message [10]. The buffer scheduling decisions have been taken according to the average contact frequency (ACF) between nodes for taking buffer scheduling decisions [11]. The message transmission status-based buffer management scheme (MTSBS) is based on priority which is decided by dissemination speed and replication count [12]. In spray and wait, the TTL-based routing (TBR) protocol implements buffer management. A contact opportunity is used to forward the message according to a priority function [13]. ST-drop is another local buffer management policy that relies only on locally available information to estimate the time and space coverage of messages in the network [14, 15]. In the weight-based buffer management policy, messages are assigned weights. These weights depend on message size, remaining time-to-live, message stay-time in the node buffer, its hop count value, and the number of replicas [16, 17]. In fuzzy-spray, the message is stored in a node and it is prioritized according to fuzzy logic. The fuzzy controller takes two inputs such as the forward transmission count (FTC) of a message and its size [18, 19].

3 Proposed Work

3.1 *Motivation and Proposed Model*

The management of buffer consists of forwarding, replication, dropping, and scheduling policy. Low-delivery ratio and high delay arise due to inefficient use of a buffer. The literature presented above has clearly expressed that dropping attacks based on the different categories of messages for OppNets are unavailable. To overcome this dropping attack, a novel technique PBMT with threshold is introduced commonly known as PBMT.

The proposed method is the advancement of PMBT by adding a one-factor threshold. In a crisis, it is a high probability that the buffer is full with high-priority message. In this condition, almost all the messages are stored in the buffer at the same time, and the remaining TTL of all the messages is almost the same. In this case, the buffer is full of high-priority messages and TTL expired at the same time. Inside the buffer, all high-priority messages consume more time in storage rather than communication which may have a chance to TTL expired before reaching the destination. Threshold avoids this stuckness by dropping the least transit message to the destination.

In PBMT with threshold, messages have some priority according to their nature. The multiple messages are allowed to transmit to nearby nodes according to their priority. The priority of the message is assigned according to the importance of the message. For example, loss of life would incur a much higher loss than the other loss. In conventional protocol, messages are transferred without considering their priority. The conventional protocols are less efficient in terms of transmission of the life-saving message which is given the highest priority message. If the node buffer is sufficiently full with distinct priority messages and a high-priority message arrives, then already stored less priority message is dropped to create space for high-priority message. By applying PBMT, the message that reaches the destination is high-priority message.

3.2 System Model

Let assume N node distributed opportunistically in the network. When any message transfers from source to destination, the scheduling and dropping policies are taken in consideration to make buffer management policy to be efficient. The first step is to set the priority of messages according to their nature, and the second step is to exchange the summary vector at the intermediate node when the buffer is full.

When a message is transferred from the source to the destination through the intermediate node. The message based on priority is transferred from the sender to that intermediate node that is in its proximity. The major advantage of this protocol is that higher priority message has less drop as compare to lower priority message. In case of FIFO, LIFO, and random buffer management policy, higher priority message has to wait for a longer duration. To overcome the longer duration of higher priority messages, a new buffer management policy was introduced. This proposed buffer management policy provides a shorter duration along with a low-dropping attack of a high-priority message. The proposed policy introduced a schedule dropping according to the threshold. The least transit probability (PLT) is the probability of a message which has very little chance to reach the destination. The least transit probability of a message depends on the remaining TTL. In this paper, both values are adjusted manually for getting a better result when compared to the conventional technique.

Algorithm 1: PBMT with threshold

1. Begin N nodes ($N0, N1, N2, N3, \ldots Nn$) in the network.
2. Source node Nk generates the message Mk and assigned priority values 1, 2, 3, 4, 5, respectively, depending on the type. 5 being the highest priority.
3. Repeat steps 4–16 while the message does not reach its destination.
4. Transfer message Mk from node Nk, keeping message with the highest priority at the top.

5. If buffer of neighbor node *Nl* is full then
6. {Pop message *Ml* from node *Nl* with lower priority than *Mk* and exchange *Mk* to *Ml*.}
7. If buffer of node *Nk* is also full and exchange of *Mk* to *Ml* is not possible
8. {Pop message *Ml* from node *Nl* with next lower priority than *Mk* and exchange *Mk* to *Ml*.}
9. If buffer of node *Nl* is full and all messages in node *Nl* are all of higher priority then
10. If the message with the least transit probability to destination is greater than the threshold value
11. {No exchange the messages will have occurred}
12. else
13. {Delete the least transit probability message}
14. else
15. Push message *Mk* from node *Nk* to node *Nl* buffer.
16. $Nl \leftarrow$ next node, $Nk \leftarrow Nl$
17. If message *Mk* has reached to destination, then
18. {Go to step}
19. Repeat steps 2 and 3 till all priority messages are transferred from *Nk* to the destination node.
20. Message delivered successfully at the destination

4 Simulation

The proposed PBMT is simulated through ONE simulator [20]. Real mobility data-trace, haggle-one-infocom 2006, is used for simulation and comparison for other conventional buffer policies FIFO and random [21]. The data-trace file contains tab-separated data fields of connection events in standard event reader syntax format. It contains five fields such as time, action, the first node, second node, and type, respectively. The first field contains simulation time which denotes the time of the event when it has occurred. The second field represents the status of the connection, and it is always "CONN." The third and fourth field denote IDs of nodes. The last field contains up and down keywords that refer to connection and disconnection.

The flowchart of proposed PBMT is given in Fig. 1, and also the various stimulation parameters are given in Table 1. The proposed method uses the length of the message for deciding priority, and one extra buffer management dropping policy is considered according to TTL to avoid the situation when the buffer is full of all messages having the highest priority.

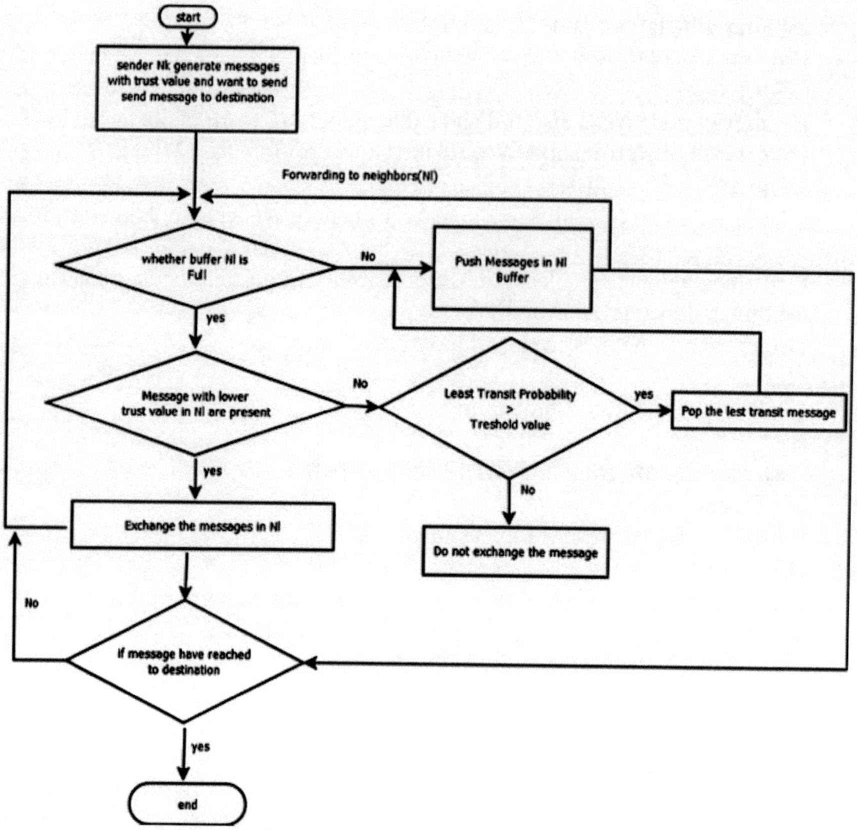

Fig. 1 Flowchart of PBMT

Table 1 Simulation parameter and its value

Parameter	Value
Communication interface	Bluetooth
Transmission range	10
Number of nodes	98
Number of contacts	170,601
Simulation time	337,418
Transmission speed	250 kbps
Message size	500 k up to 1 Mb
Buffer capacity	5 Mb
Movement model	Shortest path
Message time-to-live	300 min

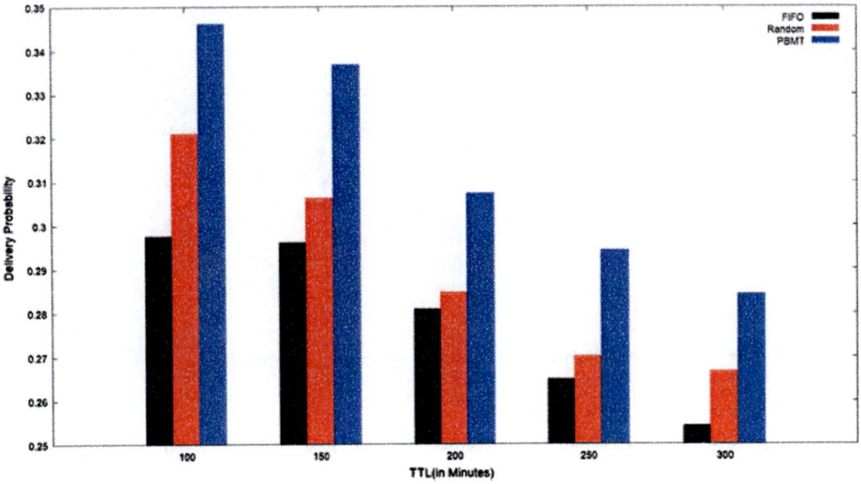

Fig. 2 Delivery probability versus TTL (in min)

4.1 Result and Discussion

Figure 2 exhibited variation of delivery probability against TTL. It is observed from the figure that the delivery probability of FIFO, random, and PBMT with threshold decreases as TTL increases. In the case of PBMT, if TTL increases, the least transit probability also increases as the duration of the remaining TTL is increases. On account of this delivery, the probability decreases as the message gets more time to be stored in the buffer. PBMT (0.3461) provides 7.85 and 16.29% higher delivery probability as compare to random (0.3209) and FIFO (0.2976).

Figure 3 exhibited variation of delivery probability against buffer size of the node. It is seen from the figure that the delivery probability of FIFO, random, and PBMT increases as buffer capacity is increases. In the case of PBMT, if the buffer size of the node increases, the least transit probability will also be increased as more messages are accommodated. On account of this, the delivery probability increases as a greater number of the message get stored in the buffer and it leads to more chances to message to reach the destination node. PBMT (0.7137) provides 2.98 and 4.35% higher delivery probability as compare to random (0.3209) and FIFO (0.2976).

Figure 4 shows the variation of average latency against TTL of the message. It found that the average latency of FIFO, random, and PBMT increases as the TTL of the message is increases. The high-TTL message will stay more time in the buffer which also substantially increases the average latency. PBMT (0.4909.9088) exhibits 4.05 and 16.81% less average latency as compare to random (5108.1033) and FIFO (5901.8884).

Figure 5 shows the variation of average latency against the buffer size of the node. It is seen from the figure that the average latency of FIFO, random, and PBMT increases as buffer capacity is increases. In the case of PBMT, if the buffer size

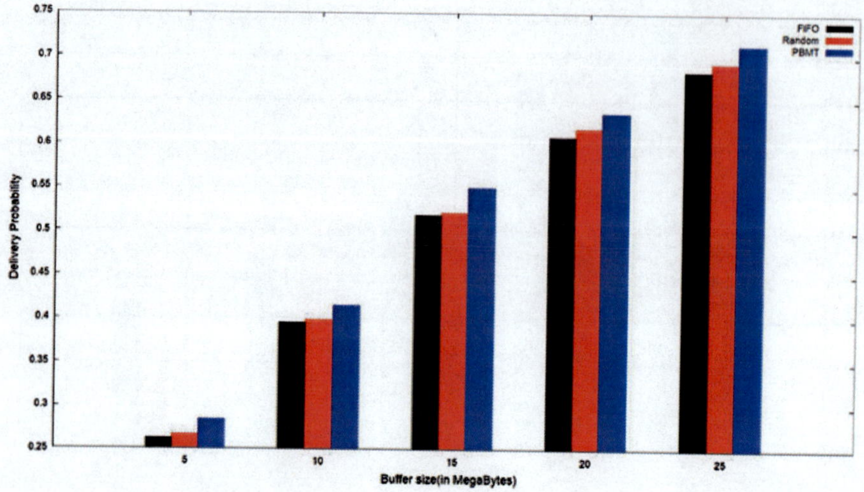

Fig. 3 Delivery probability versus buffer size (in min)

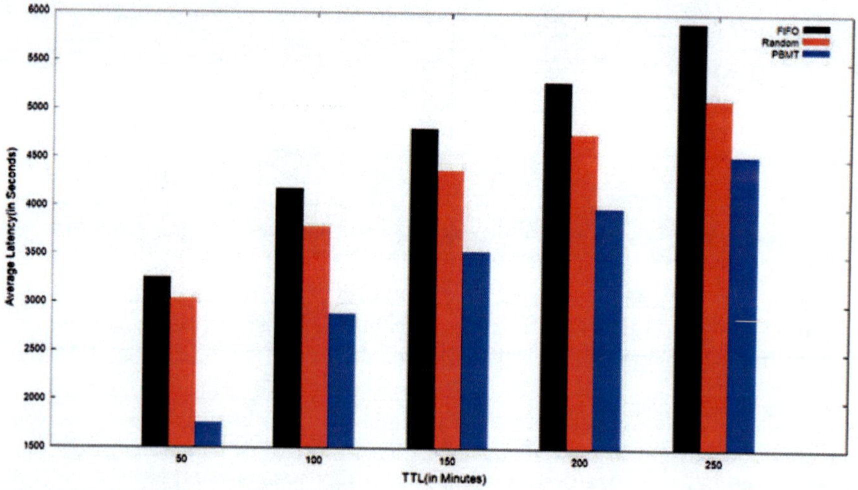

Fig. 4 Average latency versus TTL (in min)

of the node increases, average latency will be managed by the threshold value. It is observed PBMT shows the least average latency as compared to others. PBMT with a threshold (7482.6049) provides 4.54 and 2.34% less delivery probability as compare to random (7822.9013) and FIFO (0.8778.7973).

Figure 6 exhibited the variation of delivery probability when the interval between the messages is varied. It is observed from the figure that the delivery probability of FIFO, random, and PBMT increases as the message generation interval increases. If the interval between messages is increased, fewer packets are stored in the buffer. It leads less congestion and buffers management will be easy to handle. PBMT (0.3405) provides 6.10 and 14.41% higher delivery probability as compare to random (0.3209) and FIFO (0.2976).

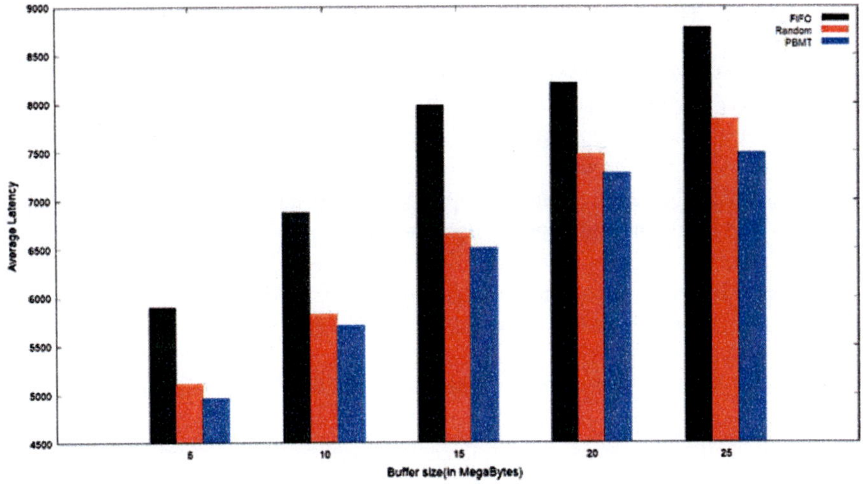

Fig. 5 Average latency versus buffer size (in min)

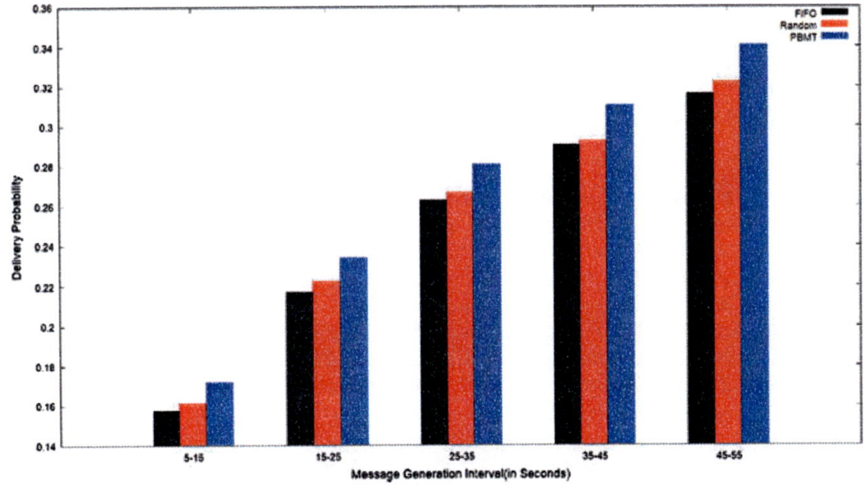

Fig. 6 Delivery probability versus message generation interval (in s)

5 Conclusion

In the proposed model, the major advantage is that it maintains the connectivity and transfers the life-saving messages with high-probability and low-average latency even though conventional connectivity does not exist. In the proposed model, buffer management policy has been addressed as its plays a crucial role in scheduling and dropping the message. The proposed model addressed the solution from messages stuck inside the buffer with all high-priority messages in standard PBMT. It introduced a threshold value that drops the message having the least probability to reach the destination. It uses the fixed value of threshold that is calculated manually by comparing standard buffer management policies. The manual selection of the least probability and threshold will lead to higher delivery probability and lower latency of the message. The proposed model fulfills the attributes which are shown in Figs. 2, 3, 4, 5, and 6. In this proposed model, least transit probability and fixed threshold values are considered manually by comparing the standard benchmark buffer policy. The threshold value can also be variable and change according to the situation. In future work, the selection of the least probable message and threshold will be calculated according to dynamic profile matrices of the given area. It also includes some other factors so that performance should be increased. This model is simulated in haggle-one-infocom 2006 data-trace. Some other data-traces can also be considered for the given proposed model.

References

1. J.A. Davis, H.A. Fagg, B.N. Levine, Wearable computers as packet transport mechanisms in highly partitioned ad-hoc networks, in *5th International Symposium on Wearable Computing*, Zurich, Switzerland (2001), pp. 141–148
2. J. Burgess, B. Gallagher, D. Jensen, B.N. Levine, MaxProp: routing for vehicle-based disruption-tolerant networks, in *25th IEEE International Conference on Computer Communications (INFOCOM) Proceedings*, Barcelona, Spain (2006), pp. 1–11
3. A. Lindgren, A. Doria, O. Schelen, Probabilistic routing in intermittently connected network. ACM SIGMOBILE Mob. Comput. Commun. Rev. **7**(3), 19–20 (2003)
4. M. Ababou, R. El Kouch, M. Bellafkih, Dynamic utility-based buffer management strategy for delay-tolerant networks. Int. J. Ad Hoc Ubiquitous Comput. (2019)
5. X. Zhang, G. Neglia, J. Kurose, D. Towsley, Performance modelling of epidemic routing. Comput. Netw. **51**(10), 2867–2891 (2007)
6. Y. Li, L. Zhao, Z. Liu, Q. Liu, N-drop: congestion control strategy under epidemic routing in DTN, in *Proceedings of the International Conference on Wireless Communications and Mobile Computing: Connecting the World Wirelessly*, Leipzig, Germany (2009), pp. 457–460
7. V.N. Soares, F. Farahmand, J.J. Rodrigues, Performance analysis of scheduling and dropping policies in vehicular delay-tolerant networks. IARIA Int. J. Adv. Internet Technol. **3**(1 and 2), 137–145 (2010)
8. F. Bjurefors, P. Gunningberg, C. Rohner, S. Tavakoli, Congestion avoidance in a datacentric opportunistic network, in *ACM ICN*, Toronto, Canada (2011), pp. 32–37
9. M. Chawla, S. Jain, Survey of buffer management policies for delay tolerant networks. J. Eng. (2014)

10. S. Rashid, Q. Ayub, M. Soperi, M. Zahid, A.H. Abdullah, Message drop control buffer management policy for DTN routing protocols. J. Wireless Pers. Commun. **72**(1), 653–669 (2013)
11. L. Tang, Y. Chai, B. Weng, Buffer management policies in opportunistic networks. J. Comput. Inf. Syst. **8**(12), 5149–5159 (2012)
12. A.T. Prodhan, R. Das, H. Kabir, G.C. Shoja, TTL based routing in opportunistic networks. J. Netw. Comput. Appl. **34**(5), 1660–1670 (2011)
13. S. Rashid, Q. Ayub, A.H. Abdullah, Reactive weight-based buffer management policy for DTN routing protocols. Wireless Pers. Commun. **3**, 993–1010 (2015)
14. M.D. Silva, I.O. Nunes, R.A.F. Mini, A.A.F. Loureiro, ST-drop: a novel buffer management strategy for D2D opportunistic networks, in *2017 IEEE Symposium on Computers and Communications (ISCC)* (2017)
15. Y. Chen, W. Yao, M. Zong, D. Wang, An effective buffer management policy for opportunistic networks, in *Collaborate Com 2016, LNICST 2016*, Beijing, China (2017), pp. 242–251
16. T. Kimura, C. Premachandra, Suppressive fair buffer management policy for intermittently connected mobile ad hoc networks. Wireless Pers. Commun. (2017)
17. S.K. Dhurandher, J. Singh, I. Woungang, J.J.P.C. Rodrigues, Priority based buffer management technique for opportunistic networks, in *2019 IEEE Global Communications Conference (GLOBECOM)* (2019), pp. 1–6
18. T. Senttawatcharawanit, S. Yamada, M.E. Haque, Message dropping policy in congested social delay tolerant networks, in *10th IEEE International Conference on Computer Science and Software Engineering*, Maha Sarakham, Thailand (2013), pp. 116–120
19. S. Abbas, K. Ahmad, Opportunistic routing protocols based on fuzzy logic: present and future directions, in ***2021 International Conference on Artificial Intelligence and Smart Systems*** *(ICAIS)* (2021), pp. 1055–1060. https://doi.org/10.1109/ICAIS50930.2021.9395955
20. A. Keränen, J. Ott, T. Kärkkäinen, The ONE simulator for DTN protocol evaluation, in *Proceedings of the 2nd International Conference on Simulation Tools and Techniques (SIMUTOOLS)* (2009)
21. J. Scott, R. Gass, J. Crowcroft, P. Hui, C. Diot, A. Chaintreau, CRAWDAD dataset Cambridge/haggle (v. 2009-05-29). Downloaded from http://crawdad.org/cambridge/haggle/20090529 (2009)

Automatic Health Speech Prediction System Using Support Vector Machine

Husam Ali Abdulmohsin

Abstract In the past couple of years, seeking for automatic disease prediction system (ADPS) on the Internet has been attempted by many users online especially after COVID-19 pandemic. This points to a new generation of medical treatment, especially when the number of internet users is growing day by day. Therefore, automatic disease prediction online applications have gained the attention of many researchers around the world. Through this work, an automatic disease prediction system depending on speech has been designed and implemented. The system aims to predict the type of disease, and the patient is suffering from depending on his voice, but this was not applicable through experiment, so the diseases were divided in to three groups, and the diagnoses were implemented accordingly. The benchmark of this work was the medical speech, transcription, and intent dataset. The features utilized in this work are the smoothness, mel-frequency cepstral coefficient (MFCC), and the spectral centroid variability (SCV) features that proved their high representation to human being medical situation in this work. The noise reduction forward–backward filter was used to remove noise from the wave files recorded online for the high noise noticed in the dataset deployed. A hybrid feature selection algorithm was designed and implemented for this work which combined the output of the genetic algorithm (GA) with the inputs of the neural network (NN) algorithm. Support vector machine (SVM), neural network, and Gaussian mixture model were utilized for classification. The highest results gained according to our design groups are 94.55 and 50.1% according to each disease, both with respect to SVM.

Keywords Automatic disease prediction · Medical speech transcription and intent dataset · Mel-frequency cepstral coefficient · Spectral centroid variability · Forward–backward filter · Hybrid feature selection algorithm · Genetic algorithm · Neural network

H. A. Abdulmohsin (✉)
Computer Science Department, Faculty of Science, University of Baghdad, Baghdad, Iraq
e-mail: husam.a@sc.uobaghdad.edu.iq

A. K. Bashir et al. (eds.), *Proceedings of International Conference on Computing and Communication Networks*, Lecture Notes in Networks and Systems 394,
https://doi.org/10.1007/978-981-19-0604-6_15

1 Introduction

Automatic disease prediction system (ADPS) has been attracted by researchers and companies in the last few years, for the fast access to medical treatment, avoid traffic and long-distance travels, avoid high expenses, and most useful to elder people, regardless of the new era the globe is passing through with the spread of COVID-19, and the periodic curfew many countries are applying to avoid the virus spread.

It is essential to understand the speech production process in human beings in order to deal with ADDAs in the most professional way. The field of phonetics studies the sounds of human speech production. Speech is produced by pushing air from lungs to the larynx (respiration), where the vocal cords may be open to allow air to pass through or may vibrate generating a sound (phonation). The lungs airflow will be shaped by the articulators in mouth and nose, who are responsible of the articulation.

Speech sounds can be considered from two different point of views, which are acoustic and articulatory phonetics. Acoustic phonetics are part of linguistics, but also is a branch of physics, where it deals specifically with acoustic and physical sound waves properties. Articulatory phonetics is how human bodies are used in speech sound production [1]. To produce speech, three mechanisms are needed. First, an energy source is needed. Anything that generates a sound requires a source of energy, and in the case of human speech sounds, the air flowing from the lungs is the source of energy. The second mechanism is the source of sound, which is the vocal cords (or vocal folds) at the larynx that produce sound. The third mechanism is the articulators where the sound will be filtered, or shaped. These articulators can be the oral cavity (mouth space), the nasal cavity (space inside and behind your nose), jaws, teeth, lips, and finally and most important the tongues [2].

Acoustic voiced sounds are generated by the air going out from the lungs passing through the vocal cords, causing vocal cords vibration, generating periodic signals. The different physical characteristics of the vocal cords for each human being generate different frequencies in the voice production that reflects the different properties existing in different human being acoustics [3]. The second member in the human speech production system is the epiglottis. The mass and elasticity characteristics of the epiglottis are different for each human being [4, 5] that also causes new features to be added to speech, regardless of the tong position and the mouth cavity that both cause different pronunciation for the same words differently pronounced by different people. The different characteristics of epiglottis differ also between genders, where we can notice that formants of females are higher in frequency than male and the spectrum of voiced sounds usually decreases in amplitude with increasing frequency. All these acoustic effects are caused by the production of speech. Therefore, it is possible to find gender specific features represented in acoustic speech signals [6].

According to the speech production process, speech features can be divided into acoustic and articulation features depending on the source of generation. If the human being is tired or feeling any pain, his articulators will reflect his medical situation, and his sound will be produced in a different way, because speaking requires a lot of energy, and as soon as the human is tired or not feeling well, the first thing that will be affected in his voice [7–9]. This is considered evidence that any human health issue will be translate through his voice, but the most important is to choose the most related features that will translate the illness, infection, pain, or any health issue.

Through this paper, the related works will be discussed in Sect. 2. The statistics and the properties of dataset bench mark deployed in the experiments of this research and the framework of the proposed method will be illustrated details in Sect. 3. Section 4 will show and discuss the highest classification results gained through this work. Section 5 will state the conclusion gained out of this work.

2 Related Work

Many researches have confirmed the existence of unique acoustic and physiologic features in human being voices that can be used to diagnose the patient symptoms, but till now, they have not reached the required classification accuracy. Some researchers have worked on ADPSs by following different approaches and mechanisms.

One of the mechanisms depended in ADPSs is depending on clinical reports to diagnose highest predicted disease that patient is suffering from [10]. Another mechanism of healthcare ADPSs is to depend on user text input (guided by users), through the question-and-answer approach. Then, the system will provide a list of top most predicted diseases [11–14]. Many researchers have used machine learning methods in diagnosing heart disease [15]. Another approach is to use natural language processing and deep learning methods to extract text that can identify some symptoms to be used in predicting the diseases [16–18] and many other researches. According to Maree Johnson survey in 2014, she stated that over 20,000 state-of-the-art works were published in the field of speech to text ADPS that deployed computational linguistics, natural language processing, human language technologies, or text mining [19].

According to the literature survey conducted through this work, the work proposed is to be considered a field is ADPS, where no linguistics, natural language processing is accomplished, and no speech is converted to text for the purpose of text mining. This work proposed an ADPS that predicts the amount of pain, and the patient is facing depending directly on his speech without ant linguistic processing.

Many challenges were faced, and one of the challenges was the psychological situation of the human being and its effect on voice. When the human being is psychologically unstable, his vowels will be differently announced when he is psychologically stable [20, 21]. The other challenge was dealing with similarity in children voices of age 3–7 years, therefore, children were avoided in the experiments applied to the method proposed in this work, and since children need adults to speak up for them. The noise and distortion found in many recordings in the utilized dataset was a challenge, and the same will be in real life.

The proposed work has some limitations and needs to be proved such as dealing with children and dealing with different languages that were not practiced through this work. Through this work, we were not able to diagnose the voice signal of each disease, but the contribution of this work was we were able to diagnose diseases into diseases that have psychological affects and those that do not and also diagnose diseases depending on the level of pain, strong, and light pain.

3 Proposed Automatic Health Prediction Speech System

The block diagram of the proposed method is shown in Fig. 1. In the following five section, the dataset and the main steps of the proposed method will be illustrated and discussed in details.

- Dataset

The dataset utilized in this work is the medical speech, transcription, intent dataset version 1. The dataset is built out of verbal descriptions in (wav) audio format that describe the medical symptoms of the patient. Each verbal description is paired with a transcription is the (csv) text format, then each are labeled according to a certain category ailment. This dataset is a one language dataset recorded in the English North American slang. It describes 25 symptoms, which are *back pain, acne, blurry vision, body feels weak, cough, ear ache, emotional pain, feeling cold, feeling dizzy,*

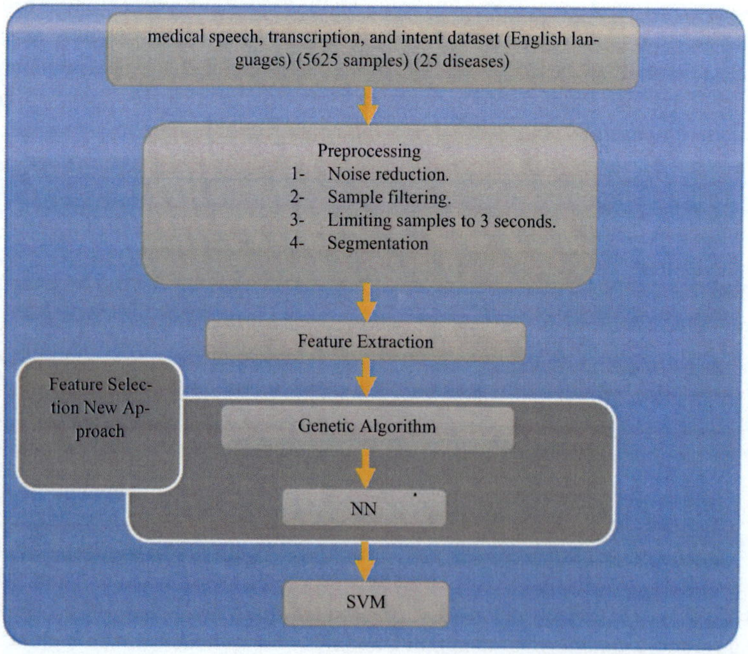

Fig. 1 Block diagram of the method proposed in this work

foot ache, hair falling out, heard to breath, head ache, heart hurts, infected wound, injury from spots, internal pain, joint pain, knee pain, muscle pain, neck pain, open wound, shoulder pain, skin issue, and stomach ache. There is a different number of recordings for each symptom, which forced us to neglect some of the recording for some symptoms to reach an equal number of recordings for each symptom which is 225 recordings for each symptom. The total number of recordings involved in this work are 5625 recording [22].

Through our work, we noticed a lot of distortion and noise in some of the wave recordings that goes back to the bad quality of microphones used in recording those samples, and the noisy environment of the place that attended the recording. This led us to pass all recordings through a forward–backward digital filter or so called the zero-phase filter, deployed in MATLAB. The reason behind using bero-phase filtering is because it preserves features filtered in the time waveform exactly where they occurred in the unfiltered signal, which at the end, does not affect the classification performance. Such kind of filters performs zero-phase digital filtering by processing the input signal, x, in both directions, in the forward direction and reverse direction. After the data are filtered in the forward direction, the filter sequence will be reversed and will run the signal back through the filter [23].

- Pre-processing

Through this step, many processes were implemented. First, the noise reduction process. The noise reduction forward–backward filter was used to remove noise from the wave files recorded online for the high noise noticed in many samples in the medical speech, transcription, and intent dataset deployed in this work. Second, the sampling filtering process. All 25 diseases represented in the dataset contained at least 225 samples, and some of the diseases were represented in more samples. So, we selected 225 samples for each disease and removed some of the samples that had high noise depending on listening all samples, and this process was time consuming. The third process was the trimming process. Most of the recordings were recorded in different time limits. Therefore, all samples were trimmed to be 3 s. The fourth process was the segmentation process. All samples were segmented according to the ratio (0.05%) of the original signal, and the overlap ratio deployed in this work was (0.025%), which provides (50%) overlap, and the total number of segments generated will be $(2n - 1)$, where n is the number of original segments.

- Feature Extraction

A feature is a measurable property established from the material being observed [24]. The most important aspect in feature extraction is extracting the most relevant features to the problem statement. In the case of this work, the features extracted were smoothness, mel-frequency cepstral coefficient (MFCC) with 12°, and the spectral centroid variability (SCV) features, which are strongly related to human being pain feeling, and were selected according to the experiments conducted on 15 types of speech features that were tested through our work, but no reasonable results were gained, only with the three types of features mentioned.

Smoothness if defined as the transaction of speech through air, as much as the speech was smooth as much as its transaction was slower, and when speech is rougher, the transaction of speech is faster [25]. The smoothness was calculated through two domains, such as first the time domain and second the spectral domain. Smoothness is calculated through Eqs. (1) and (2) [26].

$$GV_t = \frac{1}{P}\sqrt{\sum_{j=1}^{P}(\mathrm{var}_t(j))^2} \tag{1}$$

$$GV_s = \frac{1}{N}\sqrt{\sum_{i=1}^{N}(\mathrm{var}_s(i))^2} \tag{2}$$

where var_t and var_s represent the variances in time and spectral domain of the spectral feature, P is the dimension of the feature, N is the length in the time domain of the feature.

- Feature Selection

In this work, the genetic feature selection algorithm was used to filter feature groups ahead of passing it to the NN, to select the best groups of features according to a certain criterion, then pass the best selected group of features to the NN. Each group of features will be weighted, and the weight will be used in merging the groups to generate new groups. Groups with high weight will be tested, and groups with low weight will be merged, until the criterion is satisfied, the best group will be passed to the NN.

NN is used for classification purposes or used as a feature selection algorithm of type embedded and also used in data mining and machine learning [27, 28].

A filter ranking method was used through this work for three reasons. First, to filter the less relevant variables. Second, to benefit from the criteria of variable selection by order of the variable ranking techniques. Third and finally last, their simplicity and good success as reported from online applications. A ranking criterion is used to score each variable, then a threshold is fixed through experiment and used to remove variables below that threshold [24].

Feature selection methods that are applied before classification are considered filter feature selection methods thatis why ranking methods are considered filter methods. The main principle of feature selection methods is to select unique features that contains useful information of different classes in the dataset through using a basic property of that feature. This property is called feature relevance that measures the power of that feature classifying different classes [24, 29, 30].

- Feature Classification

The SVM, backpropagation NN, and the GMM classifiers were selected in this work to predict the pain of the patient.

4 Results and Discussion

- Experiment Number 1 (Exp1)

The aim of the experiment in this work was to predict the disease represented in each wave recording and classifying each recording according to the 25 different diseases represented in the dataset. Figures 2, 3, and 4 show the confusion matrix of the best accuracy classification results gained through deploying the three classifiers such as SVM, NN, and GMM, respectively. As noticed from the three confusion matrices,

0back pain	75	1	0	2	0	11	1	1	0	14	0	2	7	1	8	8	12	13	14	12	12	13	9	1	13	32.6
acne	1	119	12	23	0	0	2	3	1	1	16	1	0	0	0	0	0	1	1	1	1	0	1	11	1	60.7
blurry vision	1	21	127	21	1	1	1	1	1	1	17	2	0	1	0	3	1	1	0	0	1	0	0	14	1	58.5
body feels weak	0	24	19	126	1	0	0	1	0	0	13	1	1	0	0	0	0	1	1	1	1	0	1	14	1	61.2
cough	0	1	1	0	129	4	16	12	9	1	0	12	0	8	0	0	1	1	1	0	1	1	0	1	1	64.5
ear ache	9	0	1	1	1	99	0	0	1	8	0	0	8	1	7	12	11	9	11	14	9	10	10	1	12	42.1
emotional pain	1	0	0	0	22	0	131	11	13	1	1	11	1	10	0	1	1	0	0	0	1	1	0	0	2	63.0
feeling cold	1	1	0	1	15	0	17	144	11	0	0	13	0	9	0	0	0	1	0	1	0	1	1	0	3	65.8
feeling dizzy	0	1	0	2	16	1	23	13	135	0	1	11	1	9	1	0	0	0	0	0	1	2	0	1	1	61.6
foot ache	14	1	1	0	0	12	0	1	0	108	0	0	10	0	12	9	11	13	13	10	11	14	9	1	17	40.4
hair falling out	1	18	21	19	1	0	0	1	1	1	157	0	3	0	1	0	1	1	1	0	1	0	0	27	1	61.3
heard to breath	4	0	2	2	15	0	16	15	14	0	1	150	1	17	1	1	0	1	2	0	0	0	1	0	0	61.7
head ache	11	2	3	0	1	14	0	1	2	12	0	1	105	1	8	6	9	8	7	11	5	34	13	2	14	38.9
heart hurts	2	0	0	0	17	0	13	12	15	0	2	13	1	158	1	0	1	0	2	0	0	1	1	1	3	65.0
infected wound	8	1	1	0	1	8	0	0	1	7	0	0	13	1	116	10	7	7	7	8	6	4	7	2	7	52.3
injury from spots	7	1	0	1	0	9	0	0	2	7	0	1	11	0	7	105	18	7	12	7	5	5	8	1	6	47.7
internal pain	9	2	0	0	1	8	0	1	2	7	1	0	7	1	8	8	76	10	8	8	8	6	9	1	8	40.2
joint pain	11	1	0	1	0	9	1	1	3	7	0	1	8	1	7	8	12	78	18	7	9	4	7	1	5	39.0
knee pain	8	1	1	0	0	8	0	0	3	9	1	1	8	1	8	9	11	13	72	8	9	6	32	0	6	33.5
muscle pain	9	2	1	0	1	8	1	0	3	7	0	1	7	1	7	8	6	7	11	97	9	7	22	0	7	43.7
neck pain	12	0	1	0	1	7	0	1	2	7	0	1	9	1	8	8	11	11	11	12	111	7	9	0	7	46.8
open wound	23	1	1	0	0	7	1	1	2	8	0	1	9	2	9	7	14	16	12	10	9	96	9	0	7	39.2
shoulder pain	10	1	0	1	1	8	1	1	3	9	0	1	8	0	6	12	14	16	13	12	8	9	62	1	6	30.5
skin issue	1	25	32	24	1	0	0	1	1	1	12	1	0	2	0	2	0	1	0	0	0	0	1	144	0	57.8
stomach ache	7	1	1	1	0	11	1	3	0	9	3	0	7	0	10	8	8	9	8	6	7	4	13	1	96	44.9
	33.3%	52.9%	56.4%	56.0%	57.3%	44.0%	58.2%	64.0%	60.0%	48.0%	69.8%	66.7%	46.7%	70.2%	51.6%	46.7%	33.8%	34.7%	32.0%	43.1%	49.3%	42.7%	27.6%	64.0%	42.7%	50.1%

Fig. 2 Confusion matrix of the best classification accuracy gained from predicting 25 diseases using SVM

back pain	71	0	2	0	0	9	2	0	2	9	1	2	7	0	6	10	19	8	11	8	10	11	10	2	8	34.1
acne	1	86	31	11	0	0	2	3	2	2	23	0	2	0	2	2	2	1	0	1	2	2	1	11	1	45.7
blurry vision	0	23	56	34	0	1	3	3	3	3	25	0	3	0	2	3	1	0	0	3	2	0	1	13	0	31.3
body feels weak	1	25	33	86	0	1	0	3	1	1	12	0	2	0	2	4	2	0	0	1	2	2	1	18	1	43.4
cough	4	2	0	0	112	3	11	15	17	0	2	20	2	12	4	2	1	0	0	0	2	1	1	2	0	52.6
ear ache	9	2	1	0	3	71	0	2	0	12	1	3	8	3	7	11	14	11	15	11	13	12	11	4	8	30.6
emotional pain	0	2	1	0	25	2	167	13	19	0	2	21	2	9	2	1	1	0	0	0	3	0	1	2	1	60.9
feeling cold	0	3	1	0	24	3	8	139	16	0	1	19	2	9	3	0	0	0	0	1	0	0	1	2	1	59.7
feeling dizzy	0	4	2	1	18	2	9	13	112	2	0	23	0	12	1	4	2	1	1	1	3	1	0	0	0	52.8
foot ache	8	4	3	2	1	21	0	0	1	58	0	5	7	0	8	11	16	12	11	9	9	10	14	0	10	26.4
hair falling out	0	17	33	32	2	1	0	2	0	2	82	0	4	1	2	2	0	0	0	2	2	2	0	17	1	40.2
heard to breath	1	2	4	3	17	2	11	12	18	3	2	95	0	9	1	1	0	1	1	0	2	0	0	0	1	51.1
head ache	9	3	2	4	2	23	0	3	1	11	3	1	121	1	7	13	13	9	13	16	8	14	9	2	2	41.7
heart hurts	1	3	2	4	14	1	11	11	17	3	4	26	2	159	2	1	0	1	1	2	0	0	0	0	0	60.0
infected wound	8	2	3	9	2	8	0	0	0	8	3	2	7	1	111	15	8	8	7	8	8	9	8	2	8	45.3
injury from spots	11	3	1	1	1	7	0	0	2	9	4	0	8	0	7	66	7	12	7	9	9	9	12	0	9	34.0
internal pain	13	2	2	2	1	8	0	0	1	7	3	0	6	0	7	10	61	11	7	7	12	6	14	0	10	32.1
joint pain	13	4	2	1	1	9	0	0	2	18	4	0	0	0	7	10	31	107	8	7	9	9	12	0	7	41.0
knee pain	10	3	1	0	0	7	0	0	1	12	3	0	8	1	8	9	8	10	105	7	8	6	11	1	9	46.1
muscle pain	12	2	0	0	1	8	0	0	1	11	2	0	7	2	7	11	6	5	8	106	9	17	14	0	8	44.7
neck pain	10	3	0	0	0	9	0	1	1	22	0	1	6	1	8	9	9	8	8	6	77	9	12	1	9	36.7
open wound	18	2	0	0	0	8	0	1	2	12	0	0	7	0	7	9	7	4	8	5	14	83	12	0	8	40.1
shoulder pain	14	2	0	0	0	8	0	0	3	17	0	1	6	0	6	8	7	8	8	5	11	14	70	1	8	35.5
skin issue	0	24	45	34	1	0	0	2	2	2	45	3	2	2	1	2	1	3	0	1	2	0	0	145	1	45.6
stomach ache	11	2	0	1	0	13	1	2	1	3	3	3	6	3	7	11	9	5	6	9	8	8	10	2	114	47.9
	31.6%	38.2%	24.9%	38.2%	49.8%	31.6%	74.2%	61.8%	49.8%	25.8%	36.4%	42.2%	53.8%	70.7%	49.3%	29.3%	27.1%	47.6%	46.7%	47.1%	34.2%	36.9%	31.1%	64.4%	50.7%	43.7%

Fig. 3 Confusion matrix of the best classification accuracy gained from predicting 25 diseases using NN

back pain	34	2	3	2	1	11	1	1	2	12	0	2	19	1	8	9	15	16	14	12	14	11	9	0	9	16.3
acne	2	56	34	24	2	2	1	2	1	2	18	1	2	2	1	1	2	0	1	0	1	1	1	22	1	31.1
blurry vision	0	30	54	33	1	2	0	1	2	3	21	1	2	1	1	2	3	0	1	0	1	1	0	21	1	29.7
body feels weak	3	31	27	70	1	1	1	2	1	2	21	0	3	1	0	1	2	0	1	0	1	1	0	19	1	36.8
cough	4	2	3	3	68	4	20	18	20	2	1	28	0	11	0	1	2	1	1	1	1	1	0	0	0	35.4
ear ache	13	3	4	5	2	56	3	2	1	12	2	2	19	1	9	8	11	9	10	10	12	8	12	0	7	25.3
emotional pain	0	3	1	3	23	2	88	15	28	2	0	29	0	17	0	1	1	1	0	1	3	0	2	0	0	40.0
feeling cold	4	2	2	2	20	3	23	78	29	2	1	21	0	18	0	0	1	0	1	0	0	1	2	1	0	37.0
feeling dizzy	2	0	2	2	33	2	23	27	59	0	3	27	1	14	0	0	0	0	0	13	0	0	0	0	1	28.2
foot ache	9	0	2	2	1	22	1	3	2	58	0	3	16	2	11	12	12	16	15	9	15	16	15	1	10	22.9
hair falling out	1	32	24	24	2	3	1	1	2	0	73	1	1	3	0	2	1	0	1	1	1	1	0	32	2	34.9
heard to breath	1	2	4	4	22	3	27	24	33	0	1	67	1	20	1	0	1	0	0	0	2	1	0	1	0	31.2
head ache	12	4	2	2	4	21	2	3	1	11	31	0	54	2	15	10	11	6	12	8	8	11	7	1	12	21.6
heart hurts	0	4	4	4	21	4	24	21	31	1	1	29	0	111	0	0	1	0	0	2	1	1	0	1	0	42.5
infected wound	10	2	2	2	3	8	1	2	3	14	19	1	18	0	101	10	13	8	8	9	9	10	9	0	7	37.5
injury from spots	14	3	3	1	2	12	1	2	2	15	0	1	12	2	8	66	13	18	9	11	8	17	9	1	8	27.7
internal pain	17	4	4	3	4	11	1	3	1	11	0	0	13	3	8	20	56	15	11	12	7	11	8	1	7	24.2
joint pain	15	2	3	2	5	11	0	2	1	10	0	1	11	2	9	10	14	68	12	11	9	15	9	0	7	29.7
knee pain	13	1	4	2	3	7	1	4	1	9	0	0	10	4	8	15	11	11	65	10	11	16	9	2	7	29.0
muscle pain	14	3	2	2	2	6	0	2	1	8	1	1	9	2	9	14	15	11	14	71	9	14	13	4	7	30.3
neck pain	17	2	2	0	1	7	1	4	1	13	1	2	9	5	8	13	11	9	14	9	70	13	10	1	8	30.3
open wound	14	0	2	0	1	8	1	2	0	11	1	3	9	1	8	8	10	9	13	16	11	50	10	2	9	25.1
shoulder pain	16	0	3	0	0	8	0	3	3	11	0	0	8	0	9	11	10	14	12	10	13	11	88	1	9	36.7
skin issue	1	35	30	32	1	4	4	1	0	1	29	1	0	1	0	0	0	0	1	1	1	0	3	112	0	43.4
stomach ache	9	2	4	1	2	7	2	2	0	15	1	4	8	1	11	11	9	13	9	8	17	14	9	2	112	41.0
	15.1%	24.9%	15.1%	20.0%	28.9%	24.9%	39.1%	34.7%	20.4%	25.8%	32.4%	29.8%	39.6%	49.3%	44.9%	29.3%	24.9%	33.8%	32.0%	31.6%	31.1%	22.2%	39.1%	49.8%	49.8%	31.5%

Fig. 4 Confusion matrix of the best classification accuracy gained from predicting 25 diseases using GMM

that the classification accuracies achieved were 50.8%, 48.7%, and 31.5%, from the SVM, NN, and GMM, respectively which show low-accuracy results and are not acceptable in any ADPS, but when we analyzed the three confusion matrices, we noticed some perfect outputs, which are

1. There are some common features that relate the same recordings of the same disease, and the evidence is the results gained from Exp1. Regardless of the weak results gained, but 50.8% accuracy classification shows that there are some common features that relate the recordings, but these features need to be enhanced.
2. According to the misclassification distribution of samples in the confusion matrices, we noticed that the diseases are divided into three groups. Every disease in the same group is misclassified with diseases of the same group. After studying the diseases, we noticed that the diseases can be divided into three groups according to the amount of pain caused by that disease.
3. There are some of the diseases that affect the voice of the human, because of the amount of pain caused by that disease that makes the speech process very difficult for the patient such as painful diseases, back pain, internal pain, joint pain, knee pain, muscle pain, neck pain, open wound, shoulder pain, stomach ache, injury from spots, infected wound, head ache, ear ache, and foot ache which will be called Group 1 in Exp2, and such diseases affect the articulation phonetics. Where there are some diseases that affect the acoustic phonetic in human speech, and those can be divided in to two types. First, psychological-related diseases such as acne, blurry vision, body feels weak, hair falling out, and skin issue which will be called Group 2 in Exp2. Second, frequency-related diseases such as cough, emotional pain, feeling cold, heard to breath, heart hurts, and feeling dizzy which will be called Group 3 in Exp2.

Table 1 The groups of diseases generated after analyzing the results of Exp1

No.	Acoustic phonetic feature related disease		Articulator phonetics feature related disease
	Frequency related	Psychological related	Painful diseases
1	Cough	Acne	Back pain
2	Emotional pain	Blurry vision	Internal pain
3	Feeling cold	Body feels weak	Joint pain
4	Heard to breath	Hair falling out	Knee pain
5	Heart hurts	Skin issue	Muscle pain
6	Feeling dizzy		Neck pain
7			Open wound
8			Shoulder pain
9			Stomach ache
10			Injury from spots
11			Infected wound
12			Head ache
13			Ear ache
14			Foot ache

After implementing out experiments, we noticed that some of the diseases have been grouped together. Each group of diseases are misclassified between themselves only and not with other groups. When putting those diseases in groups as shown in Table 1, we noticed that there are some common behaviors between the diseases in the same group. Some diseases are followed with pain, some diseases cause no pain but psychological pain that affect the emotion situation of the human being, and some diseases have direct effect of the vocal folds that affect the frequency of speech. Therefore, experiment number 2 was implemented.

- Experiment Number 2 (Exp2)

The aim of this experiment is to group the diseases into three groups. Where each group of diseases share common behavior or effect on human being body of psychologically. After implementing this experiment, high-classification accuracy results were gained, through all three classifiers deployed in this work, but the SVM classifier gained the higher classification results as shown through the confusion matrix shown in Fig. 5. The classification results of the NN and GMM classifiers are shown in Figs. 6 and 7, respectively.

Fig. 5 SVM results of Exp2

Group1	3050	50	69	96.24%
Group 2	42	1056	26	93.95%
Group 3	58	19	1255	94.22%
	96.83%	93.87%	92.96%	94.55%

Fig. 6 NN results of Exp2

Group1	2972	113	72	94.14%
Group 2	91	961	35	88.41%
Group 3	89	51	1243	89.88%
	94.35%	85.42%	92.07%	90.62%

Fig. 7 GMM results of Exp2

Group1	2990	164	146	90.61%
Group 2	76	904	39	88.71%
Group 3	84	57	1167	89.22%
	94.92%	80.36%	86.44%	87.24%

5 Conclusion

Diagnosing diseases through speech is a very hard task and was impossible in the past, but now with the existence of the machine learning, everything is possible. The aim of our work was to diagnose diseases directly from human speech with any linguistic methods, but we did not reach the required results, regardless of the 50.1% accuracy classification gained with SVM classifier.

Through this work, it was noticed that diseases can be divided in many groups, such as strongly and mild pain, psychological or physical, emotional or non-emotional, and other categories. Each of the forementioned categories share similar behavior that reflects on speech. Such reflection leaves specific features related to each type of diseases.

The results gained from Exp1 can be enhanced using other feature types, or machine learning algorithms. The genetic method used gave good results in Exp2 but was a time-consuming method. So, finding a better feature selection method can speed up the ADPS.

Acknowledgements Many thanks to the medical speech, transcription, and intent dataset creators, and for the useful details.

References

1. Z. Zhang, Mechanics of human voice production and control. J. Acoust. Soc. Am. **140**(4), 2614–2635 (2016)
2. E. Phonetics, P.D.F. Trujillo, *The Study of Speech*
3. C. Anderson, *Essentials of Linguistics* (2018)
4. L.R. Rabiner, R.W. Schafer, *Digital Processing of Speech Signal* (1978)
5. J.R. Deller Jr., J.G. Proakis, J.H. Hansen, *Discrete Time Processing of Speech Signals* (Prentice Hall PTR, 1993)
6. J. Harrington, S. Cassidy, The acoustic theory of speech production, in *Techniques in Speech Acoustics* (Springer, 1999), pp. 29–56
7. P. Rose, *Forensic Speaker Identification*, vol. 1 (Taylor Francis, London, 2002), p. 380
8. J. Kreiman, G. Papçun, Voice discrimination by two listener populations. J. Acoust. Soc. Am. **77**(S1), S9–S9 (1985)
9. J.R. Elliott, *Auditory and F-Pattern Variations in Australian Okay: A Forensic Investigation* (2001)
10. X. Wang et al., Automated knowledge acquisition from clinical narrative reports, in *AMIA Annual Symposium Proceedings* (American Medical Informatics Association, 2008)
11. M. Hossain, M. Laskar, T. Rahman, Automated disease prediction system (ADPS): a user input-based reliable architecture for disease prediction. Int. J. Comput. Appl. (2015)

12. N. Dragu et al., Ontology-based text mining for predicting disease outbreaks, in *Twenty-Third International FLAIRS Conference* (2010)
13. J. Maude, Patients could provide initial differential. Br. J. Gen. Pract. (2021)
14. S. Shepperd, D. Charnock, B. Gann, Helping patients access high quality health information. BMJ **319**(7212), 764–766 (1999)
15. R. Das, I. Turkoglu, A. Sengur, Effective diagnosis of heart disease through neural networks ensembles. Expert Syst. Appl. **36**(4), 7675–7680 (2009)
16. R. Kumar et al., Disease prediction from speech using natural language processing and deep learning method, in *Congress on Intelligent Systems* (Springer, 2020)
17. C. Dreisbach et al., A systematic review of natural language processing and text mining of symptoms from electronic patient-authored text data. Int. J. Med. Inform. **125**, 37–46 (2019)
18. T. Gangavarapu et al., FarSight: long-term disease prediction using unstructured clinical nursing notes. IEEE Trans. Emerg. Top. Comput. (2020)
19. M. Johnson et al., A systematic review of speech recognition technology in health care. BMC Med. Inform. **14**(1), 1–14 (2014)
20. F.A. Shaqra, R. Duwairi, M. Al-Ayyoub, Recognizing emotion from speech based on age and gender using hierarchical models. Procedia Comput. Sci. **151**, 37–44 (2019)
21. H.A. Abdulmohsin, H.A. Wahab, A.M.J.A. Hossen, Speech emotion recognition survey. J. Mech. Contin. Math. Sci. **15**(9), 24 (2020)
22. P. Mooney, Medical speech, transcription, intent (2018), p. 1
23. F. Gustafsson, Determining the initial states in forward-backward filtering. IEEE Trans. Signal Process. **44**(4), 988–992 (1996)
24. G. Chandrashekar, F. Sahin, A survey on feature selection methods. Comput. Electr. Eng. **40**(1), 16–28 (2014)
25. G. Fant, *Acoustic Theory of Speech Production* (Walter de Gruyter, 1970)
26. P.T. Nghia et al., A measure of smoothness in synthesized speech. REV J. Electron. Commun. Math. Phys. **6**(1–2) (2016)
27. H. Yoon et al., Algorithm learning based neural network integrating feature selection and classification. Expert Syst. Appl. **40**(1), 231–241 (2013)
28. S. Ledesma et al., Feature selection using artificial neural networks, in *Mexican International Conference on Artificial Intelligence* (Springer, 2008)
29. H.A. Abdulmohsin, H.B.A. Wahab, A.M.J.A. Hossen, A new hybrid feature selection method using T-test and fitness function. CMC-Comput. Mater. Contin. **68**(3), 3997–4016 (2021)
30. R. Kohavi, G.H. John, Wrappers for feature subset selection. Artif. Intell. **97**(1–2), 273–324 (1997)

Sentiment Analysis of Political Post Classification Based on XGBoost

Ahmed Assim Nsaif and Dhafar Hamed Abd

Abstract The number of Websites and the volume of posts published on the Internet have increased dramatically in recent years. The ability to automatically assess the political polarity of a post (text) can be useful in a variety of fields including security and academics. The sentiment classification of postings, on the other hand, appears to be more complicated compared to classifying the sentiment of traditional texts. The classification procedure adopted in this research uses XGBoost algorithm and bag of word as feature extraction. To test the accuracy of the approach, the study used the confusion matrix. The proposed approach achieved 95.161 accuracy percentage.

Keywords Term frequency · XGBoost · Bag of word · Arabic article · Political · Machine learning

1 Introduction

The capacity to determine whether a text is negative, positive, or neutral in relation to a certain topic is known as opinion or sentiment classification [1, 2]. It is the ability to rate a particular text based on its overall sentiment (0 or 1), where 0 denotes an unfavorable portrayal, and 1 denotes a favorable portrayal [3]. This type of classification makes it easier to divide a collection of differing viewpoints into two groups. The use of sentiment classification to classify a large number of opinions in a great number of posts has been shown to be quite beneficial. Predicting and labeling posts can help improve the process of accessing and surfing Websites by organizing the information available to users. As a result, people can easily post and discuss their thoughts on Websites [4, 5].

A. A. Nsaif · D. H. Abd (✉)
Informatics Institute for Postgraduate Student, Baghdad, Iraq
e-mail: Dhafar.dhafar@gmail.com

D. H. Abd
Department of Computer Science, Al-Maarif University College, Anbar, Iraq

A. K. Bashir et al. (eds.), *Proceedings of International Conference on Computing and Communication Networks*, Lecture Notes in Networks and Systems 394,
https://doi.org/10.1007/978-981-19-0604-6_16

This research looks into the general classification of political posts and the labeling of these posts. Concerning the latter, the labeling of political posts cannot progress without a clear explanation of these posts. The goal of this research is to evaluate the viability of current available political datasets to subject them to sentiment classification techniques [6, 7]. More particularly, the study intended to establish XGBoost algorithm. A variety of researches have been conducted to investigate sentiment analysis at various tiers of the examined text, comprising word, phrase, and document levels. The current study's proposed approach focuses on sentiment analysis at the document level. This study has three levels of documents, which are reform, revolutionary, and conservative.

The study proposed a new method for classifying political Arabic posts into three classes: reform, conservative, and revolutionary. This classification was based on the extraction of two features, which are TF and TF-IDF. Moreover, XGBoost algorithm was used to classify these posts. The TF and TF-IDF feature extraction were used to build a vector, and XGBoost was used to identify the orientation of the post. To obtain more accurate results, this work used XGBoost algorithm with various parameters [8]. The suggested technique comprises four steps: (i) collecting data from different resources; (ii) eliminating unwanted data (pre-processing); (iii) using TF and IDF to extract words weight and build vectors; and (iv) using XGBoost algorithm to classify the post and identify its orientation. The following are the major contributions of this work:

- As a pre-trained model for post orientation, this work employed two feature extraction methods, which are TF and TF-IDF. These feature extractions were used for building vectors to feed XGBoost algorithm.
- Using a XGBoost algorithm with different parameters used to determine the orientation of political Arabic post.

Section 2 of this work discusses related works found in the literature. The process of designing the model is depicted in Sect. 3, while the description of the algorithm is provided in Sect. 4. Section 5 introduces the evaluation matrices, followed by Sect. 6, which offers the results of the experimental approach. Finally, several conclusive remarks are presented.

2 Related Works

Past studies related to sentiment classification can be classified based on the procedure used to achieve such a classification. To obtain information on document sentiment, the knowledge-based technique primarily uses language models. Hence, machine learning is going to be applied in this study to understand more about the knowledge-based method, the literature review is made in relation to this topic, and several survey papers are cited [9–11].

Lee et al. are credited with successfully applying typical machine learning processes to a movie review database [12]. The application of Naive Bayes, maximum entropy, and support vector machines enabled this historic breakthrough. They reported that the support vector machine outperformed all other machines with 82.9 accuracy rate. Their Naive Bayes classifier had an accuracy of 81.0% when it uses a unigram.

For the sentiment classification of complete documents, Hearst [13] and Sack [14] both employed cognitive linguistic models. Huettner and Subasic [15], Das and Chen [16], and Tong [17] are only among a few scholars who have investigated the manual or semi-manual creation of a discriminatory word lexicon for the grouping a passage sentiment.

Instead of an entire paragraph, Hatzivassiloglou and McKeown [18] and Turney and Littman [19] examined the classification of word orientation. To determine the semantic orientation of the holding paragraph, they used the semantic orientation of individual words or phrases. Their method involved pre-selecting a list of seed words or using linguistic heuristics to classify the passage sentiment.

3 Methodology

Political post (text) classification is widely seen as a complex process. As a result, this research offered an approach for classifying political post. With reference to the innovative compilation of political post done in this study, the suggested technique also attempted to assess the efficacy of conventional text classification algorithms. The study intended to use XGBoost algorithm and evaluated its application to the issue under investigation in this study. Gathering data from numerous Websites for the creation of initial data was the first step in the construction of the approach. Figure 1 depicts the proposed strategy.

3.1 Dataset Description

This section introduces the general concept of the political post dataset. Data were acquired in its raw form from Websites and publications [20]. As seen in Table 1, this dataset has three labels, and their total number is 206 articles.

3.2 Pre-processing

Pre-processing is a good way to guarantee that the political post (text) collected is accurate. Three independent compilations of classifiers were created to appraise modern technologies relevant to this section [21]. The first procedure was to break down the post into tokens, which are words. To train down the XGBoost, certain words such as English words and punctuation were deleted during this process. The

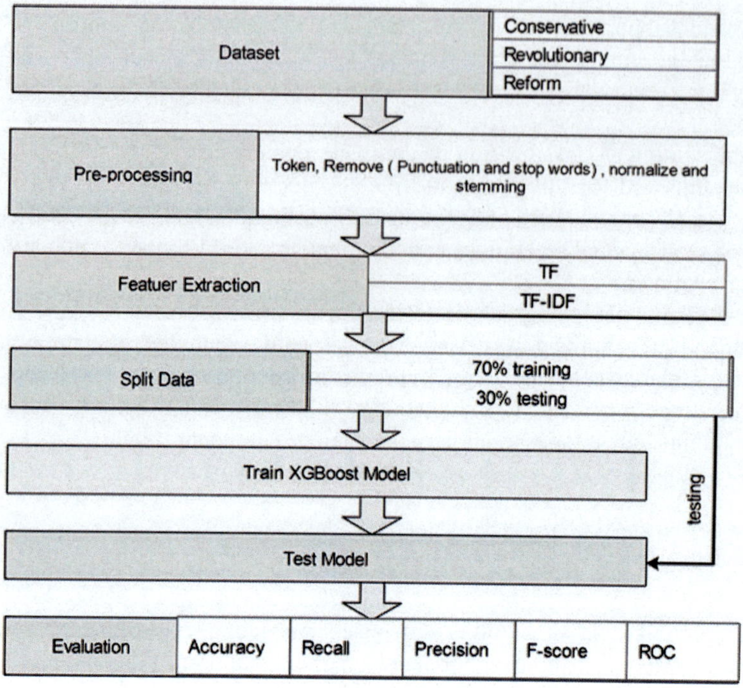

Fig. 1 Our proposed approach

Table 1 Dataset description

Political post collected	Arabic	English
80	اصلاحي	Reform
58	محافظ	Conservative
68	ثوري	Revolutionary

process of normalization Arabic texts into other forms was done in three steps [22, 23], first removes diacritics from an Arabic word as shown in Fig. 2.

The second step removes long alphabetic (ثـــــــــورة) to (ثورة), while the third step changes the alphabetic word into another as shown in Table 2.

The stop word eliminates tokens and matches the word against the stop words in the NLTK library [24]. The stemming step is the most important one in the pre-processing stage of the text analysis [25]. Stemming is used to reduce the text in the dataset by condensing a word into stem words. Two types of stemming for Arabic

Fig. 2 Diacritics must be removed

Table 2 Arabic alphabetic normalization

Input alphabetic	Output alphabetic
ا,آ,أ,إ	ا
ى	ي
ؤ , ئ	ء
ة	ه

language such as root stemming and light stemming were often used [26]. However, the present study used light stemming [27].

3.3 *Feature Extraction*

Term frequency (TF) refers to the frequency of a word in a particular political post. The word frequency in the post will be divided by the total number of words in the political post as shown in Eq. (1).

$$\mathrm{TF} = \frac{\text{frequency word in post}}{\text{total words in post}} \tag{1}$$

The TF-IDF calculated by multiply TF and IDF as shown in Eq. (2) [28].

$$\mathrm{IDF} = \log\left(\frac{\text{total posts in dataset}}{\text{number of posts with word}}\right) \tag{2}$$

The equation for calculate TF-IDF as following [29].

$$\text{TF-IDF} = \mathrm{TF} * \mathrm{IDF} \tag{3}$$

3.4 *Split Dataset*

In this phase, the dataset will be divided into training set and testing set. Table 3

Table 3 Dividing dataset

Political post	Posts for training	Posts for testing
Reform	55	25
Conservative	43	15
Revolutionary	46	22
Total	144	62

shows that 70% of the posts (144 posts) were posts for training, whereas the rest 30% posts (62 posts) were for testing set.

4 Model Description

Gradient boosting is a useful method in machine learning [30]. XGBoost is a generalized gradient boosting algorithm that has become a popular tool in supervised learning competitions. Its great predictive performance, highly optimized multicore and distributed machine implementation, and capacity to handle sparse data are the factors that give it this status. The primary idea behind XGBoost is to choose the right tree and change the parameters as illustrated in Eq. (4).

$$y_i = \sum_{k=1}^{k} f_k(x_i), \qquad f_k \in F \tag{4}$$

where f_k is the classified tree, and $f_k(x_i)$ represents the score given by the k tree to the ith observation in features x_i. The input and output of XGBoost algorithm are training sets of political post, with the input belonging to TF and TF-IDF attributes $(x_1, x_2, x_3, \ldots, x_n)$ and the result for the political post (labels) $(y_1, y_2, y_3, \ldots, y_m)$. $x_i \in$ is the vector, and $y_i \in$ {label}, then it becomes (x_n, y_m). The XGBoost algorithm is described below with the set of weight w_i for estimating the accurate value of (y) [31].

XGBoost algorithm

Input:

$S = \{(x_i, y_i) | x_i \in R^n, y_i \in m,\ i \in \{1,2,\ldots,N\}\}$ –training set;

$Z = \{z_i | z_i \in R^m, i \in \{1,2,\ldots,t\}\}$ –Test set;

Initialization:

$Y \leftarrow \emptyset$

Computation:

for $z_i \in Z$ **do**

$r \leftarrow$ calculate f_k;

$y \leftarrow$ predicted class by used r on z_i;

$C \leftarrow C \cup \{y\}$

Output:

$C = \{y | y_i \in N\ i \in \{1,2,\ldots,t\}\}$ - Predicted set.

Table 4 Equations for evaluation our model

Metric	Equation
Accuracy (A)	$\frac{(\text{TP+TN})}{(\text{TP+TN+FP+FN})}$
Recall (R)	$\frac{\text{TP}}{(\text{TP+FN})}$
Precision (P)	$\frac{\text{TP}}{(\text{TP+FP})}$
$F1$-score (F)	$2 * \frac{\text{Recall} * \text{Precesion}}{\text{Recall+Precesion}}$

5 Evaluation Metric

Several measurements were employed to examine the performance of the proposed model as shown in Table 4. These measurements were for accuracy, recall, precession, and F-score. All these measurements were based on the confusion matrix. The recall was utilized to check the correctness of each class. The precision was used to incorrectly classify the political dataset. The $F1$ scores are calculated with the use of both recall and precision.

6 Results

Experimentations were carried out to evaluate the effectiveness of the proposed model for optimizing XGBoost algorithm parameters. During training, a category sample was separated from the other samples and placed in a different class. Unknown samples were grouped together in the category of those with the highest classification function value. Table 5 shows the parameters of XGBoost algorithm. The testing method was carried out using Python programming.

The dataset of this study includes 206 political posts. These posts were divided into training set and testing set as shown in Table 3. Tables 6 and 7 show the confusion matrix for both TF and TF-IDF techniques to give robust results about each class.

Table 5 XGBoost algorithm parameters

Parameters (P)	Value (V)
Base score	0.5
Booster	Gbtree
Gamma	0
Learning rate	0.3
Max depth	6
Estimator	100
Alpha	0
Random	0

Table 6 XGBoost algorithm using TF

True label	Predicted label		
	Conservation	Reform	Revolutionary
Conservation	14	1	0
Reform	1	23	1
Revolutionary	1	1	20

Table 7 XGBoost algorithm using TF-IDF

True label	Predicted label		
	Conservation	Reform	Revolutionary
Conservation	15	0	0
Reform	1	23	1
Revolutionary	0	1	21

Tables 6 and 7 show the confusion matrix for both feature extractions; TF and TF-IDF. For TF technique, conservation class achieved higher accuracy compared to other classes, where errors were only found in one post. TF-IDF feature as a consecration class also achieved higher accuracy than the others with 15 correct posts out of 15. Finally, conservative class in both feature extractions outperformed other classes with three posts for TF-IDF, and five posts for TF; hence, TF-IDF was better than TF. Tables 8 and 9 show other evaluations for both features extraction.

Table 9 shows the TF-IDF feature extraction technique with XGBoost algorithm. This technique achieved 95.161% accuracy. For precession, the higher-class was the reform. For recall, the conservation class achieved 100%. For F-score, the conservation class achieved 0.97, which is higher than other classes. Finally, conservation

Table 8 TF vector with XGBoost

Label	Precision (p)	Recall (r)	$F1$-score	Avg. accuracy (%)
Conservation	0.88	0.93	0.90	91.935
Reform	0.92	0.91	0.92	
Revolutionary	0.95	0.91	0.93	

Table 9 TF-IDF vector with XGBoost

Label	Precision	Recall	$F1$-score	Avg. accuracy (%)
Conservation	0.94	100	0.97	95.161
Reform	0.96	0.92	0.94	
Revolutionary	0.95	0.95	0.95	

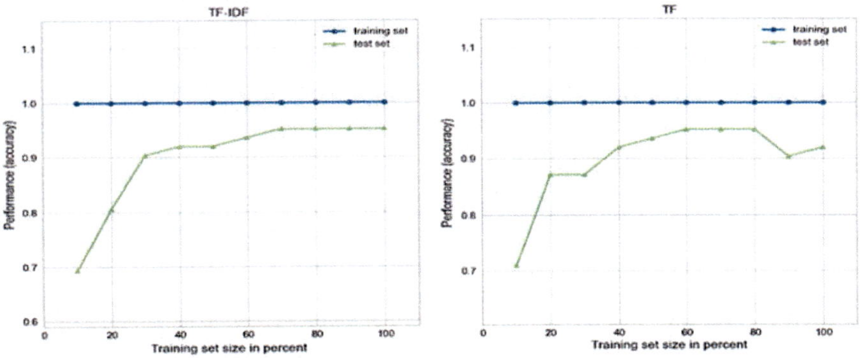

Fig. 3 Feature extraction in both case training then testing

class achieved higher accuracy value compared to other classes. For more details, see Fig. 3.

The proposed model needs to measure the carves between training and testing and see the difference between them in both feature extraction techniques (TF, TF-IDF). Figure 3 shows the learning carves for TF and TF-IDF with XGBoost algorithm and compares between them to identify which one is better. As shown in Fig. 3, both techniques were good in the training set with 100% accuracy. Yet, for testing the TF-IDF, one should start from a low point then goes up and to reach 95% approximately. On the other hand, TF starts from a low point then the end of learning is not very well that is because TF-IDF takes into account the relation between the post and the dataset to find the value that measure this relation. Figure 4 shows precision and recall for both feature extractions.

Figure 4 shows two feature extractions; each one of them was taken to learn XGBoost algorithm. For TF, both classes, conservative and revolutionary, achieved 0.99%, reform class, on the other hand, achieved 0.96, which is the least class compared to others. For TF-IDF feature extraction technique, both conservative and

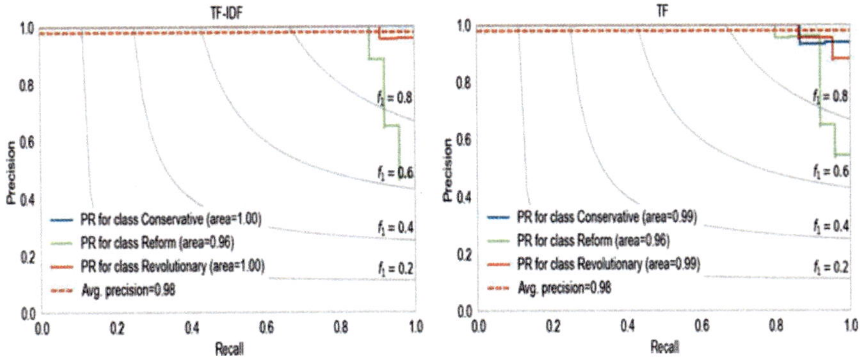

Fig. 4 Precision-recall

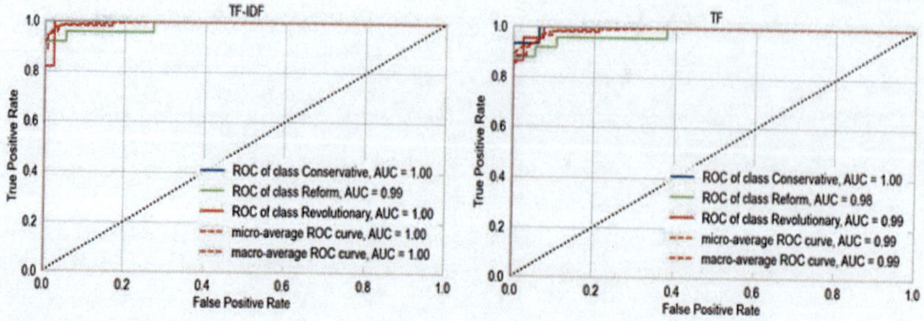

Fig. 5 ROC model

revolutionary achieved 100%, while reform achieved 0.96% with the same value for TF feature extraction. The ROC model in our empirical analysis is depicted in Fig. 5.

Figure 5 shows two techniques for feature extraction, which are the TF and the TF-IDF. TF-IDF showed high value in both conservative and revolutionary classes with 100%. Reform class achieved 0.99%, while micro and macro achieved 100% in both terms. In term of TF, the conservative class achieved 100%, which is the highest among other classes. However, the revolutionary class achieved 0.99%, reform achieved 0.98%, and the micro and macro in TF achieved 0.99% value. Finally, TF-IDF was proved to be the best feature extraction technique to be used with XGBoost algorithm.

6.1 Discussion

The first experiment in this study was conducted to compare the TF feature extraction with TF-IDF as used with the XGBoost algorithm. Tables 8 and 9 show the precision, recall, and *F*-score of this experiment when using the TF and TF-IDF. As shown in Fig. 3, it can be observed that the proposed method has higher testing set to TF-IDF. As shown in Tables 6 and 7, the accuracy of the TF-IDF feature extraction was higher than that of TF. To verify the accuracy of the proposed model, TF and TF-IDF of the corpus were compared. As shown in Fig. 5, the value for TF-IDF was larger than the predicted level of TF. Generally, compared with the TF and TF-IDF, the proposed method achieved lower classification error rate. It has been observed through experiments conducted in the present study that the proposed model has achieved the best result (95.161%), which is better than the results of TF.

7 Conclusions

This research focused on using XGBoost to a unique feature set from the proposed dataset that was derived from political posts. This work presented evidence to back up the claim that an XGBoost algorithm is responsive to the features that are taken from our dataset. The study aims were to identify how the size of the feature affected the classification efficiency and accuracy of the used XGBoost algorithm, by picking the most important features. This work used the TF-IDF and TF to generate the vector, where TF-IDF showed a high accuracy level (95.161%). Vector size can be reduced in future works via selecting the best words by means of feature selection technique.

References

1. B. Pang, L. Lee, Opinion mining and sentiment analysis. Found. Trends® Inf. Ret. **2**(1–2), 1–135 (2008)
2. B. Liu, Sentiment analysis and subjectivity, in *Handbook of Natural Language Processing*, vol. 2, no. 2010 (2010), pp. 627–666
3. B. Liu, Sentiment analysis and opinion mining. Synth. Lect. Hum. Lang. Technol. **5**(1), 1–167 (2012)
4. C. Tan, L. Lee, J. Tang, L. Jiang, M. Zhou, P. Li, User-level sentiment analysis incorporating social networks, in *Proceedings of the 17th ACM SIGKDD International Conference on Knowledge Discovery and Data Mining* (ACM, 2011), pp. 1397–1405
5. D.H. Abd, A.R. Abbas, A.T. Sadiq, Analyzing sentiment system to specify polarity by lexicon-based. Bull. Electr. Eng. Inform. **10**(1), 283–289 (2021)
6. R. Xia, C. Zong, S. Li, Ensemble of feature sets and classification algorithms for sentiment classification. Inf. Sci. **181**(6), 1138–1152 (2011)
7. J.K. Alwan, A.J. Hussain, D.H. Abd, A.T. Sadiq, M. Khalaf, P. Liatsis, Political Arabic articles orientation using rough set theory with sentiment lexicon. IEEE Access **9**, 24475–24484 (2021)
8. T. Chen, C. Guestrin, XGBoost: a scalable tree boosting system, in *Proceedings of the 22nd ACM SIGKDD International Conference on Knowledge Discovery and Data Mining* (2016), pp. 785–794
9. J.A. Balazs, J.D. Velásquez, Opinion mining and information fusion: a survey. Inf. Fusion **27**, 95–110 (2016)
10. W. Medhat, A. Hassan, H. Korashy, Sentiment analysis algorithms and applications: a survey. Ain Shams Eng. J. **5**(4), 1093–1113 (2014)
11. K. Ravi, V. Ravi, A survey on opinion mining and sentiment analysis: tasks, approaches and applications. Knowl.-Based Syst. **89**, 14–46 (2015)
12. B. Pang, L. Lee, S. Vaithyanathan, Thumbs up?: sentiment classification using machine learning techniques, in *Proceedings of the ACL-02 Conference on Empirical Methods in Natural Language Processing*, vol. 10 (Association for Computational Linguistics, 2002), pp. 79–86
13. M.A. Hearst, Direction-based text interpretation as an information access refinement, in *Text-Based Intelligent Systems: Current Research and Practice in Information Extraction and Retrieval* (1992), pp. 257–274
14. W. Sack, On the computation of point of view, in *AAAI* (1994), p. 1488
15. A. Huettner, P. Subasic, Fuzzy typing for document management, in *ACL 2000 Companion Volume: Tutorial Abstracts and Demonstration Notes* (2000), pp. 26–27

16. S. Das, M. Chen, Yahoo! for Amazon: extracting market sentiment from stock message boards, in *Proceedings of the Asia Pacific Finance Association Annual Conference (APFA)*, vol. 35 (Bangkok, Thailand, 2001), p. 43
17. Z. Wang, V.J.C. Tong, P. Ruan, F. Li, Lexicon knowledge extraction with sentiment polarity computation, in *2016 IEEE 16th International Conference on Data Mining Workshops (ICDMW)* (IEEE, 2016), pp. 978–983
18. V. Hatzivassiloglou, K.R. McKeown, Predicting the semantic orientation of adjectives, in *Proceedings of the 35th Annual Meeting of the Association for Computational Linguistics and Eighth Conference of the European Chapter of the Association for Computational Linguistics* (Association for Computational Linguistics, 1997), pp. 174–181
19. P.D. Turney, M.L. Littman, Unsupervised learning of semantic orientation from a hundred-billion-word corpus. arXiv preprint cs/0212012 (2002)
20. D.H. Abd, A.T. Sadiq, A.R. Abbas, PAAD: political Arabic articles dataset for automatic text categorization. Iraqi J. Comput. Inform. **46**(1), 1–10 (2020)
21. D.H. Abd, A.T. Sadiq, A.R. Abbas, Classifying political Arabic articles using support vector machine with different feature extraction, in *International Conference on Applied Computing to Support Industry: Innovation and Technology* (Springer, 2019), pp. 79–94
22. A. Oussous, A.A. Lahcen, S. Belfkih, Impact of text pre-processing and ensemble learning on Arabic sentiment analysis, in *Proceedings of the 2nd International Conference on Networking, Information Systems & Security* (ACM, 2019), p. 65
23. D.H. Abd, A.T. Sadiq, A.R. Abbas, Political articles categorization based on different Naïve Bayes models, in *International Conference on Applied Computing to Support Industry: Innovation and Technology* (Springer, 2019), pp. 286–301
24. N. Hardeniya, J. Perkins, D. Chopra, N. Joshi, I. Mathur, *Natural Language Processing: Python and NLTK* (Packt Publishing Ltd, 2016)
25. M. Mustafa, A.S. Eldeen, S. Bani-Ahmad, A.O. Elfaki, A comparative survey on Arabic stemming: approaches and challenges. Intell. Inf. Manag. **9**(02), 39 (2017)
26. R. Abooraig, S. Al-Zu'bi, T. Kanan, B. Hawashin, M. Al Ayoub, I. Hmeidi, Automatic categorization of Arabic articles based on their political orientation. Digit. Investig. **25**, 24–41 (2018)
27. D.H. Abd, W. Khan, K.A. Thamer, A.J. Hussain, Arabic light stemmer based on ISRI stemmer, in *International Conference on Intelligent Computing* (Springer, 2021), pp. 32–45
28. C.C. Aggarwal, C. Zhai, A survey of text classification algorithms, in *Mining Text Data* (Springer, 2012), pp. 163–222
29. D.H. Abd, A.T. Sadiq, A.R. Abbas, Political Arabic articles classification based on machine learning and hybrid vector, in *2020 5th International Conference on Innovative Technologies in Intelligent Systems and Industrial Applications (CITISIA)* (IEEE, 2020), pp. 1–7
30. R. Mitchell, E. Frank, Accelerating the XGBoost algorithm using GPU computing. PeerJ Comput. Sci. **3**, e127 (2017)
31. M. Khalaf et al., An application of using support vector machine based on classification technique for predicting medical data sets, in *International Conference on Intelligent Computing* (Springer, 2019), pp. 580–591

Applications of Deep Learning Approaches in Speech Recognition: A Survey

Sameer I. Ali Al-Janabi and Ali Azawii Abdul Lateef

Abstract Automated speech recognition (ASR) appeared to be a driving force for a variety of machine learning (ML) techniques, include to ubiquitously utilized discriminative learning, Bayesian learning, hidden Markov model, adaptive learning, and structured sequence learning. Although machine learning utilize ASR as a large scale, it can reasonable application to thoroughly test viability for a given procedure and to motivate unused issues emerging from intrinsically consecutive and discourse energetic nature. Also, although ASR is accessible commercially for a few applications used in this research through the limitation and research gaps that the researcher try to access high accuracy of these systems. The advance technology from new ML techniques appears incredible guarantee to progress the literature review in ASR innovation. This study gives reader with a diagram of present-day ML methods as used within the relevant and current as significant for ASR future systems and research. The study goal is to promote advanced cross-pollination between ML and ASR communities more than has hither to occurred.

Keywords Machine learning · Automated speech recognition · Speech identification · Speech to text

1 Introduction

Speech is the foremost characteristic and viable strategy of communication between human creatures. Speech identification aimed to decipher speech to text [1]. It may be a standard classification issue where discourse signals got to be mapped to or recognized as words. Therefore, it is not conceivable to work with discourse reports

S. I. Ali Al-Janabi (✉)
Collage of Islamic Science, University of Anbar, Anbar, Iraq
e-mail: isl.samir.ia2012@uoanbar.edu.iq

A. A. A. Lateef
Human Resources Department, University of Anbar, Anbar, Iraq
e-mail: Aliazawii@uoanbar.edu.iq

A. K. Bashir et al. (eds.), *Proceedings of International Conference on Computing and Communication Networks*, Lecture Notes in Networks and Systems 394,
https://doi.org/10.1007/978-981-19-0604-6_17

in case they are recorded as sound signals. Hence, discourse acknowledgment has gotten to be a vital zone of research [2, 3].

There are numerous challenges which make real-time speech recognition a difficult issue. Different possible pronunciations are a few of these challenges. There is a considerable misfortune in precision when we move from a controlled exploratory setup to genuine life circumstances. Despite this, automated speech recognition system has copious utilization in correspondence, human–machine interfacing and control of machines among others.

2 Speech Recognition Techniques

Recently, there has been a broad applications of deep learning approaches and neural networks to perform speech recognition leading to critical new outcomes. The slant started two decades back, when modern comes about were accomplished utilizing hybrid ANN-HMM schemes. These schemes appropriated the utilize of neural network (NN) with as it was one layer of covered up units having nonlinear actuation capacities to foresee probabilities over well states from brief windows of acoustic coefficients [4]. The Neural Network is effective approach that can speak to complex nonlinear capacities but at that time, not one or the other the computation control nor the preparing calculations that were accessible, were progressed sufficient for preparing NN with numerous covered up layers. So, cross-breed ANN-HMM schemes might not supplant the exceptionally fruitful combination of HMMs with acoustic models based on Gaussian mixtures.

Malla et al. proposed a system which can recognizing feeling within the discourse from the speech signals. This system done based on the most recent studies within speech emotion recognition (SER) field schemes using neural network convolutional related with the issue and give an ideal solution. The details of framework proposed are dataset, stage of extraction feature, and classification task that assist a help within the usage and assessing the framework. This framework will help the conclusion clients in emotion acknowledgment from discourse flag and making AI more vigorous by utilizing neural organize convolutional, encouraging a colossal nearness within the future system [5].

Dhande and Shaikh studied how the epochs playing a vital part within preparing databases. The epochs number chooses whether the information over trained or not. Results depend on database prepare. Speech recognition broadly utilized application in these days. Deep learning based on speech recognition has changed the viewpoint of the world to see at the innovation. The proposed system is based on speech recognition with deep learning approach where there are sound files and content transcripts within the datasets. The sound records are prepared with the acknowledgment show, and transcript contents are prepared by a dialect demonstrate. The dataset that is used in this architecture is made up of pieces of sounds taken from three diverse situation, to be specific clean, white clamor and persistent noise [6].

Tarunika et al. used k-nearest neighbor and deep neural network for recognition of emotion from speech particularly terrifying state of intellect. The field of applications of the framework is primarily concerned over the healthcare sectors. The establishment of this inquire has primary firm applications in field of palliative care. Beneath most exact result, the signals of caution are made by cloud. Numerous crude information is collected beneath extraordinary accentuation methods. After that the acoustic voice signals are changed over to wave shape, discourse level highlight extraction feeling classification, existing database acknowledgment, alarm flag creation through cloud is the grouping of steps to be followed [7].

Yousefi and Hansen proposed a block-based CNN design to address discourse covering modeling in streams sound with outlines as brief as 25 ms. The proposed engineering is strong for: (i) shifts in arrange enactments dispersion due to changing in arrange parameters amid preparing, (ii) nearby varieties from input highlights caused by extraction highlight, natural commotion, or room interference. Moreover, examine substitute input highlights counting ghostly greatness, MFCC, MFB, and pyknogram impact on both computational time and classification execution [8].

Tzirakis et al. presented a modern method for persistent emotion recognition from discourse. The proposed system comprised from a convolutional neural arrange (CNN), which extricates highlights from crude flag, and stacked on beat a two-layer long short-term memory (LSTM), to consider the relevant data within information. In terms of concordance relationship coefficient, our show essentially outflanks the state-of-the-art strategies for RECOLA database [9].

Arif and Puji developed the framework of existing speech recognition Indonesian that has a precision and still not great for unconstrained speech recognition. The framework is prepared utilizing HMM-GMM acoustic show. In this ponder, unconstrained discourse information collected in Indonesian for duration 14 h and discourse acknowledgment framework execution was progressed by supplanting acoustic show with a neural network-based demonstrate. The utilized neural networks topology are time delay neural network, deep neural network, and convolutional neural network [10].

Zakiah and Lestari propose advancement iterative acoustic models by using an extra unlabeled speech corpus. They used unlabeled information for revamp acoustic models by utilizing segment's translations created by already directed created ASR. For more urge solid translation, we utilized four ASRs with four sorts of profound learning-based acoustic models (CNN, TDNN, DNN, and LSTM) and chosen fragments with reliable transcripts given by models or fragments with completely understanding names [11].

Nugroho et al. discussed gender voice identification for Javanese individuals who are handled utilizing mel recurrence cepstral coefficient extracted features, at that voice classification point is done utilizing deep learning technique combined with singular value decomposition strategy in decreasing information delivered measurements. The dataset used for building this approach divided into two parts: the first part is 70% of dataset used for training model and the second part is the remaining 30% of the dataset used for testing model information the comes about of the inquire for appear profound learning method's precision is (97.78%) higher than calculated relapse strategy (95.56%) and SVM (93.33%). Discourse acknowledgment investigate appears profound learning, and SVD strategy can be utilized for performing discourse acknowledgment with high precision degree 93.33% [12].

Agrawal and Ganapathy offer a deep variational model-based method for learning modulation filters. They formulate filter learning problem in a deep unsupervised generative modeling framework, in which variational autoencoder convolutional filters capture voice modulations significant. In combined spectro-temporal domain, the spectrogram properties for voice identification for process and train are used two-dimensional modulation filters and deep variational networks, respectively. Several voice recognition studies are carried out on a series of challenges that include reverberation (REVERB Challenge), noise addition with reverberation (REVERB Challenge) (CHiME-3), noise addition with artifacts channel (Aurora-4). The modulation filter learning framework beats baseline properties and a range of current noise-resistant front ends in these suggested tests (average relative improvements of over the baseline features 7.5 and 20% in Aurora-4 and CHiME-3 databases, respectively). In addition, the proposed method has been demonstrated to be beneficial in semi-supervised automatic voice recognition systems. By employing 30% labeled training data, for example, on the Aurora-4 database, a relative improvement of 25% over the baseline system was discovered [13].

The metaheuristic algorithm pigeon inspired optimization (PIO) technique was introduced by Waris and Aggarwal used to optimize weight matrix for DNN model. This heuristic method is used to optimize the weight matrix. DNN training time is reduced because of this, and the system's recognition rate improves. The weight matrix optimization result is tested on phoneme recognition TIMIT database [14].

Liu et al. offer a two-module deep representation learning system that is local–global aware. To learn local representation, for example, time frequency CNN (TFCNN) is one module includes a multi-scale CNN. Framework with dense connections for several blocks is another module to learn deep and shallow global knowledge. Each block in this structure is a fully functional CapsNet that has been enhanced by a new routing algorithm [15].

A convolutional neural network (CNN) architecture is proposed by Saheaw et al. In order to compare it with long short-term memory (LSTM), the Thai language speech dataset turn-on and off by seven types from electrical applications. The process of reducing noise and silence from the front and the back audio is completed by 14 classes. According to tests findings, the proposed long short-term memory has best accuracy [16].

Han et al. studied and quickly explained the principles and categories, methodologies, and applications of transfer learning, as well as the application of speech emotion identification, before noting the important areas that require more investigation [17].

Table 1 shows the summary of most work in literature review with the used methods and its achievements.

Table 1 Summary of deep learning approaches applied in speech recognition

Authors	Approach	Year
Malla et al. [5]	CNN model which can recognizing feeling within the discourse from the speech signals. Typically done based on the most recent studies within speech emotion recognition (SER) field schemes using neural network convolutional related with the issue and give an ideal solution	2020
Dhande and Shaikh [6]	Studied how the epochs playing a vital part within preparing databases. The epochs number chooses whether the information over trained or not. Results depend on database prepare	2019
Tarunika et al. [7]	Used *k*-nearest neighbor and deep neural network for recognition of emotion from speech particularly terrifying state of intellect. The field of applications of the framework is primarily concerned over the healthcare sectors	2018
Yousefi and Hansen [8]	Proposed a block-based CNN design to address discourse covering modeling in streams sound with outlines as brief as 25 ms. The proposed engineering is strong for: (i) shifts in arrange enactments dispersion due to changing in arrange parameters amid preparing, (ii) nearby varieties from input highlights caused by extraction highlight, natural commotion, or room interference	2021
Tzirakis et al. [9]	Presented a modern method for persistent emotion recognition from discourse. The proposed system comprised from a convolutional neural arrange (CNN), which extricates highlights from crude flag, and stacked on beat a two-layer long short-term memory (LSTM), to consider the relevant data within information	2018
Arif and Puji [10]	Developed the framework of existing speech recognition Indonesian that has a precision and still not great for unconstrained speech recognition. The framework is prepared utilizing HMM-GMM acoustic show	2020
Zakiah and Lestari [11]	Propose advancement iterative acoustic models by using an extra unlabeled speech corpus. They used unlabeled information for revamp acoustic models by utilizing segment's translations created by already directed created ASR	2020
Nugroho et al. [12]	Discussed gender voice identification for Javanese individuals who are handled utilizing mel recurrence cepstral coefficient extracted features, at that voice classification point is done utilizing deep learning technique combined with singular value decomposition strategy in decreasing information delivered measurements	2019

(continued)

Table 1 (continued)

Authors	Approach	Year
Agrawal and Ganapathy [13]	Offer a deep variational model-based method for learning modulation filters. They formulate filter learning problem in a deep unsupervised generative modeling framework, which variational autoencoder convolutional filters capture voice modulations significant	2019
Waris and Aggarwal [14]	Metaheuristic algorithm pigeon inspired optimization (PIO) technique that used to optimize weight matrix for DNN model. This heuristic method uses to optimize the weight matrix	2018
Liu et al. [15]	Offer a two-module deep representation learning system that is local–global aware. To learn local representation, for example, time frequency CNN (TFCNN) is one module includes a multi-scale CNN	2020
Saheaw et al. [16]	The Thai language speech dataset turn-on and off by seven types from electrical applications using a CNN architecture contrasted to LSTM	2020
Han et al. [17]	Study and quickly explain the principles and categories, methodologies, and applications of transfer learning, as well as the application of speech emotion identification	2019

3 Challenges

High reliability and stable detection are still difficult to achieve due to the intricacy of speech recognition system. The following are the key reasons: (1) The context of speech, such as the speaking scene, the speaker's manner of speaking, and the speaker's age, gender, and speaking behaviors, all influence human audio generation. (2) Speech data gathering is difficult, and it must account for ambient noise. (3) Emotion is a personal experience, and there is no proper statement of emotion. (4) The capacity of the human that defines the data to perceive emotion has an impact on the annotation of the emotion data. Annotation is time-consuming since it depends on the whole display of speech information. As a result, the lot of public speaking emotion corpora that have been annotated is restricted.

4 Conclusions and Future Works

The field of deep learning has seen quick advance and lead to critical enhancements in different areas. In this survey, we have given a brief instructional exercise and outline of deep learning procedures and models within the domain of speech recognition. Recently, acoustic models based on CNNs and DBNs have effectively supplanted Gaussian blends and have been illustrated to work very well for expansive lexicon assignments. Additionally, there has been the thought of killing preparing stages, utilizing one unified neural organize to attain end-to-end discourse acknowledgment. To this conclusion, RNNs are presently being tested with but require much computation control for preparing. The utilization of RNNs for acoustic system inside a hybrid DNN-HMM framework as compared to the utilization of RNNs for end-to-end speech recognition utilizing CTC misfortune work and a dialect demonstrate has had blended responses. Deep learning holds the control to work with crude inputs and learn wealthy representations whereas disposing of difficult handling stages. With quick progression of computational advances, deep learning will as it was developed within the future.

In the future, greater datasets will be used to test deeper CNN models for speech analysis. We believe that using the raw signal, we can achieve superior results for various speech analysis tasks. When creating a new model, however, we must stick to the core principles of kernel size and pooling size.

References

1. A. Kumar, S. Verma, H. Mangla, A survey of deep learning techniques in speech recognition, in *Proceedings of IEEE 2018 International Conference on Advances in Computing, Communication Control and Networking, ICACCCN 2018* (2018), pp. 179–185
2. E. Trentin, M. Gori, A survey of hybrid ANN/HMM models for automatic speech recognition. Neurocomputing **37**(1–4), 91–126 (2001)
3. L. Deng, X. Li, Machine learning paradigms for speech recognition: an overview. IEEE Trans. Audio Speech Lang. Process. **21**(5), 1060–1089 (2013)
4. H. Bourlard, N. Morgan, *Connectionist Speech Recognition—A Hybrid Approach* (1994)
5. S. Malla, A. Alsadoon, S.K. Bajaj, A DFC taxonomy of speech emotion recognition based on convolutional neural network from speech signal, in *CITISIA 2020—IEEE Conference on Innovative Technologies in Intelligent Systems and Industrial Applications, Proceedings* (2020)
6. G. Dhande, Z. Shaikh, Analysis of epochs in environment based neural networks speech recognition system, in *Proceedings of the International Conference on Trends in Electronics and Informatics, ICOEI 2019*, no. Icoei (2019), pp. 605–608
7. K. Tarunika, R.B. Pradeeba, P. Aruna, Applying machine learning techniques for speech emotion recognition, in *2018 9th International Conference on Computing, Communication and Networking Technologies* (2018), pp. 1–5
8. M. Yousefi, J.H.L. Hansen, Block-based high performance CNN architectures for frame-level overlapping speech detection. IEEE/ACM Trans. Audio Speech Lang. Process. **29**, 28–40 (2021)
9. P. Tzirakis, J. Zhang, W. Schuller, *End-to-End Speech Emotion Recognition Using Deep Neural Networks* (Department of Computing, Imperial College London, London, UK Chair of Embedded Intelligence for Health Care and Wellbeing, University of Augsburg, Germany, 2018), pp. 5089–5093
10. D.A. Rahman, Indonesian spontaneous speech recognition system using deep neural networks (2020), pp. 2020–2022
11. I. Zakiah, D.P. Lestari, Iterative deep learning-based acoustic models using transcription agreement from multi-models automatic speech recognitions, in *2020 7th International Conference on Advanced Informatics: Concepts, Theory and Applications. ICAICTA 2020* (2020), pp. 1–5
12. K. Nugroho, E. Noersasongko, Purwanto, Muljono, H.A. Santoso, Javanese gender speech recognition using deep learning and singular value decomposition, in *Proceedings—2019 International Seminar on Application for Technology of Information and Communication: Industry 4.0: Retrospect, Prospect, and Challenges, iSemantic 2019* (2019), pp. 251–254

13. P. Agrawal, S. Ganapathy, Modulation filter learning using deep variational networks for robust speech recognition. IEEE J. Sel. Top. Signal Process. **13**(2), 244–253 (2019)
14. A. Waris, R.K. Aggarwal, Optimization of deep neural network for automatic speech recognition, in *Proceedings of the International Conference on Inventive Research in Computing Application. ICIRCA 2018*, no. Icirca (2018), pp. 524–527
15. J. Liu, Z. Liu, L. Wang, L. Guo, J. Dang, Speech emotion recognition with local-global aware deep representation learning, in *ICASSP, IEEE International Conference on Acoustics, Speech, and Signal Processing. Proceedings*, May 2020, vol. 2020 (2020), pp. 7174–7178
16. W. Saheaw, S. Jaiyen, A. Hanskunatai, Thai voice recognition for controlling electrical appliances using long short-term memory, in *2020 IEEE 7th International Conference on Industrial Engineering and Applications. ICIEA 2020* (2020), pp. 697–700
17. Z. Han, H. Zhao, R. Wang, Transfer learning for speech emotion recognition, in *Proceedings of 5th IEEE International Conference on Big Data Security on Cloud, BigDataSecurity 2019, 5th IEEE International Conference on High Performance and Smart Computing, HPSC 2019, 4th IEEE International Conference on Intelligent Data and Security, IDS 2019* (2019), pp. 96–99

Support-Based High Utility Mining with Negative Utility Values

Pushp and Satish Chand

Abstract High utility itemset mining (HUIM) aims at knowledge discovery from the datasets by finding patterns that have high utility values. Most of the existing algorithms suffer from the drawback of generating huge number of results that overwhelm the decision-making process for industry applications. Also, the real-life datasets often consist of items that have both positive and negative utility values in order to represent the profit and losses, respectively. In this paper, we propose a novel mining algorithm that maps closely to the real-life applications by producing only a reasonable number of outputs based on a support measure, from the datasets that have both positive and negative utility values. Several experiments are undertaken to test the efficacy of the proposed approach. Empirical evaluation suggests that the proposed approach is highly efficient for dense datasets.

Keywords Knowledge discovery · Data mining · High utility itemset mining

1 Introduction

Knowledge discovery in datasets has attracted the attention of the research community in the last decade. Most of the data mining algorithms are designed for extracting imperceptible knowledge from large datasets. The mined information reflects the trends and patterns in the underlying database and can be useful in various paradigms, depending on the use case. One of the earliest applications of data mining is frequent pattern mining (FPM) that aims at discovering the frequently occurring patterns from the customer transaction datasets, which is also referred to as *market-basket analysis*. The mined outputs enhance the decision-making process and aid in business growth-related operations like designing of cross marketing strategies, customer classification and market segmentation.

Pushp (✉) · S. Chand
Jawaharlal Nehru University, New Delhi, India
e-mail: srapushp@gmail.com

S. Chand
e-mail: schand@mail.jnu.ac.in

A. K. Bashir et al. (eds.), *Proceedings of International Conference on Computing and Communication Networks*, Lecture Notes in Networks and Systems 394,
https://doi.org/10.1007/978-981-19-0604-6_18

The commonly used techniques for FPM are based on the *downward closure property*, which states that all the supersets of a non-frequent itemset are also non-frequent. This property is intuitively correct as it is based on the recurrence of an item in a database or the *support* measure. Recent advances in knowledge discovery have given rise to more sophisticated mining tasks. One of the emerging areas in this field is mining the high utility itemsets, where utility is a user-defined parameter, that can hold an aesthetic or a quantitative value. The information mined using the FPM approach is indicative of the most frequently occurring patterns in a dataset; however, it may not completely imply its usefulness in terms of the measures like profit. Example, for a set of sales data in a retail store over a week, while bread and butter can be the most frequently sold items that are produced as an output by FPM techniques, the items sold at highest profit might be rare that are not produced as an output by FPM. The high utility itemset mining addresses this challenge by extracting those patterns from a database that have utilities higher than a user-defined threshold. The mined patterns are referred to as high utility itemsets (HUIs). The utility measure for transaction datasets is defined as the profit made by selling an item which is the product of the profit per unit of the item and the quantity of that item within a transaction. Mining the high utility patterns alone leads to a large number of outputs being produced, which can easily overwhelm the decision-making process in an organisation. So, for the mining results to be suitable for real-life applications, it is essential to design algorithms that take into account both, support and utility measure. These mined outputs are therefore, representatives of those itemsets that occur atleast with a frequency corresponding to a minimum support value within a database.

Moreover, the real-life datasets often contain items that have negative utility values. This is because certain business decisions are strategically designed to sell some items at losses, which thereby hold a negative profit value. For example, in order to enhance the sale of a newly launched product, another product can be tagged along with it, to be sold for free or at a discounted price.

In this paper, we design an algorithm to effectively address the real-life data mining requirements. The main contribution of the proposed study is the design of an efficient algorithm for mining the frequently occurring HUIs from the databases with items holding negative utility values.

2 Problem Statement

Here, we introduce the problem of mining frequently occurring HUIs from the databases with items holding negative utility value, by using a supporting example. Consider a sample dataset consisting of three items and six transactions as given in Table 1. Every transaction consists of a unique identifier, *TID*, and the quantity of individual items p, q, r. The unit profit of each item is also provided. It can be observed from this table that item (q) holds a negative utility value.

Table 1 Transaction database

TID	p	q	r
TID_1	1	0	0
TID_2	4	0	0
TID_3	7	0	1
TID_4	2	2	0
TID_5	1	1	0
TID_6	0	1	10

Item	Unit profit
p	2
q	−3
r	7

The *utility* of an item is the product of the quantity and it's per unit profit. Here, the utility of the item (p) in TID_1 is *quantity* $(p, TID_1) \times profit\,(p) = 1 \times 2 = 2$ units. The *total utility* of an item is defined as the sum of its utilities across all transactions in the database. Therefore, the utility of item (p) in the database is $\sum_{i=1}^{6} util(p)_{TID(i)}$, which computes to 30. For utility threshold 25, the item (p) (with utility value 30) is a high utility item. The *support* of an item is defined as the count of the number of transactions in which the item is present. For the dataset in Table 1, the support of item (p) is 5 as it occurs in transactions $TID_1, TID_2 \ldots TID_5$. If the support threshold is set to 4, the item (p) would be considered as a frequent item. The item (q) holds a utility value of -12 and a support of 3, so, it is a low utility, low frequency item.

It may be noticed that even though the item (q) holds a negative utility value, it is still possible to have combinations of items with item (q), that qualify as having high utility. For example, consider the itemset (q, r) that holds a combined utility value of 73, which is higher than the specified minimum utility threshold of 25, and is therefore a high utility item.

Given this premises, the objective is to discover those combinations of the items that have utility value and support higher than the pre-defined threshold values for the utility and support, respectively.

3 Related Work

Several algorithms exist in the literature for FPM based on a given minimum support threshold, *minsup*. The most widely applicable algorithms for FPM are the apriori algorithm [1] and the FP tree [6] algorithm. The apriori algorithm [1] is based on the *downward closure property*, which utilises the anti-monotone trend of the frequent patterns and enables efficient pruning of the search space. It first scans the database and generates the candidates in a level by level fashion by combining items from the previous level. This algorithm simultaneously prunes the candidates to effectively mine the correct results.

The FP tree [6] algorithm stores the information regarding the frequency of occurrence of an itemset in a tree like structure and then explores the tree in a depth first search fashion to mine the high frequency items. These algorithms are however not suitable to mine the high utility itemsets, primarily because the high utility itemsets do not hold the downward closure property. The non-compliance of utility mining techniques with the downward closure property is justifiable as the utility of an item is dependant on both, the quantity and per unit profit. Therefore, the supersets of a non-high utility item may or may not have high utility.

In order to avoid the combinatorial explosion in the generation of candidate itemsets for HUIM, it is essential to establish an ordering between the patterns, which allows for pruning of non-promising candidates. A close resemblance to the downward closure property is introduced in the two-phase algorithm [11], called the transaction weighted utilisation (TWU)-based property. The property states that for an itemset X, if $TWU(X) < minutility$, then (X) and all its supersets are low utility itemsets, where the transaction weighted utilisation of (X) is defined as the total summation of transaction utilities of all those transactions in which the itemset (X) is present. Other techniques to efficiently organise the search space include using tree structures and utility lists [2, 7, 10]. The utility lists efficiently store the utility information of every item in form of transaction ID, utility value and the remaining utility of the transaction. HUI-miner [10] uses utility lists to find HUIs by recursively exploring the extensions of single itemsets. An improvement to the HUI-miner [10] is introduced in *FHM* [5] which uses an EUCS structure to store the TWU values of pair of itemsets and improves the pruning of non-potential candidates. A few recent studies make use of heuristics [12] and distributed computing [8] to mine HUIs.

Mining of HUIs from the databases with items holding the negative utility values is a complicated task, as the negative utility value can decrease the transaction weighted utilisation, that may result in incorrect pruning of the search space. This can underestimate or overestimate the utilities of items which can lead to erroneous results. There are only two algorithms in the literature that account for negative utility values. The authors in [3] discuss the *HUINIV* which is based on two-phase algorithm [11] and overestimates HUIs by considering only the positive values of the items. The second algorithm is *FHN* [9] that is based on the utility lists [10] and the EUCS structure proposed in *FHM* [5]. The FHN maintains separate records for positive and negative utility items and deploys the TWU-based pruning using only the positive utility values. However, for large input datasets, FHN produces a huge volume of HUIs which can undermine the usability of the generated results. To the best of our knowledge, no algorithm exists in the literature to address the problem of finding frequent HUIs from the databases where the items hold negative utility values. In this paper, we introduce the support-based mining with negative utility (SMNU) algorithm to efficiently mine the high utility items while taking into consideration the item support and negative utility items.

4 Proposed SMNU Algorithm

In this section, we introduce our proposed SMNU algorithm that is inspired by *FHN* [9]. The main procedure of SMNU scans the input database D, to compute the transaction weighted utilisation (TWU) and support of the single itemsets. As suggested in [9], only the itemsets with positive utility values are used for computing TWU (line: 9). SMNU takes into account the frequency of itemsets unlike the FHN. For every item in the database D, if the item is present in a transaction T_j, then its support is incremented (line: 5). An important point regarding the computation of the support is that the support measure of the items is incremented regardless of the fact if the item holds a negative or a positive utility value. This is because the support of an item should reflect the total frequency of its occurrence in D. After computation of support, transaction weighted utilisation of single items is computed by taking only the positive utility values into consideration. The utility lists of the single items are then formed (line: 15), and the search procedure (2) is called (line: 17), to recursively produce the candidate items.

Algorithm 1 SMNU main procedure

INPUT:
D: Database of size N
OUTPUT:
UL: Set of Utility Lists L_i

1: $TWU(i) = 0$
2: $Sup(i) = 0$
3: **for all** Single Items $i \in D$ **do**
4:     **for all** $T_j = 1$ to N **do**
5:         $Sup(i) \leftarrow Sup(i) + 1$
6:         **if** $Utility(i) < 0$ **then**
7:             $NI^* : NegativeItemsets \leftarrow (i)$
8:         **else**
9:             $TWU(i) = TWU(i) + TU(T_j)$
10:         **end if**
11:         $I^* \leftarrow (i)$
12:     **end for**
13: **end for**
14: **for all** $i \in I^*$ **do**
15:     Form $UtilityList\ UL_i$
16: **end for**
17: SEARCH($\emptyset$, I^*, $MinUtilThresh$, UL)

The search procedure (2) computes the sum of positive and negative utilities of each item and checks the sum against the minimum utility threshold (line: 4). Additionally, in order to ensure that only the frequent HUIs are produced as outputs, a check against the minimum support value $SThrsh$ is implemented (line: 4). If the input itemset passes these two checks, it is produced as a HUI (line: 5). Further, all the combinations of itemsets are explored sequentially only if the sum of utilities and remaining utilities for an itemset is higher that the minimum utility (line: 7). A noteworthy point is that the check against the minimum support value is implemented

Algorithm 2 Search procedure

INPUT:
I: ItemSet
ExtI: Extension ItemSets of *I*
MThrsh: Minimum Utility Threshold
SThrsh: Minimum Support Threshold
UL: Set of Utility Lists
OUTPUT:
HUIMs: Set of HUIs

1: **for all** $I_x \in ExtI$ **do**
2: $\quad SumUtil = SumUtilp + SumUtiln$
3: $\quad SumRUtil = SumRUtilD$
4: $\quad$ **if** $SumUtil > MThrsh \ \& \ Sup(i) > SThrsh$ **then**
5: $\qquad$ HUI $\leftarrow I_x$
6: $\quad$ **end if**
7: $\quad$ **if** $SumUtil + SumRUtil > MThrsh \ \& \ Sup(i) > SThrsh$ **then**
8: $\qquad ExtI_x \leftarrow \emptyset$
9: $\qquad$ **for all** $I_y \in ExtI$ **do**
10: $\qquad\quad I_{xy} \leftarrow I_x \cup I_y$
11: $\qquad\quad UL_{xy} \leftarrow Ext_x, Ext_y$
12: $\qquad\quad$ **if** $UL_{xy}.iutil > MThrsh$ **then**
13: $\qquad\qquad ExtI_x \leftarrow ExtI_x \cup I_{xy}$
14: $\qquad\quad$ **end if**
15: $\qquad$ **end for**
16: $\qquad$ SEARCH(I_x, $ExtI_x$, *MThrsh*, *SThrsh*, UL)
17: $\quad$ **end if**
18: **end for**

here because, if an itemset is non-frequent, then all its supersets will also be non-frequent as per the *downward closure property*. This allows for pruning of the search space, as the extensions of the non-frequent itemsets need not be explored. The qualified candidates after the check (line: 7) are used to perform the join operation between the utility lists of the extensions (line: 9 to line: 15). The search procedure is called (line: 16) with the formed extensions to recursively explore all the valid itemsets in a depth first fashion.

The main contribution of SMNU is the computation of the support count of each itemset and new pruning conditions based on the support measure, which helps to minimise the search space and improves efficiency. In the next section, we perform several experiments to validate the performance of SMNU against the state-of-the-art algorithms.

5 Experimental Evaluation

SMNU is the first algorithm in the literature to mine HUIs based on a support measure from datasets where items can have negative utility values also. Therefore, to test the efficiency of our method, we carry out extensive experiments on various datasets that

Table 2 Statistics of datasets

Dataset	Transactions	No of items	Average length
Retail	88,162	16,470	10,30
Mushroom	8416	119	23
Chess	3196	75	37

include *retail*, *mushroom* and *chess*. The summary of characteristics of these datasets is provided in Table 2. These datasets are available in the SPMF library [4]. First, we compare the performance of SMNU and FHN, when a very small support threshold supplied to SMNU. Then, we vary the minimum support threshold and compare the performance of SMNU with FHN, for iterations by varying the minimum utility threshold. The comparison of candidate count and HUI count between SMNU and FHN is provided in Figs. 1 and 2, respectively. For the convenience of representation, the candidate count in Fig. 1 is represented in multiples of 100.

For the *retail* dataset, a minimum support of 0.01 is set, which means that only those itemsets should be produced as an output by SMNU, that occur in atleast 1% of the total transactions in the input dataset. For the *mushroom* and *chess* datasets, the minimum support threshold is set to 0.1, which is 10% of the total transactions. As it is evident from Fig. 1, for dense dataset (*retail*), SMNU reduces the candidate count by almost a factor of seven as compared to FHN. Also, as shown in Fig. 2, the total number of HUIs for *retail* dataset is about three times higher for FHN as compared to SMNU. These candidate itemsets and HUIs are reflective of the itemsets when the support threshold is set to 0.01. This implies that only one third of the HUIs in the entire database occur atleast in 10% of the transactions. The remaining HUIs as produced by FHN, occur in less than 10% of the transactions. In real-life applications, analysing such itemsets manually can consume a lot of irrelevant time and manpower for large datasets. Similar observations can be found for the dataset *mushroom* from Figs. 1 and 2, where the candidates and HUI count differ by almost a factor of two for a relatively higher support value of 0.1. For smaller datasets like *chess*, it can be observed that the difference between the two algorithms for candidate and HUI count is negligible. So, it can be established that the performance of SMNU is highly efficient for large and dense datasets.

The comparisons for memory and time requirements between SMNU and FHN is presented in Figs. 3 and 4, respectively. It can be observed from these figures that the difference between FHN and SMNU is significant for large and dense datasets like *retail* and *mushroom*.

The comparisons for memory, time and HUI count are shown in Figs. 5, 6 and 7 by varying the minimum support threshold from 0.1 to 0.7. Both the algorithms have been executed on the three datasets with varying minimum utility thresholds.

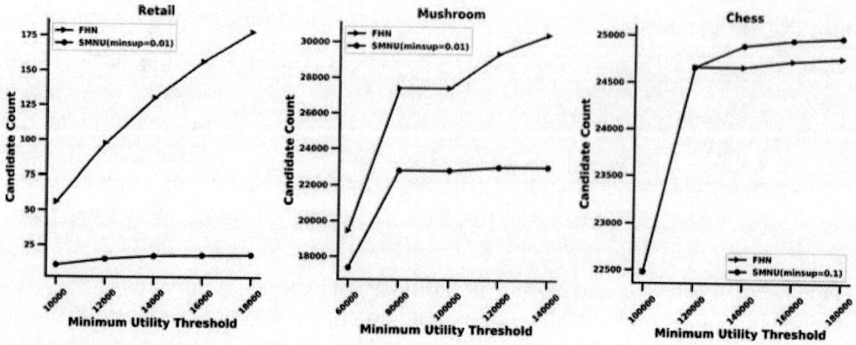

Fig. 1 Candidate count

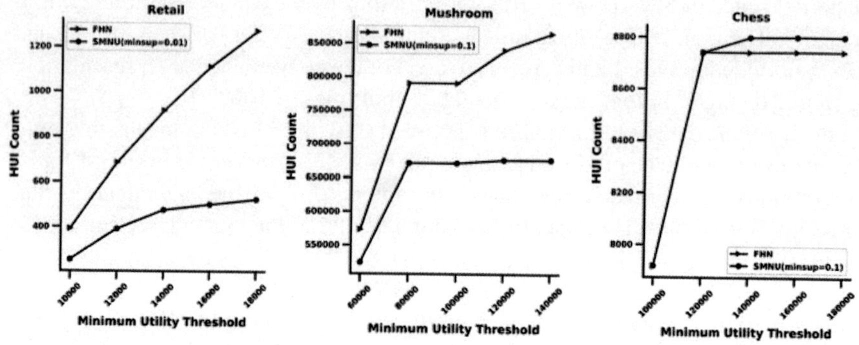

Fig. 2 High utility itemset count

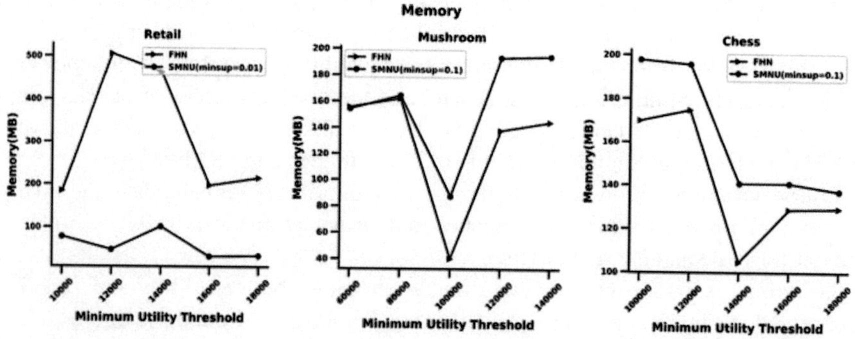

Fig. 3 Memory (MB)

It is evident from Fig. 5 that for the *retail* dataset, the SMNU outperforms the FHN algorithm in terms of memory, time and HUI count for all runs of the varying support values. Similar trends have been observed for the *mushroom* dataset as shown in Fig. 6. The results on *chess* dataset as shown in Fig. 7, are comparable for FHN and SMNU algorithms.

So, the SMNU guarantees an optimal count of HUIs to be produced as output and is also more scalable than the FHN algorithm for large and dense datasets.

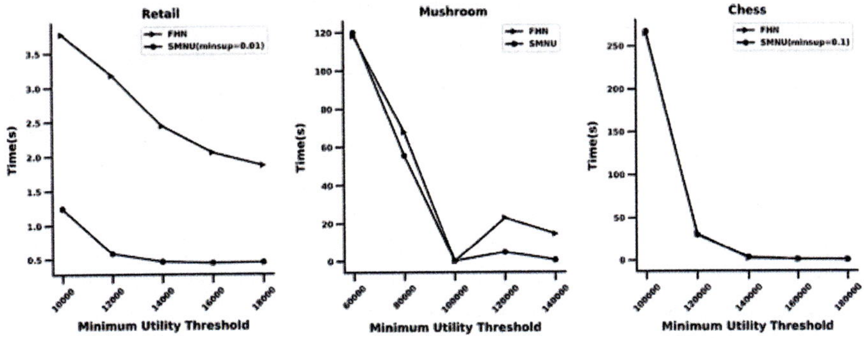

Fig. 4 Time (s)

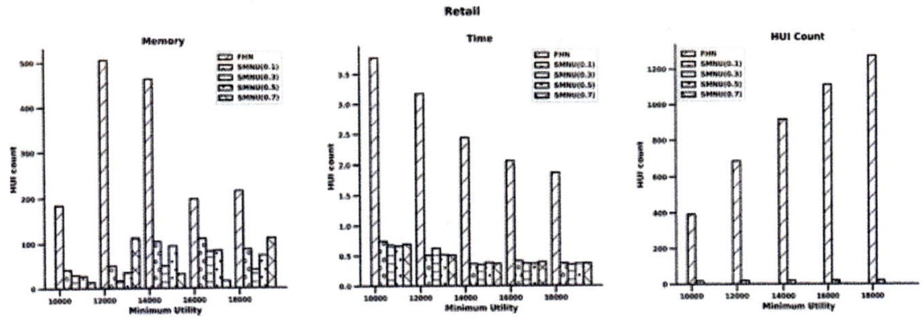

Fig. 5 Retail

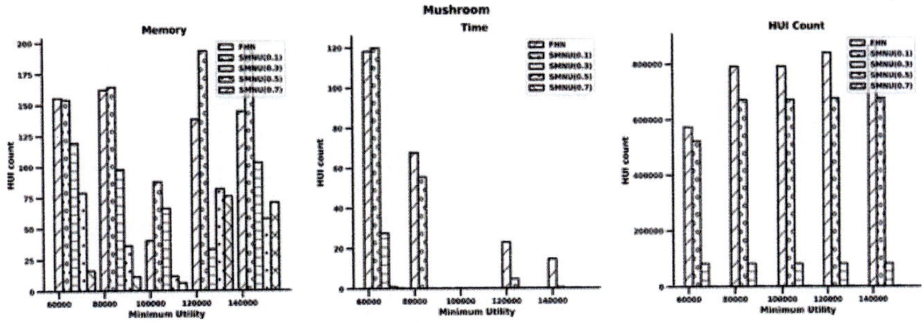

Fig. 6 Mushroom

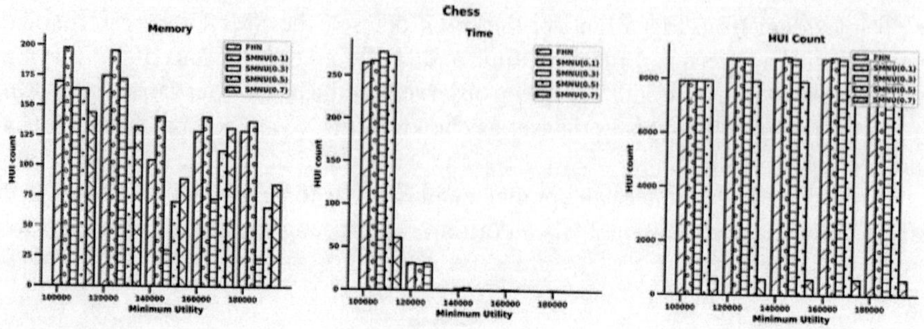

Fig. 7 Chess

6 Conclusion

In this paper, we have presented an algorithm, called SMNU, for support-based mining of HUIs from the databases with items that have negative utility values. The SMNU produces HUIs that can be directly utilised for real-life applications, unlike for most of the other mining algorithms where the support or negative utility values are not taken into consideration. The experimental results show that SMNU is more scalable and requires less execution time as well as memory for large datasets. However, for small and sparse datasets, the proposed approach has similar performance as the approaches that do not use support.

In future, we aim to propose optimisations to further enhance the performance of SMNU for large as well as small datasets. We also aim to design a novel method to generate association rules for the mined HUIs.

References

1. R. Agrawal, R. Srikant et al., Fast algorithms for mining association rules, in *Proceedings of the 20th International Conference on Very Large Data Bases, VLDB*, vol. 1215 (Citeseer, 1994), pp. 487–499
2. Y. Baek, U. Yun, H. Kim, J. Kim, B. Vo, T. Truong, Z.-H. Deng, Approximate high utility itemset mining in noisy environments. Knowl.-Based Syst. **212**, 106596 (2021)
3. C.-J. Chu, V.S. Tseng, T. Liang, An efficient algorithm for mining high utility itemsets with negative item values in large databases. Appl. Math. Comput. **215**(2), 767–778 (2009)
4. P. Fournier-Viger, *SPMF: A Java Open-Source Data Mining Library*. Philippe-fournier-viger.com. (2021)
5. P. Fournier-Viger, C.-W. Wu, S. Zida, V.S. Tseng, FHM: faster high-utility itemset mining using estimated utility co-occurrence pruning, in *International Symposium on Methodologies for Intelligent Systems* (Springer, 2014), pp. 83–92
6. J. Han, J. Pei, Y. Yin, Mining frequent patterns without candidate generation. ACM SIGMOD Rec. **29**(2), 1–12 (2000)
7. X. Han, X. Liu, J. Li, H. Gao, Efficient top-k high utility itemset mining on massive data. Inf. Sci. **557**, 382–406 (2021)

8. S. Kumar, K.K. Mohbey, High utility pattern mining distributed algorithm based on spark RDD, in *Computer Communication, Networking and IoT* (Springer, 2021), pp. 367–374
9. J.C.-W. Lin, P. Fournier-Viger, W. Gan, FHN: an efficient algorithm for mining high-utility itemsets with negative unit profits. Knowl.-Based Syst. **111**, 283–298 (2016)
10. M. Liu, J. Qu, Mining high utility itemsets without candidate generation, in *Proceedings of the 21st ACM International Conference on Information and Knowledge Management* (2012), pp. 55–64
11. Y. Liu, W.-K. Liao, A. Choudhary, A two-phase algorithm for fast discovery of high utility itemsets, in *Pacific-Asia Conference on Knowledge Discovery and Data Mining* (Springer, 2005), pp. 689–695
12. W. Song, C. Zheng, C. Huang, L. Liu, Heuristically mining the top-k high-utility itemsets with cross-entropy optimization. Appl. Intell. 1–16 (2021)

Comparative Analysis of Regressor Models on Non-invasive Blood Glucose Dataset

Neha Tuniya, Mahesh Parihar, Shital Patil, Komal Lawand, and Hemalata Nawale

Abstract Diabetes affects more than 285 million people globally according to estimates by the International Diabetes Federation. A compact and non-invasive monitoring system can measure blood glucose continuously without posing many problems and easy to use for the diabetic population. In this paper, a near-infrared optical sensor-based non-invasive system has been designed and calibrated against conventional invasive glucose measuring techniques. A synthetic dataset of a sufficiently large population has been developed in a suitable range as defined by the Medical Council of India. Upon development of the laboratory prototype of this device, by analyzing the variation in voltages received after reflection of incident light in the cases the approximate glucose level of the individual is going to be predicted using statistical models. A compact framework for non-invasive blood glucose measurement has been designed and tested successfully for a set of 5000 populations of fasting as well as random blood glucose samples of different patients. This paper emphasizes on different approaches using basic regression methods and classical machine learning algorithms like support vector machines, K-nearest neighbor, random forest with their hybrid regression methods for the predictions of blood glucose. Here, classical and hybrid machine learning algorithms and techniques have been applied as a measuring tool, to the large synthetic datasets to come out with laboratory-based prototypes in an attempt to automate the analysis of large and complex patient data

M. Parihar
College of Engineering Pune, Pune, Maharashtra, India

N. Tuniya (✉)
MmM Ltd, Navi Mumbai, Maharashtra, India
e-mail: tneha49@gmail.com

S. Patil
Ramrao Adik Institute of Technology, Mumbai, India
e-mail: shital.patil@rait.ac.in

K. Lawand
Vidyalankar Institute of Technology, Mumbai, India

H. Nawale
Sandip University, Nashik, India

A. K. Bashir et al. (eds.), *Proceedings of International Conference on Computing and Communication Networks*, Lecture Notes in Networks and Systems 394,
https://doi.org/10.1007/978-981-19-0604-6_19

attributes. Also, this work envisages performance optimization using grid search CV and randomized search to find the best suitable values for the hyperparameters and its impact on improvising accuracy of algorithmic models leading to more accurate prediction of blood glucose values.

Keywords Non-invasive technique · Glucose level · Machine learning techniques

1 Introduction

The techniques of machine learning have been successfully employed in assorted applications including medical diagnosis. By developing classifier system, machine learning algorithm may immensely help to solve the health-related issues which can assist the physicians to predict and diagnose diseases at an early stage. We can ameliorate the speed, performance, reliability, and accuracy of diagnosing on the current system for a specific disease by using the machine learning classification algorithms [1]. Applying machine learning and data mining methods in diabetes mellitus (DM) research is a key approach to utilizing large volumes of available diabetes-related data for extracting knowledge. The severe social impact of the specific disease renders DM one of the main priorities in medical science research, which inevitably generates huge amounts of data. Undoubtedly, therefore, machine learning and data mining approaches in DM are of great concern when it comes to diagnosis, management, and other related clinical administration aspects. Hence, in the framework of this study, efforts were made to review the current literature on machine learning and data mining approaches in diabetes research [2].

The second chapter discusses sensor system setup along with the hardware used in it. It also emphasizes on the synthetic data generation extrapolated from actual data collected from sample population of diabetic patients. Third chapter discusses the implementation of ML algorithms, their analysis, and performance comparison for better accuracy.

2 Device Setup and Synthetic Data

In this paper, a near-infrared optical sensor is used to measure the blood glucose level which is linearly calibrated against the voltage indicated by the sensor. This has been compared and calibrated against conventional invasive methods taken for same patients population. The range of the dataset has slowly been increased 100-fold to 500 different patients initially belonging to different lifestyle with different socioeconomic conditions. The dataset is collected from different blood donation camps, pathology laboratories, and hospitals as well. This ensures that dataset consists of healthy (from blood donation camps) as well as unhealthy people (from hospitals and pathology laboratories). Using extrapolation methods, a sufficiently large synthetic

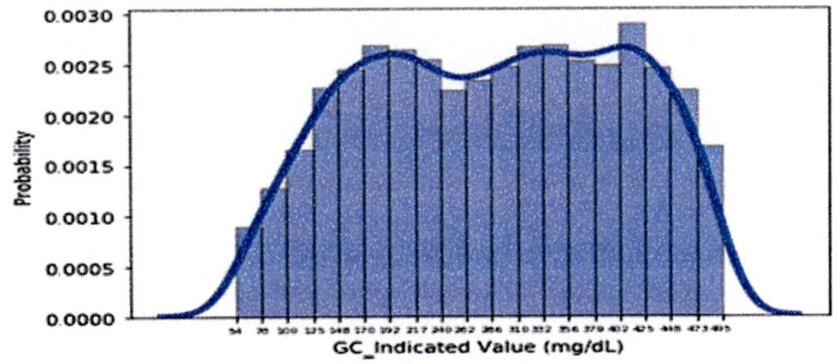

Fig. 1 Normalized distribution of dataset

dataset of within maximum range of different types of diabetics, viz. has been developed. Figure 1 shows the normalized distribution of dataset for various values of glucose levels. A blood glucose level less than 70 mg/dL called as hypoglycemia as well as more than 200 mg/dL called as hyperglycemia has been considered while selecting the sample data space. Extrapolated dataset is divided into the ratio of 15:60:25 with respect to different types of diabetics, viz. hypoglycemia, normal range, and hyperglycemia, respectively, as per recommendation by the group of physicians.

3 Hardware Setup

The experimental circuit is set up using near-infrared (NIR) spectral range to measure the blood glucose. A sensor-based system is used to detect blood glucose non-invasively using near-infrared (NIR) radiation using spectroscopic reflection analysis. It describes the principle of glucose measurement using NIR method. The device consists of an infrared LED having a wavelength of 900–1100 nm as emitter. The intensity of reflected light is used to determine the blood sugar level. The hardware and software requirement of the setup consists of Arduino microcontroller, IDE software, and computer.

The processing is done by the microcontroller, and the level of glucose is to be displayed on the computer. The hardware part also consists of infrared light emitting diode (IR LED), a photodiode sensor, after amplification components and microcontroller. After performing second-order regression analysis, it is found that glucose level (Y) in real time in terms of voltage intensity (x) is given by Eq. 1

$$Y = 1.8375x^2 + 18.945x + 41.288 \tag{1}$$

Figure 2 shows the regression analysis of glucose data showing the linear relationship between voltage indicated and glucose.

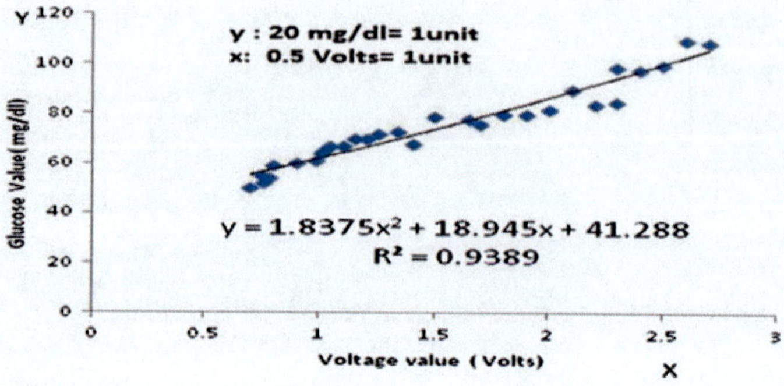

Fig. 2 Regression analysis of glucose data with sensor output

4 Methodology

To carry out regression analysis for predicting glucose, dataset is collected from 5000 patients. The dataset consists of information about various tests done on diabetes patients using two different methods. Regression algorithms are applied to calculate the glucose value for different values of voltage index. The blood glucose level measured by the optical device is further used to automate the analysis of large and complex patient data attributes using basic regression and machine learning techniques. Initially, basic regression techniques such as linear regression and polynomial regression of various degrees are implemented along with some classical methods like decision trees, random forest, support vector regressor, and k-nearest neighbor. To further investigate the model, some hybrid methods are used. These include methods like ridge regression, Lasso regression, elastic net, and gradient boosting regressor. Also, performance optimization using grid search CV and random search CV is implemented on ridge regression. Finally, some advance methods like XGBoost, LightGBM, Catboost, artificial neural network, and some stacked methods such as LBGM + ANN + RidgeCV, XGBoost + ANN + LassoCV are implemented. Hardware mentioned above is used to calculate the blood glucose level for training purpose. The combined results of all methods are listed in Table 1.

5 Analysis, Results, and Discussion

Constructed system determines a method for the prediction of blood glucose level for human using non-invasive methods. Various basic and classical machine learning techniques along with some hybrid techniques are used to correctly predict the glucose level in the body. Dataset consists of 5000 patients along with their glucose level collected by prick method and optical method. Using this dataset, machine learning algorithms are trained and optimization is done by using advance and hybrid

Table 1 Accuracy results of the all methods

Method	Mean squared error	Mean absolute percentage error	Accuracy ($r2$ score)
Linear regression	5.290553703	41.9839390	0.999469
Polynomial regression degree 2	2.653233685	27.74397314	0.99957
Polynomial regression degree 3	2.635144804	26.51472235	0.99958
Polynomial regression degree 4	1.396237706	18.10051644	0.99972
Decision tree	0.000488	0.096429	0.99999
Random forest	0.0027259	0.77733	0.99999
Support vector regressor	43.88772	97.2438	0.99017
K-nearest neighbor	0.016158	2.01658556	0.99999
Ridge regression	3.322519	35.17507	0.99946
Lasso regression	3.585725	38.329494	0.99942
Elastic net	454.714012	432.769912	0.95248
Gradient boosting regressor	0.641002	12.337249	0.999939
Grid search CV	1.97199	24.86016	0.99946
Randomized search	1.97199	24.8596	0.99946
XGBoost	0.639	12.2	0.999938
LightGBM	0.3893	11.2057	0.999956
Catboost	0.9298	16.8893	0.99972
Artificial neural network	1.9939	21.9658	0.99946
LBGM + ANN + RidgeCV	0.4331	11.8764	0.999959
XGBoost + ANN + LassoCV	0.547	11.5862	0.999944

methods. The performance is evaluated step by step and discussed below in detail. Figure 3 shows the graphs of all methods.

5.1 Basic Methods

Basic methods include linear regression model and polynomial regression of various degrees of polynomial. Initially, linear regression is applied on the 5000 data points to get the baseline results. To further experiment with the dataset polynomial regression is applied with various degrees of polynomials. Table 1 shows the results obtained from basic methods. The results show that as the degree of polynomial increases it

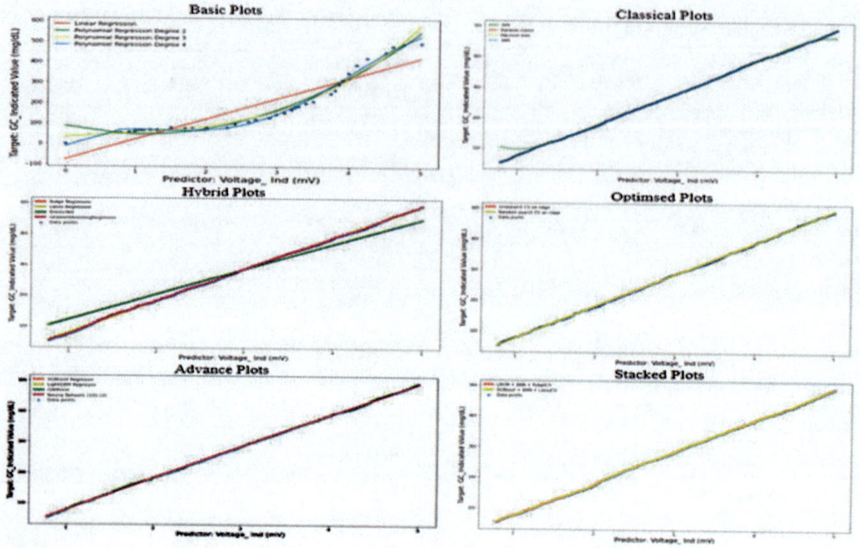

Fig. 3 Plots of all methods

becomes more accurate but this is with respect to the training data points which can result in overfitting as well. Even though the model is performing well on the training data as the degree of polynomial increases, it may not perform well with testing data. Therefore, this results in overfitting. Thus, an optimum degree of polynomial has to be chosen carefully where the model has minimum testing and training error and is free from extreme bias and variance conditions (high bias, low variance or low bias, and high variance).

5.2 *Classical Methods*

Apart from the basic methods, some classical methods were applied. These methods include random forest, decision tree, K-nearest neighbor, and support vector regressor. The results of these methods are shown in Table 1. Results are very accurate for decision tree, random forest, and KNN, whereas in SVR at both the starting and ending the curve gets deviated. This shows that at very low and very high voltages the system is not able to predict the glucose level accurately. Therefore, support vector regressor is not the appropriate model for this particular application.

5.3 Hybrid Methods

These methods include ridge regression, Lasso regression, elastic net, and gradient boosting regressor. The results of these methods are illustrated in Table 1. From the results, we can see that introducing L2 penalty in ridge regression and L1 penalty in Lasso regression has not affected the accuracy of linear regression to certain extent. The accuracy of linear regression, ridge and Lasso regression is almost the same. However, when both the penalties are introduced in elastic net, the accuracy is reduced to a certain extent. This is because it has to underfit the data as shown in the figure. Among all the hybrid methods, gradient descent has the highest accuracy.

5.4 Optimized Methods

To obtain the ideal parameters and optimize the performance of the algorithm, optimization techniques like grid search CV and randomized search are performed on ridge regression to obtain the optimum value of lambda. The results of the optimization are shown in Table 1. It can be seen that even though after applying grid search CV and randomized search, the accuracy of the model is not changed. It means that in ridge regression, initially, before applying optimization techniques, the lambda selected was already optimal. Therefore, there is no change in the accuracy after applying optimization techniques.

5.5 Advance Methods

In this project, four advance methods are implemented to predict the glucose level which include XGBoost, Catboost, LightGBM, and artificial neural network. The results obtained are shown in Table 1.

It can be seen that the accuracy of XGBoost is highest, and ANN is least among the four advanced techniques.

5.6 Stacked Methods

In this project, two combinations are applied LBGM + ANN + RidgeCV and XGBoost + ANN + LassoCV. Both stacked regressors give out great results are shown in Table 1. These combinations are chosen because these regressors individually worked well and stacking them will make the results even better. Both stacked regressors give out great results. The MSE of model 1 is lesser but MAPE is slightly higher. But the model accuracy is also higher than that model 2; therefore, model 1 can be considered as better.

6 Conclusion

In this paper, several machine learning techniques are applied from basic to advance methods to check which model is best suited to predict the glucose level. Then classical methods like kNN, decision trees, random forest, and SVR are explored which show that decision trees perform better among them. Further, hybrid methods like ridge regression, Lasso regression, elastic net, and gradient boosting methods are implemented to compare the accuracy of model in which gradient boosting edges over others. To obtain the optimized parameters, some optimization methods like grid search CV and random search CV are applied on ridge regression but it does not have much impact on accuracy. To extend further, advance methods like XGBoost, LightGBM, Catboost, artificial neural network are applied on dataset to explore better accuracy of the model and stacked methods like LBGM + ANN + RidgeCV as well as XGBoost + ANN + LassoCV are performed on the dataset wherein latter works better than the former. Comparing the results of all these methods, even though other methods give fruitful results, the accuracy of classical methods mainly decision tree, random forest, and K-nearest neighbor is higher than others and hence stands the most appropriate models for glucose level prediction.

7 Future Scope

Synthetic data will be validated against the actual collected data for adding correction factor in the calibration equation of the sensor, and also entire exercise of ML algorithms will be implemented on actual dataset for model validation. Sensor will be equipped to compute the levels of hemoglobin, and ML algorithms will be applied to predict the values of glucose as well as hemoglobin at the same time as a part of better calibration of the designed glucose sensor system.

References

1. A. Choudhury, D. Gupta, A survey on medical diagnosis of diabetes using machine learning techniques, in *Proceedings of Recent Developments in Machine Learning and Data Analytics* (2018)
2. I. Kavakiotis, O. Tsave, A. Salifoglou, N. Maglaveras, I. Vlahavas, I. Chouvarda, Machine learning and data mining methods in diabetes research. Comput. Struct. Biotechnol. J. (2017)
3. S. Coster, M.C. Gulliford, P.T. Seed, J.K. Powrie, R. Swaminathan, Monitoring blood glucose control in diabetes mellitus. Health Technol. Assess. **4**(12) (2000)

4. K. Lawand, M. Parihar, S.N. Patil, Design and development of infrared LED based non invasive blood glucometer, in *IEEE India Council International Conference 2015 (INDICON)*, Jan 2016, vol. 3, no. 2 (2015), pp. 1–6
5. Q. Zou, K. Qu, Y. Luo, D. Yin, Y. Ju, H. Tang, Predicting diabetes mellitus with machine learning techniques (2018)
6. O. Rodriguez, A generalization of ridge, lasso and elastic net regression to interval data, in *Proceedings of Tilburg and the 2013 Conference of the International Federation of Classification Societies (IFCS)* (2013)
7. T. Chen, C. Guestrin, XGBoost: a scalable tree boosting system (2016)
8. G. Ke, Q. Meng, T. Finley, T. Wang, W. Chen, W. Ma, Q. Ye, T.-Y. Liu, LightGBM: a highly efficient gradient boosting decision tree. Neural Inf. Process. Syst. **30** (2017)
9. L. Prokhorenkova, G. Gusev, A. Vorobev, A.V. Dorogush, A. Gulin, CatBoost: unbiased boosting with categorical features, in *Proceedings of 32nd Conference on Neural Information Processing Systems*, Montréal, Canada (2018)
10. J. Bergstra, Y. Bengio, Random search for hyper-parameter optimization. J. Mach. Learn. Res. (2012)
11. P. Bhatt, H. Prosper, S. Sekmen, C. Stewart, Optimizing event selection with the random grid search. Comput. Phys. Commun. **228** (2018)
12. S. Dzeroski, B. Zenko, *Is Combining Classifiers with Stacking Better Than Selecting the Best One?* (Department of Intelligent Systems, Jozef Stefan Institute, Ljubljana, Slovenia)
13. J.H. Friedman, Greedy function approximation: a gradient boosting machine. Ann. Statist. **29** (2001)
14. B. Pavlyshenko, Using stacking approaches for machine learning models, in *IEEE Second International Conference on Data Stream Mining and Processing* (2018)
15. J.T. Hancock, T.M. Khoshgoftaar, Catboost for big data: an interdisciplinary review (2020)

Intelligent Parking System Using Automated License Plate Recognition and Face Verification

Quang Than Van, Huong Nguyen Van, Linh Duong Vu Hoang, Thanh Nguyen Ngoc, Vinh Luong Duc, Duong Le Dai, Tan Nguyen Bao, Sang Nguyen Quang, Linh Nguyen Van, Anh Dang Trung, Cong Le Thanh, and Ho Nguyen Manh

Abstract In this paper, we design a ticketless parking system using automated license plate recognition (ALPR) technology and face verification technology. Cameras are deployed to get the license plate and the driver's face information at the entrance; then, this information is analyzed and stored in a database to verify the vehicle's owner at the exit. Based on high-precision technologies, our system brings outstanding benefits about safety, time, and convenience compared with manual ones. These advantages demonstrate the ability of this solution to be applied not only to parking lots but also to toll stations on highways.

1 Introduction

Parking is an essential part of the transportation system. In Vietnam, there are 4.3 million cars and tens of millions motorbikes in circulation, followed by the enormous demand for parkings. Parking is the mandatory facility in any commercial centers, supermarkets, office buildings, and some urban areas. Besides an essential service, it also becomes a great source of income for investors. With indisputable importance contributed to transport infrastructure, parking requires practical solutions to optimize costs and improve its operating efficiency. However, the manual parking solution based on ticket information and vehicle matching to certify the ownership instead of driver's information remains inconvenient, time-consuming, and unsecured. As a result, we are motivated to design a modern parking system using the

Q. T. Van · H. N. Van (✉) · L. D. V. Hoang · T. N. Ngoc · V. L. Duc · D. L. Dai · T. N. Bao · S. N. Quang · L. N. Van · A. D. Trung · C. L. Thanh · H. N. Manh
Viettel Solutions, Hanoi, Vietnam
e-mail: huongnv75@viettel.com.vn

Q. T. Van
e-mail: quangtv13@viettel.com.vn

A. K. Bashir et al. (eds.), *Proceedings of International Conference on Computing and Communication Networks*, Lecture Notes in Networks and Systems 394,
https://doi.org/10.1007/978-981-19-0604-6_20

most updated technologies. This paper introduces a ticketless parking solution with the confirmation method by attaching the vehicle's license plate information to the owner's facial information. Specifically, we have designed a ticketless parking system architecture that can operate the whole process automatically. To do that, we have built artificial intelligence (AI) models based on existing researches and adapted them to our system, including:

- The first AI model used to detect 4-wheel and 2-wheeler license plates achieved 99.2% and 99.5% accuracy on 4-wheelers and 2-wheelers, respectively.
- The second model is built to recognize the characters on the registration plate, helping the system identify the license plate. The model has an accuracy of 96.1% for cars and 85.7% for motorcycles.
- The third model is trained on the masked face dataset to detect the masked face features; this model achieves 95% average precision on the WIDER FACE Easy evaluation dataset.
- The fourth model is also trained on the masked face dataset to extract the masked face features; this model achieves 99.8% accuracy on the LFW evaluation dataset.

In the following part, Sect. 2 demonstrates the literature reviews in this area. Section 3 describes how the AI models are built with license plate recognition model in Sect. 3.1; face detection model in Sect. 3.2; and face verification model in Sect. 3.3. In Sect. 4, we illustrate our proposed solution with system processing flow in Sect. 4.1 and system development in Sect. 4.2. Finally, we demonstrate some experiment results and discussion in Sect. 5, and conclusion is in Sect. 6.

2 Literature Review

In 2016, Subhashini et al. designed a system that automatically identifies vehicles [1]. This system authorizes vehicle owner identification to secure highly restricted areas like housing, defense, military, parliament, etc. Using image processing technology, both the number plate recognition and the vehicle owner recognition processes are implemented in MATLAB. The system contains an embedded section that controls the opening of the security gate and the communicated mean. In 2019, Persada et al. implemented face and vehicle license plates identification for the smart parking system at the parking lot in the Telkom University area [2]. The SSIM algorithm is applied as the core in the face recognition process, gaining an accuracy rate of 76.67% on 30 samples of vehicles with the highest SSIM value 0.83.

In [3], Joshi et al. proposed a parking system using optical character recognition (OCR) and radio frequency identification (RFID). After using OCR to read a vehicle's license plate number and save that data to the database, this system controls the vehicle by the authorized TAG containing license plate information. In [4], Kamaruzaman et al. introduced a ticketless parking system called PARKEY. It is based on license plate recognition technology and built on Raspberry Pi 3B+.

3 Artificial Intelligence Model Construction

3.1 License Plate Recognition

Currently, there are multiple methods of license plate recognition researched with numerous scientific articles and open-source code. The license plate detection model in this project is built based on the implementation of Silva and Jung [5]: Unrestricted scenario uses optical character recognition (OCR) to identify the characters on the license plate. In this work, the open-source code mentioned above is applied to train the model based on the dataset we created.

Dataset Preparation: We collect and label the training dataset consisting of 10,000 images containing license plates of motorbikes and cars in Vietnam in various sizes, shapes, as well as under different lighting conditions and view angles. In addition, we also create a dataset of 4000 license plate images for testing purposes.

Training: Based on ALPR in unconstrained scenarios GitHub repository [5], we re-train the model on our dataset. The model's output includes 26 letters from A to Z and 10 digits from 0 to 9. We train the model using Adam optimizer (with weight decay at 0.0005, batch size at 32 on 3 NVIDIA GeForce RTX2080 Ti GPUs (11 GB)). The initial learning rate value is 0.01, decreasing 0.01 times every 10 epochs. In the end, the training process terminates after 40 epochs.

Testing: We test the models on a dataset including car and motorbike license plate images with 2000 items for each category. Concerning the license plate detection model, our accuracy reaches 99.2% for the car image set and 99.5% for the motorbike one. However, the results go down to 96.1 and 85.7% in license plate recognition (see Table 1).

3.2 Face Detection

The face detection model in this project is built based on the research of Deng et al. [6], which is a single-stage methods object detection algorithm. The model's architecture

Table 1 License plate recognition system accuracy

Test dataset	Image	Rate of detection (%)	Rate of recognition (%)
Car license plate	2000	99.2	96.1
Motorbike license plate	2000	99.5	85.7

used for image feature extraction is based on the ResNet50 [7] backbone. In addition, some simplifications of the loss function are made in order to save computational resources.

Dataset Preparation: We train and test the model on the WIDER FACE dataset [8]. This dataset consists of 32,203 images and 393,703 bounding boxes of faces in various sizes, poses, expressions, situations, and lighting conditions. The WIDER FACE dataset is split into training (40%), validation (10%), and testing (50%) subsets by randomly sampling from 61 scene categories. Based on EdgeBox's detection rate, the author groups this dataset into three sub-datasets corresponding to easy, medium, and hard.

Training: For loss head, we use multitask loss, including softmax loss for face/not face classification and smooth-L1 loss for coordinate regression. Compared to the original paper [6], we drop dense regression loss so as to simplify the loss function. For data augmentation, since there are about 20% small faces in the WIDER FACE training set [8], we randomly cut square arrays from the original image and resize these arrays to 640 $\times$ 640 to generate large training faces. More specifically, square arrays are clipped from the original image with a random size between [0:3.1] of the shorter edge in the original image. We train the model using an SGD optimizer (with momentum at 0.9, weight decay at 0.0005, batch size at 24 out of 3 NVIDIA GeForce RTX2080 Ti GPUs (11 GB)). The initial learning rate value is 0.001, decreases by 0.01 every 5 epochs, and divides by 10 at the 55th and 68th epochs. Finally, the training process terminates after 80 epochs.

Testing: We assess the model on the WIDER FACE test set [8]. Box voting is applied on the union set of predicted face boxes using an IoU threshold at 0.4. Results are shown in Table 2. Good accuracy is observed (>88).

3.3 *Face Verification*

To verify whether two faces belong to the same person or not, we construct a model with ResNet50 [7] backbone to extract the masked facial feature vectors based on the research of Deng et al. [9]. Then, we normalize vectors and compute the dot product of these two vectors. If it is greater than or equal to 0.5, two faces belong to the same person. Otherwise, two faces belong to two different people.

Table 2 Face detection average precision

WIDER FACE test set [8]	Average precision (%)
Easy	95
Medium	93
Hard	88

Dataset Preparation: In this session, the MS1M-ArcFace dataset which is generated from MS-Celeb-1M dataset [10] is utilized to train the model. In our work, we create a program to generate masked face data from this dataset. For each original face image, two more masked photographs are generated with randomly selected mask styles. As a result, we obtain a dataset which includes 15.3 million images belonging to 93,431 people. Each person is identified by an ID that contains many images of both non-masked faces and masked faces.

Training: We use mis-classified vector guided softmax loss (MV-softmax loss) [11] for loss head. And for data augmentation, we align the face image and resize it to 112×112 before normalizing the data with mean of [0.5, 0.5, 0.5] and std of [0.5, 0.5, 0.5]. In addition, we also use the horizontal random flip technique to increase the amount of data. We train the model with an SGD optimizer (with momentum at 0.9, weight decay at 0.0005, batch size at 128) on 3 NVIDIA GeForce RTX2080 Ti GPUs (11 GB). The learning rate initialization value is 0.1, decreasing 10 times every 8 epochs. The training process terminates after 20 epochs.

Testing: For evaluating our model, we test the training results with three different datasets LFW [12], CFP-FP [13], and AgeDB-30 [14]. The test properties and results are represented in Table 3. A good accuracy (>98%) is observed between our trained model and the references. Consequently, we use this model for the following steps.

4 The Proposed Solution

4.1 System Processing Flow

Figure 1 presents the flow of our system, which are operated with several steps as follows:

At entry gate:

- Step 1: When a vehicle enters the entry gate, two cameras simultaneously capture the license plate's image and the driver's face image.
- Step 2: The system analyzes the images to get the characters of the license plate and extract facial feature vector containing condensed face information.

Table 3 Face verification accuracy

Test datasets	Identity	Image	Accuracy (%)
LFW [12]	5749	13,233	99.8
CFP-FP [13]	500	7000	98.8
AgeDB-30 [14]	568	16,488	98.2

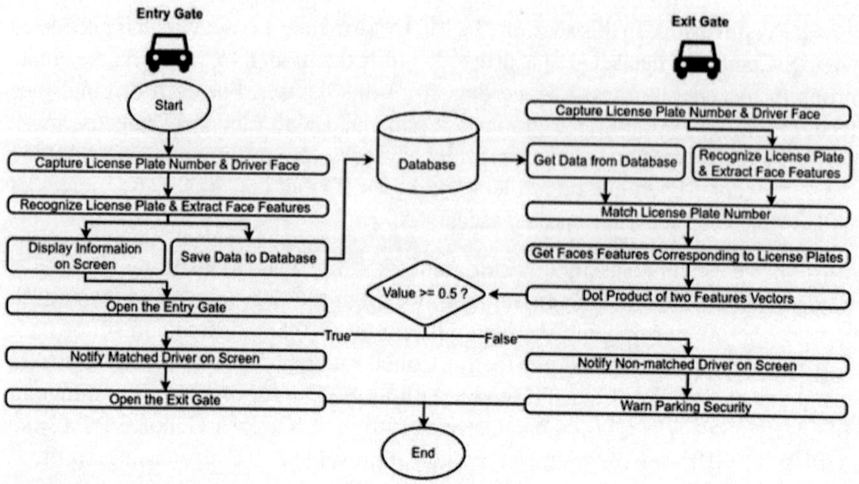

Fig. 1 System processing flow

- Step 3: The system saves the information in step 2 into the database and displays it on the screen to notify the driver.
- Step 4: The barrier at the entry gate is opened.

At exit gate:

- Step 5: When a vehicle is taken out, the system repeats the process from step 1 to step 2 to get the vehicle's license plate information and the face features vector of the driver. Simultaneously, the system gets data from the database.
- Step 6: The system matches the license plate number between the one on the exit gate and database.
- Step 7: Two facial feature vectors are obtained corresponding to license plates in Step 6.
- Step 8: The system computes the dot product of these two vectors. If the result is greater than or equal to 0.5, the system concludes that this case is the same person; otherwise, the conclusion is not the same.
- Step 9: If the result of step 8 is the same person, the screen shows a message of successful confirmation, and the barrier is opened. In contrast, the system will warn the parking security and display a non-matched message on the screen.

4.2 System Development

As introduced in Fig. 2, our ticketless parking system has two different parts that will work together to form the product. At each gate, two IP cameras record images with a resolution of 1920 * 1080 and are utilized to get the license plate's image and the

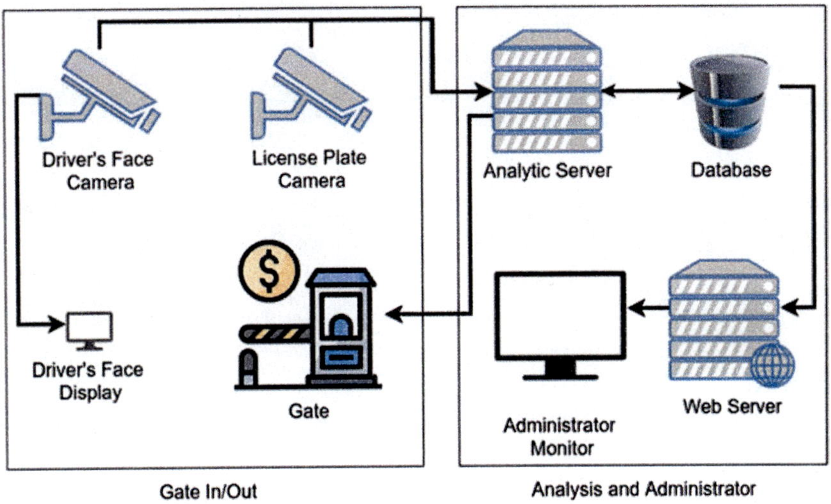

Fig. 2 Our proposed ticketless parking system architecture

driver's face. Then, a Raspberry Pi 4 is used to display images on an 10 inch LCD screen in order to help the system interact with drivers before the barrier is opened. The administrator section includes the analytics server, image storage, and web applications. The web application provides user interaction functions that include viewing parking log history, searching the license plate to view entry and exit information, as well as statistics of the amount of vehicles entering and leaving each day.

5 Result and Discussion

To investigate the precision and efficiency of the system, we have deployed the system at the parking lot with a capacity of 700 vehicles. During a week of experimentation, the system recorded 2480 vehicle visits. Table 4 shows the results with license plates. The rate of detection is calculated by the number of license plates detected divided by the total number. The number of correctly recognized license plates divided by total number of detected license plates is the recognition rate. The experiment results are not as good as the model testing results in Sect. 3.1. Most license plates are not detected or misidentified because they are so filthy that it is not human-readable.

Table 5 shows the results on the face; face detection rate is determined by the number of detected faces divided by the total number of faces; the number of correctly verified faces divided by the number of detected faces is the verification rate. In this experiment, many drivers wear helmets and goggles, so facial recognition and verification rates are worse than when evaluating the model in Sects. 3.2 and 3.3.

Table 4 Experimental plate result

Amount of plates	Success detection	Rate of detection (%)	Success recognition	Rate of recognition (%)
2480	2428	97.9	2309	95.1

Table 5 Experimental face result

Amount of faces	Success detection	Rate of detection (%)	Success verification	Rate of verification (%)
2480	2472	99.7	2414	97.7

The average time each vehicle stops in front of the barrier is 2 s at the entrance gate and 3 s at the exit gate. Those results guarantee that our proposed system can be well applied in parking lots.

6 Conclusions

In the twenty-first century, when technology is changing rapidly, AI applications permeate our life. The proposed solution in this paper integrates multiple AI technologies to provide the most convenient solution. Built on top of the latest research and popular open source, our solution solves the problems in the traditional parking system as well as provides new features benefiting users. Although there exists limitations such as an extensive server system required to build this product, optimization for deployment to embedded IoT devices will help solve this problem in the future.

References

1. G. Subhashini, M.E. Markhi, R. Abdulla, Automatic car park management system using face and vehicle registration recognition, in *The Fourth International Conference on Technological Advances in Electrical, Electronics and Computer Engineering (TAEECE2016)*, Sept 2016, p. 48
2. R.P. Persada, S. Aulia, D. Burhanuddin, H. Sugondo, Automatic face and VLP's recognition for smart parking system. Telkomnika (Telecommun. Comput. Electron. Control) **17**(4), 1698–1705 (2019). https://doi.org/10.12928/telkomnika.v17i4.11746
3. Y. Joshi, P. Gharate, C. Ahire, N. Alai, S. Sonavane, Smart parking management system using RFID and OCR, in *2015 International Conference on Energy Systems and Applications* (2015). https://doi.org/10.1109/icesa.2015.7503445
4. M.A.M.B. Kamaruzaman, N.R.M. Nasir, PARKEY: ticket-less parking system using license plate recognition approach. J. Phys. Conf. Ser. **1860**(1), 012006 (2021). IOP Publishing. https://doi.org/10.1088/1742-6596/1860/1/012006
5. ALPR in Unconstrained Scenarios Github Repository. Available at: https://github.com/sergiomsilva/alpr-unconstrained.git

6. J. Deng, J. Guo, Y. Zhou, J. Yu, I. Kotsia, S. Zafeiriou, Retinaface: single-stage dense face localisation in the wild. arXiv preprint arXiv:1905.00641. https://doi.org/10.1109/CVPR42600.2020.00525
7. K. He, X. Zhang, S. Ren, J. Sun, Deep residual learning for image recognition, in *Proceedings of the IEEE Conference on Computer Vision and Pattern Recognition* (2016), pp. 770–778. https://doi.org/10.1109/cvpr.2016.90
8. S. Preuss, A. Demchuk Jr., M. Stuke, Appl. Phys. A
9. J. Deng, J. Guo, N. Xue, S. Zafeiriou, ArcFace: additive angular margin loss for deep face recognition, in *Proceedings of the IEEE/CVF Conference on Computer Vision and Pattern Recognition* (2019), pp. 4690–4699. https://doi.org/10.1109/cvpr.2019.00482
10. Y. Guo, L. Zhang, Y. Hu, X. He, J. Gao, MS-Celeb-1M: a dataset and benchmark for large-scale face recognition, in *European Conference on Computer Vision*, Oct 2016 (Springer, Cham, 2016), pp. 87–102. https://doi.org/10.1007/978-3-319-46487-9_6
11. X. Wang, S. Zhang, S. Wang, T. Fu, H. Shi, T. Mei, Mis-classified vector guided softmax loss for face recognition, in *Proceedings of the AAAI Conference on Artificial Intelligence*, Apr 2020, vol. 34, no. 07, pp. 12241–12248. https://doi.org/10.1609/aaai.v34i07.6906
12. G.B. Huang, M. Ramesh, T. Berg, E. Learned-Miller, Labeled faces in the wild: a database for studying face recognition in unconstrained environments. Techn. Rep. **5**(6), 8 (2007)
13. S. Sengupta, J.C. Chen, C. Castillo, V.M. Patel, R. Chellappa, D.W. Jacobs, Frontal to profile face verification in the wild, in *2016 IEEE Winter Conference on Applications of Computer Vision (WACV)*, Mar 2016 (IEEE, 2016), pp. 1–9. https://doi.org/10.1109/WACV.2016.7477558
14. S. Moschoglou, A. Papaioannou, C. Sagonas, J. Deng, I. Kotsia, S. Zafeiriou, AgeDB: the first manually collected, in-the-wild age database, in *Proceedings of the IEEE Conference on Computer Vision and Pattern Recognition Workshops* (2017), pp. 51–59. https://doi.org/10.1109/CVPRW.2017.250

A Review Study of Smart Vehicle Seat Sensor for Real-Time Postural Analysis

Praneeth Kumar Reddy Dendi, Yagna Gurjala, Sylvia Bhattacharya, and Jason S. Metcalfe

Abstract The era of autonomous driving is gradually transitioning from humans as drivers to humans as passengers. With the advancement of artificial intelligence (AI), humans are switching roles from driver to passenger. Most of the early researches are driver central. But now, it is essential to also study passenger data to understand human states in an autonomous vehicle as a passenger. Smart sensor monitoring systems can play a major role in assessing human and AI dynamics in an autonomous vehicle. Several research studies have utilized electroencephalography, electrocardiography, or electromyography technologies to study human states. But there is a huge gap in the practical applicability of such a sensor in everyday life. Hence, this paper conducts an extensive review of smart seat sensors for easy implementation in a car.

Keywords Driver monitoring system · Smart cars · Sensors · Algorithm · Machine learning model

1 Introduction

According to National Highway Traffic Safety Administration (NHTSA), in 2018, about two thousand eight hundred people were killed and four hundred thousand people were injured due to the driver being distracted in the U.S.A. Since then, a lot of

P. K. R. Dendi (✉) · Y. Gurjala
Department of Computer Science, Kennesaw State University, Marietta, GA, USA
e-mail: pdendi@students.kennesaw.edu

Y. Gurjala
e-mail: ygurjala@students.kennesaw.edu

S. Bhattacharya
Electrical Engineering Technology, Kennesaw State University, Marietta, GA, USA
e-mail: sbhatta6@kennesaw.edu

J. S. Metcalfe
US Devcom Army Research Laboratory, Human Research and Engineering Directorate, Aberdeen Proving Ground, MD, USA
e-mail: jason.s.metcalfe2.civ@mail.mil

A. K. Bashir et al. (eds.), *Proceedings of International Conference on Computing and Communication Networks*, Lecture Notes in Networks and Systems 394,
https://doi.org/10.1007/978-981-19-0604-6_21

research has been conducted in minimizing accidents. The driver monitoring system is an essential intelligent tool that provides information about the driver and driving state. It is traditionally used to monitor the alertness of the driver. Driver drowsiness and distracted driving have been important factors that cause road accidents and lead to severe physical injuries, deaths, and significant economic losses [1]. Besides this, features such as navigation also became an enabler for lack of focus in drivers. Key sources of distraction are using mobile phones, handling kids, pets, applying makeup, and so on. The usage of mobile phones to talk or text while driving is one of the most common reasons for distracted driving and leads to accidents. Apart from the challenges mentioned above, drivers are working for a long time than what is recommended. In such cases, driver fatigue and drowsiness is the primary cause for accidents as they typically reduce the decision-making capability and the perception level of the driver which in turn affects the ability of the driver to control the vehicle. Driver fatigue can cause traffic accidents by degrading the driver's alertness and performance.

Drinking alcohol and speeding also top the list of driver behaviors leading to car crashes. Driving when intoxicated is extremely dangerous and should be completely avoided. Though it might not be the number one cause of car accidents, it is the deadliest. Drivers may feel tempted to increase their speed, but it is good to remain within the posted speed limits. Excessive lane changes, speeding over the limit, and aggressive behaviors are dangerous. Another factor would be trusting, the emotions and comfort of a passenger changes when there is trust in the driver. In some situations, the passenger would sit on the edge of the seat although the speed of the vehicle is limited. As there are many issues involved, systems to monitor driver behavior have become a priority for vehicles.

There are various approaches for driver and driving behavior monitoring [2–5]. Advanced features are used in automobile devices such as the driver monitoring system [6, 7], also known as driver attention monitor, a vehicle safety system to alert and warn the driver if needed and eventually apply the brakes. The driver monitoring system includes a CCD camera placed on the steering column which helps to track the face, via infrared LED detectors. Advanced onboard software collects data points from the driver and creates an initial baseline of the driver's normal state. If there is a situation where the driver is not paying attention to the road and a dangerous situation is detected, the system will warn the driver with warning sounds, flashing lights. If the driver does not take any action, then the vehicle will apply the brakes. The software can also determine whether the driver is blinking more if the eyes are narrowing or closing. Several steps have been taken to ensure that future vehicles adopt driver monitoring features like mandatory application of drowsiness and attention, detection, and mandatory application of drowsiness (including distraction) recognition. Driver monitoring systems helps in enhancing both driver and passenger safety.

An area that has seen a persistent and long-term flurry of research and development is that of autonomous vehicle technologies. In both military and civilian domains, there is a clear intention to replace human drivers with advanced automated systems. This is, of course, causing a shift toward humans experiencing vehicles from the primary perspective of being a passenger. While decades of research have provided huge amounts of data regarding vehicle drivers, very little is known about how people experience being a passenger in a vehicle that they do not control as well as how their changing states are manifest in both anticipatory and reactive responses at all levels from physiological to behavioral. Whether driven by another human or automation, passengers have regular opportunities to make decisions regarding ongoing driving performance, and in many cases, these decisions lead to both beneficial and harmful attempts at interventions (e.g., a verbal warning to a human driver or a take over from automated cruise control). Such passenger behaviors may critically impact vehicle behavior and, more importantly, when and how they occur is likely to be subject to a variety of factors such as individual decision-making style, driving preferences, level of fatigue, risk tolerance, and personality. In particular, human decisions regarding cooperative behaviors such as reliance and compliance are believed to be significantly impacted by the amount of trust held in the paired agent, whether human or automated. Moreover, endeavors to understand the individual factors that lead to passenger states like trust and stress will likely inform the design of intelligent vehicle systems that are capable of responding to, if not also anticipating, individual passenger needs and preferences. By enabling vehicle understanding of passenger states and intentions, we may begin to develop technologies that support and address human autonomy challenges and solutions that undergird the capabilities for achieving transformational overmatch in the evolving future of complex multi-domain operations.

The current generation is becoming more flexible and is willing to adopt different technologies like wearable sensors [8, 9]. Biomedical signals such as electroencephalography (EEG), electrooculography (EOG), electrocardiography (ECG), electromyography (EMG), and eye tracking also provide information regarding driver state. EEG records the electrical activity of the human brain and is considered an effective technique in detecting driving fatigue or distraction [10, 11]. On the other hand, ECG is used to check the heart's rhythm and electrical activity, EMG helps in measuring the electrical activity produced by skeletal muscles. Sensors attached to the skin are used to detect the electrical signals produced by the heart. Research has already investigated using ECG and EMG to detect driver's drowsiness [12–14]. As the EOG signal helps in investigating the abnormalities of the outermost layer of the retina, it is found to be a promising drowsiness detector [15]. Eye tracking is also one of the important techniques in collecting evidence about the driver [16].

2 Sensor Types in Car

2.1 Importance of Seat Sensors and Their Types

Although biomedical signals provide a lot of information about driver monitoring, and seat sensors also turned out to be one of the best ways of providing a piece of big information when combined with biomedical signals. Exploiting the position change caused by drowsiness or fatigue during driving can be predicted by seat sensors [17, 18]. Especially, in an autonomous world, analyzing human sitting posture can provide information about autonomous driving styles and can predict human preference and comfort level. Several studies were focusing on the analysis of real-time seating posture in a car. Some studies are focused based on video analysis for posture diagnosis while some work focuses on the body part deformation. On the other hand, some studies are concentrated especially on car drivers. But all these studies turned out to be not suitable for daily use because these studies necessarily involve some equipment or the help of a clinical specialist. As a solution, simple sensors embedded on a car seat turned out to be a more efficient way of predicting human behavior. A sensor can be defined as an appliance that detects changes in physical or electrical or other quantities and by this means, generally, produces an electrical or optical signal output as an acknowledgment of the change in that specific quantity. Many wearable or unwearable sensors have been developed to classify human postures. Few wearable sensors are light in weight for humans which helps in the measurement of various activities, but wearable sensors are still difficult to be set on the human body or make humans feel uncomfortable and safety measures also must be taken so that body does not get harmed. In recent generations, most vehicles will have seat position sensors provided. There are different types of sensors. It is primarily classified into analog and digital sensors. Sensors that produce continuous analog output signals are called analog sensors. There are various types of analog sensors like accelerometers, pressure sensors, light sensors, sound sensors, temperature sensors, and so on. The sensors that are mostly used in the postural analysis research are pressure sensors, accelerometers, gyroscopes, force sensors, proximity sensors, textile sensors, and so on.

Analog sensors that are used to detect changes in position, velocity, orientation, shock, vibration, and tilt by sensing motion are called accelerometers. Some models automatically detect proper/improper sitting postures by making use of accelerometer readings from some human spinal points and a Web camera which established relationships of human body frames and proper sitting posture [19, 20]. On the other hand, pressure sensors are those that are used to measure the amount of pressure applied to a sensor. These are used for many automotive, medical, industrial, consumer, and building devices, which depend on accurate and stable pressure measurements to operate reliably. As more industries rely on pressure sensors to monitor and control their applications, demand for these technologies has greatly increased (Fig. 1).

Pressure sensors are always a good way to analyze the human seating postures [21] which in turn helps to classify the good and bad postures [22]. Initially, when the drive starts, the driver adopts a good driving position which can be considered as the reference position. By analyzing the pressure distribution change of the driver's body, it is possible to detect fatigue and drowsiness state which in turn indicates the change in the driver's sitting position [23, 24]. There are many studies where the pressure sensor seat is used; a pressure sensor seat on a chair to identify the sitting postures [25, 26]. Besides determining the change in seating posture, seat sensors also

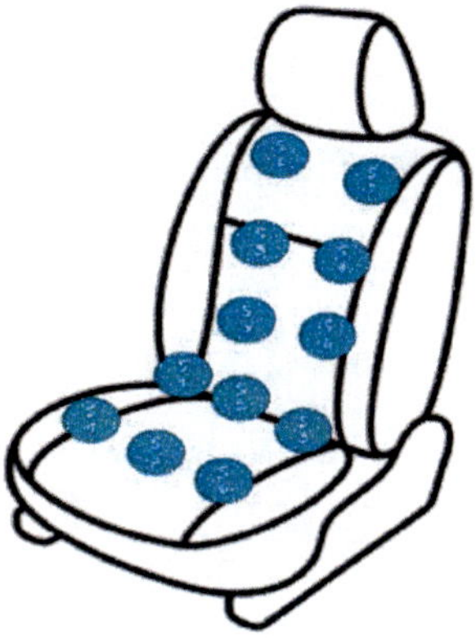

Fig. 1 Car seat sensor

recognize driving style (aggressive acceleration, hard braking) from the passenger posture. Some car seats do not provide enough comfort for drivers/passengers which in turn puts additional strain on the spine. Some other cars force driver bodies into unhealthy positions to operate the vehicle. Though it might not always be the car's fault, it might be simply how we sit. Bad driving postures could result in chronic pain [27].

There are other approaches to find the posture of the passengers, concerning the back seat angle. In this approach, the sitting posture is found as follows when a person moves left vision sensor finds how inclined the person is to the back seat, in the equivalent way when a person moves forward, the sensors find the angle between the person and back seat sensor. But these angles are not efficient when compared with pressure sensors on the seat. Monitoring the driver's posture provides the information for evaluating the driver's attention, operational intention, seating posture and position, and so on.

Advantages: When a person sits in a static position the sensor's pressure value has a high impact in recognizing the posture of the person. Even a camera can be used for posture analysis, but seat sensors are efficient and accurate to capture the sitting postures.

Disadvantages: When a person is in the dynamic state, i.e., in a driving/moving state, the road conditions can also impact the pressure value to increase and decrease. If the number of postures increases, the accuracy of posture identification by sensor decreases. As the passengers have their seat belts when they are on ride the change in pressure value from one posture to another is not too high. Here, the seat belt restricts the movement of the passenger.

Safety is the topmost priority for the U.S. Department of Transportation. Loss of life is unacceptable on roadways. Though 2019 Fatality Data Shows Continued Annual Decline in Traffic Deaths compared to 2018, 2020 Fatality Data Shows Increased traffic fatalities. While Americans drove less in 2020 due to pandemics, NHTSA estimates show that an estimated 38,860 people died in motor vehicle traffic

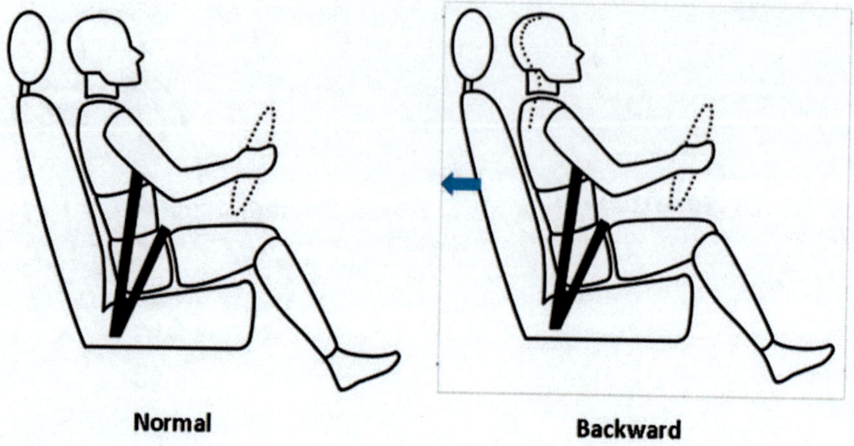

Fig. 2 Backward posture

crashes. The analysis shows that the main behaviors that drove this increase included impaired driving, speeding, and failure to wear a seat belt. There are many reasons why a car accident may occur. It might be due to driver negligence, bad weather, poor road conditions, and also the negligence of third parties. According to National Highway Traffic Safety Administration (NHTSA), it is found that somewhere between 94 and 96% of motor vehicle accidents are caused by some type of human error. NHTSA identifies that hard braking, excessive speeding, sudden lane change, distracted driving, and aggressive/reckless driving are several different events why car accidents occur.

Excessive Speeding: Speeding endangers the lives of everyone on the road. When a driver is traveling too fast, it becomes difficult to slow down and react to conditions on the road. Severe injuries or maybe death results due to collisions at higher speeds. Seat sensors help in analyzing if a vehicle is speeding. For instance, considering the driver/passenger's normal posture, an increase in speed makes the person apply pressure on the seat-back sensors as shown in Fig. 2.

Hard Braking: Hard braking is an event when more force than normal is applied to the vehicle's brake and increases the risk of accident or injury to the driver and passengers and others sharing the road. In such cases, the driver/passenger posture moves forward from normal. Accelerometer helps in the detection of harsh braking which is now included in most GPS devices. Whenever hard barking occurs, the posture of the driver/passenger moves forward shown in Fig. 3.

Lane Change: 9% of all accidents in the United States of America are caused due to reckless lane changes. Crashes that occur in the left lane are more serious than those that occur in the right. Left-lane crashes often result in severe injuries and fatalities. Seat sensors help in detecting the person's posture changing to left and right as shown in Fig. 4.

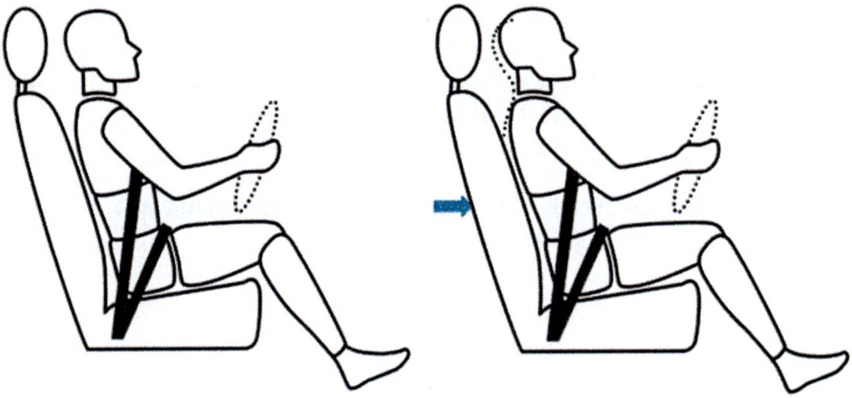

Fig. 3 Forward posture

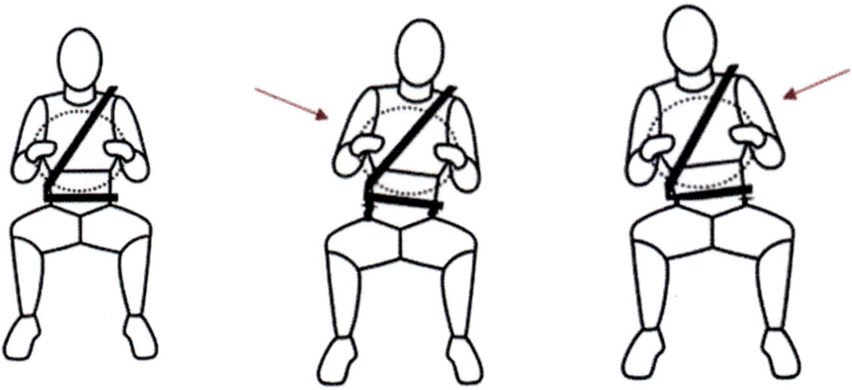

Fig. 4 Lane change postures

3 Physics of Real-Time Postural Changes in Car

Objects are in motion all around us. Everything involves motion. Even when we are resting, our heart moves blood through our veins. Likewise, inertia is something that keeps a vehicle moving until some force slows it down or stops it. This force can be own action of applying a brake, the conditions on the road, an object on the road like a tree, or if the driver is not paying enough attention, even another vehicle. Imagine a car is traveling at 65 mph before you stop suddenly. Once the car has stopped, the driver and passengers will continue moving at 65 mph because of inertia. For instance, if the pressure value of the sensors on the left side of the seat increases, it indicates the person or an object is applying more force on those sensors. Here, the driver is moved more to the left side. If the value of the pressure in some areas in the seat becomes null, it indicates that there is a gap between the driver's body and the seat, which means there is no physical contact.

For instance, if P is the probability of the change in a person's posture, A being the pressure values of the left seat sensors and B being the pressure values of the right seat sensors, if the probability of A is greater than 75%, then it means that person is moving toward left. On the other hand, if the probability of B is greater than 75%, then it means that person is moving toward the right side.

$$P(A) \geq 0.75 = \text{Person's Posture Moving left} \tag{1}$$

$$P(B) \geq 0.75 = \text{Person's Posture Moving Right} \tag{2}$$

When it comes to driving, when the vehicle takes the right to turn, the person's force is applied on the left-side seat pressure sensors due to inertia. In vice versa, when the vehicle takes a left turn, the person's force is applied on the right-side seat pressure sensors. Not only it helps in classifying the left, right sitting postures but also normal, forward, backward postures, and so on [28].

4 Methods and Techniques to Implement Seat Sensors Machine Learning Models

Knowledge of the posture of a seated person is useful in many ways. Postures can be predicted using machine learning algorithms [29]. As machine learning comprises regression, classification, and deep learning models, to get accurate results, it is important to choose the right machine learning model. Various machine learning techniques such as support vector machines, multinomial regression, boosting, neural networks, and random forest are applied to automatically identify the user's position which will be discussed in this paper [30]. There are several other effective techniques like sensor calibration, data representation, and dynamic time warping-based classification proposed to improve the recognition rate of sitting postures [31] (Fig. 5).

The commonly used machine learning algorithms in this area are as follows:

4.1 Support Vector Machine (SVM)

Support vector machines (SVMs) [32] are the most widely used and one of the highest-performing classifiers. It is a classifier and regression algorithm and makes predictions based upon the given data. SVM is one of the best algorithms for multi-

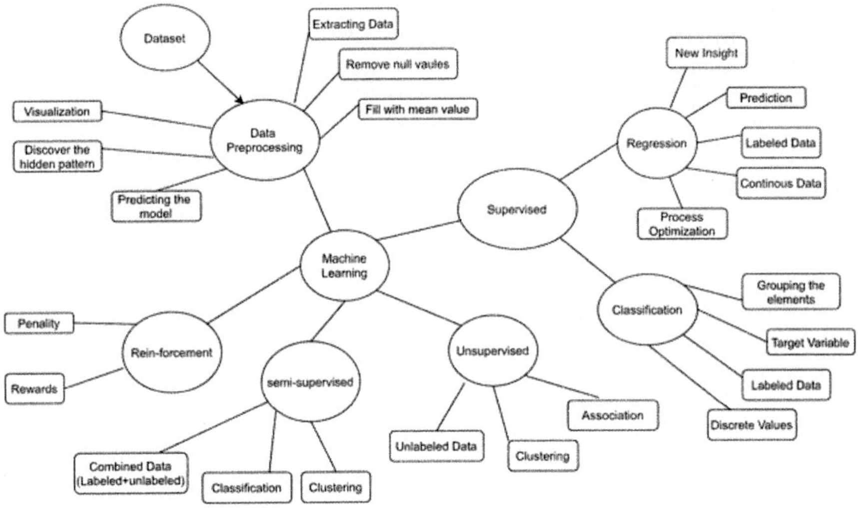

Fig. 5 Machine learning classification

class classifiers since more than two postures need to be classified. SVM's with a linear, Gaussian radial basis function and so on were used to classify different sitting postures. Many of the studies showed that the SVM has the best accuracy compared to other algorithms like Gaussian naive Bayes and AdaBoost [33]. SVM has many hyperparameters tuning which can be used for optimizing the accuracy.

4.2 Multinominal Regression

Multinomial logistic regression (MLR) was introduced by McFalden. Multinomial regression is an extension of logistic regression that adds help in multi-class classification problems and uses maximum likelihood estimation to evaluate the probability of categorical membership. MLR is also a decent algorithm that helps in classifying the sitting postures but not with the best accuracy compared to other algorithms.

4.3 Boosting

Though machine learning is undoubtedly one of the powerful techniques in AI, there are times when ML models are weak learners. Boosting is a technique that takes several weak models and combines them into a stronger one. Three different boosting algorithms are AdaBoost, gradient boosting, and XG Boost.

4.4 Neural Networks

As neural networks mimic the operations of a human brain to recognize the relationships between vast amounts of data, they are used in a variety of applications. Classification accuracy through the convolutional neural networks algorithm is superior to conventional machine learning techniques such as ANN and DNN.

4.5 Random Forest

Random forest is also a Machine Learning technique that is used to solve regression and classification problems. It consists of many decision trees. Random forests merge multiple decision trees which produce more accurate and stable predictions. Random forest always proved to be the best way in classifying the sitting postures even when compared with SVM and other algorithms.

5 Application of Seat Sensors

Sitting posture analysis is widely applied in many day-to-day applications such as biomedical, education, and health care domains.

Algorithm name	Accuracy (%)	No. of sensors	No. of positions
Support Vector Machine	97	Seat (4)	7
SVM	93.90	Seat (8 * 8)	9
Random Forest	90.90	2-armrests	7
		4-backrest	
		10-seat pan	
AdaBoost	87	10-seat pan	15
Naive Bayes	82	Seat (6 * 8)	9
		Backrest (2 * 8)	

5.1 Find the Seat Occupancy

There exists much research to find seat occupancy using seat sensor data. Seat occupancy helps in airbag sensing, driver departure, seat belt alarm sensors, automatic taxi fares, and so on. The seat sensors are also used to find whether the seat is occupied or not. If there is no person in the seat, the airbag will not open. Some area which it applicable are driver seat and passenger seat. When the pressure values of the seat are more than the threshold, it is classified as seat occupied else the seat is empty.

5.2 *Finding the Human Body Posture*

Body posture helps in determining how the person is sitting on the chair. Like the person is resting to the back seat or not and other postures are leaning to left, leaning to the right, leaning forward, cross the legs, left leg crosses, right leg crossed. All these postures can be identified by seat pressure sensors and the pressure numeric value of those sensors [34].

5.3 *Finding the Drivers Active State Using Seat Sensors*

As we know if any person drives for a long time, the person may feel drowsy and stressed. So, the person must take a break to prevent the accident. Here, the person state can be defined by how long the seat sensors are in an active state (pressure on seat sensors).

6 Research Gap and Future Scope

Using seat sensor data to determine the driving state in a real-time car is still an emerging topic with research using postural information. The major research gap is that several studies have been conducted reporting on driver drowsiness and fatigue [35, 36]; however, there are limited studies on seat sensor's response during other miscellaneous states such as drowsiness or fatigue.

For postural variation as an indicator for drowsiness while driving, the research must keep an eye on each participant as it varies from person to person. The primary extracted feature must be normal posture which is considered as base posture. Considering normal posture, other postures can be predicted easily. Once the features have been obtained, a neural network should be trained with all the recorded data. The time required to train the neural network is dependent on the quantity of the dataset. Even if the dataset is large, it may be taking hours or days to train but will be extremely accurate. If the researchers plan to develop a real-time system, then random forest or SVM should be used as the primary classifier with the neural network as a baseline for accuracy.

7 Conclusion

This paper is not exhaustive but gives a contemporary view of the methods and technologies used to determine the human state. The literature review clearly states that seat sensors are more effective and easy to use in studying the body posture of human subjects in a real-world car compared to other sensors. Seat sensor information can be easily obtained and is non-intrusive to the driver or passenger when compared with other biomedical signals. Postural movement in a car can provide precise valuable information about driving state, passenger's comfort level, external environment, etc. The features from the postural changes are used to build different prediction and classification models. The overall system is capable of supporting healthy AI and human passenger dynamics. The information can be fed to the AI driver to continuously improve its driving based on passenger preferences.

References

1. M.Q. Khan, S. Lee, A comprehensive survey of driving monitoring and assistance systems. Sensors **19**(11), 2574 (2019)
2. H.-B. Kang, Various approaches for driver and driving behavior monitoring: a review, in *Proceedings of the IEEE International Conference on Computer Vision Workshops* (2013), pp. 616–623
3. S. Kanarachos, S.-R.G. Christopoulos, A. Chroneos, Smartphones as an integrated platform for monitoring driver behaviour: the role of sensor fusion and connectivity. Transp. Res. Part C Emerg. Technol. **95**, 867–882 (2018)
4. Q. Wang, J. Yang, M. Ren, Y. Zheng, Driver fatigue detection: a survey, in *6th World Congress on Intelligent Control and Automation*, vol. 2 (IEEE, 2006), pp. 8587–8591
5. M. Stork, J. Skala, P. Weissar, R. Holota, Z. Kubik, Various approaches to driver fatigue detection: a review, in *2015 International Conference on Applied Electronics (AE)* (2015), pp. 239–244
6. Y. Dong, Z. Hu, K. Uchimura, N. Murayama, Driver inattention monitoring system for intelligent vehicles: a review. IEEE Trans. Intell. Transp. Syst. **12**(2), 596–614 (2011)
7. T. Brandt, R. Stemmer, A. Rakotonirainy, Affordable visual driver monitoring system for fatigue and monotony, in *2004 IEEE International Conference on Systems, Man and Cybernetics (IEEE Cat. No.04CH37583)*, vol. 7 (2004), pp. 6451–6456
8. M.M. Rodgers, V.M. Pai, R.S. Conroy, Recent advances in wearable sensors for health monitoring. IEEE Sens. J. **15**(6), 3119–3126 (2015)
9. N. Jalloul, Wearable sensors for the monitoring of movement disorders. Biomed. J. **41**(4), 249–253 (2018)
10. Y. Ma, B. Chen, R. Li, C. Wang, J. Wang, Q. She, Z. Luo, Y. Zhang, Driving fatigue detection from EEG using a modified PCANet method. Comput. Intell. Neurosci. **2019** (2019)
11. S. Barua, M.U. Ahmed, S. Begum, Classifying drivers' cognitive load using EEG signals, in *pHealth* (2017), pp. 99–106
12. M.A. Serhani, H.T. El Kassabi, H. Ismail, A. Nujum Navaz, ECG monitoring systems: review, architecture, processes, and key challenges. Sensors **20**(6), 1796 (2020)
13. A. Tjolleng, K. Jung, W. Hong, W. Lee, B. Lee, H. You, J. Son, S. Park, Classification of a driver's cognitive workload levels using artificial neural network on ECG signals. Appl. Ergon. **59**, 326–332 (2017)
14. R. Fu, H. Wang, Detection of driving fatigue by using noncontact EMG and ECG signals measurement system. Int. J. Neural Syst. **24**(03), 1450006 (2014)
15. T.C. Chieh, M.M. Mustafa, A. Hussain, S.F. Hendi, B.Y. Majlis, Development of vehicle driver drowsiness detection system using electrooculogram (EOG), in *2005 1st International Conference on Computers, Communications, & Signal Processing with Special Track on Biomedical Engineering* (IEEE, 2005), pp. 165–168
16. B.-G. Lee, W.-Y. Chung, Driver alertness monitoring using fusion of facial features and biosignals. IEEE Sens. J. **12**(7), 2416–2422 (2012)

17. S. Lee, M. Kim, H. Jung, D. Kwon, S. Choi, H. You, Effects of a motion seat system on driver's passive task-related fatigue: an on-road driving study. Sensors **20**(9), 2688 (2020)
18. A. Sahayadhas, K. Sundaraj, M. Murugappan, Detecting driver drowsiness based on sensors: a review. Sensors **12**(12), 16937–16953 (2012)
19. W.Y. Wong, M.S. Wong, Detecting spinal posture change in sitting positions with tri-axial accelerometers. Gait Posture **27**(1), 168–171 (2008)
20. J. Ma, H. Kharboutly, A. Benali, F. Benamar, M. Bouzit, Joint angle estimation with accelerometers for dynamic postural analysis. J. Biomech. **48**(13), 3616–3624 (2015)
21. M. Zhao, G. Beurier, H. Wang, X. Wang, Exploration of driver posture monitoring using pressure sensors with lower resolution. Sensors **21**(10), 3346 (2021)
22. J. Roh, H.-J. Park, K.J. Lee, J. Hyeong, S. Kim, B. Lee, Sitting posture monitoring system based on a low-cost load cell using machine learning. Sensors **18**(1), 208 (2018)
23. M. Ding, T. Suzuki, T. Ogasawara, Estimation of driver's posture using pressure distribution sensors in driving simulator and on-road experiment, in *2017 IEEE International Conference on Cyborg and Bionic Systems (CBS)* (2017), pp. 215–220
24. I. Teyeb, O. Jemai, M. Zaied, C.B. Amar, Towards a smart car seat design for drowsiness detection based on pressure distribution of the driver's body, in *ICSEA 2016* (2016), p. 230
25. D. Bibbo, M. Carli, S. Conforto, F. Battisti, A sitting posture monitoring instrument to assess different levels of cognitive engagement. Sensors **19**(3), 455 (2019)
26. L. Xu, G. Chen, J. Wang, R. Shen, S. Zhao, A sensing cushion using simple pressure distribution sensors, in *2012 IEEE International Conference on Multisensor Fusion and Integration for Intelligent Systems (MFI)* (2012), pp. 451–456
27. G. Liang, J. Cao, X. Liu, Smart cushion: a practical system for fine-grained sitting posture recognition, in *2017 IEEE International Conference on Pervasive Computing and Communications Workshops (PerCom Workshops)* (2017), pp. 419–424
28. K. Kamiya, M. Kudo, H. Nonaka, J. Toyama, Sitting posture analysis by pressure sensors, in *2008 19th International Conference on Pattern Recognition* (2008), pp. 1–4
29. Y.M. Kim, Y. Son, W. Kim, B. Jin, M.H. Yun, Classification of children's sitting postures using machine learning algorithms. Appl. Sci. **8**(8), 1280 (2018)
30. R. Zemp, M. Tanadini, S. Plüss, K. Schnüriger, N.B. Singh, W.R. Taylor, S. Lorenzetti, Application of machine learning approaches for classifying sitting posture based on force and acceleration sensors. BioMed Res. Int. **2016** (2016)
31. W. Xu, M.-C. Huang, N. Amini, L. He, M. Sarrafzadeh, ecushion: a textile pressure sensor array design and calibration for sitting posture analysis. IEEE Sens. J. **13**(10), 3926–3934 (2013)
32. C. Cortes, V. Vapnik, Support-vector networks. Mach. Learn. **20**(3), 273–297 (1995)
33. C. Ma, W. Li, R. Gravina, G. Fortino, Posture detection based on smart cushion for wheelchair users. Sensors **17**(4), 719 (2017)
34. S. Matuska, M. Paralic, R. Hudec, A smart system for sitting posture detection based on force sensors and mobile application. Mob. Inf. Syst. **2020** (2020)
35. B. Alshaqaqi, A.S. Baquhaizel, M.E.A. Ouis, M. Boumehed, A. Ouamri, M. Keche, Driver drowsiness detection system, in *2013 8th International Workshop on Systems, Signal Processing and Their Applications (WoSSPA)* (IEEE, 2013), pp. 151–155
36. V. Saini, R. Saini, Driver drowsiness detection system and techniques: a review. Int. J. Comput. Sci. Inf. Technol. **5**(3), 4245–4249 (2014)

A Proxy Re-signcryption Scheme with Delegation Property

Abdulrahman Alamer

Abstract Recently, proxy re-encryption (PRE) cryptosystem has been increasingly suggested in many fields, including the cloud and fog computing paradigm. In the PRE paradigm, in which the only confidentiality of security features of the transmitted message is guaranteed, the proxy entity can re-encrypt the message without exposing the corresponding plaintext. However, other security features such as integrity and authenticity are not guaranteed with using PRE cryptosystem. Thus, in a true assumption, the proxy entity is indicated as a semi-trusted entity, mostly in fog computing where fog nodes, such as the road side unit (RSU), are considered untrusted. The malicious RSU fog node can easily modify the integrity of the encrypted message. This paper proposes a proxy re-signcryption scheme with delegation property (PRSCD), in which the security features of data integrity, confidentiality, and authenticity are completely guaranteed during the transmission process. Thus, the PRSCD is proved in CCA on security models. An analysis of security illustrates that the PRSCD is able to prevent a malicious RSU fog node from performing a number of security challenges.

Keywords Proxy re-encryption · Confidentiality · Integrity · Authentication · Signcryption · Proxy re-signcryption

1 Introduction

Proxy re-encryption (PRE) [1] has been designed as a special form of public-key encryption, which allows a semi-trusted entity to perform a re-encryption (RE) process on a transformation ciphertext message utilizing a permit re-encryption key (REK) without knowning the underlying original content of the message while re-encrypting a ciphertext. Therefore, many PRE schemes have been introduced against chosen-ciphertext attacks (CCA) and replayable chosen-ciphertext attacks (RCCA) [2–4].

A. Alamer (✉)
Department of Information Technology and Security, Jazan University, Jazan, Saudi Arabia
e-mail: amalameer@jazanu.edu.sa

A. K. Bashir et al. (eds.), *Proceedings of International Conference on Computing and Communication Networks*, Lecture Notes in Networks and Systems 394,
https://doi.org/10.1007/978-981-19-0604-6_22

With these useful properties, PRE has been suggested for use in protecting many Internet of Things (IoT) application scenarios in cloud or fog computing network [5, 6]. However, security and privacy challenges still prevent the PRE from being fully used on these IoT application scenarios [7, 8]. As an example, because of the less verifiable property of the PRE paradigm, a malicious proxy entity can easily play with the integrity of a transmitted ciphertext [9]. Thus, following the RE phase, the proxy entity might forward a forged ciphertext message to the recipient [10]. With no guarantee of protecting the ciphertext integrity, the PRE technique may not be recommended for various IoT application scenarios [6].

Although significant works have been conducted using a digital signature scheme, together with a separate a public-key encryption scheme in order to guarantee ciphertext confidentiality and integrity [11–13], these works offer neither sufficient nor practical suggestions for IoT application scenarios since they are built on the assumption of a sign-then-encrypt scheme [14]. This is because the cost involved in this approach is actually the combined cost involved in having to sign the message then encrypt and re-encrypt it [15].

Motivated by this potential issue, this paper presents a design for a proxy re-signcryption scheme with delegation property (PRSCD), which is a new promising paradigm in public-key cryptography. This is considered as an enhancement of the PRE technique in order to simultaneously guarantee confidentiality, integrity, and authentication for each transmitted ciphertext message in the IoT network. In the PRSCD technique, the proxy entity can concurrently perform the re-encryption and re-signature in one process without the requirement to access the corresponding plaintexts. The receiver then has the ability to authenticate the sender and proxy entity by verifying the received ciphertext message. The received ciphertext message can be considered as a valid message only if the receiver validly authenticates both the sender and the proxy on the ciphertext message. Therefore, PRSCD can be effectively utilized as a security scheme for protecting many types of IoT applications. A review of current research suggests that no other proxy re-signcryption schemes dealing with delegation property have yet been designed. This work concentrates on the design of the first such PRSCD method with proofs of a CCA-security in the proposed security model.

1.1 Related Work

Proxy re-encryption (PRE) has been designed to prohibit a semi-trusted proxy from knowing the underlying message content while performing RE process [16]. It is categorized into two forms: a bidirectional PRE (B-PRE) method and a unidirectional PRE (U-PRE) method. In the B-PRE, the RE process is implemented in both directions using the same REK [17]. In the U-PRE, RE process is implemented in only one direction using one REK [1].

Therefore, authors in [18] introduced the first U-PRE method to supply public verifiability. In addition, Weng et al. [19] presented a new U-PRE method without random oracles. Paul et al. [20] proposed the first PRE method in CCA-secure without pairings under the problem of computational Diffie–Hellman. Yao et al. [21] presented an identity-based conditional PRE scheme with CCA-secure for secure cloud data sharing. Wang et al. [22] introduced a PRE with additional benefits of a noninteractivity and collusion resistance and RCCA-secure. Authors in [23] introduced the PRE method with a delegatable verifiability feature (PREDV) that is defend against RCCA to achieve delegatable verifiability. Many PRE methods have been studied for addressing many of security issues that arise from different application backgrounds in real-life practice [24, 25]. According to B-PRE and U-PRE concepts in CCA-secure and RCCA-secure, many works have been proposed for supporting different network security [26–28].

However, it seems that these works still need significant improvement regarding the computational costs and communication overhead used by them to achieve security requirements including integrity, confidentiality, and authentication for every transmitted ciphertext message in the IoT network. Thus, these works are not practical for use in the IoT paradigm.

1.2 Contributions

Motivated by the above-mentioned issues, this paper proposes a proxy re-signcryption scheme that supports a delegation feature (PRSCD) with proofs of the CCA-secure in the standard model. Due to the fact, there is no work yet is proposed a proxy re-signcryption method with a delegation feature designed up to this end, this paper selects the most dynamic PRE methods in [18, 19] as benchmarks. The contributions of this work are listed as follows.

- An efficient proxy re-signcryption method that supports a delegation feature (PRSCD) is proposed with CCA-security proofs in the proposed security models.
- The proposed PRSCD scheme is used to enhance the PRE security features. The PRSCD can achieve integrity, confidentiality, and authentication for first and second level ciphertexts while RE process.
- Security analysis is performed to display the ability of the PRSCD in achieving the security features of confidentiality, integrity, and authentication that can prevent any potential threats.
- Evaluation is performed to display the PRSCD efficiency compared with PRE methods in [18, 19].

1.3 Paper Organization

The rest of the paper is ordered as the following. In Sect. 2, the preliminaries and security models of PRE are presented. Section 3 first presents the security models of

PRSCD, and then the PRSCD in detail is presented followed with security proofs. The performance evaluation is presented in Sect. 4 and conclusions in Sect. 5.

2 Preliminaries

2.1 Bilinear Pairings

Let $\mathcal{G}$ and $\mathcal{G}_T$ be two cyclic groups with the same prime order p. $\hat{e}$ is an admissible bilinear pairing $\hat{e} : \mathcal{G} \times \mathcal{G} \rightarrow \mathcal{G}_T$ if the following functions are met:

- Bilinearity: For all $a_1, a_2 \in Z_p^*$ and $\mathcal{Q}, \mathcal{V}, \mathcal{R} \in \mathcal{G}$, where
  - $\hat{e}(\mathcal{Q} + \mathcal{V}, \mathcal{R}) = \hat{e}(\mathcal{Q}, \mathcal{R})\hat{e}(\mathcal{V}, \mathcal{R})$.
  - $\hat{e}(\mathcal{Q}, \mathcal{R} + \mathcal{V}) = \hat{e}(\mathcal{Q}, \mathcal{R})\hat{e}(\mathcal{Q}, \mathcal{V})$.
  - $\hat{e}(\mathcal{Q}^{a_1}, \mathcal{V}^{a_2}) = \hat{e}(\mathcal{Q}, \mathcal{V})^{a_1 a_2} = \hat{e}(\mathcal{Q}^{a_2}, \mathcal{V}^{a_1})$.
- Non-Degeneracy: $\hat{e}(\mathcal{Q}, \mathcal{V}) \neq 1_{\mathcal{G}_T}$.
- Computability: $\hat{e}(\mathcal{Q}, \mathcal{V})$ is computed.

2.2 Complexity Assumptions

The complexity problem of the discrete logarithms that are assumed to be related to the PRSCD is listed as follows:

Definition 1 *Computational Diffie–Hellman (CDH) Problem. Let $a_1, a_2 \in Z_q^*$ and g is $\mathcal{G}$ generator. Thus, with given $g, g^{a_1}, g^{a_2} \in \mathcal{G}$, the problem of CDH is how to compute $g, g^{a_1}, g^{a_2} \in \mathcal{G}$.*

Definition 2 *Decisional Bilinear Diffie–Hellman (DBDH) Problem. Let $a_1, a_2, a_3 \in Z_q^*$ and g is $\mathcal{G}$ generator. Thus, with given $g, g^{a_1}, g^{a_2}, g^{a_3} \in \mathcal{G}$, the problem of DBDH is to decide whether $f = \hat{e}(g, g)^{a_1 a_2 a_3}$, where $f \in \mathcal{G}_T$.*

Definition 3 *Collision-resistant (CR) hash function. Given $a_1, a_2 \in Z_p^*$, it is infeasible for an adversary with CR hash $\mathcal{H}$ function to discovery $a_1 \neq a_2$ with $\mathcal{H}(a_1) = \mathcal{H}(a_2)$.*

2.3 Model of Single-Hop U-PRE

Definition 4 *(Single-hop U-PRE). A single-hop U-PRE method consists of the following probabilistic polynomial-time (PPT) algorithms.*

- Initiate $(\gamma) \rightarrow (\mathcal{P})$. Given γ as a system security parameter, the authorized server executes the Initiate algorithm to output the $\mathcal{P}$ as public parameters of the system.
- $K\text{Gen}(\mathcal{P}) \rightarrow (pk_i, sk_i)$. Given the system public parameters $\mathcal{P}$, the authorized server will run the KGen algorithm to output the public and private keys (pk_i, sk_i) for each registered user i.
- $RK\text{Gen}(pk_2, sk_1) \rightarrow (rk_{1\rightarrow 2})$. Given a sender's sk_1 and the delegated user's pk_2, the authorized server will run the RKGen algorithm to generate the REK $rk_{1\rightarrow 2}$.
- $\text{Enc}(pk_1, m) \rightarrow (\alpha)$. Given a plan text $m \in \{0, 1\}^*$ and pk_1, the sender 1 will run the Enc algorithm to generate a first level α ciphertext under pk_1.
- $R\text{Enc}(rk_{1\rightarrow 2}, \alpha) \rightarrow (\alpha')$. Given the α and REK $rk_{1\rightarrow 2}$, the proxy entity runs the REnc algorithm to generate the second level α' ciphertext under the pk_2.
- $\text{Dec}(\alpha', sk_2) \rightarrow (m)$. Given the second level α' ciphertext and a delegated user's sk_2, the receiver user 2 runs Dec algorithm to output the message m.

Correctness. For any message $m \in \{0, 1\}^*$, (pk_1, sk_1), $(pk_2, sk_2) \leftarrow K\text{Gen}(1^\gamma)$, the following conditions must hold:

$$rk_{1\rightarrow 2} \leftarrow (pk_1, RK\text{Gen}(pk_2, sk_1)) \tag{1}$$

$$\alpha' \rightarrow \text{Dec}\left(sk_2, R\text{Enc}\left(rk_{1\rightarrow 2}, \underbrace{\text{Enc}(pk_1, m)}_{\alpha}\right)\right) = m \tag{2}$$

2.4 Security Models for U-PRE

RCC-Security for U-PRE. The U-PRE scheme generates two levels of ciphertexts. The first ciphertext level (1cl) is known as the original ciphertext that is generated from the encrypted message m, and the second ciphertext level (2cl) is generated from the re-encrypted the original 1cl. Hence, the RCC-security for the U-PRE scheme is defined from these two cases.

2.4.1 CCA – 1

The security challenge of the 1cl.

Initiate: The challenger $\mathcal{S}$ server generates the $\mathcal{P}$ from choosing (γ).

Phase 1: The adversary $\mathcal{D}$ generates a set of queries $\varrho = \{q_1, \ldots, q_n\}$. Each $q_i \in \varrho$ is considered as one of the $\Phi = (\mathcal{O}_{pk}, \mathcal{O}_{sk}, \mathcal{O}_{rk}, \mathcal{O}_{re}, \mathcal{O}_{dec})$ oracles, in which they model the ability of an adversary, as follows:

- Public key $\mathcal{O}_{pk}$ generation oracle: Given $(\mathcal{P})$, the $\mathcal{S}$ runs the $KGen(\mathcal{P})$ algorithm to output (pk, sk). The pk will be given to $\mathcal{D}$ and (pk, sk) will be recorded in $\mathbf{T}_{pk}$ table. For any $q_j \in \varrho$ oracle query that involves pk_i, we require that (pk_i, sk_i) can be found in $\boldsymbol{T}_{pk}$, otherwise the oracle will return 0.
- Private key $\mathcal{O}_{sk}$ generation oracle: Given pk_i, $\mathcal{S}$ returns "1" to $\mathcal{D}$ if sk_i with respect to pk_i is existing in $\boldsymbol{T}_{sk}$, otherwise outputs "0".
- REK $\mathcal{O}_{rk}$ generation oracle: Given (pk_i, pk_j), the $\mathcal{S}$ returns $rk_{i \to j} = RKGen(pk_j, sk_i)$ to $\mathcal{D}$ as well as records it in $\boldsymbol{T}_{rk}$ table where $pk_i \neq pk_j$.
- RE $\mathcal{O}_{re}$ generation oracle: Given $(rk_{i \to j}, C_i)$, $\mathcal{S}$ returns $C_i^{'} = REnc(RKGen(sk_i, pk_j), pk_i, C_i)$ to $\mathcal{D}$ as the 1cl and $C_i^{'}$ as the 2cl.
- Decryption $\mathcal{O}_{\text{dec}}$ generation oracle: Given $(pk_i, C_i^{'})$, $\mathcal{S}$ outputs the $m = \text{Dec}(sk_i, C_i)$ to $\mathcal{D}$.

Challenge: The $\mathcal{S}$ randomly selects $\mathcal{B} \in \{0, 1\}^*$ and returns ciphertext $C_i^* = \text{Enc}(m_{\mathcal{B}}, pk_i^*)$. It then forwards C_i^* ciphertext to $\mathcal{D}$ as a challenge.

Phase 2: This is the same as Phase 1 but the $\mathcal{S}$ will reject the following queries:

- For any public key $pk_j = pk_i^*$, $\mathcal{D}$ cannot be allowed to generate queries on $\mathcal{O}_{sk}(pk_i^*)$.
- For any $pk_j = pk_i^*$, $\mathcal{D}$ cannot be allowed to generate queries on $\mathcal{O}_{sk}(pk_j)$ when (pk_i^*, pk_j, C_i^*) has been queried to $\mathcal{O}_{rk}$.
- For any pk_j, $\mathcal{D}$ cannot be allowed to concurrently generate queries on $\mathcal{O}_{sk}(pk_j)$ and $\mathcal{O}_{re}(pk_j, pk_i^*, C_i^*)$.
- For any pk_j, $\mathcal{D}$ cannot be allowed to concurrently generate queries on $\mathcal{O}_{sk}(pk_j)$ and $\mathcal{O}_{re}(pk_j, pk_i^*, C_i^*)$.
- For the first level ciphertext $C_i \leftarrow \mathcal{O}_{re}(C_i^*, pk_j, pk_i^*)$, $\mathcal{D}$ cannot be allowed to generate the decryption queries on $\mathcal{O}_{\text{dec}}(pk_j, C_i) \in \{0, 1\}^*$.

Guess: The $\mathcal{D}$ returns a guess $\mathcal{B}^* \in \{m_0, m_1\}^*$, and it will win the game only if $\mathcal{B}^* = \mathcal{B}$.

Theorem 1 *The U-PRE method is $CCA-1$ secure if all efficient adversaries $\mathcal{D}$ are specified as above. $\mathcal{D}$'s advantage $(\text{Adv}_{\text{PRE}}^{\text{CCA}-1}(1^\gamma))$ is negligible.*

Proof In the U-PRE scheme, the $\mathcal{D}$'s advantage $\text{Adv}_{U-\text{PRE}}^{\text{CCA}-1}(1^\gamma)$ against the CCA $-$ 1 is formulated as $|\text{Pr}[b = b'] - 1/2|$ due to the $\mathcal{D}$'s advantage running in two phases, guess (∂) and query (ϱ), as is formulated in:

$$\text{Adv}_{U-\text{PRE}}^{\text{CCA}-1}(1^\gamma) = \left| Pr\left[\mathcal{B}^* = \mathcal{B} \,\middle|\, \begin{matrix} \varrho; \\ \partial. \end{matrix} \right] - \frac{1}{2} \right| \tag{3}$$

where ϱ and ∂ can be calculated as:

- $\varrho = (pk_i^*, (m_0, m_1), st) \leftarrow \mathcal{D}_{\varrho}^{\Phi}(\mathcal{P})$
- $\partial = \mathcal{B}^* \leftarrow \mathcal{D}_{\partial}^{\Phi}(\mathcal{P}, C^*, st)$

st denotes the $\mathcal{D}$'s internal state information and $\Phi = \left(O_{pk}, O_{sk}, O_{rk}, O_{re}, O_{dec}\right)$. Thus, for any time (t), $\mathcal{D}$, respectively, makes queries $q_i \in \varrho$ to Φ oracles, such that $\mathfrak{D}\left(t \cdot \text{Adv}_{U-\text{PRE}}^{\text{CCA}-1}(1^{\gamma})\right) \leq \varepsilon$. Therefore, no polynomial-time algorithm $\mathfrak{D}$ with an advantage ε and running time t is to solve the $DBDH$ and CDH problems.

2.4.2 CCA-2

The security challenge of the second level ciphertext.

Phase 1: This is a similar case to that in the security challenge of the first level ciphertext (CCA-1).

Challenge: The $\mathcal{S}$ randomly selects $\mathcal{B} \in \{0, 1\}^*$ and computes $C' = R\text{Enc}\left(rk_{i \to j}, \text{Enc}\left(m_{\mathcal{B}}, pk_i^*\right)\right)$ and then forwards C^* to $\mathcal{D}$ as a challenge.

Phase 2: This is the same as Phase 1; however, the $\mathcal{S}$ will return 0 in the following queries:

- If $pk_j = pk_i^*$ and pk_j has been queried to $\mathcal{O}_{sk}$.
- If $\left(pk_j, C_i\right) = \left(pk_i^*, C_i'\right)$ has been queried to $\mathcal{O}_{dec}$.

Guess: Giving a guess $\mathcal{B}^* \in \{m_0, m_1\}^*$, $\mathcal{D}$ will win the game only if $\mathcal{B}^* = \mathcal{B}$.

Theorem 2 *The $U-PRE$ method is a* CCA $-$ 2 *secure if all efficient $\mathcal{D}$ are defined as above. The $\mathcal{D}$'s advantage* $\left(\text{Adv}_{U-\text{PRE}}^{\text{CCA}-2}(1^{\gamma})\right)$ *is negligible.*

Proof The $\mathcal{D}$'s advantage $\text{Adv}_{U-\text{PRE}}^{\text{CCA}-2}(1^{\gamma})$ against the CCA $-$ 2 is defined as $|Pr\left[\mathcal{B}' = \mathcal{B}\right] - 1/2|$ due to the $\mathcal{D}$'s advantage using (ϱ) and (∂), which is similar to that in the security challenge of the first level ciphertext (CCA $-$ 1).

3 The Proposal

The proposed PRSCD scheme addresses the integrity and verification security issues related to the general concept of the U-PRE scheme. Hence, PRSCD can protect 1cl and 2cl ciphertexts from multiple security threats. The PRSCD is explained as follows.

3.1 Model of Single-Hop Unidirectional PRSCD

Definition 5 *(Single-hop U-PRSCD). A single-hop U-PRSCD method is consisted of PPT algorithms, as follows:*

- Initiate$(\gamma) \rightarrow (\mathcal{P})$. This algorithm takes γ as input of security parameter and outputs the $\mathcal{P}$ as the public parameters of the system. This algorithm is executed only by an authorized trust (AT) server.
- $K\mathrm{Gen}(\mathcal{P}) \rightarrow (pk_i, sk_i)$. This algorithm takes the $\mathcal{P}$ as input and outputs the public and private keys (pk_i, sk_i) for each registered user i. This algorithm also runs only by the AT-server.
- $RSK\mathrm{Gen}(sk_i, pk_j) \rightarrow (rsk_{i\rightarrow j})$. This algorithm takes the user j's pk_j and the user i's sk_i, and then outputs the re-signcryption key $rsk_{i\rightarrow j}$. This algorithm is executed by the user i.
- $\mathrm{Signc}(pk_i, sk_i, m_i) \rightarrow (\alpha)$. This algorithm takes the $m_i \in \{0, 1\}^*$ and the pk_i, sk_i as input and generate a signcrypted ciphertext $\alpha_i = \{c_i, \sigma_i\}$ under the public key pk_i, where c_i is the encrypted message with its corresponding signature σ_i. This algorithm is also executed by the user i
- $\mathrm{ReSignc}(rsk, sk_p, \alpha_i) \rightarrow (\alpha_i^{'})$. This algorithm takes the (α_i), the proxy-server's private key (sk_p) and the re-signcryption key $rsk_{i\rightarrow j}$ as input and produces a re-signcrypted ciphertext $\alpha_i^{'} = \{c_i^{'}, \sigma_i^{'}\}$ under the pk_j. This algorithm is only executed by the authorized proxy entity that has authority to hold the re-signcryption key $rsk_{i\rightarrow j}$.
- $Ver(\sigma_i^{'}, pk_i, pk_p) \rightarrow (1, 0)$. This algorithm takes the $\sigma_i^{'}$, sender's pk_i and the proxy-server's pk_p as input and outputs 1 if the $\sigma_i^{'}$ are generated by a valid pk_i, pk_p, otherwise it outputs 0.
- $\mathrm{Dec}(\alpha_i^{'}, sk_j) \rightarrow (m_i^{'})$. This algorithm takes the $\alpha_i^{'}$ ciphertext and the sk_j as input and outputs the message $m_i^{'} = m_i$. Note that the verification $\mathrm{PRSCD_{Ver}}$ and decryption $\mathrm{PRSCD_{Dec}}$ algorithms are both performed by the user j holding the pk_j, sk_j who was delegated by the user i holding pk_i.

Correctness. For any message $\sigma_i^{'} = \{\sigma_i^{'}, c_i^{'}\}$, $(sk_i, pk_i) \leftarrow K\mathrm{Gen}(\gamma)$, and $(sk_j, pk_j) \leftarrow K\mathrm{Gen}(\gamma)$, the following conditions must be held:

$$\mathrm{Ver}(\sigma', \sigma) \rightarrow (c_i, c_i^{'}) = \begin{cases} 1, \text{ if } (pk_i, pk_p) \text{ valid} \\ 0, \text{ otherwise} \end{cases} \tag{4}$$

$$c' \rightarrow \mathrm{Dec}\left[\mathrm{ReSignc}\left(\underbrace{\mathrm{Signc}(m, pk_i)}_{c}, rk_{i\rightarrow j}\right), sk_j\right] = m \tag{5}$$

3.2 The Detailed PRSCD

The proposed PRSCD scheme contains from a set of algorithms:

3.2.1 Initiate (γ)

In this algorithm, the AT-server tacks γ as input of security parameter in order to output the $\mathcal{P}$ public parameters $\mathcal{P}$ of the system, as follows:

- Choosing bilinear groups $(\mathcal{G}, \mathcal{G}_T)$ with the same q prime order.
- Choosing g $\in \mathcal{G}$ as a generator.
- Choosing three CR hashes $\mathcal{H}_1 : \{0, 1\}^* \rightarrow Z_q^*$, $\mathcal{H}_2 : \mathcal{G} \rightarrow \{0, 1\}^{256}$ and $\mathcal{H}_3 : \{0, 1\}^{256} \rightarrow \mathcal{G}$.
- Setting $\hat{e} : \mathcal{G} \times \mathcal{G} \rightarrow \mathcal{G}_T$.
- Outputing $\mathcal{P} = \big(q, \mathrm{g}, \hat{e}, \mathcal{G}, \mathcal{G}_T, \mathcal{H}_1, \mathcal{H}_2, \mathcal{H}_3\big)$.

3.2.2 KGen($\mathcal{P}, ID_i$)

In this algorithm, the AT-server tacks $\mathcal{P}$ and user i's identity ID_i as input in order to output the public and private keys for each user i as follows:

- Selecting $x_i \in Z_q^*$ randomly as a private key for user i.
- Computing $X_i = g^{x_i}$ as a corresponding user i's public key.
- Computing $Q_i = \mathcal{H}_1(ID_i, \text{salt})$ as a user i's pseudo-identity [29, 30].

  In a secure way, sends (X_i, x_i, Q_i) to end user i.

3.2.3 RSKGen(x_i, X_j)

In this algorithm, the user i's takes hes private key x_i and public key $\big(X_j\big)$ of user j as input in order to generate $rsk_{i\rightarrow j} = (X_j)^{\lambda/x_i}$ where $\lambda \in Z_q^*$ as a secure agreement information between sender i and receiver user j. Subsequently, $\big(rsk_{i\rightarrow j}\big)$ will be forwarded to the proxy-server in a secure way.

3.2.4 Signc (X_i, x_i, m_i.)

In this algorithm, the user i's takes hes public and private key x_i, X_i, and message $m_i \in \{1, 0\}^*$ as input in order to generate a signcryption (α_i) ciphertext under (X_i). This algorithm runs by first selecting a random value $r_i \in Z_q^*$, then computing the following:

- $C_i = (X_i)^{r_i}$
- $L_i = g^{r_i}$
- $K_i = \mathcal{H}_2(L_i) \oplus m_i$
- $R_i = \mathcal{H}_3(K_i)$
- $\sigma_i = (R_i \cdot C_i)^{x_i}$
- $\alpha_i = (C_i, K_i, \sigma_i)$

3.2.5 ReSignc ($rsk_{i\rightarrow j}, x_p, \alpha_i$)

In this algorithm, the proxy-serve takes hes private key x_p and the signcryption ciphertext (α_i) as input in order to generate a re-signcryption ciphertext $\left(\alpha_i^{'}\right)$ under the $\left(X_j\right)$, as the following:

- $C_j^{'} = (C_i) \cdot rsk_{i\rightarrow j}$
- $\sigma_i^{'} = (C_i^{'})^{x_p} \cdot \sigma_i$
- $\alpha_i^{'} = \left(C_i^{'}, K_i, \sigma_i^{'}\right)$

3.2.6 Ver ($\sigma_i^{'}$)

This Ver algorithm runs by the receiver j who holds the public key X_j to authenticate the sender user i with X_i and the proxy-server with X_p. The receiver j determines the X_i, X_p and computes $R_i = \mathcal{H}_3(K_i)$ to verify if $\sigma_i^{'}$ holds true value by verifying the following:

$$\hat{e}\left(g, \sigma_i^{'}\right) \stackrel{?}{=} \hat{e}\left(X_p, C_i^{'}\right)\hat{e}(X_i, (R_i \cdot C_i)) \tag{6}$$

3.2.7 Dec ($x_j, \sigma_i^{'}$)

This Dec algorithm also runs by the receiver j who holds the public key X_j. If the output from Eq. (6) holds, the receiver j will decrypt the ciphertext $\alpha_i^{'}$ as the following:

- $L_i^{'} = (C_i^{'})^{\frac{1}{x_j\lambda}} = (g^{r_i x_j \lambda})^{\frac{1}{x_j\lambda}} = g^{\frac{r_i x_j \lambda}{x_j \lambda}} = g^{r_i}$
- $m' = L_i^{'} \oplus K_i$

Correctness. For an honest verification of the above Eq. (6), the verifier must have the verifiability on the sender's X_i and the proxy-server's X_p. The correctness of Eq. (6) can be simulated under the $DBDH$ assumption as follows:

$$\begin{aligned}\hat{e}\left(g, \sigma_i^{'}\right) &= \hat{e}(g, \left(E_i^{'})^{x_{px}} \cdot \sigma_i\right) = \hat{e}(g, \left(C_i^{'})^{x_p}\right)\hat{e}(g, \sigma_i) \\ &= \hat{e}(g, \left(C_i^{'})^{x_p}\right)\hat{e}(g, (R_i \cdot C_i)^{x_i}) \\ &= \hat{e}\left(g^{x_p}, C_i^{'}\right)\hat{e}(g^{x_i}, (R_i \cdot C_i)) \\ &= \hat{e}\left(X_p, C_i^{'}\right)\hat{e}(X_i, (R_i \cdot C_i))\end{aligned}$$

3.3 Security Models for PRSCD

The PRSCD scheme generates two levels of ciphertexts. The first ciphertext level (1cl) is known as the original ciphertext that is generated from the signcrypted message m, and the second ciphertext level (2cl) is generated from the re-signcrypted the original 1cl. Hence, the RCC-security for the PRSCD scheme is defined from these two cases.

Theorem 3 *The PRSCD is CCA-secure at the* 1cl *only if the PRSCD is CCA-1 secure under the $DBDH$ and CDH assumption.*

Proof Assuming $\mathcal{D}$ is an adversary with ability of breaking the CCA-1 security of the PRSCD, and $\mathcal{F}$ is an algorithm that is designed for breaking the CCA-1 security of the PRSCD using $\mathcal{D}$. Therefore, $\mathcal{F}$ and $\mathcal{D}$ will challenge each other through investigating the CCA-1 security game, as follows:

Initiate (γ): $\mathcal{F}$ runs Initiate(γ) algorithm to output the $\mathcal{P}$.

Phase 1: For all $\mathcal{D}$'s queries, $\mathcal{F}$ answers $\mathcal{D}$ as follows:

- In $\mathcal{O}_{pk}$ oracle, $\mathcal{F}$ runs KGen to output sk, pk. Then, $\mathcal{F}$ queries its own $\mathcal{O}_{pk} \in \Phi$ oracle for $PRSCD$ to output pk in $PRSCD$. Thus, $\mathcal{F}$ performs the following:
  - Selects $x_i, a, b \in Z_p^*, w_i \in \{-1, 0, 1\}$
  - If $w_i = -1$, sets $pk_i = g^{x_i}$
  - If $w_i = 0$, sets $pk_i = g^{bx_i}$
  - If $w_i = 1$, sets $pk_i = g^{ax_i}$
    Subsequently, $\mathcal{F}$ records (pk_i, x_i, w_i) in $\boldsymbol{T}_{pk}$, and outputs pk_i for $\mathcal{D}$.
- In $\mathcal{O}_{sk}$ oracle, if $w_i = -1$, $\mathcal{F}$ searches the corresponding (x_i, pk_i, w_i) in $\boldsymbol{T}_{pk}$ and then returns $sk_i = x_i$ as the private key to $\mathcal{D}$. Otherwise, $\mathcal{F}$ returns a random $\mathcal{B} \in \{0, 1\}^*$.
- In $\mathcal{O}_{rsk}$ oracle, $\mathcal{F}$ seeks for (pk_i, x_i, w_i) and $\left(pk_j, x_j, w_j\right)$ in $\boldsymbol{T}_{pk}$ to output $rsk_{i \to j}$ as a delegation from user i to user j. $\mathcal{F}$ then returns $rsk_{i \to j}$ to $\mathcal{D}$ depending on the following conditions:
  - $sk_i = x_i$ if $Pr\left[w_i = -1\right] \vee Pr\left[w_i = w_j\right]$, returns $rk_{i \to j} = pk_j^{\lambda/x_i}$.
  - $sk_i = ax_i \wedge sk_j = bx_j$ if $Pr\left[w_i = 1\right] \wedge Pr\left[w_j = 0\right]$, returns $rk_{i \to j} = pk_j^{b\lambda/ax_i}$.
  - $sk_i = ax_i \wedge sk_j = x_j$ if $Pr\left[w_i = 1\right] \wedge Pr\left[w_j = -1\right]$, returns $rk_{i \to j} = pk_j^{\lambda/ax_i}$.
  - $sk_i = bx_i \wedge sk_j = ax_j$ if $Pr\left[w_i = 0\right] \wedge Pr\left[w_j = 1\right]$, returns $rk_{i \to j} = pk_j^{a\lambda/bx_i}$.
  - $sk_i = bx_i \wedge sk_j = x_j$ if $Pr\left[w_i = 0\right] \wedge Pr\left[w_j = -1\right]$, returns random $\mathcal{B} \in \{0, 1\}^*$.

- In $\mathcal{O}_{\text{Signc}}$ oracle, $\mathcal{F}$ fetches the ciphertext α_i and queries its own re-signcryption $\mathcal{O}_{rsk} \in \Phi$ oracle to output a re-signcrypted $\alpha_i^{'}$ ciphertext that is identical to $\left(rsk_{i\to j}, sk_p, \alpha_i\right)$. $\mathcal{F}$ then proceeds to execute $\alpha_i^{'}$ as the following:
  - $C_i^{'} = (C_i) \cdot rsk_{i\to j} = X_i^{r_i} \cdot X_j^{\lambda/x_i} = g^{x_i r_i} \cdot g^{x_j\lambda/x_i} = g^{x_i r_i x_j \lambda/x_i} = g^{r_i x_j \lambda}$
  - $\sigma_i^{'} = (C_i^{'})^{x_p} \cdot \sigma_i = g^{x_p r_i x_j \lambda} \cdot (R_i \cdot C_i)^{x_i}$

  $\mathcal{F}$ then returns $\alpha_i^{'}$ to $\mathcal{D}$. For a noticing, the $C_i^{'} = g^{r_i x_j \lambda}$ ciphertext is blinded by a secret value λ between user i and user j as well as a random number $r_i \in Z_p^*$, hence no information about λ and r_i values can be guessed or leaked to the adversary $\mathcal{D}$.

- $\mathcal{O}_{\text{ver}}$: In the verification oracle $\mathcal{O}_{ver}$, $\mathcal{F}$ seeks for $\left(pk_s, pk_p\right)$ in $\boldsymbol{T}_{pk}$ to generate a verification $\mathcal{O}_{\text{ver}}$ process in order to authenticate the sender s and proxy-server p. $\mathcal{F}$ then returns the result of $PRSCD \cdot \text{Ver}\left(\sigma_i^{'}\right)$ to $\mathcal{D}$.
- $\mathcal{O}_{\text{dec}}$: In the decryption oracle $\mathcal{O}_{\text{dec}}$, $\mathcal{F}$ seeks for $\left(C_i^*\right)$ in $\boldsymbol{T}_{pk}$ corresponding to pk_i. $\mathcal{F}$ returns m_i to $\mathcal{D}$ depending on the following conditions:
  - If the C_i^* is the first level of ciphertext $C_i^* = C_i$, $\mathcal{F}$ seeks receiver i's sk_i in $\boldsymbol{T}_{pk}$ corresponding to pk_i. $\mathcal{F}$ then queries $\mathcal{O}_{\text{dec}}$ to obtain m corresponding to pk_i in PRSCD and then returns m to $\mathcal{D}$.
  - If the C_i^* is the second level of ciphertext $C_i^* = C_i^{'}$, $\mathcal{F}$ seeks delegated user j's sk_j in $\boldsymbol{T}_{pk}$ corresponding to pk_j. $\mathcal{F}$ then queries $\mathcal{O}_{dec}$ to obtain m corresponding to pk_j in PRSCD and outputs m_i to $\mathcal{D}$.

Challenge: If $\mathcal{D}$ is decided that Phase 1 is over, it will return pk_i and $m_0, m_1 \in \{0, 1\}^*$ for the purpose of challenging. $\mathcal{F}$ recovers tuple $\left(pk_i^*, x_i^*, w_i^*\right)$ and then queries its own challenge Φ oracle for $\left(x_i^*, pk_i^*, w_i^*\right)$ and $\left(m_0^*, m_1^*\right)$ to output the challenge C_i^* ciphertext and finally returns C_i^* as the ciphertext challenge to $\mathcal{D}$.

Phase 2: This is similar to Phase 1 but with the CCA-1 security game restrictions.

Guess: $\mathcal{D}$ continues querying and $\mathcal{F}$ answers to $\mathcal{D}$'s queries as in Phase 1, since $\mathcal{D}$ has to follow the CCA-1 security game restrictions.

Output: $\mathcal{D}$ returns a guess $\mathcal{B}^* \in \{0, 1\}^*$ such that if $\mathcal{B} = \mathcal{B}^*$, $\mathcal{F}$ will return "1", otherwise, will return "0". However, when $\mathcal{B} = \mathcal{B}^*$, the following theorem will be obtained.

Theorem 4 *The PRSCD is an CCA-2 secure at 2cl only if the PRSCD is an CCA-2 secure under the $DBDH$ and CDH problems.*

Proof Similar with Theorem 3 proof, in which if $\mathcal{D}$ is an adversary with ability of breaking the CCA-2 security of the PRSCD and $\mathcal{F}$ is an algorithm that is designed to break the CCA-2 security of the $PRSCD$ through using $\mathcal{D}$. Therefore, $\mathcal{F}$ and $\mathcal{D}$ will be challenging each other through investigating the following CCA-2 security game.

Initiate: $\mathcal{F}$ executes Initiate (γ) algorithm to output the $\mathcal{P}$ as system parameters.

Phase 1: For all $\mathcal{D}$'s queries, $\mathcal{F}$ answers in a similar way to Theorem 3 proof.

Challenge: If $\mathcal{D}$ is decided the Phase 1 is over, it will output pk_i^* and $m_1^*, m_0^* \in \{0, 1\}^*$ for the purpose of challenging. $\mathcal{F}$ then queries its own challenge Φ oracle with $(pk_i^*), m_i^*0, m_1^*)$ to output the C_i^* as a ciphertext challenge. Thus, $\mathcal{F}$ outputs C_i^* to $\mathcal{D}$ as a ciphertext challenge.

Phase 2: This is similar to Phase 1 but with the CCA-2 security game restrictions.

Guess: $\mathcal{D}$ continues querying and in contrast $\mathcal{F}$ answers to $\mathcal{D}$'s queries as described in Phase 1 if $\mathcal{D}$ is following the CCA-2 security game restrictions.

Output: If $\mathcal{D}$ returns $\mathcal{B} = \mathcal{B}^*$, $\mathcal{F}$ will return "1", otherwise, returns "0".

Here, the explanation of the PRSCD is completed, the following theorem will begin to analyze the PRSCD simulation.

Theorem 5 *The proposed PRSCD is an CCA-secure at all ciphertext levels (CCA-1, CCA-2) through fulfilling most of the security feature requirements such as confidentiality, integrity and authentication since the underlying PRSCD is an CCA-secure with assumptions of $DBDH$ and CDH problems.*

Proof Assuming $\mathcal{F}$ is an algorithm obtains $(g, g^{a_1}, g^{a_2} \in \mathcal{G})$ and $(Q \in \mathcal{G}_T)$ with zero-knowledge of $a_1, a_2, a_3 \in Z_q^*$, in which $\mathcal{F}$ aims to compute $g^{a_1 a_2 a_3} \in \mathcal{G}$ in order to find if $Q = e(g, g)^{a_1 a_2 a_3} \in \mathcal{G}_T$. Thus, $\mathcal{F}$ can break the CDH and $DBDH$ assumptions. Thus, if $\mathcal{F}$ is a polynomial-time algorithm with an advantage ε and has ability to solve the CDH and $DBDH$ problems in $(\mathcal{G}, \mathcal{G}_T)$ with:

$$\begin{aligned}&\left|\Pr\left[\mathcal{F}\left(g, g^{a_1}, g^{a_2}, g^{a_3}, e(g, g)^{a_1 a_2 a_3}\right) = Q\right]\right.\\ &\quad \left.-\Pr\left[\mathcal{F}\left(g, g^{a_1}, g^{a_2}, g^{a_3}, Q\right) = 1\right]\right| \geq \varepsilon \end{aligned} \tag{7}$$

$$\left|\Pr\left[\tau + O(t)\left(q_{pk} + q_{rk} + q_{dec}\right)\right]\right| \leq \varepsilon \tag{8}$$

The (ε, τ)-CDH is holded only if there is no ε algorithm with τ time can compute the CDH problem where t is the maximum time to execute $g^{a_1 a_2 a_3} \in \mathcal{G}$. Therefore, since the message m is encrypted and signed in the (CCA-1, CCA-2) with assumptions of $DBDH$ and CDH problems, it will be difficult to be computed in a polynomial time, as shown in Theorem 5's proof, the proposed PRSCD scheme can guarantee the most important security features such as simultaneous confidentiality, integrity, and authentication.

4 Performance Evaluation

This subsection demonstrates the performance evaluation of the proposed $PRSCD$ scheme from two aspects: computational and communication overhead.

4.1 Computational Overhead

The computational overhead focuses on counting complicated cryptographic operations, including bilinear pairing in $\hat{e}$, scalar exponentiation in $\mathcal{G}$ and multiplication in $\mathcal{G}_T$. To illustrate the computational overhead of the proposed $PRSCD$ scheme, the evaluation is executed on a MacBook pro with CPU-i7 Intel core, 8 GB RAM, and @2.9 GHz. The theoretic number of the cryptography is executed by the MIRACL core cryptographic library. The Weil pairing is executed on $y = x^3 + 1$ elliptic curve over $\mathcal{F}_p$ with a generator g is 512 bits. The notations that are used in this computational overhead are denoted as $E_{\mathcal{G}}$, $E_{\mathcal{G}_T}$ and BP, which represent the time cost of a scalar exponentiation operation in $\mathcal{G}$, the time cost of exponentiation operation in $\mathcal{G}_T$ and the time cost of computing a bilinear pairing operation in $\hat{e}$, respectively. Table 1 demonstrates the cryptographic operation costs for each process in the proposed PRSCD. In addition, Table 3 shows the calculation running time for the $E_{\mathcal{G}}$, $E_{\mathcal{G}_T}$, BP operations in milliseconds. On the basis of the computational cost value results, the PRSCD is compared with other PRE schemes in [18, 19], as displayed in Table 2.

Table 1 PRSCD computational cost

Key and rekey-signcryption generation		
Phases	Operations	Run time (ms)
$KGen$	$n(E_{\mathcal{G}})$	$n(1.0)$
$RSKGen$	$E_{\mathcal{G}}$	1.0
Signcryption and re-signcryption		
Phases	Operations	Run time (ms)
$Signc$	$3(E_{\mathcal{G}})$	3(1.0)
$ReSignc$	$E_{\mathcal{G}}$	1.0
Verification and decryption		
Phases	Operations	Run time (ms)
Ver	$3BP$	3(2.9)
Dec	$E_{\mathcal{G}}$	1.0

Table 2 PRSCD overhead analysis

Schemes	Computational overhead	First level ciphertext length size	Second level ciphertext length size
Libert and Vergnaud [18]	$11E_{\mathcal{G}} + 8BP + 2E_{\mathcal{G}_T}$	$\lvert\sigma\rvert + 2\lvert\mathcal{G}\rvert + \lvert\mathcal{G}_T\rvert + \lvert sv\rvert$	$\lvert\sigma\rvert + 4\lvert\mathcal{G}\rvert + \lvert\mathcal{G}_T\rvert + \lvert sv\rvert$
Weng et al. [19]	$5E_{\mathcal{G}} + 5BP + 4E_{\mathcal{G}_T}$	$3\lvert\mathcal{G}\rvert + \lvert\mathcal{L}\rvert + \lvert Z_p^*\rvert$	$\lvert\mathcal{L}\rvert + \lvert Z_p^*\rvert + 2\lvert\mathcal{G} + \lvert\mathcal{G}_T\rvert$
Proposed scheme	$6E_{\mathcal{G}} + 3BP$	$\lvert\mathcal{L}\rvert + 2\lvert\mathcal{G}\rvert$	$\lvert\mathcal{L}\rvert + 2\lvert\mathcal{G}\rvert$

Table 3 Cryptographic operation running time

	Descriptions	Execution time (ms)
$E_{\mathcal{G}}$	Exponentiation in $\mathcal{G}$	1.0
$M_{\mathcal{G}_T}$	Exponentiation in $\mathcal{G}_T$	0.2
BP	Pairing operation	2.9

4.2 Communication Overhead

The length size of the ciphertext message is considered as a crucial value that can affect on the successful of any cryptographic scheme. Thus, it is crucial to display the communication overhead efficiency of the PRSCD scheme. The PRSCD communication overhead is analyzed from two phases: the length size of the first ciphertext level, which denotes the communication from the sender-to-proxy and the length size of the second ciphertext level, which denotes the communication from the proxy-to-receiver. Here, $|\mathcal{G}|$ denotes the length size of an element in $\mathcal{G}$ while $|\mathcal{L}|$ denotes the length size of an element in $\{1, 0\}$. Assuming that the length size for each scalar expsonentiation in $\mathcal{G}$ is 160 bits and the ciphertext message in $|\mathcal{L}|$ is 256 bits (Table 3).

Sender-to-proxy. The size of the ciphertext length $|\boldsymbol{\alpha}_i|$ that is generated by the sender to send a message m as a ciphertext $\boldsymbol{\alpha}_i$ to the proxy-server is $|\boldsymbol{\alpha}_i| = (|\mathcal{L}| + 2|\mathcal{G}|)$. Hence, the size of the first ciphertext level $|\boldsymbol{\alpha}_i| = 320 + 256$ bits.

Proxy-to-receiver. When the proxy-server receives the ciphertext $\boldsymbol{\alpha}_i$, it will perform the re-signcryption $ReSignc$ algorithm to output a second level $\boldsymbol{\alpha}_i^{'}$ ciphertext. Thus, the size of the second ciphertext level $\boldsymbol{\alpha}_i^{'}$ is $\left|\boldsymbol{\alpha}_i^{'}\right| = (|\mathcal{L}| + 2|\mathcal{G}|) = 320 + 256$ bits.

Table 2 shows a comparison of the communication overhead with PRE schemes in [18, 19]. Here, $|\sigma|$ is the length size of the LV08 one-time signature method and $|sv|$ is the length size of the key verification. A summary of the above evaluations, the proposed PRSCD provides an efficient protocol that has a lower computation and communication cost that is suitable for narrow bandwidth and terminals with limited resources. Therefore, the PRSCD is capable of being used in any application scenario, including IoT, cloud, and fog computing.

5 Conclusions

This paper presents a unidirectional proxy re-signcryption method with a feature of delegation process (PRSCD). This paper proves that the PRSCD is CCA-security through the proposed security models. Thus, the evaluation results show that the PRSCD is able to increase the security features compared with other PRE schemes. However, the proposal in this paper has a limited on only achieving a CCA-secure single-hop U-PRSCD method. Therefore, in future works, designing a multi-hop U-PRSCD method will be concentrated.

References

1. M. Blaze, G. Bleumer, M. Strauss, Divertible protocols and atomic proxy cryptography, in *International Conference on the Theory and Applications of Cryptographic Techniques*. (Springer, 1998), pp. 127–144
2. R. Guo, G. Yang, H. Shi, Y. Zhang, D. Zheng, O 3-r-cp-abe: an efficient and revocable attribute-based encryption scheme in the cloud assisted IoMT system. IEEE Internet Things J. **8**(11), 8949–8963 (2021)
3. G. Mamidisetti, R. Makala, C. Anilkumar, A novel access control mechanism for secure cloud communication using SAML based token creation. J. Ambient Intell. Human. Comput. 1–19 (2020)
4. P. Dutta, W. Susilo, D.H. Duong, P.S. Roy, Collusion-resistant identity-based proxy re-encryption: lattice-based constructions in standard model. Theoret. Comput. Sci. **871**, 16–29 (2021)
5. Alamer, Security and privacy-awareness in a software-defined fog computing network for the internet of things, Opt. Switching Netw. 100616 (2021)
6. H.-Y. Lin, Y.-M. Hung, An improved proxy re-encryption scheme for IoT-based data outsourcing services in clouds. Sensors **21**(1), 67 (2021)
7. S.M. Patil, B. Purushothama, Non-transitive and collusion resistant quorum controlled proxy re-encryption scheme for resource constrained networks. J. Inf. Secur. Appl. **50**, 102411 (2020)
8. Alamer, S. Basudan, P.C. Hung, A secure tracing method in fog computing network for the IoT devices, in *Proceedings of the 12th International Conference on Management of Digital EcoSystems*, pp. 104–110 (2020)
9. Alamer, Y. Deng, X. Lin, Secure and privacy-preserving task announcement in vehicular cloud, in *2017 9th International Conference on Wireless Communications and Signal Processing (WCSP)*. (IEEE, 2017), pp. 1–6
10. Alamer, Y. Deng, G. Wei, X. Lin, Collaborative security in vehicular cloud computing: a game theoretic view. IEEE Netw. **32**(3), 72–77 (2018)
11. R. Nidhya, S. Shanthi, M. Kumar, A novel encryption design for wireless body area network in remote healthcare system using enhanced RSA algorithm, in *Intelligent System Design*. (Springer, 2021), pp. 255–263
12. Alamer, An efficient group signcryption scheme supporting batch verification for securing transmitted data in the internet of things. J. Ambient Intell. Human. Comput. 1–18 (2020)
13. S. Basudan, Lega: a lightweight and efficient group authentication protocol for massive machine type communication in 5G networks. J. Commun. Inf. Netw. **5**(4), 457–466 (2020)
14. Alamer, S. Basudan, An efficient truthfulness privacy-preserving tendering framework for vehicular fog computing. Eng. Appl. Artif. Intell. **91**, 103583 (2020)
15. N.N. Mohamed, Y.M. Yussoff, M.A. Saleh, H. Hashim, Hybrid cryptographic approach for internet of hybrid cryptographic approach for internet of things applications: a review. J. Inf. Commun. Technol. **19**(3), 279–319 (2020)
16. X. Yu, C. Xu, B. Dou, Y. Wang, Multi-user search on the encrypted multimedia database: lattice-based searchable encryption scheme with time-controlled proxy re-encryption. Multimedia Tools Appl. **80**(2), 3193–3211 (2021)
17. J. Hou, M. Jiang, Y. Guo, W. Song, Efficient identity-based multibit proxy re-encryption over lattice in the standard model. J. Inf. Secur. Appl. **47**, 329–334 (2019)
18. Libert, D. Vergnaud, Unidirectional chosen-ciphertext secure proxy re-encryption. IEEE Trans. Inf. Theory **57**(3), 1786–1802 (2011)
19. J. Weng, M. Chen, Y. Yang, R. Deng, K. Chen, F. Bao, CCA secure unidirectional proxy re-encryption in the adaptive corruption model without random oracles. Sci. China Inf. Sci. **53**(3), 593–606 (2010)
20. Paul, S.S.D. Selvi, C.P. Rangan, A provably secure conditional proxy re-encryption scheme without pairing. IACR Cryptol. ePrint Arch. **2019**, 1135 (2019)

21. S. Yao, R.V.J. Dayot, H.-J. Kim, I.-H. Ra, A novel revocable and identity-based conditional proxy re-encryption scheme with ciphertext evolution for secure cloud data sharing. IEEE Access **9**, 42801–42816 (2021)
22. Q. Wang, W. Li, Z. Qin, Proxy re-encryption in access control framework of information-centric networks. IEEE Access **7**, 48417–48429 (2019)
23. X. Lin, R. Lu, Proxy re-encryption with delegatable verifiability, in *Australasian Conference on Information Security and Privacy*. (Springer, 2016), pp. 120–133
24. J. Lai, Z. Huang, M.H. Au, X. Mao, Constant-size CCA-secure multi-hop unidirectional proxy re-encryption from indistinguishability obfuscation. Theoret. Comput. Sci. **847**, 1–16 (2020)
25. S. Prasad, B. Purushothama, CCA secure and efficient proxy re-encryption scheme without bilinear pairing. J. Inf. Secur. Appl. **58**, 102703 (2021)
26. Manzoor, A. Braeken, S.S. Kanhere, M. Ylianttila, M. Liyanage, Proxy re-encryption enabled secure and anonymous IoT data sharing platform based on blockchain. J. Netw. Comput. Appl. **176**, 102917 (2021)
27. Ge, W. Susilo, J. Baek, Z. Liu, J. Xia, L. Fang, A verifiable and fair attribute-based proxy re-encryption scheme for data sharing in clouds. IEEE Trans. Dependable Secure Comput. (2021)
28. W. Li, H. Xiong, Efficient proxy re-encryption scheme for e-voting system. KSII Trans. Internet Inf. Syst. **15**(5) (2021)
29. Alamer, S. Basudan, X. Lin, A privacy-preserving incentive framework for the vehicular cloud, in *2018 IEEE International Conference on Internet of Things (iThings) and IEEE Green Computing and Communications (GreenCom) and IEEE Cyber, Physical and Social Computing (CPSCom) and IEEE Smart Data (SmartData)*. (IEEE, 2018), pp. 435–441
30. Alamer, A secure anonymous tracing fog-assisted method for the internet of robotic things. Library Hi Tech (2020)

Deep Learning-Based COVID-19 Detection Using Lung Parenchyma CT Scans

Zeynep Kaya, Zuhal Kurt, Şahin Işık, Nizameddin Koca, and Sümeyye Çiçek

Abstract During the outbreak of the COVID-19 pandemic, it is important to improve early diagnosis using effective ways in order to lower the risks and further spread of the viruses as early as possible. This is also important when it comes to appropriate treatments and the reduction of mortality rates. In this respect, computer tomography (CT) scanning is a useful technique in detecting COVID-19. The present paper, as such, is an attempt to contribute to this process by generating an open-source, CT-based image dataset. This dataset contains the CT scans of lung parenchyma regions of 180 COVID-19 positives and 86 COVID-19 negative patients, all from Bursa Yuksek Ihtisas Training and Research Hospital. The experimental studies demonstrate that this dataset is effectively utilized deep learning-based models for diagnostic purposes. Firstly, a smart segmentation mechanism based on the k-means algorithm is applied to this dataset as a pre-processing stage. Then, the performance of the proposed method is evaluated using InceptionV3 and Xception convolutional neural networks, yielding a 96.20% and 96.55% accuracy rate and 95.00% and 95.50% F1-score, respectively. These state-of-the-art models are observed to detect COVID-19 cases faster and more accurately. In addition, the fine-tuning stage of the convolutional neural network (CNN) features sufficiently improves this accuracy rate. For these features, the support vector machine (SVM) classifier is used, resulting in remarkable 96.76% accuracy rate and 95.81% F1-score. The implications of the proposed method are immense both for present-day applications as well as future developments.

Z. Kaya
Department of Electrical and Electronics Engineering, Eskisehir Osmangazi University, Eskisehir, Turkey

Z. Kurt (✉)
Department of Computer Engineering, Atilim University, Ankara, Turkey
e-mail: zuhal.kurt@atilim.edu.tr

Ş. Işık
Department of Computer Engineering, Eskisehir Osmangazi University, Eskisehir, Turkey

N. Koca · S. Çiçek
Department of Internal Medicine, University of Health Sciences, Bursa, Turkey
e-mail: nizameddin.koca@sbu.edu.tr

A. K. Bashir et al. (eds.), *Proceedings of International Conference on Computing and Communication Networks*, Lecture Notes in Networks and Systems 394,
https://doi.org/10.1007/978-981-19-0604-6_23

Keywords Lung parenchyma · Deep learning · COVID-19 detection · CT image dataset · Fine-tuning · K-means · Support vector machine

1 Introduction

A pandemic is known as an infectious disease that spreads rapidly across a large region and affects a large number of people. The pandemic with the most deaths in human history was the black death, which took the lives of an estimated 25–200 million individuals in the fourteenth century [1]. The term "pandemic" can into use with the Spanish Flu [2], which resulted in the deaths of 17–50 million people between 1918 and 1920. On March 11, 2020, the World Health Organization (WHO) announced COVID-19 as a pandemic after its first appearance in the Wuhan province of China, and its spread worldwide since December 2019. Marked by symptoms such as sore throat, headache, fever, runny nose, coughing, etc., this disease is easily transmitted among individuals and it weakens the immune system [3] to the point of multiple organ failures and eventual deaths [4]. Up to August 5, 2021, some 200,174,883 cases of COVID-19 were recorded by the WHO, and the death toll reached 4,255,892 with a rapid increase of cases all over the world. As for Turkey, these figures stood at 5,822,487 cases and 51,767 deaths on the same date [5].

According to the Chinese government, the COVID-19 cases were diagnosed using real-time polymerase chain reaction (RT-PCR), which consumes a lot of time and suffers from high false negative rates [6]. Most of the RT-PCR-positive patients either show no symptoms or only minimal symptoms. The most important exposure determining the prognosis in COVID-19 disease appears in the lungs. As a result, planning treatments based on diagnosing COVID-19-triggered pneumonia in the early stages can reduce mortality by allowing the disease to be intervened before it progresses: However, it can be difficult to make a definitive diagnosis and avoid delays in starting treatment. To make up for this disadvantage, thorax computer tomography (CT) can be used as a more reliable and faster method, allowing for speedy intervention since all medical centers are equipped with such facilities.

The deep learning-based approach is one of the most effective techniques utilized by biomedical applications. With this fast and efficient method, many diseases are diagnosed with high-accuracy rates. CT images are used by radiologists to determine which patients are COVID-19 positive or negative. An efficient automated deep learning-based system can prevent misdiagnosis and delayed treatment. Several works [7–10] have developed such methods to identify COVID-19 cases using CTs with a high-accuracy rate. Additionally, open-access COVID-19 chest X-ray image datasets [11] and the COVID-19 radiography database [12] are available for diagnostic purposes. These datasets are also utilized in [13], and they include three types of chest images: COVID-19, pneumonia, and normal. Each image is pre-processed before being trained with deep learning models. In this pre-processing stage, the dataset is reconstructed using the fuzzy technique and stacking technique. Then, the developed dataset is trained using the MobileNetV2 and SqueezeNet deep learning models, and the models are classified using the SVM method as in [13].

The integration of deep learning to detect the presence of COVID-19 in X-ray images was proposed by exploiting transfer learning in [14]. The network activation layers are shown as an additional contribution to identify the areas that the model considered in order to generate predictions and improve the interpretability of the predictions. Furthermore, an algorithm was introduced in [14] that can be used by radiologists to immediately pinpoint the X-ray areas worthy of further investigation. In [15], an automated deep learning-based approach was introduced for the detection of COVID-19 infection in chest X-rays. The main contribution of this paper was to overcome the lack of sensitivity of RT-PCR, and the chest X-ray images were utilized to detect and diagnose of COVID-19. An extreme version of inception (Xception) algorithm was used in the same work, since the algorithm can be trained with the weights of networks for large datasets, as well as for fine-tuning the weights of pre-trained networks on small datasets. In [16], an intelligent COVID-19 infection diagnosis model was proposed based on the convolutional neural networks (CNNs) and machine learning techniques. This model ensures an end-to-end learning scheme that can directly learn discriminative features from the input chest CT X-ray images and eliminate the handcrafted feature engine. The developed model in the stated work can be used to assist field specialists, physicians, and radiologists in the decision-making process. Aims of this study are to reduce the misdiagnosis rates and to apply the proposed model as a retrospective evaluation tool.

A new deep learning-based computer-aided design (CAD) scheme for chest X-ray radiography images was developed in [17]. The scheme can detect and classify the images into 3 classes, namely COVID-19 pneumonia, other community-acquired non-COVID-19 pneumonia, and normal (non-pneumonia) cases. Image processing algorithms were applied to remove the majority of diaphragm regions in this study, as opposed to directly using the original chest X-ray images to train the deep learning models. This pre-processing stage may help to significantly improve the models' performance and robustness in detecting COVID-19 cases and in distinguishing them from other community-acquired non-COVID-19 infected pneumonia cases. A well-trained VGG16-based CNN model as a transfer learning model was selected in this CAD scheme [17].

Modern artificial intelligence (AI) and machine learning (ML) technology are recently employed to deal with challenges during the outbreak and spread of the COVID-19 pandemic. In [18], a comprehensive review of studies is offered concerning the model and technology for this purpose, including the types of AI and ML and those of datasets. The final performance of each model and the compromises and commutations of modern techniques are also presented to give a general insight into their robustness. A deep learning-based system that combines the CNN and long short-term memory (LSTM) networks to automatically detect COVID-19 from X-ray images was introduced in [19], where CNN was used for feature extraction and LSTM to classify COVID-19 based on those features. The 2D CNN and LSTM layout features are combined to improve the classification accuracy. The dataset used was collected from multiple sources, and also pre-processing was performed to reduce the noise.

In [20], texture is one of the main visual attributes present in chest X-ray images. Some popular texture descriptors and also a pre-trained CNN model were used to extract features from chest X-ray images. Multiple features were extracted using different texture descriptions; then, different fusion techniques were used to take advantage of each descriptor's strength, both on early and late fusion modes. In [9], a novel CAD system is proposed based on deep learning to classify COVID-19 infection versus other atypical and viral pneumonia-related diseases. Here, ten well-known pre-trained CNNs were used to diagnose infections related to the COVID-19. It can be deduced that the deep learning method can help radiologists in diagnosing infections related to COVID-19. Another deep learning model was proposed for the automatic diagnosis of COVID-19 in [9]. The developed model has an end-to-end architecture without using any feature extraction methods, and it requires raw chest X-ray images to enable diagnosis. Such tests performed after 5–13 days are found to be positive in recovered patients, concluding that they may continue to spread of the virus; hence, the need for more accurate methods to provide predictions. In [8], the COVIDNet-CT is introduced as a deep convolutional neural network architecture tailored specifically for this purpose using chest CT images and a machine-driven design exploration approach. A CT image dataset derived from CT imaging data was introduced and collected from the China National Center for Bioinformation (CNCB). The decision-making behavior of the model was investigated, and an interpretability-driven performance validation was carried out regarding its predictions. A new network architecture was proposed in [10]. Binary class (COVID-19 and no-findings) and multi-class (COVID-19, no-findings, and pneumonia) classification were carried out with capsule networks. COVID-19 and other images of different sizes have been pre-processed to be input data to capsule networks. A novel model that is different from the existing capsule networks was proposed.

Despite the promising results reported in the previous studies, many issues have not been well investigated in regards to how to train deep learning models for estimation of COVID-19 with optimal recognition rates; for instance, whether applying image pre-processing algorithms can help to improve the performance and robustness of the deep learning models. To better address some of the challenges or technical issues, the present study applies a set of experiments on effective deep neural network models, namely InceptionV3 and Xception. After determinate the best set of parameters, these models are used to detect COVID-19 among the patients. The proposed models allow us to explore the critical visual factors associated with COVID-19 infection after mapping the characteristic of COVID-19 with large-size filters. The trained models not only improve the decisions of experts but also minimize costly procedures required to diagnose COVID-19.

Furthermore, the importance and originality of this study are that it integrates the pre-processing step before mapping the raw image data to CNN features. For this purpose, it applies a simple but effective pre-processing step to the obtained an open-sourced dataset—Turkish COVID-19. The usefulness of this dataset is confirmed by a senior radiologist, who has been diagnosing and treating COVID-19 patients since the outbreak of this pandemic. Prior to training, a smart data augmentation procedure is followed to tackle overfitting and underfitting when training a typical CNN algorithm. After data augmentation, experimental studies are conducted to further demonstrate that this dataset is useful for developing AI-based diagnosis models of COVID-19. After conducting simulations on this custom dataset, the findings of the proposed methods are reported as 96.20% and 96.55% for InceptionV3 and Xception models, respectively. Moreover, a superior performance 96.76% is obtained with fine-tuning approach, which refers to feature-engineering of CNN architectures.

This paper is organized as follows. First, the Turkish COVID-19-CT dataset is introduced in Sect. 2, where the potential of the proposed deep NN models is investigated in terms of detecting COVID-19. In Sect. 3, the experimental results are reported and compared with other studies. Finally, conclusions and comments appear in Sect. 4.

2 Methods and Tools

2.1 Proposed Methodology

An overview of the COVID-19 detection framework appears in Fig. 1. The proposed approach includes the following stages:

- Segment the lung region from the input image;
- Crop the region of interest (ROI) from the segmented image;
- Train the dataset using pre-trained deep learning models; and
- Obtain the results by using SVM classifiers.

2.2 Deep Learning Models

The present work proposes a qualitative case study methodology to investigate the use of lung parenchymal images in detecting COVID-19 among patients. It is believed that this research will provide some insights into the importance of CNN features for this purpose. The segmented lung regions are encoded with reduced size of features after applying the common CNN methods. The key motivation is that spatio-temporal CNN features could be used in discriminative analyzes of COVID-19 features in the CTs of patients with and without infection.

In the field of medical image analysis, deep learning models are popularly used for segmentation and classification. There is a growing body of studies emphasizing the importance of using CNN architectures in COVID-19 detection based on a comprehensive literature review carried out for this paper. Typically, a CNN architecture consists of three major components: (i) convolution layers, (ii) fully connected layers, and (iii) other parameters such as activation function and optimizer. As mimics of

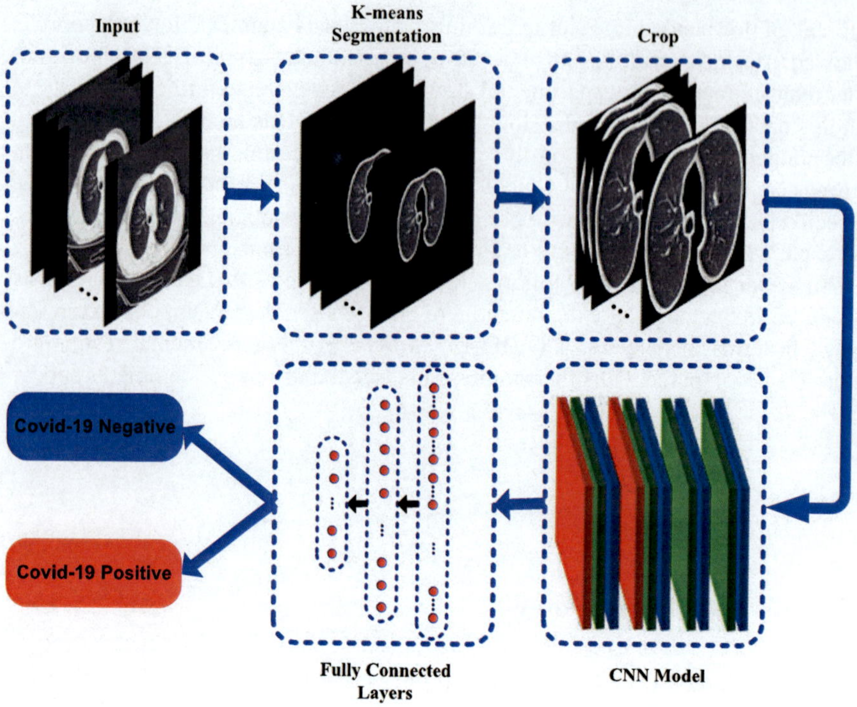

Fig. 1 An overview of the framework for the proposed approach

the human learning mechanism, these massive convolutional filters are optimized through feed-forward and feed-backward stages. Therefore, the key backbone of the learning stage is based on the gradient descent method, which seeks to find optimum global weights by moving in the direction of the steepest descent. However, investigating a minimization of cost function depends on various factors. The first parameter is the learning rate, which is a steering control parameter as an acceleration rate to determine a global solution. The second parameter is regularizing the residuals with a smart optimizer.

Meanwhile, the success of a CNN method relies on the combination of a good design and appropriate component (hyperparameters, filter size, etc.) selection. This important point was first noticed in the ImageNet 2012 competition, where the AlexNet model [21] achieved a better score over traditional approaches. Shortly afterward, the VGG16 [22] emerged by modifying the depth and size of filters in AlexNet. In the family of CNN design, the various architectures have been developed with data-driven purposes. After systematically reviewing the performance of the current CNN methods, it can be observed that the InceptionV3 [23] and Xception [24] models are efficient for classification tasks (https://keras.io/api/applications/). The model selection is based on the accuracy (top-1 and top-5) and the number of parameters that are utilized in the memory.

Table 1 Parameter details about Xception and InceptionV3

Parameter name	Value
Epochs	100
Batches	8, 16
Input length	224 × 224 × 3
Learning rate	1e−4
Activation function	Softmax
Loss function	Categorical-cross entropy
Optimization function	Adam

The Xception architecture has 36 convolutional layers that appear before to two fully connected layers. 4096-dimensional features are extracted from each image in these layers. These models both involve depth-wise separable convolutions. The performance of Xception is slightly better than InceptionV3 model as about 2% accuracy rate. Moreover, the parameter sizes of the Xception model are 22,855,952, whereas for the InceptionV3 model, they are 23,626,728, which means that Xception is faster than InceptionV3. Inspired by the fine-tuning strategy, these pre-trained CNN architectures are modified by freezing the classification head of the network and training the remaining layers. The parameter settings of two models are given in Table 1.

Moreover, the data augmentation process is applied to avoid overfitting and underfitting, as well as enhancing the generalizability of the trained models. For this purpose, the most popular data augmentation methods are employed. Technically, random rotation (90° angles), shear transformation (0.1 ratios), zoom transformation (0.1 ratios), and horizontal flip operations are applied to the data. The images are normalized within the 0–1 range by multiplying with 1/255. The batch size is defined with regard to the memory of the GPU capacity.

3 Experimental Study

3.1 Dataset and Pre-processing

To build the proposed methodology, a dataset of Turkish COVID-19 cases is compiled to include lung parenchymal CT images obtained from the University of Health Sciences, Bursa Yüksek Ihtisas Training and Research Hospital as an affiliate of the Ministry of Health, Republic of Turkey. In compliance with the national procedures, this study was approved by the ethics committee of the university.

The dataset of our study includes 266 lung parenchymal CT scans in total, 180 COVID-19 pneumonias, and 86 normal (non-COVID-19) patients, all acquired from the stated university between March 29th, 2020 and October 20th, 2020. The demographic characteristics of COVID-19 and normal patients are given in Table 2. There

Table 2 Demographic characteristics of patients with COVID-19 and without

	COVID-19	Normal
Male	104 (58%)	45 (52%)
Female	76 (42%)	41 (48%)
Age	21 ± 92	19 ± 81

is no significant difference in terms of sex and age between the two groups. According to the treatment program of COVID-19 researchers at the hospital, the clinical classification of COVID-19 patients is categorized as mild, moderate, and severe. In this study, the COVID-19 dataset includes 49 milds, 53 moderate, and 78 severe patients.

A total of 9729 lung parenchymal cutting/cross-section images are obtained from the CT images for COVID-19. 70% of the data (6810) is used for training, 10% (973) for validation, and 20% (1946) for testing. The images were resized at a resolution of 480 × 640 pixels. Figure 2 shows the visualization of CT images of the two groups (COVID-19 and non-COVID-19). Detecting the presence of the coronavirus among patients is one of the most challenging tasks in medical image processing-based diagnostic systems. If AI-based, these systems can enhance the speed, precision, and effectiveness of the decisions made by clinicians [10, 18]. However, implementing an effective deep learning algorithm on lung parenchymal CT images can face certain difficulties, all of which can be addressed using a convenient pre-processing method which is widely known to improve the accuracy [17]. Also lung region segmentation is needed. The crucial stage of lung detection is based on removing the redundant noises, and the background of the digital imaging and communications in medicine (DICOM) images does not include any potentially risky regions of the lungs.

Detecting the presence of the coronavirus among patients is one of the most challenging tasks in medical image processing-based diagnostic systems. If AI-based, these systems can enhance the speed, precision, and effectiveness of the decisions

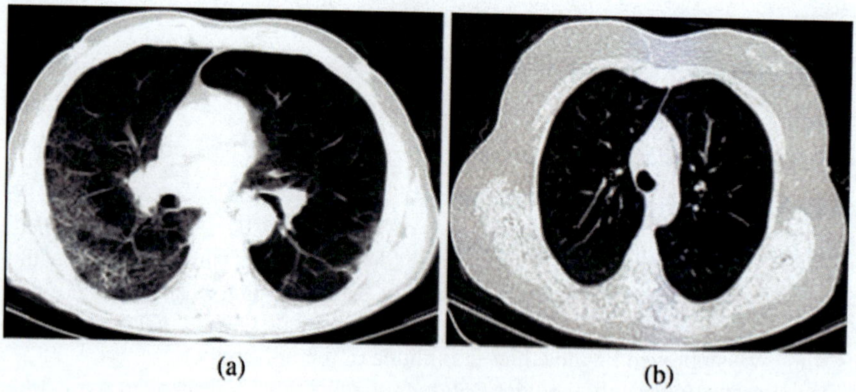

Fig. 2 Samples of chest CT images from the COVID-19 dataset, **a** COVID-19-positive case, **b** COVID-19-negative case

made by clinicians [10, 18]. However, implementing an effective deep learning algorithm on lung parenchymal CT images can face certain difficulties, all of which can be addressed using a convenient pre-processing method which is widely known to improve the accuracy [17]. Also lung region segmentation is needed. The crucial stage of lung detection is based on removing the redundant noises, and the background of DICOM images does not include any potentially risky regions of the lungs (Fig. 1).

$$F1\text{-Score} = \frac{2 \times \text{TP}}{2 \times \text{TP} + \text{FP} + \text{FN}} = \frac{2 \cdot \text{Precision} \times \text{Recall}}{\text{Precision} + \text{Recall}} \tag{1}$$

The $F1$-score rate is obtained on the basis of the true positive (TP), true negative (TN), false positive (FP), and false negative (FN) rates. For this purpose, the estimated and actual values are utilized to compute these statistical values. From its characteristic, the F1-score is the harmonic mean of precision and recall, as shown in Eq. (1); thus, the extreme values are removed from the model output. This is a more convenient evaluation measurement for imbalanced class distribution. In the equation, precision refers to TP/(TP + FP), and recall is represented with TP/(TP + FN). The confusion matrices and the overall accuracy results of each three model are drawn in Fig. 3. For a detailed performance inspection, the F1-score and area

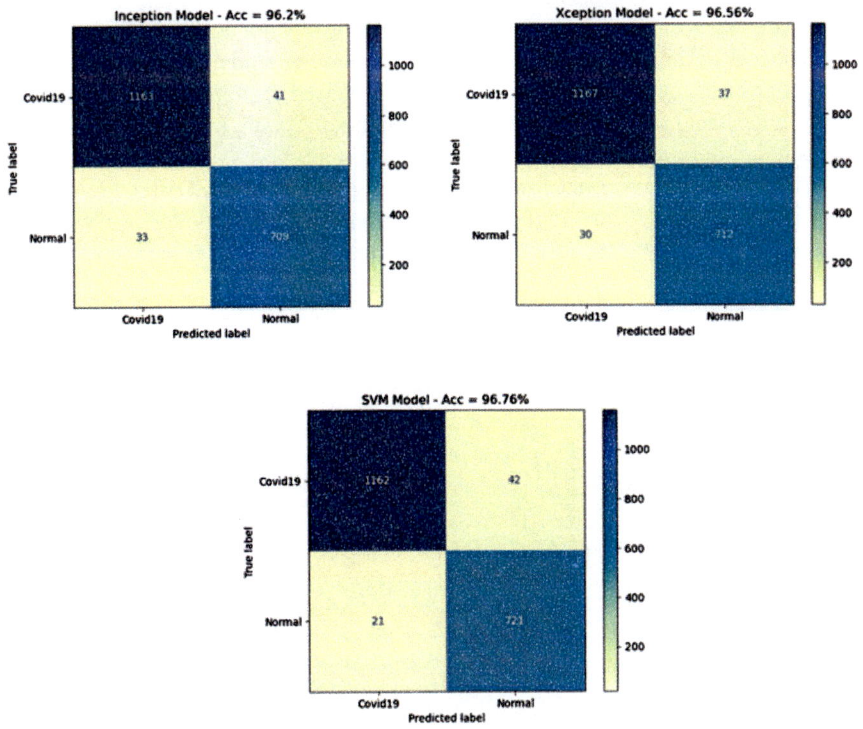

Fig. 3 The confusion matrices and % accuracy results

Table 3 The overall average performances for each classifier

Method	Acc (%)	F1-score	ROC AUC
InceptionV3	96.20	0.9500	0.9607
Xception	96.55	0.9550	0.9644
SVM fine-tune	96.76	0.9581	0.9684

under the receiver operating characteristic curve (ROC AUC) rates of each classifier are given in Table 3.

From the overall scores given in Table 3, it can be noted that fine-tuning the features of the Xception model gives superior results. When comparing Xception and InceptionV3, it can be stated that the Xception model presents more steady results for all the evaluation metrics. In the case of fine-tuning, the SVM model is trained on the features returned from the last fully connected layer of the Xception model. Technically, the $1 \times 100{,}352$ feature vectors are calculated from the training (6810 samples) and validation (973 samples) sets. Later, the chi-squared feature selection method is applied to retain meaningful features while eliminating the redundant ones. After feature selection, the only 4096×3 dimensional distinguishable features are extracted from each sample. In total, the size of the training matrix becomes $7783 \times 12{,}288$ and $1946 \times 12{,}288$ for the test matrix. The SVM parameter is determined with a squared hinge loss function, and the regularization parameter is 0.05. The AUC score of SVM is demonstrated in Fig. 4.

The accuracy results of the proposed method are compared with the recently published works. The details about these methods and our study are summarized in Table 4. For each method, the type of imaging, focused region, number of cases, models, and related accuracy scores are highlighted. Although each method is a state-of-the-art approach in terms of its own specific purposes, we will review the weakness and robustness of each method in terms of detecting COVID-19. When monitoring the accuracy scores, at first, it appears that the deepest model (ResNet-50)

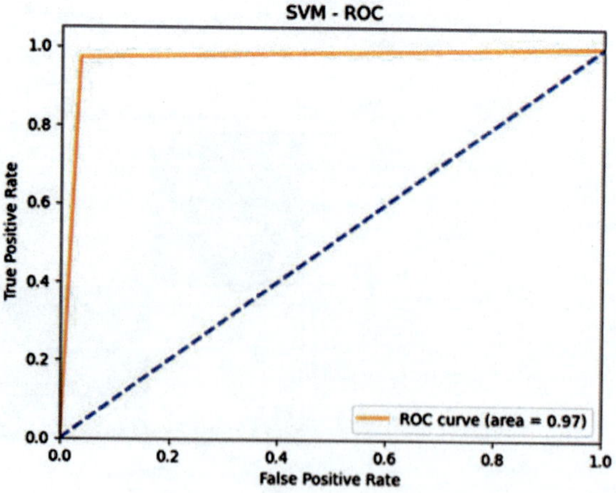

Fig. 4 ROC curve performance of SVM fine-tuning

Table 4 The detailed comparison performance of the proposed method with available methods

Study	Type of images	View	Dataset	Method	Acc (%)
Narin et al. [26]	X-ray	PA chest	50 COVID-19 50 no-findings	ResNet-50	98.00
Nour et al. [16]	X-ray	PA chest	219 COVID-19 1341 no-findings 1345 pneumonia	CNN model	98.97
Yang et al. [29]	CT	Lung parenchyma	216 COVID-19 463 no-findings	DenseNet-169	98.00
Toraman et al. [10]	X-ray	PA chest	1050 COVID-19 1050 no-findings 1050 Pneumonia	CapsNet	84.22
Zheng et al. [27]	CT	Lung parenchyma	313 COVID-19 229 no-findings	DeCovNet	90.01
Song et al. [25]	CT	Lung parenchyma	88 COVID-19 86 no-findings	VGG16	84.00
Song et al. [25]	CT	Lung parenchyma	88 COVID-19 86 no-findings	DenseNet	82.00
Song et al. [25]	CT	Lung parenchyma	88 COVID-19 86 no-findings	ResNet-50	86.00
Song et al. [25]	CT	Lung parenchyma	88 COVID-19 86 no-findings	DRE-Net	86.00
Gunraj et al. [8]	CT	Lung parenchyma	325 COVID-19 740 Pneumonia	InceptionNet	89.50

(continued)

Table 4 (continued)

Study	Type of images	View	Dataset	Method	Acc (%)
Ozturk et al. [9]	X-ray	PA chest	127 COVID-19 500 no-findings	DarkCovidNet	98.08
Proposed	CT images	Lung parenchyma	180 COVID-19 86 no-findings	Xception	96.55
Proposed	CT images	Lung parenchyma	180 COVID-19 86 no-findings	InceptionV3	96.20
Proposed	CT images	Lung parenchyma	180 COVID-19 86 no-findings	SVM fine-tune	**96.76**

has better performance; however, when training with large batch sizes, it consumes a significant amount of CPU or GPU memory. The ideal scenario is to have a high rate of detection while offering economy in terms of memory use.

As shown in Table 4, the findings of our study validate the results obtained by the previous researches. Following the presented results, the finding of the method of Song et al. [25] is contrary to the study of Narin et al. [26], which suggests that ResNet-50 yields higher discriminative scores for detecting COVID-19 positive cases. Nonetheless, these variations can also be attributed to different degrees of complexity concerning the datasets. For example, the study by Nour et al. [16] indicated that the overall recognition rate was 98.97%, much higher than that of previously reported scores in the work of Toraman et al. [10] and Zheng et al. [27]. Contrary to these two, [16] developed a modest CNN architecture by enhancing the number of convolutional layers and large number of filters, which is similar to VGG16 in terms of functional details. These improved results suggest that using a large number of weights and layers achieves higher recognition performance. The discrimination score (86.00%) observed in the investigation of DRE-Net by Song et al. [25] is far below those obtained in DarkCovNet by Ozturk et al. [9], which accounts to 98.08%. To elaborate, in [25], the DRE-Net was built on a pre-trained ResNet-50 [28] by integrating feature pyramid network (FPN) in order to extract the K-dimension features from each sample image. After comparing the dataset size of each method, one can say that the generalizability of classification performance is high for Toraman et al. [10] since the number of positive COVID-19 cases (1050) and negative cases (1050) are higher than other studies.

Upon inspecting the results obtained by the proposed methods, it can be stated that the performance of InceptionV3 (96.20%), Xception (96.55%), and SVM fine-tuning (96.76%) compete with top scores. The most interesting aspect of this experimental stage is that the fine-tuning of Xception model not only boosts the recognition scores but also reduces the memory required for label estimation. With the fine-tuning concept, the spatial discrimination between classes is converted to the minimum number of features, which is usually in the $1 \times 12{,}288$ dimensional vector form.

Furthermore, we have compared the generalizability of our study in terms of population. One can say that the number of COVID-19 positive and negative samples are 180:86, which is above the average number when compared with some existing works. What is not surprising is that the VGG16 model negatively impacts of the performance of COVID-19 detection, which means inadequate potential to capture the variations between positive and negative cases when taking the CT scans of lung parenchyma regions as reference. Also, as a fact of overfitting, after a certain limit within the layers and filters, an increase in the filters' depth or the numbers of layers fails to improve the performance of any CNN model. This can be observed from the results of ResNet-50, DenseNet-169, InceptionV3, and Xception models. As a characteristic of CNN-based probability estimation, the accuracy could not improve after converging to a global optimum point even with the best set of tuned parameters.

4 Conclusion

The purpose of the current study is to investigate the capability of efficient CNN models for COVID-19 detection. After empirical evaluations, InceptionV3, Xception, and an SVM classifier in the fine-tuning layer can be considered as reliable predictors to detect the presence of COVID-19 cases.

Also, the most apparent finding of this study is that the pre-processing stage significantly boosts the performance. Prior to segmenting the lung parenchyma regions, the performance of the system is evaluated at around 86% accuracy rate. This weakness is caused by the remaining redundant information on the CT scans. Therefore, the pre-processing stages are certainly necessary to detect the presence of COVID-19 among patients. In this regard, focusing on the most susceptible regions is a practical way for the purpose of reliable measurements with the deep learning concept. The following conclusions can be drawn from the present study:

- Using a parameter-free model, namely CNN, yields effective results. However, the top of a CNN architecture consists of a blind activation function, typically Softmax. Thus, it fails to provide a line or a plane that separates the two classes.
- Using an SVM classifier in the fine-tuning layer and focusing on the CNN features show that it is possible to set up an effective model with promising performances. The features of CNN are extracted and put forward by a determined classifier, such as SVM, random forest (RF), or decision tree (DT).
- The running time of the chosen method is important for real-time applications. After analyzing the cost, it turns out that the execution time of InceptionV3 and Xception model is about 640 milliseconds (ms) each iteration. This duration can

be determined when doing the experiments with a standard computer (Intel(R) Xeon(R) CPU E5-1620 v3 with 3.50 GHz CPU, 24 GB memory, and 4 GB GPU).
- All of the obtained quantitative results support the robustness of the proposed methods for COVID-19 detection.

References

1. P. Ziegler, The black death. Faber & Faber (2013)
2. A. Trilla, G. Trilla, C. Daer, The 1918 "Spanish flu" in Spain. Clin. Infect. Dis. **47**(5), 668–673 (2008)
3. M.S. Razai, K. Doerholt, S. Ladhani, P. Oakeshott, Coronavirus disease 2019 (COVID-19): a guide for UK GPs. BMJ **368** (2020).
4. L. Li, L. Qin, Z. Xu, Y. Yin, X. Wang, B. Kong, J. Bai, Y. Lu, Z. Fang, Q. Song, K. Cao, D. Liu, G. Wang, Q. Xu, X. Fang, S. Zhang, J. Xia, J. Xia, Artificial intelligence distinguishes COVID-19 from community acquired pneumonia on chest CT. Radiology (2020)
5. W. COVID (2020). 19: USA. https://covid19.who.int/region/amro/country/us. Last accessed 2021/08/06
6. WHO, Report of the WHO-China joint mission on coronavirus disease 2019 (COVID-19). https://www.who.int/publications/i/item/report-of-the-who-china-joint-mission-on-coronavirus-disease-2019-(covid-19). Last accessed 2021/08/06
7. A.A. Ardakani, A.R. Kanafi, U.R. Acharya, N. Khadem, A. Mohammadi, Application of deep learning technique to manage COVID-19 in routine clinical practice using CT images: results of 10 convolutional neural networks. Comput. Biol. Med. **121**, 103795 (2020)
8. H. Gunraj, L. Wang, A. Wong, COVIDNET-CT: a tailored deep convolutional neural network design for detection of COVID-19 cases from chest CT images. Front. Med. **7** (2020)
9. T. Ozturk, M. Talo, E.A. Yildirim, U.B. Baloglu, O. Yildirim, U.R. Acharya, Automated detection of COVID-19 cases using deep neural networks with X-ray images. Comput. Biol. Med. **121**, 103792 (2020)
10. S. Toraman, T.B. Alakus, I. Turkoglu, Convolutional capsnet: a novel artificial neural network approach to detect COVID-19 disease from X-ray images using capsule networks. Chaos Solitons Fractals **140**, 110122 (2020)
11. J.P. Cohen, COVID-19 chest X-ray dataset or CT dataset, Git Hub: https://github.com/ieee8023/covid-chestxray-dataset (2020)
12. MCT Rahman, A. Khandakar, COVID-19 radiography database, Kaggle. https://www.kaggle.com/tawsifurrahman/COVID-19-radiography-database/data. Last accessed: 2021/08/06
13. M. Toğaçar, B. Ergen, Z. Cömert, COVID-19 detection using deep learning models to exploit Social Mimic Optimization and structured chest X-ray images using fuzzy color and stacking approaches. Comput. Biol. Med. **121**, 103805 (2020)
14. L. Brunese, F. Mercaldo, A. Reginelli, A. Santone, Explainable deep learning for pulmonary disease and coronavirus COVID-19 detection from X-rays. Comput. Methods Programs Biomed. **196**, 105608 (2020)
15. N.N. Das, N. Kumar, M. Kaur, V. Kumar, D. Singh, Automated deep transfer learning-based approach for detection of COVID-19 infection in chest X-rays. IRBM (2020)
16. M. Nour, Z. Cömert, K. Polat, A novel medical diagnosis model for COVID-19 infection detection based on deep features and Bayesian optimization. Appl. Soft Comput. 106580 (2020)
17. M. Heidari, S. Mirniaharikandehei, A.Z. Khuzani, G. Danala, Y. Qiu, B. Zheng, Improving the performance of CNN to predict the likelihood of COVID-19 using chest X-ray images with preprocessing algorithms. Int. J. Med. Inform. **144**, 104284 (2020)
18. S. Lalmuanawma, J. Hussain, L. Chhakchhuak, Applications of machine learning and artificial intelligence for Covid-19 (SARS-CoV-2) pandemic: a review. Chaos, Solitons Fractals 110059 (2020)

19. M.Z. Islam, M.M. Islam, A. Asraf, A combined deep CNN-LSTM network for the detection of novel coronavirus (COVID-19) using X-ray images. Inform. Med. Unlocked **20**, 100412 (2020)
20. R.M. Pereira, D. Bertolini, L.O. Teixeira, C.N. Silla Jr, Y.M. Costa, COVID-19 identification in chest X-ray images on flat and hierarchical classification scenarios. Comput. Methods Programs Biomed. **194**, 105532 (2020)
21. A. Krizhevsky, I. Sutskever, G.E. Hinton, Imagenet classification with deep convolutional neural networks. Adv. Neural. Inf. Process. Syst. **25**, 1097–1105 (2012)
22. K. Simonyan, A. Zisserman, Very deep convolutional networks for large-scale image recognition. arXiv preprint arXiv:1409.1556 (2014)
23. C. Szegedy, V. Vanhoucke, S. Ioffe, J. Shlens, Z. Wojna, Rethinking the inception architecture for computer vision, in *Proceedings of the IEEE Conference on Computer Vision and Pattern Recognition*, pp. 2818–2826 (2016)
24. F. Chollet, Xception: deep learning with depthwise separable convolutions, in *Proceedings of the IEEE Conference on Computer Vision and Pattern Recognition*, pp. 1251–1258 (2017)
25. Y. Song, S. Zheng, L. Li, X. Zhang, X. Zhang, Z. Huang, J. Chen, R. Wang, H. Zhao, Y. Chong, J. Shen, Y. Zha, Y. Yang, Deep learning enables accurate diagnosis of novel coronavirus (COVID-19) with CT images. IEEE/ACM Trans. Comput. Biol. Bioinform. (2021)
26. A. Narin, C. Kaya, Z. Pamuk, Automatic detection of coronavirus disease (COVID-19) using X-ray images and deep convolutional neural networks. Pattern Anal. Appl. 1–14 (2021)
27. C. Zheng, X. Deng, Q. Fu, Q. Zhou, J. Feng, H. Ma, W. Liu, X. Wang, Deep learning-based detection for COVID-19 from chest CT using weak label. MedRxiv (2020)
28. K. He, X. Zhang, S. Ren, J. Sun, Deep residual learning for image recognition, in *Proceedings of the IEEE Conference on Computer Vision and Pattern Recognition*, pp. 770–778 (2016)
29. X. Yang, X. He, J. Zhao, Y. Zhang, S. Zhang, P. Xie, COVID-CT-dataset: a CT scan dataset about COVID-19 (2020)

Spatiotemporal Location Privacy Preservation in 5G-Enabled Sparse Mobile Crowdsensing

MingChu Li, Qifan Yang, Xiao Zheng, and Liqaa Nawaf

Abstract With the increasing popularity of 5G communications, smart cities have become one of the inevitable trends in the development of modern cities, and smart city services are the foundation of 5G smart cities. Sparse mobile crowdsensing (SparseMCS), as a new and informative urban service model, has attracted the attention of many researchers. Generally, the data required for a sensing task often has a high spatial and temporal correlation, which means that the data uploaded by users need to carry their location information, which may cause serious location privacy issues. The existing location privacy protection mechanism usually only pays attention to the location information of the user's travel and ignores that people's daily travel often has a fixed pattern. The attacker can use long-term observation and prior knowledge to infer the victim's travel mode and analyze its location information. To achieve efficient, robust, and private data sensing, we built a SparseMCS framework with the following three elements: (1) We train the data adjustment model offline on the server-side and solve the position mapping matrix; (2) Design a noise-sensitive data reasoning algorithm improves the accuracy of data; (3) Combining differences and spatiotemporal location privacy to protect the user's location information and travel mode. Experiments based on real datasets prove that our 5G-supported sparse mobile crowdsensing framework provides more comprehensive and effective location privacy protection.

Keywords 5G · Mobile crowdsensing · Location privacy · Differential privacy · Spatiotemporal phenomena

M. Li (✉) · Q. Yang
School of Software, Dalian University of Technology, Dalian, China
e-mail: mingchul@dlut.edu.cn

Q. Yang
e-mail: dekusmash_yqf@mail.dlut.edu.cn

X. Zheng
School of Computer Science and Technology, Shandong University of Technology, ZiBo, China
e-mail: xiao_zheng0910@163.com

L. Nawaf
Cardiff Metropolitan University, Cardiff, UK
e-mail: LLLNawaf@cardiffmet.ac.uk

A. K. Bashir et al. (eds.), *Proceedings of International Conference on Computing and Communication Networks*, Lecture Notes in Networks and Systems 394,
https://doi.org/10.1007/978-981-19-0604-6_24

1 Introduction

5G is the current mainstream new-generation mobile communication technology and an essential part of the next-generation information infrastructure [1]. The high-quality information services of 5G provide a good communication foundation for the construction of smart cities and industrial Internet of Things [2–6]. Mobile crowdsensing systems can provide city services for the smart city systems, traffic information, weather information, and other services system. Therefore, mobile crowdsensing (MCS) has developed rapidly in recent years and has become a significant computing paradigm in smart city data sensing scenarios. MCS plays a crucial role in collecting ambient temperature, traffic flow [7], noise [8], and air quality [9] in inter-city areas. In mainstream MCS, the publisher will launch a data sensing task for a specific target area. Service providers screen and recruit mobile users according to task requirements and perform tasks in the target area. However, large-scale data collection tasks such as urban tasks require many users to cover all target points. Therefore, urban tasks often require much budget, and target points are often missed due to uneven population distribution, and data redundancy may also occur in densely populated areas.

One solution is *Sparse Mobile Crowdsensing* (SparseMCS), which combines historical records and sensing data in nearby areas to infer task demand data in unperceived areas [10]. In SparseMCS, users need to report their location and time when uploading data, bringing considerable risks to user privacy [11]. Therefore, designing an effective privacy protection mechanism for the system can attract and retain more participants. In order to enable the MCS server to distinguish the data uploaded by each user, the privacy protection mechanism designed according to anonymity usually needs to retain the mapping information between the user's real identity and the anonymous information. If the server is attacked, users will face personal severe privacy risks. In contrast, according to the obfuscated design mechanism, it can usually be configured in a lightweight manner on a mobile device, thereby avoiding the hosting of accurate information. Therefore, we design a location privacy protection mechanism based on confusion.

Researchers have conducted extensive research on location privacy in location-based services. These two mechanisms are usually considered to protect user location privacy [12]: (a) The user protects privacy by making location tracking and personal identity impossible to associate by anonymous means; (b) Remapping the location to change the location information released by the user.

Cloaking is a prevalent obfuscation technology. The user can hide the actual location in multiple fine-grained stealth areas instead of one or several specific areas or units. However, when the adversary has some knowledge of the target user, the effect of cloaking may be significantly reduced [13]. For example, when the adversary learns that the target user is a doctor and that the user's cloaking area covers a hospital or other medical facilities, it is easy to locate the target.

In response to this problem, we use *differential privacy* [14] to ensure that the probability of accurate location mapping to different locations is approximate.

In location-based service (LBS), we usually use the distance between the actual location and the confusion location to measure data loss. Unlike it, the ultimate goal of SparseMCS is to collect target data, so the data loss in the system is determined by the difference between the actual location and the confusion location of the target data. According to this feature, we can think that in SparseMCS, as long as the target data difference between the actual location and the confusing location is slight, the user can theoretically map the actual location to a very far place. Therefore, we should redesign the location privacy protection mechanism according to the characteristics of SparseMCS.

Even if differential privacy is used, the user cannot control the range of the adversary's estimation of his location, which is an inference error [15]. Therefore, we added a *distortion privacy* [15] mechanism to control inference errors. Distortion privacy controls the adversary's estimated range by controlling the expected distance between the adversary's inferred location and the actual location. Applying distortion privacy requires the presumption of prior knowledge possessed by the adversary. Distortion privacy is to limit the inference attack to the preset inference error. The adversary cannot achieve a better inference error within the preset prior knowledge range than the optimal error. However, we cannot know the adversary's prior knowledge, so distortion privacy is not a powerful privacy protection mechanism. It needs to be used in conjunction with other privacy protection methods to provide more comprehensive protection.

However, the above LPPMs only consider the user's exact location information and do not realize that the location change of mobile users is a complex combination of time and space [16]. For example, "Alice went to a certain supermarket last week" (this behavior may occur more than once) and "Bob travels between A address and B address" (this behavior may occur every working day). In this article, we call it spatiotemporal location. We do not know whether the differential privacy mechanism can simultaneously guarantee a certain level of privacy in spatiotemporal locations. Therefore, we have introduced the privacy goal of the spatiotemporal location to ensure that users' daily travel patterns can be protected.

The main contributions of our work are:

- In order to provide more comprehensive location privacy protection, we propose a privacy protection framework that includes three privacy mechanisms. (a) Differential privacy guarantees the geographic indistinguishability; (b) Distortion privacy limits the adversary's optimal estimation error on prior knowledge; (c) Spatiotemporal location privacy guarantees the privacy of user behavior patterns.
- In order to improve the reliability and efficiency of the system, we designed a noise-aware reasoning algorithm to improve the data accuracy of the unperceived area.
- We validated our framework using real-world temperature datasets. The results show that our method, while providing a higher level of location privacy protection, limits the error within the range of 10^{-2}.

2 Related Work

With the rollout of 5G networks, the 5G environment integrates numerous location-based services, and mobile group awareness is one of them. Mobile crowdsensing is a data collection service that uses mobile devices to collect environmental data in the urban environment (for example, noise, air quality, temperature information, traffic flow) by hiring users distributed in different locations in the city. However, due to the large sensing area or limited budget, there may not be enough users to complete the sensing task. Wang et al. [10] proposed sparse mobile crowdsensing to solve this problem. Both MCS and SparseMCS can be regarded as a kind of LBS. Recruited users often need to expose their location to the task organizer, which involves serious location privacy issues.

Currently, most location privacy protection mechanisms mainly use two technologies: anonymity and obfuscation [17]. However, these two technologies will significantly reduce the strength of privacy protection when facing adversaries with prior knowledge [13]. In response to this problem, Andrés et al. [13] introduced the concept of differential privacy into location privacy protection to prevent attacks from adversaries with prior knowledge. According to the survey results of Pournajaf et al. [18], the current location privacy protection technology in MCS is mainly obfuscation technology. Many researchers combine MCS with edge computing. Putra et al. [19], Li et al. [20], and others have studied the location privacy protection mechanism in this environment. The DU-Min-$\epsilon\delta$ [21] proposed by Wang et al. realizes location privacy protection in the SparseMCS environment. Compared with this algorithm, our work considers the time dimension of location information and realizes the protection of user travel patterns.

3 Sparse Mobile Crowdsensing Concepts

3.1 *Sparse Mobile Crowdsensing*

3.1.1 Computation Paradigm

As shown in Fig. 1 (Basic), when monitoring temperature changes in a target city is started, the city will be divided into multiple fine-grained target areas. The user will collect the temperature data of the current area and upload the collected temperature data, identity information, and location information to the server. The server will use real-time sensor data and historical data to infer temperature information in areas that the user has not reached.

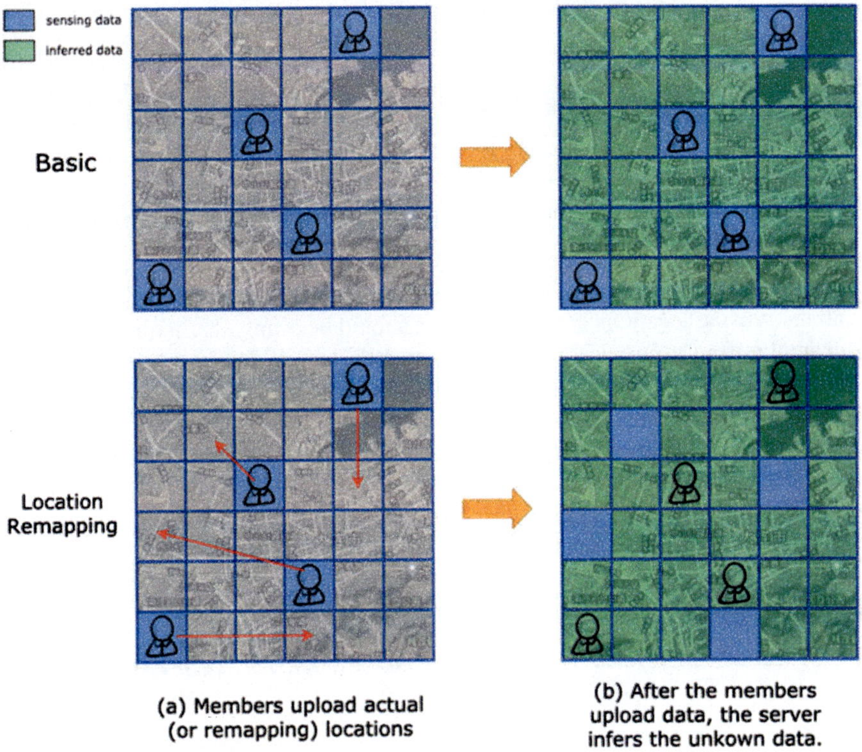

Fig. 1 Basic data collection is in sparse mobile crowdsensing, and data collection with location privacy protection added

3.1.2 Data Collection

The user will start the data collection task at the current location. In order to be able to verify identity and verify data, the server will require the user to upload identity information. At the same time, the target data collected by the user should have location information to achieve complete semantic functions, so location information should also be uploaded. Therefore, the simultaneous exposure of the user's identity information and location information to the task organizer will cause serious privacy risks. The SparseMCS system needs to enable strict LPPM to reduce the user's risk. However, LPPM will adjust the location information of the target data, which destroys the semantic function of the target data to a certain extent. The destruction of semantic functions will reduce the accuracy of target data.

3.1.3 Data Inference

In our work, we use compressed sensing as our data inference algorithm [22]. Candès and Plan [23] have proved that recovering an unknown low-rank matrix can uniformly sample a small number (less than the size of the matrix) with noisy entries. The recovery error is proportional to the noise level. In other words, there are two inherent assumptions in the use of compressed sensing algorithms to achieve data inference:

- Uniform distribution: The compressed sensing algorithm requires that sampled data are evenly distributed in the sampling space. That is to say, in SparseMCS, all the sensing locations in the target area should be evenly distributed. If not, for example, if no user exists in a specific area during all the sensing periods, it is impossible to infer the missing data in that area.
- Weak noise environment: When the sampling items do not carry noise and meet uniform sampling, the missing data in the matrix can be accurately inferred. When the sampling items carry noise, the total inference error is proportional to the noise level. That is to say, the smaller the noise carried in the sampling items, the higher the accuracy of the inference results.

3.2 *Location Privacy-Preserving Framework for SparseMCS*

The sensing data uploaded by the user in SparseMCS should include the target data and the actual location. Figure 1 (Location Remapping) shows that using obfuscation technology to add noise to the location information can reduce the user's privacy risk. However, this method will bring about data quality loss because the actual location and the target data of the confusing location may be different. In response to this problem, we designed a location privacy-sensitive SparseMCS framework composed of two parts: location remapping and data adjustment.

Figure 2 shows the location privacy protection framework of SparseMCS we designed, which consists of two parts: the server-side and the mobile user side. Before the task starts, the server will realize the data adjustment function and generate the location mapping probability matrix in the offline state according to the historical data. The data adjustment function is based on learning the relationship between the historical data in any two regions. This function will reduce the quality loss of the target data caused by the noise caused by the location remapping. By adjusting the probability matrix item $[i, j]$ (the probability that location i is mapped to location j), we can ensure that the adversary cannot accurately infer the user's actual location even if he gets the matrix.

Before performing the task, the user saves the data adjustment function and the mapping matrix on the mobile device. The task execution process is as follows: First, the user adjusts to the confusion location according to the current cycle and actual location according to the location mapping probability matrix (step M1). Subsequently, according to the actual location and the confused location, the original

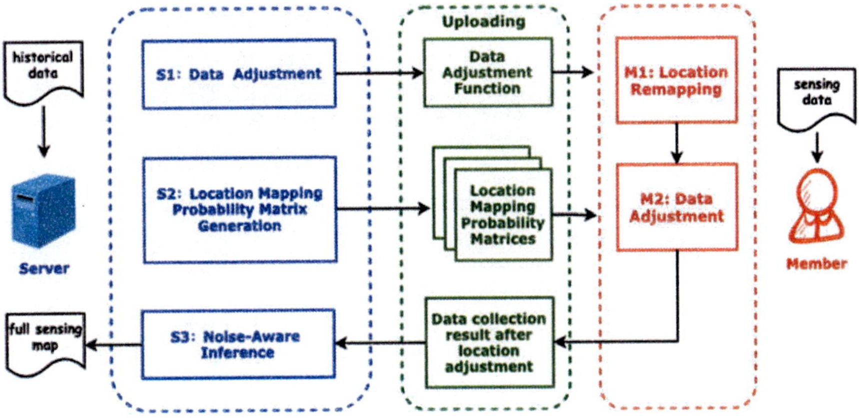

Fig. 2 Location privacy-preserving framework for SparseMCS

data is adjusted to the adjusted data with noise using the data adjustment function (step M2). The user uploads the adjustment data and the confused location to the server, and then the server combines the historical data to infer a complete sensing map (step S3).

4 Differential and Distortion and Spatiotemporal Location Privacy

This section introduces the privacy protection concept we applied in SparseMCS. Our privacy protection mechanism focuses on Bayesian attacks. Distortion privacy cannot resist Bayesian attacks, but it can effectively limit the optimal attack model based on Bayesian inference to minimize errors. The major notations are summarized in Table 1.

4.1 Differential Location Privacy

The purpose of introducing differential privacy is to bind the improvement of the posterior knowledge acquired by the adversary to the prior knowledge [14]. Differential privacy will make the probability of mapping from the real location r to any confusion location r^* similar.

Table 1 Notations

$\mathcal{R}$	Sensing task target area, $\mathcal{R} = \{r_1, r_2, \ldots, r_n\}$
P	Location mapping probability matrix
r	Real location $r \in \mathcal{R}$
r', r^*	Confusion locations $r', r^* \in \mathcal{R}$
$\tilde{r}$	Adversary inferred location $\tilde{r} \in \mathcal{R}$
$P(r^*\|r)$	The probability of location r mapped to location r^*
$\tilde{\beta}$	Adversary's inference attack
$\rho(\tilde{r}, r)$	The distance between $\tilde{r}$ and r^*
$\eta_u(r)$	The location distribution of target user
T	The time period for the user to release locations
$\mathcal{O}$	The user's observable location sequence
$\mathcal{S}$	User-defined sensitive areas

Definition 1 (*ϵ-Differential Privacy*) Assuming that the $\mathcal{R}$ is divided into multiple fine-grained areas r, then the P satisfies ϵ-Differential Privacy iff:

$$P(r^*|r) \leq e^{\epsilon} \cdot P(r^*|r'), \qquad \forall r, r', r^* \in \mathcal{R} \tag{1}$$

where ϵ represents the privacy budget.

4.2 *Distortion Location Privacy*

Although differential privacy limits the adversary's information gain, users still cannot determine how close the adversary's estimated location is to its actual location, that is, how small the adversary's inference error is. In order to limit the inference error, we adopt distortion privacy [15]. This method can ensure that the adversary's optimal attack inference error will be greater than a particular value for a given user's public location distribution information.

4.2.1 Attack Inference Error

The attack inference error can be obtained by the following equation:

$$\sum_{r^* \in \mathcal{R}} P(r^*|r) \sum_{\tilde{r} \in \mathcal{R}} \tilde{\beta}(\tilde{r}|r^*) \cdot \rho(\tilde{r}, r) \tag{2}$$

We assume that the location distribution η_u is partially disclosed (for example, social network check-ins [24]). Furthermore, the adversary obtains the distribution, and he can minimize the expected reasoning error on η_u to achieve the optimal attack.

$$\arg\min_{\tilde{\beta}} \sum_{r^* \in \mathcal{R}} \eta_u(r) \sum_{r^* \in \mathcal{R}} P(r^*|r) \sum_{\tilde{r} \in \mathcal{R}} \tilde{\beta}(\tilde{r}|r^*) \cdot \rho(\tilde{r}, r) \tag{3}$$

4.2.2 Definition of Distortion Location Privacy

Definition 2 (*δ-Distortion Privacy*) The location mapping probability matrix P satisfies δ-Distortion Privacy iff:

$$\sum_{r^* \in \mathcal{R}} \eta_u(r) \sum_{r^* \in \mathcal{R}} P(r^*|r) \sum_{\tilde{r} \in \mathcal{R}} \tilde{\beta}(\tilde{r}|r^*) \cdot \rho(\tilde{r}, r) \geq \delta \tag{4}$$

where δ is the lower bound of privacy disclosure acceptable to users.

The disclosed location distribution η_u does not always summarize the adversary's prior knowledge. The distortion privacy only makes a mild assumption and cannot contain some extreme situations.

4.3 Spatiotemporal Location Privacy

Figure 3 vividly shows the relationship and difference between the spatial dimension, time dimension, and space-time dimension of location privacy. Differential and distortion privacy only realizes the case of privacy protection in the spatial dimension, and it is not clear whether it can provide location privacy protection in the spatial and temporal dimensions. Therefore, we use spatiotemporal location privacy [25] to extend location privacy protection to the spatial and temporal dimensions.

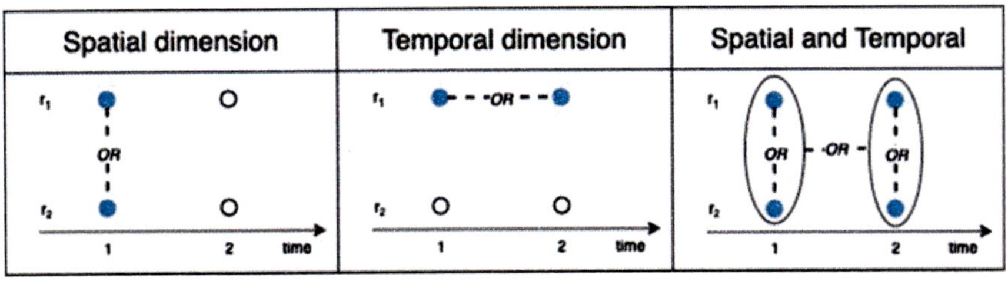

Fig. 3 Dimensional analysis of location privacy. Spatial dimension: privacy refers to a sensitive area including location r_1 and r_2; temporal dimension: privacy refers to accessing location r_1 at time point 1 or 2; spatial and temporal dimension: the user's sensitive area consists of locations r_1 and r_2 at time point 1 and 2

Definition 3 (*ϵ-Spatiotemporal Location Privacy*) Suppose that in period $T = \{1, 2, \ldots, t\}$ and area $\mathcal{R}$. The user sets the sensitive area $\mathcal{S}$, and generates the observation sequence $\mathcal{O} = \{r^1, r^2, \ldots, r^t\}$, $\forall r^i \in \mathcal{R}$. $\mathcal{O}$ satisfies the ϵ-Spatiotemporal Location Privacy iif:

$$P(\mathcal{O}|\mathcal{S}) \le e^{\epsilon} \cdot P(\mathcal{O}|\neg\mathcal{S}) \tag{5}$$

where $\neg\mathcal{S}$ is the complementary set of $\mathcal{S}$, and $U = \mathcal{S} \cap \neg\mathcal{S}$ represents all possible sensitive areas in the target area.

4.4 *Combined Location Privacy-Preserving Mechanism*

In this section, we will briefly describe the advantages of combining the above three privacy concepts. For differential privacy, it is difficult for the adversary to predict the correct location of the user accurately. For distortion location privacy, the adversary's prediction should keep a certain distance from the correct location even if the guess is wrong. For spatiotemporal location privacy, it is difficult for adversaries to analyze the user's travel habits through long-term observation. Therefore, we combine these three privacy concepts to provide users with more comprehensive location privacy protection.

5 Location Privacy Protection Mechanism with Minimal Data Loss

5.1 *Data Quality Requirements for Location Mapping*

Recall that the prerequisites for data inference introduced in Sect. 3.1.3 include the average distribution of sampled data and a weak noise environment [23]. However, the introduction of a location privacy mechanism will destroy these two premises:

(a) *Uniform distribution of confusion locations*: In real life, users may be evenly distributed within the city, but the location distribution may be very uneven after location mapping. For example, suppose that no user's location may map to location i. Then, the value of the ith row in the sensing matrix will be lost, and the ith row data will not be restored during the inference process.
(b) *Weak noise environment*: After the location remapping process, the target data submitted by the user corresponds to the original location rather than the confusing location. Although the resulting error can be reduced through data adjustment, the uploaded target data is still inaccurate. Therefore, we need to generate a matrix P that can minimize the loss of data quality.

5.2 *Optimal Location Mapping Matrix Generation*

In order to meet the challenge mentioned in Sect. 5.1, we designed an optimization problem to generate the position mapping probability matrix.

5.2.1 Objective: Minimize Data Loss

We denote the data loss produced by this process as a matrix L. The entry $L[r, r^*] \in L$ records the residual standard deviation between the original data and the adjusted data. Intuitively, the less data loss caused by the location mapping process, the better. We hope to find a mapping matrix P that can minimize the overall expectation of the data loss caused by the process, denoted as $\bar{L}$. Our optimization goal is to minimize $\bar{L}$.

$$\bar{L} = \sum_{r \in \mathcal{R}} \eta_u(r) \cdot \sum_{r^* \in \mathcal{R}} L[r, r^*] \cdot P[r, r^*] \tag{6}$$

5.2.2 Location Mapping Probability Matrix Generation

According to Definition 5, spatiotemporal location privacy should protect a sensitive area in the time period, and differential and distortion location privacy is protected for the location at the time point. Therefore, we generate a matrix P^t that satisfies differential and distortion location privacy at each time point in the sensing cycle T. $\mathcal{P} = \{P^1, P^2, \ldots, P^t\}, \forall t \in T$ satisfies the spatiotemporal location privacy in the period. In order to minimize the expectation of data loss and ensure the location mapping probability matrix $\mathcal{P}$ of differential privacy, distortion privacy, and spatiotemporal location privacy.

(a) *Constraint 1*: *ϵ-Differential Privacy*: The first constraint is ϵ-Differential Privacy, which is implemented by Eq. 8.
(b) *Constraint 2*: *δ-Distortion Privacy*: The second constraint is δ-Distortion Privacy, which is implemented by Eqs. 9 to 10. Because Eq. 4 in Definition 2 contains an optimization problem (Eq. 3), we cannot directly use it as a constraint. Therefore, according to Shokri's work [15], we convert Eq. 4 into Eqs. 9 and 10.
(c) *Constraint*: *ϵ-Spatiotemporal Location Privacy*: The third constraint is ϵ-Spatiotemporal Location Privacy, which is implemented by Eq. 5. At each time t in the period T, a P^t that meets the requirements is generated. P^t meets the requirements of spatiotemporal location privacy, which means that the probability of the user appearing in the sensitive area and not appearing in the sensitive area within this period is similar. According to Definition 5, we can abstract the user's presence in the sensitive area S in T into a boolean expression. Assuming that the period $T = \{1, 2\}$, the sensitive area $\mathcal{S} = \{r_1, r_2\}, \forall r_i \in \mathcal{R}$, the user's appearance in the sensitive area during this period can be abstracted as $[(u^1 = r_1) \vee (u^1 = r_2)] \wedge [(u^2 = r_1) \vee (u^2 = r_2)]$. In the expression, $u^t = r_i$

means that the user is at the location of r_i at time t. Abstracting the period T and the sensitive area $\mathcal{S}$ and converting the Boolean expression into a probability expression, Eq. 11 can be obtained.

The optimization problem established based on the above content is as follows:

$$\underset{P^t}{\arg\min} \quad \bar{L}(P^t) = \sum_{r \in \mathcal{R}} \eta_u(r) \sum_{\substack{r^* \in \mathcal{R} \\ t \in T}} L\left[r, r^*\right] \cdot P^t\left(r^* \mid r\right) \tag{7}$$

$$\text{s.t.} \quad P^t(r^*|r) \leq e^{\epsilon} \cdot P^t(r^*|r'), \forall r, r', r^* \in \mathcal{R}, \forall t \in T \tag{8}$$

$$\sum_{r \in R} \eta_u(r) P^t\left(r^*|r\right) d(\tilde{r}, r) \geqslant x^t\left(r^*\right), \forall \tilde{r}, r^* \in \mathcal{R}, \forall t \in T \tag{9}$$

$$\sum_{r^* \in \mathcal{R}} x^t(r^*) \geq \delta, \forall t \in T \tag{10}$$

$$\prod_{t \in T} \eta_u(r) \sum_{r^* \in R} P^t\left(r^*|r\right) \leqslant e^{\epsilon} \prod_{t \in T} \eta_{\mu}\left(r'\right) \sum_{r^* \in R} P^t\left(r^*|r'\right), \quad \forall r \in \mathcal{S}^*, r' \in \mathcal{S}_i, \forall \mathcal{S}_i \in \neg \mathcal{S}^* \tag{11}$$

$$\sum_{r \in \mathcal{R}} \eta_u(r) \cdot P^t(r^*|r) = 1/\mathcal{R}, \forall r^* \in \mathcal{R}, \forall t \in T \tag{12}$$

$$\sum_{r^* \in \mathcal{R}} P^t(r^*|r) = 1, \forall r \in \mathcal{R}, \forall t \in T \tag{13}$$

$$P^t(r^*|r) \geq 0, \forall r, r^* \in \mathcal{R}, \forall t \in T \tag{14}$$

5.3 Noise-Aware Inference Algorithm

Compressed sensing algorithms require a weak noise environment, but in order to ensure location privacy, we have violated this condition. In order to solve this problem, we designed a noise-aware mechanism to sample data with a small amount of noise with a higher weight. The following equation obtains the data loss expectation corresponding to each mapping location:

$$\bar{L}_{\cdot, r^*} = \sum_{r \in \mathcal{R}} \eta_u(r) \cdot P(r^*|r) \cdot L[r, r^*] \tag{15}$$

According to the expected data loss of each mapping location, we obtain the sampling weight corresponding to each mapping location through the following equation:

$$\omega_{r^*} = \omega_0 + (1 - \omega_0) \cdot \frac{\bar{L}_{\max} - \bar{L}_{\cdot, r^*}}{\bar{L}_{\max} - \bar{L}_{\min}} \tag{16}$$

where, $\bar{L}_{max}$ and $\bar{L}_{min}$, respectively, represent the largest and smallest data loss expectations in all mapping locations. We denote the sampling weight of the mapping location corresponding to the expected maximum data loss as ω_0. According to the experiment in Sect. 6, we recommend setting the weight to 0.75.

6 Evaluation

6.1 Configuration Environment

6.1.1 Baseline

We use three baselines that implement differential location privacy protection. Under the same level of differential privacy, we will show that our method will additionally protect user behavior patterns with similar data quality loss.

(a) *Self*: *Self* [26] algorithm provides a higher probability for location self-mapping. Formally, the location mapping matrix generated by this algorithm satisfies differential privacy:

$$P_{i,j} = \begin{cases} \alpha\, e^{\epsilon}, & \text{if } i = j, \\ \alpha, & \text{o.w.} \end{cases}$$

(b) *Laplace*: The Laplacian mechanism [13] completes privacy protection by adding Laplacian noise to the actual data. This method is more inclined to map locations to neighbor locations.

(c) *DU-Min-$\epsilon\delta$*: This method [21] constructs a linear optimization problem to achieve local location differential privacy in a sparse crowdsensing environment.

6.1.2 Evaluation Environment

We used SensorScope [27] open-source actual temperature sensing data as our experimental dataset. They deployed temperature sensors on the EPFL campus, covering an area of 300 m $\times$ 500 m. We divide it into 100 sub-areas with 30 m $\times$ 50 m, of which 57 contain temperature sensors (that is, contain real data). The data collection lasted for a week, the sensing period was 30 min, the first day's data was used as the training data adjustment function, and the location mapping matrix was solved, and the rest were used as tests (Table 2).

Table 2 Evaluation parameters

	Default	Description
k	4	Number of sensing data collected in each cycle
ϵ	ln 4	Differential privacy budget
c	3	Number of cycles users perform sensing tasks
ω	0.75	Basic sampling weight

6.1.3 Experimental Parameters

We assign different privacy budgets, ϵ the number of sensing data collected by participants in each sensing cycle, k, and the number of cycles that participants perform sensing tasks, c, as experimental independent variables. ϵ is usually customized by the user. For convenience, we set it from $\ln 2$ to $\ln 8$. The publisher generally determines k based on the budget held and the quality of the data required. The service provider will set c based on the user's travel mode and expected data quality.

6.1.4 Data Quality Metric

We use the *Mean Absolute Error (MAE)* to calculate the data loss of the inferred data compared to the real data. Every time we modify the experimental parameters, we perform five repeated experiments and take the average value. The data loss caused by the location privacy protection mechanism (LPPM) is defined as follows:

$$L_{\mathrm{MAE}}(\mathrm{LPPM}) = \mathrm{MAE}(\mathrm{LPPM}) - \mathrm{MAE}(\text{No-Privacy})$$

6.2 *Experimental Performance*

Our experimental results show that our work can provide more effective location privacy protection at a lower cost of data loss. Compared with the baseline algorithm, our work provides more comprehensive location privacy protection, and the additional data loss generated on this basis is also controlled within the range of 10^{-2}. When the number of task cycles c is small, the data quality loss caused by our work can be further controlled within 10^{-3}.

We measured how the target data quality loss changes with the privacy budget ϵ. From Fig. 4, we can see that as the privacy budget increases (the intensity of privacy protection decreases), data quality loss will decrease. In general, our work is better than the Laplace and self versions under the same conditions. When the privacy budget is small, the error level of our work is similar to that of DU-Min-$\epsilon\delta$.

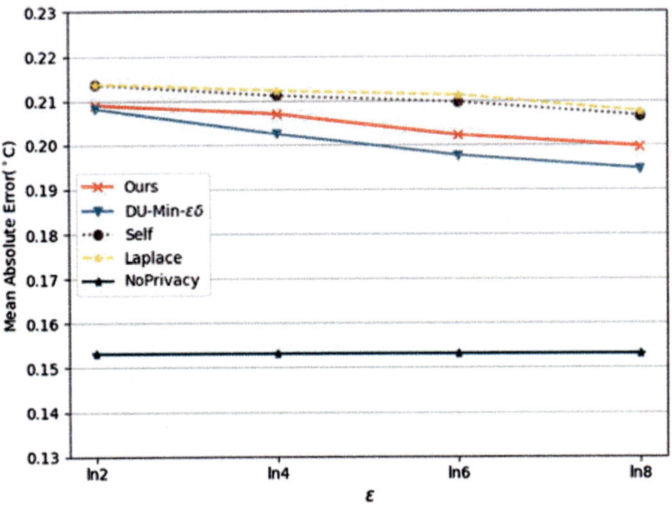

Fig. 4 MAE changes with ϵ

Figure 5 shows how the target data quality varies with the number of sensor data k collected in each cycle. From Fig. 5, we can see that the target data quality loss will decrease with the increase of k, even if we change the number of cycles c for the user to perform sensing tasks. In addition, due to the data noise caused by LPPM, the data quality loss decreases more and more slowly and cannot reach the level of no privacy. Moreover, our algorithm is superior to *Laplace* and *Self* in terms of data loss. Compared with DU-Min-$\epsilon\delta$, the error is also controlled within 10^{-2}. More importantly, we provide more comprehensive location privacy protection.

Figure 6 shows the loss of data quality when the number of cycles c of the user performing the sensing task is 2 and 3. Since the number of constraints in the optimization problem for generating the optimal location mapping probability matrix is the factorial of c, the complexity of the problem will become higher when c is larger, so we only calculated the cases where c is 2 and 3. However, to ensure the quality of the data in the SparseMCS environment, we usually do not need the same user to perform multiple sensing cycles. In future work, we will further reduce the complexity of the optimization problem to adapt to more scenarios.

In order to verify the impact of the basic sampling weight on the data quality loss, we change the weight and calculate the data quality loss under different privacy budgets when k and c are the default values. Figure 7 shows the results of the experiment. **Under different privacy budgets, we find that the data error is the smallest when ω_0** is 0.75.

Fig. 5 MAE changes with the amount of sensing data collected by participants in each cycle

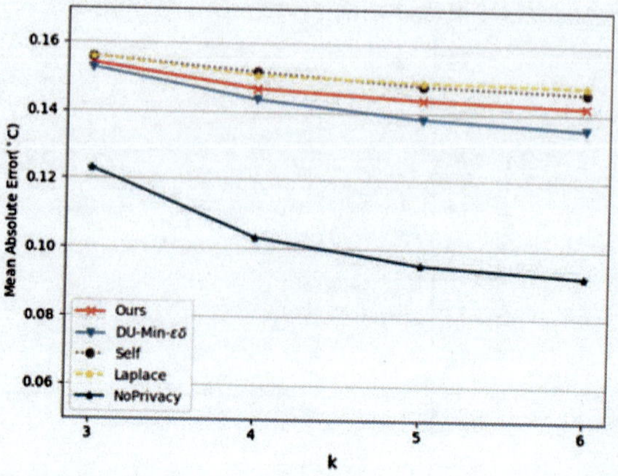

(a) $c = 2$

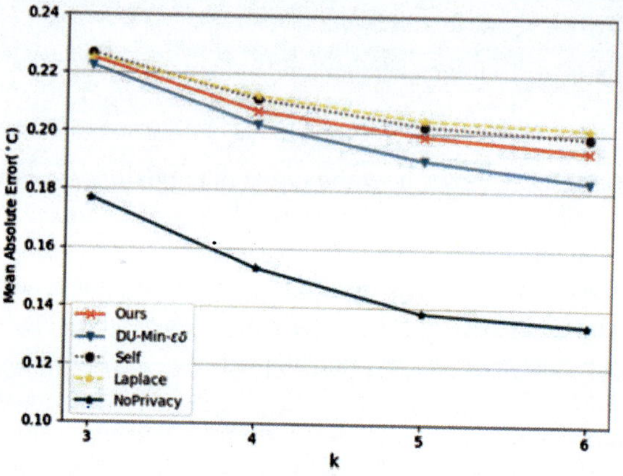

(b) $c = 3$

7 Conclusions

This paper presents a spatiotemporal and differential location privacy protection mechanism for 5G-enabled sparse mobile crowdsensing. It considers the level of location privacy protection required by users, the protection of travel modes, the ability to resist attacks from attackers with prior solid knowledge, and the loss of data quality due to location mapping. In particular, users can use this framework to develop personalized location privacy protection based on their travel mode. Experiments based on real data verify the effectiveness of the framework.

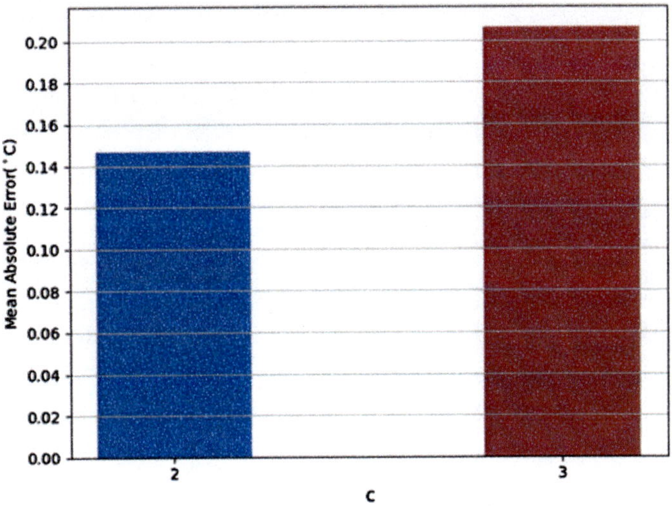

Fig. 6 MAE changes with c

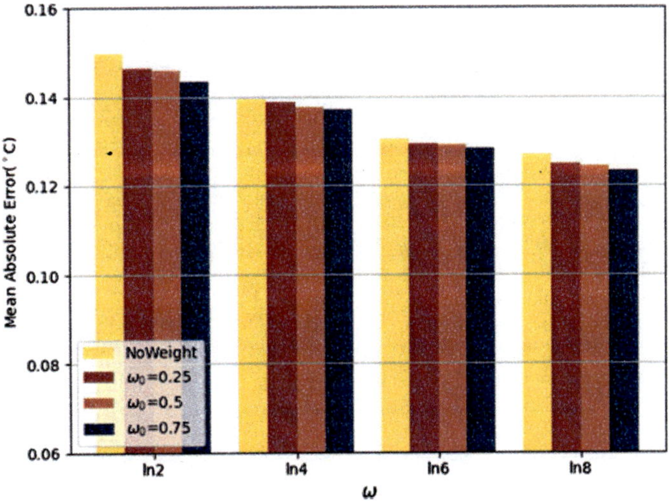

Fig. 7 MAE changes with ω

Acknowledgements This work was support from the National Science Foundation of China (Grant No. 61877007), Fundamental Research Funds for the Central Universities (Grant No. DUT20GJ205), and the Shandong National Science Foundation of China (Grant No. ZR202103040468).

References

1. L. Tan, H. Xiao, K. Yu, M. Aloqaily, Y. Jararweh, A blockchain-empowered crowdsourcing system for 5G-enabled smart cities. Comput. Stand. Interfaces **76**, 103517 (2021). [Online]. https://doi.org/10.1016/j.csi.2021.103517
2. S.B. Shah, C. Zhe, F. Yin, I.U. Khan, S. Begum, M. Faheem, F.A. Khan, 3D weighted centroid algorithm & RSSI ranging model strategy for node localization in WSN based on smart devices. Sustain. Cities Soc. **39**, 298–308 (2018). [Online]. Available: https://www.sciencedirect.com/science/article/pii/S2210670717312982
3. S.B.H. Shah, L. Wang, M.E. Haque, M.J. Islam, A. Carie, N. Kumar, Lifetime improvements of smart sensors maintenance protocol in prospect of IoT-based Rampal power plant, in *2020 16th International Conference on Mobility, Sensing and Networking (MSN)* (2020), pp. 260–267
4. S.B.H. Shah, Z. Chen, S.H. Ahmed, F. Yin, M. Faheem, S. Begum, Depth based routing protocol using smart clustered sensor nodes in underwater WSN, in *Proceedings of the 2nd International Conference on Future Networks and Distributed Systems*, Amman, Jordan, 26–27 June 2018, ed. by A. Abuarqoub, B. Adebisi, M. Hammoudeh, S. Murad, M. Arioua (ACM, 2018), pp. 53:1–53:7. [Online]. https://doi.org/10.1145/3231053.3231119
5. M. Faheem, R.A. Butt, B. Raza, M.W. Ashraf, M.A. Ngadi, V.C. Gungor, A multi-channel distributed routing scheme for smart grid real-time critical event monitoring applications in the perspective of industry 4.0. Int. J. Ad Hoc Ubiquitous Comput. **32**(4), 236–256 (2019). [Online]. https://doi.org/10.1504/IJAHUC.2019.103264
6. M. Faheem, R.A. Butt, R. Ali, B. Raza, M.A. Ngadi, V.C. Gungor, CBI4.0: a cross-layer approach for big data gathering for active monitoring and maintenance in the manufacturing industry 4.0. J. Ind. Inf. Integr. **24**, 100236 (2021). [Online]. Available: https://www.sciencedirect.com/science/article/pii/S2452414X21000364
7. Y. Zhu, Z. Li, H. Zhu, M. Li, Q. Zhang, A compressive sensing approach to urban traffic estimation with probe vehicles. IEEE Trans. Mob. Comput. **12**(11), 2289–2302 (2013)
8. R.K. Rana, C.T. Chou, S.S. Kanhere, N. Bulusu, W. Hu, Ear-phone: an end-to-end participatory urban noise mapping system, in *Proceedings of the 9th International Conference on Information Processing in Sensor Networks, IPSN 2010*, Stockholm, Sweden, 12–16 Apr 2010, ed. by T.F. Abdelzaher, T. Voigt, A. Wolisz (ACM, 2010), pp. 105–116. [Online]. https://doi.org/10.1145/1791212.1791226
9. D. Hasenfratz, O. Saukh, S. Sturzenegger, L. Thiele, Participatory air pollution monitoring using smartphones. Mob. Sens. (2012)
10. L. Wang, D. Zhang, Y. Wang, C. Chen, X. Han, A. M'hamed, Sparse mobile crowdsensing: challenges and opportunities. IEEE Commun. Mag. **54**(7), 161–167 (2016)
11. J.E. Dobson, P.F. Fisher, Geoslavery. IEEE Technol. Soc. Mag. **22**(1), 47–52 (2003). [Online]. https://doi.org/10.1109/MTAS.2003.1188276
12. J. Krumm, A survey of computational location privacy. Pers. Ubiquitous Comput. **13**(6), 391–399 (2009)
13. M.E. Andrés, N.E. Bordenabe, K. Chatzikokolakis, C. Palamidessi, Geo-indistinguishability: differential privacy for location-based systems, in *2013 ACM SIGSAC Conference on Computer and Communications Security, CCS'13*, Berlin, Germany, 4–8 Nov 2013, ed. by A. Sadeghi, V.D. Gligor, M. Yung (ACM, 2013), pp. 901–914
14. C. Dwork, Differential privacy, in *33rd International Colloquium on Automata, Languages and Programming, ICALP 2006*, Proceedings, Part II, Venice, Italy, 10–14 July 2006, ed. by M. Bugliesi, B. Preneel, V. Sassone, I. Wegener. Lecture Notes in Computer Science, vol. 4052 (Springer, 2006), pp. 1–12
15. R. Shokri, Privacy games: optimal user-centric data obfuscation, in Proc. Priv. Enh. Technol. **2015**(2), 299–315 (2015). [Online]. https://doi.org/10.1515/popets-2015-0024
16. Y. Cao, Y. Xiao, L. Xiong, L. Bai, PriSTE: from location privacy to spatiotemporal event privacy, in *35th IEEE International Conference on Data Engineering, ICDE 2019*, Macao, China, 8–11 Apr 2019 (IEEE, 2019), pp. 1606–1609

17. V. Primault, A. Boutet, S.B. Mokhtar, L. Brunie, The long road to computational location privacy: a survey. IEEE Commun. Surv. Tutor. **21**(3), 2772–2793 (2019)
18. L. Pournajaf, D.A. Garcia-Ulloa, L. Xiong, V.S. Sunderam, Participant privacy in mobile crowd sensing task management: a survey of methods and challenges. SIGMOD Rec. **44**(4), 23–34 (2015)
19. K.T. Putra, H. Chen, Prayitno, M.R. Ogiela, C. Chou, C. Weng, Z. Shae, Federated compressed learning edge computing framework with ensuring data privacy for PM2.5 prediction in smart city sensing applications. Sensors **21**(13), 4586 (2021). [Online]. https://doi.org/10.3390/s21134586
20. M. Li, Y. Li, L. Fang, ELPPS: an enhanced location privacy preserving scheme in mobile crowd-sensing network based on edge computing, in *19th IEEE International Conference on Trust, Security and Privacy in Computing and Communications, TrustCom 2020*, Guangzhou, China, 29 Dec 2020–1 Jan 2021, ed. by G. Wang, R.K.L. Ko, M.Z.A. Bhuiyan, Y. Pan (IEEE, 2020), pp. 475–482. [Online]. https://doi.org/10.1109/TrustCom50675.2020.00071
21. L. Wang, D. Zhang, D. Yang, B.Y. Lim, X. Han, X. Ma, Sparse mobile crowdsensing with differential and distortion location privacy. IEEE Trans. Inf. Forensics Secur. **15**, 2735–2749 (2020)
22. L.T. Nguyen, J. Kim, B. Shim, Low-rank matrix completion: a contemporary survey. IEEE Access **7**, 94215–94237 (2019)
23. E.J. Candès, Y. Plan, Matrix completion with noise. Proc. IEEE **98**(6), 925–936 (2010)
24. D. Yang, D. Zhang, Z. Yu, Z. Yu, Fine-grained preference-aware location search leveraging crowdsourced digital footprints from LBSNs, in *The 2013 ACM International Joint Conference on Pervasive and Ubiquitous Computing, UbiComp '13*, Zurich, Switzerland, 8–12 Sept 2013, ed. by F. Mattern, S. Santini, J.F. Canny, M. Langheinrich, J. Rekimoto (ACM, 2013), pp. 479–488. [Online]. https://doi.org/10.1145/2493432.2493464
25. Y. Cao, Y. Xiao, L. Xiong, L. Bai, M. Yoshikawa, PriSTE: protecting spatiotemporal event privacy in continuous location-based services. Proc. VLDB Endow. **12**(12), 1866–1869 (2019)
26. S. Agrawal, J.R. Haritsa, A framework for high-accuracy privacy-preserving mining, in *Proceedings of the 21st International Conference on Data Engineering, ICDE 2005*, Tokyo, Japan, 5–8 Apr 2005, ed. by K. Aberer, M.J. Franklin, S. Nishio (IEEE Computer Society, 2005), pp. 193–204
27. F. Ingelrest, G. Barrenetxea, G. Schaefer, M. Vetterli, O. Couach, M. Parlange, Sensorscope: application-specific sensor network for environmental monitoring. ACM Trans. Sens. Netw. **6**(2), 17:1–17:32 (2010)

Artificial Intelligence Techniques Applied on Renewable Energy Systems: A Review

Ali Azawii Abdul Lateef, Sameer I. Ali Al-Janabi, and Omar Azzawi Abdulteef

Abstract Renewable energy is gaining traction as an efficient alternative source of energy; it is considerably safer and healthier than traditional energy, and it has greatly contributed to this area. However, there are still several areas that need improvement in order to meet this rapidly expanding technology. AI technology can evaluate the previous, improve the current, and predict what will happen. As a result, AI will fix the majority of these issues. AI is complicated, but it lowers error and aspires for better precision, making energies more intelligent. This paper presents an overview of commonly utilized artificial intelligence (AI) techniques in sustainable sources of energy applications. AI is applied in practically every form of energy for design, optimization, prediction, administration, transmission, and regulation (wind, solar, geothermal, hydro, ocean, bio, hydrogen, and hybrid). Throughout this aspect, the purpose of this study is to highlight the AI techniques utilized in the field of renewable energy.

Keywords Artificial intelligence · Renewable energy · Solar energy

1 Introduction

Renewable energy resources (RE) offer immense potential and can satisfy today's global energy needs. It could increase energy production industry variety, provide long-term sustainable supply, and lower local and global emissions. It can indeed give financially appealing choices for meeting specific power service requirements

A. A. A. Lateef (✉)
Human Resources Department, University of Anbar, Anbar, Iraq
e-mail: Aliazawii@uoanbar.edu.iq

S. I. Ali Al-Janabi
College of Islamic Science, University of Anbar, Anbar, Iraq
e-mail: isl.samir.ia2012@uoanbar.edu.iq

O. A. Abdulteef
Ministry of Education, Anbar Directorate, Planning Department, Anbar, Iraq

A. K. Bashir et al. (eds.), *Proceedings of International Conference on Computing and Communication Networks*, Lecture Notes in Networks and Systems 394,
https://doi.org/10.1007/978-981-19-0604-6_25

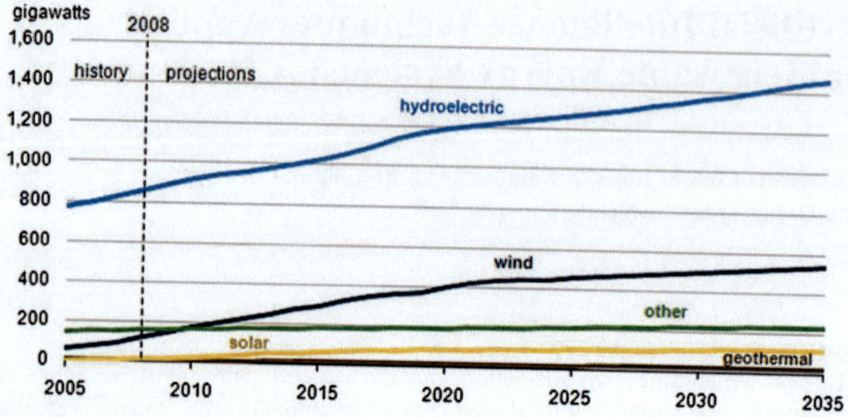

Fig. 1 Global installed power generation capacity by energy [2]

(mainly in developing nations and rustic regions), as well as opportunities for local component production. For design, improvement, rating, operation, distribution, and legislation, AI is employed in practically every kind of renewable energy. Throughout this scenario, the purpose of this study is to highlight the AI techniques utilized in the field of renewable energy [1].

Between 2008 and 2035, existing hydroelectric power generation is predicted to grow faster than other renewable energy sources. Implemented solar power generation, on the other hand, is expected to develop at the fastest rate over the forecast period. In comparison with the rest, as indicated in Fig. 1 [2].

Due to rising computational capacity, tools, and data collection, artificial intelligence (AI) is becoming more prevalent in many sectors of renewable energy systems (REs). The present approaches for design, control, and maintenance in the energy business have been shown to produce somewhat erroneous outcomes. Furthermore, the use of artificial intelligence (AI) to execute these activities has improved accuracy and precision, and it is currently at the forefront.

AI has been one of the most popular areas of research in recent decades, owing to its ability to automate systems for improved quality and productivity [3]. Through training techniques with a set of sophisticated instructions, it allows them to learn, reasoning, and decide in the same way that humans do.

Additionally, the use of AI in the digitalization of energy systems has been classified as having significant capability to improve in power system network continuity, stability, dynamic responsiveness, and other critical developments [4]. Nowadays, AI is being used to integrate components of the power system such as design [4], forecasting, control, optimization, maintenance, and security [5–7].

2 Renewable Energy Types

2.1 *Solar Energy*

Solar power can be generated physically using photovoltaic (PV) cells or implicitly by gathering and concentrating solar power (CSP) to generate steam, which will be used to operate a turbine to generate electricity. The photovoltaic effect, which leads to the idea that photons of light push electrons into a higher energy state, is used to directly generate electricity from solar radiation. Although photovoltaics were first used to power spacecrafts, there are several PV power generating usage in ordinary living, including grid-independent homes, water utilization pumps, e-mobility, wayside emergency phones, and remote sensing [8, 9].

2.2 *Wind Energy*

Wind is a renewable energy source that is pure, cheap, and easily accessible. Wind turbines collect the air's energy and turn it into electricity every day across the globe. Wind energy is becoming more essential in terms of how we power our world—in a clean, sustainable way. Wind as a key source of energy has been used for ages by converting its dynamic power into electricity using windmills and wind turbines. [10, 11].

2.3 *Hydroelectric Energy*

Hydroelectric energy is generated when water is coming thru a dam (hydroelectric electricity is created while water is coming through with a dam) (the dam can be opened or closed to varying degrees to control water flow and to produce the amount of electricity needed, based on demand). Water goes into an intake behind the dam, where it powers turbine blades. A turbine spins a generator to generate electricity. The amount of electricity produced is proportional to the distance, and the water drops as well as the volume of water that flows through the system. Energy can also be supplied to households, industries, and companies via lengthy electric wires. Hydroelectric power is the most frequently used renewable resource, responsible for nearly 16% of electricity generated by renewable use [12].

2.4 *Ocean Energy*

Sea energy refers to a broad variety of technical systems for using a number of transformation techniques to create electricity from the ocean. It is a new industry, with the first commercial units being installed in 2008 and 2009. Although the huge source of renewable energy has yet to be used on a large scale, the ocean energy sector plays an important role to make a big difference to the supply of electricity to coastal countries and people [13].

3 Artificial Intelligence (AI)

Artificial intelligence (AI) allows a computer, robot, or device to mimic human cognitive behavior. The basic goal of artificial intelligence is to improve computer functions that are involved in human cognition, such as thinking, learning, and problem-solving. AI is particularly useful for digitizing cognitive capacities; and a common use of AI is facial recognition. Research on the application of artificial intelligence approaches to power and renewable energy systems is now underway. Artificial neural networks, fuzzy logic, and knowledge-based systems are now the most widely utilized and effective of these techniques. The AI techniques can make predictions better, faster, and more practical than any of the traditional methods. On the other side, inherently, noisy data from renewable energy procedures are a great candidate for handling with AI systems [14].

Even if comprehending the intricate thought of a human mind is a difficult topic to tackle, AI aspires to understand human thought in order to create intelligent beings capable of solving complex problems. The advancement of AI has decreased the strain of manual computation [15].

4 AI Techniques Applied in Renewable Energy

For design, optimization, estimate, management, distribution, and policy, AI is employed in practically every type of renewable energy (wind, solar, hydro, ocean, and hybrid). People's attention has been drawn to renewable energy as the environment has deteriorated, and conventional supplies have been depleted. Wind power has been quickly expanding in many locations, particularly in Europe, as a non-polluting renewable energy source. In Spain, for example, wind power generation accounts for 4% of global energy consumption.

Figure 2 shows a simplified display of several sorts of renewable power sources and AI technologies [16–18].

Lalot employed artificial neural networks to detect the solar detectors' timing constraints. Two factors properly explain the static behavior of a flat plate collector,

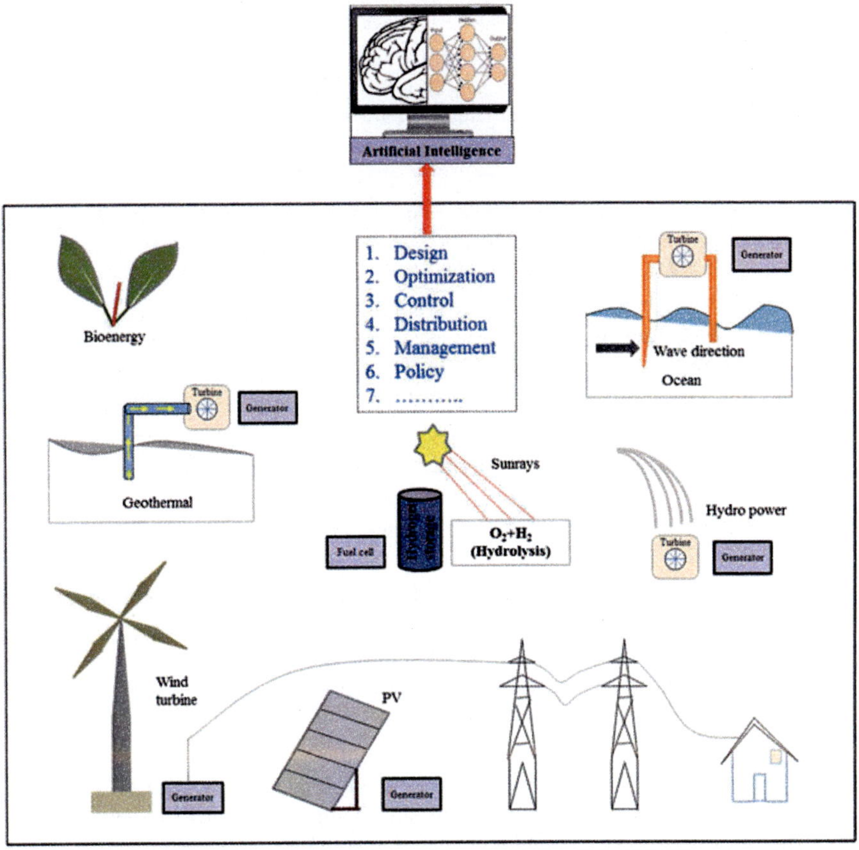

Fig. 2 Diagram depicting the use of artificial intelligence in various RE sources [18]

whereas two additional parameters are required to clearly explain the dynamic behavior. When a second-order process was investigated, however, the network's discrimination ability was not very great. Collectors have been demonstrated to be regarded third-order systems. To correctly determine pure third-order systems, a radial basis function (RBF) neural network is used. Based on number of learning steps, the Euclidean distance between the collectors and their models was computed to validate the neural network. Finally, neural networks were proven to be capable of discriminating collectors with similar parameters: the suggested network identified a difference of 2% for one parameter [19].

Veerachary and Yadaiah used an artificial neural network (ANN) to find the best operating point for a photovoltaic (PV) system. The ANN controller is trained using a gradient descent technique to identify the maximum power point of a solar cell array and the integrated system's gross mechanical power operation. Solar insolation is the essential input to the neural network, and the converter chopping ratio corresponding to the maximum power output of the PV cells or gross mechanical energy production of the integrated PV system is the output parameter. For centrifugal and volumetric pump loads, the ANN forecasts had an error of less than 2% and 7%, respectively [20]. A full survey of the uses of NN in power electronics is presented in [21]. Many particular control and system identification examples are given. Additional AI technologies, such as fuzzy logic, metaheuristic approaches, and so on, have not been touched on. Although [22] goes into greater detail about these strategies, it focuses on illustrative examples instead of an in-depth study of AI algorithms. Bose [22] presents a comprehensive explanation of metaheuristic approaches for MPPT in photovoltaic (PV) systems. The AI techniques utilized to PV systems are covered in [14], which is focused solely on the PV application.

Using a genetic algorithm, Senjyu et al. created an ideal configuration of power generating systems in isolated islands with RE (GA). This technique can be used to figure out how many solar panels, wind turbine generators, and battery configurations are best. Diesel generators, wind turbine generators, a photovoltaic system, and batteries make up the generating system. In compared to diesel generators alone, the proposed technique can minimize operation costs by around 10% [23].

Dufo-Lopez et al. [24] present a revolutionary genetic algorithm-optimized technique for controlling stand-alone hybrid renewable electrical systems with hydrogen storage. RE resources (wind, PV, and hydro), batteries, fuel cells, an AC generator, and electrolyze make up the optimal hybrid system.

Lopez and Agustin created the hybrid optimization by genetic algorithms (HOGAs), a program that designs a PV-diesel system using a genetic algorithm (GA) (sizing and operation control of a PV-diesel system). C ++ was used to create the software. A HOGA-optimized PV-diesel system is compared to a stand-alone PV-only system dimensioned using a traditional design method based on available energy under worst-case scenarios. The need and sun irradiation are the same in both circumstances. The computational findings demonstrate the PV-hybrid system's cost-effectiveness. HOGA is also compared to a commercial program for hybrid system optimization [25].

Mabel and Fernandez [26] used a neural network with feed-forward backpropagation to estimate wind power over a three-year period from seven wind farms. The prediction accuracy of the BPNN is commendable (the test set had an RMSE of 0.0065, and the training set had an RMSE of 0.0070.). The performance of three distinct forms of ANN approaches (BPNN, RBFNN, and adaptive linear element network (ADALINE)) has been investigated in the calculation of wind velocity data from two separate locations [27].

For wind power estimation, Kariniotakis et al. [28] ANN (recurrent high-order neural networks) was employed in a more advanced form. The ANN model's performance is compared to that of the Naive Bayes (NB) technique. When compared to the NB, the ANN has the lowest RMSE. For the years 1993–1997, the BPNN approach was employed to anticipate wind speed in the Marmara [29].

Damousis and Dokopoulos proposed fuzzy approaches for wind speed and power estimates utilizing multiple GA algorithms (real coded GA and binary coded GA). Data about wind energy from a faraway site were obtained utilizing wireless modems and analyzed using the fuzzy methodology, which produced 29.7% and 39.8% accurate accuracy results than the permanent technique for the following hour and lengthy, accordingly [30].

Solar energy applications have also used several evolutionary AI approaches [31, 32]. GA in solar tracking was suggested by Mashohor et al. [31] for increased PV system performance. The best GA-solar system has an initial size of the population of 100, 50 epochs, and mutation and crossover chances of 0.7 and 0.001, accordingly. The low-standard deviation (1.55) in production yield also demonstrates the system's efficiency. GA is used to design a solar water heating system that is as efficient as possible. The plate gathering area has been actually developed with the GA set to 63 m, resulting in a solar portion value of 98% [31].

Kumar et al. employed GA to track the highest point of power of a PV array coupled to a battery. The GA's effectiveness is compared to that of the perturb and observe (PO) algorithm. The boost converter produces a 400 V line voltage [32].

The employment of GA in parameter adjustment of the hidden layer by Monteiro et al. resulted in improved prediction efficiency (RMS 0.0432). The GA + HISIMI model (RMSE 283.89) approach is compared to the BPNN (RMSE 286.11) and conventional persistence (RMSE 445.48) methods [33].

O'Sullivan et al. employed PSO to optimize the size of a hybrid RE system in order to make it more cost effective [34]. In the operation optimization of a hybrid RE system, an upgraded GA is applied, which outperforms the classic GA method [35]. The bee method is used to improve the performance characteristics of a hybrid RE system (net present cost (NPC), cost of energy (COE), and generation cost (GC)) [36].

Table 1 shows the summary of most work in literature review with the used methods and its achievements.

5 Comparative Analysis

The models discussed above each have their own unique qualities and can behave effectively in a variety of settings. Artificial neural network (ANN) models are effective in the photovoltaic (PV) field and can provide improved long-term prediction outcomes. They are frequently utilized as feed to time-series models, since ARMA helps them get improved result.

The tenacity models are the most straightforward time-series algorithms. In terms of very simple prediction, they can outperform several other algorithms. Despite their inconsistency in prediction accuracy, they are commonly employed in practice. In the last thirty years, the majority of research on time-series modeling techniques has been done by academics.

New artificial intelligence-based models such as neural network models and fuzzy logic models have been developed. Algorithms that use a vast amount of historical data for modeling input, such as wind energy consumption algorithms and fuzzy logic systems, can produce precise short-term predictions.

Table 1 Summary of AI techniques applied in renewable energy

References	Methods	Description	Achievement
Lalot [19]	ANNs/RBF	A radial basis function was used to identify temporal characteristics of solar collectors using ANNs (RBF)	For one parameter, the suggested network identified a difference of 2%
Veerachary and Yadaiah [20]	ANN	An artificial neural network (ANN) was used to find the best operating point for a photovoltaic (PV) system	For centrifugal and volumetric pump loads, the ANN forecasts had an error of less than 2% and 7%, respectively
Senjyu et al. [23]	Genetic algorithm (GA)	Using a genetic algorithm, developed an optimal configuration of power generating systems in isolated islands with RE (GA)	In comparison with diesel generators alone, the proposed technique can cut operation costs by around 10%
Dufo-Lopez et al. [25]	Hybrid optimization by genetic algorithms (GAs)	Produced hybrid optimization by genetic algorithms (HOGAs), a tool for designing a PV-diesel system using a genetic algorithm (GA) (sizing and operation control of a PV-diesel system)	The PV-hybrid system's economic benefits are demonstrated by the computational findings
Mabel and Fernandez [26]	Feed-forward backpropagation neural network (BPNN)	BPNN is used to evaluate wind power from seven wind farms during a three-year timeframe	The BPNN has a respectable prediction accuracy (RMSE 0.0070 for the training set and 0.0065 for the test set)
Kariniotakis et al. [28]	Advanced version of ANN	For wind power estimation, an upgraded form of ANN was implemented	In comparison with the NB, the ANN has the smallest RMSE
Damousis and Dokopoulos [30]	Fuzzy methods using the two GA algorithms	For wind speed and power estimate, developed fuzzy approaches employing the multiple GA algorithms (real coded GA and binary coded GA)	The fuzzy method outperforms the persistent method by 29.7% and 39.8% for the next hour and long-term predictions, respectively
Mashohor et al. [31]	Genetic algorithm (GA)	GA in solar tracking is recommended for increased PV system performance	The system's efficiency is also demonstrated by the low standard deviation (1.55) in generation gain

(continued)

Table 1 (continued)

References	Methods	Description	Achievement
Atia et al. [37]	Genetic algorithm (GA)	GA is used to develop a solar water heating system that is as efficient as possible	With the GA set to 63 m, the plate catcher region has been improved, resulting in a solar fraction value of 98 percent
O'Sullivan et al. [34]	Particle swarm optimization (PSO)	PSO is used to optimize the size of a hybrid RE system	Improve the cost-effectiveness of the hybrid RE system
Lalot [19]	ANNs/ RBF	The identification was done using a radial basis function. Temporal characteristics of solar collectors using ANNs (RBF)	For one parameter, the suggested network showed a difference of 2%
Veerachary and Yadaiah [20]	ANN	An artificial neural network (ANN) was used to determine the best operating point of a photovoltaic (PV) system	For centrifugal and volumetric pump loads, the ANN predictions had an error of less than 2% and 7%, respectively
Senjyu et al. [23]	Genetic algorithm (GA)	Using a genetic algorithm, produced an optimal configuration of power generating systems in isolated islands with RE (GA)	In comparison with diesel generators alone, the proposed technique can cut operation costs by around 10%
Dufo-Lopez et al. [25]	Hybrid optimization by genetic algorithms (GAs)	Produced hybrid optimization by genetic algorithms (HOGAs), a tool that designs a PV-diesel system using a GA (sizing and operation control of a PV-diesel system)	The computational findings demonstrate the PV-hybrid system's cost-effectiveness

Raw data input is handled well by neural networks, which also have significant learning and training capabilities. When it comes to reasoning difficulties, fuzzy logic models surpass others, but their learning and adapting abilities are subpar. Fuzzy logic and neural networks were merged in new approaches to get good results. Because these strategies are dependent on varied settings, meaningful comparisons of all of them are difficult, and data collecting is a difficult undertaking. However, there are some comparisons and similar studies that prove that artificial-based algorithm outperforms other approaches in terms of short-term prediction.

6 Challenges

Based on existing AI advances, the implementation of AI technology in RE is projected to encounter the following significant obstacles:

Reliability needs to be enhanced even more. While AI technology implemented to energy systems has achieved a high rate of issue and defect detection, it still falls short of actual application requirements. At this time, AI can only be utilized as a supplement to traditional methods of work.

There is a need to upgrade infrastructure. The use of AI is dependent on a large number of data samples, high-computer power, and global network interaction. The supporting capability and degree of necessary infrastructure assets, such as big data, are, however, important considerations.

7 Conclusion and Future Directions

Through the previous review of all technologies used in the fields of renewable energy, it is very important to develop these techniques and work to spread these techniques because of a great benefit in producing electric power without environmental harmful. The important role of artificial intelligence techniques and their effective role in developing electricity production using renewable energy techniques. After reviewing most of the techniques used in these areas, it is very important to focus on the deep learning and machine learning techniques to improve work in renewable energy and production electricity.

Advances in currently accessible AI approaches are extremely likely to be seen in the coming years. There appears to be a data large disparity in the economy right now, but with the rise of IOT's solutions, the implementation of a wide range of sensors, adaptive streaming supplied by drones for monitoring purpose, and NLP techniques, the problem of a lack of data is likely to fade away.

It is worth noting that, among all AI techniques, neural networks (NNWs) are now receiving the most attention for future applications.

References

1. M. Asif, T. Muneer, Energy supply, its demand and security issues for developed and emerging economies. Renew. Sustain. Energy Rev. **11**(7), 1388–1413 (2007)
2. U.S. Briefing, International energy outlook 2013. US Energy Inf. Adm. **506**, 507 (2013)
3. J.J. Bryson, The past decade and future of AI's impact on society. Towar. New Enlight. 150–185 (2019)
4. S. Zhao, F. Blaabjerg, H. Wang, An overview of artificial intelligence applications for power electronics. IEEE Trans. Power Electron. (2020)
5. V.S.B. Kurukuru, F. Blaabjerg, M.A. Khan, A. Haque, A novel fault classification approach for photovoltaic systems. Energies **13**(2), 308 (2020)

6. M.A. Khan, A. Haque, V.S.B. Kurukuru, Performance assessment of stand-alone transformer-less inverters. Int. Trans. Electr. Energy Syst. **30**(1), e12156 (2020)
7. S. Sahoo, T. Dragicevic, F. Blaabjerg, Cyber security in control of grid-tied power electronic converters-challenges and vulnerabilities. IEEE J. Emerg. Sel. Top. Power Electron. 1–15 (2020)
8. J.M. Carrasco et al., Power-electronic systems for the grid integration of renewable energy sources: a survey. IEEE Trans. Ind. Electron. **53**(4), 1002–1016 (2006)
9. M. Liserre, T. Sauter, J.Y. Hung, Future energy systems: integrating renewable energy sources into the smart power grid through industrial electronics. IEEE Ind. Electron. Mag. **4**(1), 18–37 (2010)
10. T. Burton, N. Jenkins, D. Sharpe, E. Bossanyi, *Wind Energy Handbook*. Wiley (2011)
11. J.F. Manwell, J.G. McGowan, A.L. Rogers, *Wind Energy Explained: Theory, Design and Application*. Wiley (2010)
12. A. Blakers, M. Stocks, B. Lu, C. Cheng, A review of pumped hydro energy storage. Prog. Energy (2021)
13. M. Esteban, D. Leary, Current developments and future prospects of offshore wind and ocean energy. Appl. Energy **90**(1), 128–136 (2012)
14. A. Mellit, S.A. Kalogirou, L. Hontoria, S. Shaari, Artificial intelligence techniques for sizing photovoltaic systems: a review. Renew. Sustain. Energy Rev. **13**(2), 406–419 (2009)
15. R.S. Michalski, J.G. Carbonell, T.M. Mitchell, *Machine Learning: An Artificial Intelligence Approach*. Springer Science and Business Media (2013)
16. I. Sanchez, Short-term prediction of wind energy production. Int. J. Forecast. **22**(1), 43–56 (2006)
17. R.O.S. Juan, J. Kim, Utilization of artificial intelligence techniques for photovoltaic applications. Curr. Photovolt. Res. **7**(4), 85–96 (2019)
18. S.K. Jha, J. Bilalovic, A. Jha, N. Patel, H. Zhang, Renewable energy: present research and future scope of artificial intelligence. Renew. Sustain. Energy Rev. **77**, 297–317 (2017)
19. S. Lalot, Identification of the time parameters of solar collectors using artificial neural networks, in *Proceedings of Eurosun*, (2), pp. 1–6 (2000)
20. M. Veerachary, N. Yadaiah, ANN based peak power tracking for PV supplied DC motors. Sol. energy **69**(4), 343–350 (2000)
21. B.K. Bose, Neural network applications in power electronics and motor drives—an introduction and perspective. IEEE Trans. Ind. Electron. **54**(1), 14–33 (2007)
22. B.K. Bose, Artificial intelligence techniques in smart grid and renewable energy systems—some example applications. Proc. IEEE **105**(11), 2262–2273 (2017)
23. T. Senjyu, D. Hayashi, A. Yona, N. Urasaki, T. Funabashi, Optimal configuration of power generating systems in isolated island with renewable energy. Renew. Energy **32**(11), 1917–1933 (2007)
24. R. Dufo-Lopez, J.L. Bernal-Agustín, J. Contreras, Optimization of control strategies for stand-alone renewable energy systems with hydrogen storage. Renew. Energy **32**(7), 1102–1126 (2007)
25. R. Dufo-López, J.L. Bernal-Agustín, Design and control strategies of PV-diesel systems using genetic algorithms. Sol. Energy **79**(1), 33–46 (2005)
26. M.C. Mabel, E. Fernandez, Analysis of wind power generation and prediction using ANN: a case study. Renew. Energy **33**(5), 986–992 (2008)
27. G. Li, J. Shi, On comparing three artificial neural networks for wind speed forecasting. Appl. Energy **87**(7), 2313–2320 (2010)
28. G.N. Kariniotakis, G.S. Stavrakakis, E.F. Nogaret, Wind power forecasting using advanced neural networks models. IEEE Trans. Energy Convers. **11**(4), 762–767 (1996)
29. A. Öztopal, Artificial neural network approach to spatial estimation of wind velocity data. Energy Convers. Manag. **47**(4), 395–406 (2006)
30. I.G. Damousis, P. Dokopoulos, A fuzzy expert system for the forecasting of wind speed and power generation in wind farms, in *PICA 2001. Innovative Computing for Power-Electric Energy Meets the Market. 22nd IEEE Power Engineering Society. International Conference on Power Industry Computer Applications (Cat. No. 01CH37195)*, pp. 63–69 (2001)

31. S. Mashohor, K. Samsudin, A.M. Noor, A.R.A. Rahman, Evaluation of genetic algorithm based solar tracking system for photovoltaic panels, in *2008 IEEE International Conference on Sustainable Energy Technologies*, pp. 269–273 (2008)
32. P. Kumar, G. Jain, D.K. Palwalia, Genetic algorithm based maximum power tracking in solar power generation, in *2015 International Conference on Power and Advanced Control Engineering (ICPACE)*, pp. 1–6 (2015)
33. C. Monteiro, T. Santos, L.A. Fernandez-Jimenez, I.J. Ramirez-Rosado, M.S. Terreros-Olarte, Short-term power forecasting model for photovoltaic plants based on historical similarity. Energies **6**(5), 2624–2643 (2013)
34. M.J. O'Sullivan, K. Pruess, M.J. Lippmann, State of the art of geothermal reservoir simulation. Geothermics **30**(4), 395–429 (2001)
35. J. Zeng, M. Li, J.F. Liu, J. Wu, H.W. Ngan, Operational optimization of a stand-alone hybrid renewable energy generation system based on an improved genetic algorithm, in *IEEE PES General Meeting*, pp. 1–6 (2011)
36. B. Tudu, S. Majumder, K.K. Mandal, N. Chakraborty, Optimal unit sizing of stand-alone renewable hybrid energy system using bees algorithm, in *2011 International Conference on Energy, Automation and Signal*, pp. 1–6 (2011)
37. D.M. Atia, F.H. Fahmy, N.M. Ahmed, H.T. Dorrah, Optimal sizing of a solar water heating system based on a genetic algorithm for an aquaculture system. Math. Comput. Model. **55**(3–4), 1436–1449 (2012)

A Personalized Healthcare Platform for Monitoring Mental Health of a Person During COVID

C. Jyotsna, J. Amudha, and Sreedevi Uday

Abstract Stress is the human body's response to various factors such as mental, physical, or emotional pressure. Coronavirus Disease 2019 pandemic has disrupted the mental health of most people worldwide. Stress plays a crucial role in corona virus disease patients during their medication period. Therefore, a remote mental health monitoring system has become a necessity. The physiological data captured using body sensors can provide rich information about the stress experienced by a person. Paper proposes a personalized stress indicator for monitoring a person's mental health through a personalized healthcare platform. The physiological data from body sensors such as the galvanic skin response sensor, electrocardiogram module, and accelerometer module are sent in real-time to an Internet of things platform, 'ThingSpeak.' In the ThingSpeak platform, MATLAB analysis is performed to calculate the baseline threshold value of each user. Then, the stress percentage is evaluated based on the data rate above the threshold. The stress percentage is displayed on an output channel of the ThingSpeak platform. It enables remote monitoring of patients' mental health by sending the health updates to the doctor or caretaker through email.

Keywords Stress · Galvanic skin response · Activity recognition · IoT · Accelerometer · ThingSpeak · COVID-19

1 Introduction

Stress has become so common in our daily life, which can be termed as the response of our body to various conditions such as mental, physical or emotional pressure from different environments like home, worksite or public places. Stress is a mechanism

C. Jyotsna (✉) · J. Amudha · S. Uday
Department of Computer Science and Engineering, Amrita School of Engineering, Amrita Vishwa Vidyapeetham, Bengaluru, India
e-mail: c_jyotsna@blr.amrita.edu

J. Amudha
e-mail: j_amudha@blr.amrita.edu

A. K. Bashir et al. (eds.), *Proceedings of International Conference on Computing and Communication Networks*, Lecture Notes in Networks and Systems 394,
https://doi.org/10.1007/978-981-19-0604-6_26

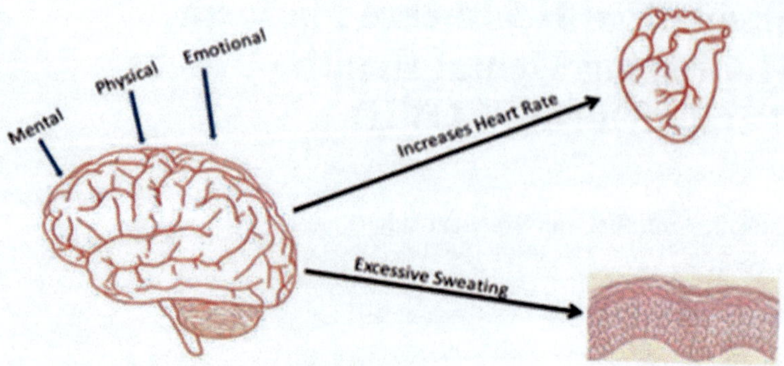

Fig. 1 Role of sympathetic nervous system in a stressful state

of our body to keep the balance of our system in a 'flight or fight' response. If it is pertaining for too long period, it can result in causing adverse effects on our body. When a person has stress, his sympathetic nervous system gets activated, and activates the physiological signals in our body. Figure 1 represents the role of the sympathetic nervous system in physiological changes. It includes increased heart rate, sweating, pulse, blood pressure, dilation in pupil diameter, faster breathing, muscle tension, dry mouth, increased blood sugar as shown in Fig. 1.

During the pandemic situation due to Coronavirus Disease 2019 (COVID-19), most people are mentally and physically affected [1, 2]. Therefore, it has become a necessity to regularly monitor a person's mental health. COVID-19 has affected the everyday life of many people. Facing new realities like unemployment, work from home, homeschooling, and reduced income has mentally affected most people. COVID-19 has disrupted crucial mental health services worldwide when they are most needed [3, 4].

Stress plays a vital role in the recovery period of patients with various medical conditions, especially in COVID patients, cancer patients, and heart patients [5]. Stress can trigger cancer cells and contribute to their growth in the human body. Cardio patients should not get affected mentally. There are chances of shooting up blood pressure and elevated heart rate during stressful situations, which is not desirable for them [6, 7]. It can be beneficial to know everyone's stress status regularly. It helps to take various preventive measures to cope with the adverse effects and improve their health status [8].

Today's one of the most advanced technology is the Internet of Things (IoT), which connects computing objects via the internet and allows their management, enabling them to receive and send data for various purposes like monitoring, controlling or performing its analysis [9]. Paper proposes a personalized stress indicator (PSI), which monitors the physiological signals GSR, HR, and activity data of the user and displays the stress percentage on IoT platform. The physiological signals galvanic skin response (GSR), heart rate (HR), and activity data are sensed and transmitted to the IoT platform with the microcontroller's help. Then, the computation is performed on the cloud, and stress percentage is evaluated and sent to the user through their email.

In the previous study [10], we could extract physiological parameters such as GSR [11] and HR [10] and indicated stress levels. Always we cannot consider the changes in physiological signals as an indication of stress. When a person is physically active, there can be changes in his physiological signals. Therefore, it can make a false prediction of stress. Recognizing the activity using an accelerometer module helps to avoid those wrong predictions.

The paper is organized as follows: Sect. 2 discusses various methodologies used for stress analysis, including the physiological parameters and techniques used. Also, it provides an insight into the previous study on physiological measures for the detection of stress. Section 3 gives the system overview and explains the role of the accelerator module in better predicting stress percentage. Section 4 provides a view of the hardware components and the software used to implement the work for performing stress analysis. It also shows the results obtained at various stages of the development of this work, the outcome of the computation, and a representation of the classification made by the machine learning algorithm. Finally, Sect. 5 shows the conclusion of this research and the possible future enhancements of this work.

2 Related Works

This section reviews the research study on the technologies used for remote health monitoring and related work performed for the detection of stress, its classification, and its involvement in the study of various medical conditions. Several physiological parameters and emerging technologies like IoT and machine learning are being used in this field to analyze the stress level.

There are multiple measures to detect and monitor the mental health of a person. It can be measured using subjective and objective measures. Subjective measures are standardized and well-tested questionnaires that a person can fill in to check his mental health. However, the subjective measures cannot be collected in real-time and can be psychologically biased. On the other hand, physiological measures are objective and can be collected in real time. Since unobtrusive and involuntary data collection is possible with physiological measures, it has been widely used in mental health monitoring [12]. Various sensing technologies help to monitor mental health by sensing physiological parameters like eye measures, brain signals, heart rate, dermal activity, facial expression, sound, temperature, respiratory rate (BPM), and blood pressure (BP).

Panigrahy et al. developed a stress detector using a GSR signal. The classification of GSR data into stressed or relaxed conditions is performed with the help of supervised binary classifiers [11]. The results found that the J48 classifier performs better for the classification of GSR data.

Luay et al. developed a tool for monitoring a person's mental health while doing his daily activities [13]. The tool continuously monitors physiological signals like skin temperature, GSR, and heart rate variability (HRV). The recorded data will send to the cloud database through Wi-Fi. A user interface has been used for visualization and further analysis.

Basjaruddin et al. developed a model to measure the stress level by monitoring the physiological signals HR, GSR, body temperature, and oxygen saturation [14]. The sensed data has been processed and sent to the cloud platform by the microcontroller. Based on the features of the physiological signals, a fuzzy-based model is designed to determine the stress level.

Sabrina et al. had considered physiological signals such as galvanic skin response, HRV, peripheral capillary oxygen saturation and proposed an artificial intelligence-based stress detection system [15]. The K-nearest neighbor (KNN) and support vector machine (SVM) algorithms have been used for the prediction of stress levels.

Jyotsna et al. developed a method to detect the stress level of students by monitoring the eye measures like pupil diameter and blink frequency [16]. The data has been collected using the eye tracker. Mathematical questions have been used as stimuli to induce stress and observed changes in those parameters as indications of stress. Statistical analysis has been done to validate the obtained result.

Hassanalieragh et al. focused on highlighting challenges in the remote health monitoring procedures like sensing, analytics, and visualization using IoT in home and work environments and the need to address those challenges before its actual integration into the clinical practice [17].

Multimodal data collection and analysis always helps to increase the prediction accuracy. Various data like digital, physical, psychometric, physiological, and environmental data helps to monitor a patient. Multiple data collection helps in the real-time monitoring of patients. The GSR and HR are considered as significant measures in predicting stress [18]. The IoT serves as a healthcare platform that helps to track a patient's health conditions constantly.

In the previous work [10], two wearable modules, smart band, and a chest strap module were used to indicate the stress score. The smart band module uses a GSR sensor and an Arduino LilyPad microcontroller for measuring the galvanic skin response. The chest strap module gives the HR data using an ADS1292R ECG module and an Arduino Uno microcontroller. The measured GSR and HR data are then fed to an IoT platform using an ESP8266 Wi-Fi module for storage and further analysis.

3 System Architecture of Personalized Stress Indicator (PSI)

The proposed system personalized stress indicator (PSI) enables constant tracking of mental health conditions of a person with the help of Thingspeak, an IoT platform. The user's physiological signals GSR and HR are sensed and sent to the Thingspeak platform for finding their stress percentage. The PSI aims at improving the accuracy

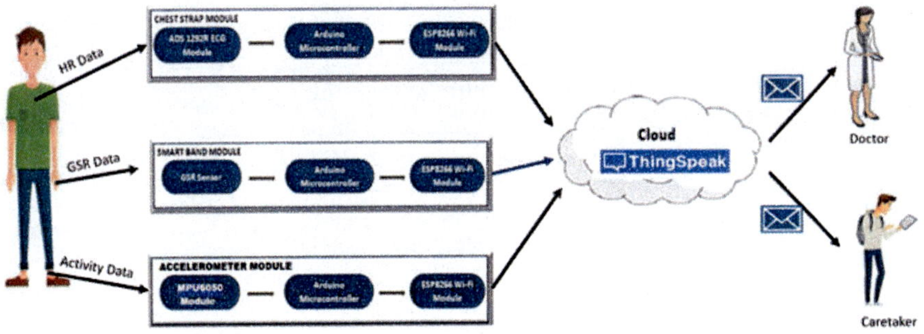

Fig. 2 System architecture of personalized stress indicator

of the stress percentage prediction by introducing an accelerometer module. When a person is engaged in physical activity, there are chances of experiencing excessive sweating and increased heart rate. Therefore, data collection during this period can lead to an erroneous calculation of the stress percentage. An accelerometer module is developed to avoid this scenario, which gives the activity profile of the person. This activity data helps in improving the result by ignoring the GSR and HR signal during the activity period. PSI uses all three modules: a smart band module worn over the wrist, a chest strap module worn around the chest, and an accelerometer module worn over the ankle. The system is designed to measure three parameters GSR, HR, and the activity data from the subject using Arduino microcontroller and transmit those signals to the ThingSpeak IoT platform using the Wi-Fi module. Figure 2 shows the system architecture of PSI.

ThingSpeak is open-source software that allows users to connect with Internet-enabled devices. The ThingSpeak platform provides the provision to create channels for storing the data sent to the platform. Channel is created so that several fields can be enabled in each channel, where each field serves as storage space for each parameter. The data sent from the three wearable modules are stored in the respective fields of the enabled channel for each parameter. The analysis is performed on this data for the calculation of stress percentage.

3.1 Experimental Setup and Visual Stimulus

An experimental setup was created to conduct the research study. An isolated room was chosen to provide a calm environment for the subject. Fourteen healthy subjects (age range $= 30 \pm 10$) voluntarily participated in the research study. The methodology of this experiment was explained to each subject and obtained a written consent for their voluntary participation. Each participant was given the three wearable bands and a headset, which they wore according to the instructions provided. A short video was played on the laptop, with a combination of calm and stressful scenes. A short animation movie [10] served as an introductory video, followed by a calm video scene and a stressful video scene. The world's most relaxing film [10] is selected as the calm video. The physiological parameter readings of the person are taken during this period. Reading during the introductory video was discarded because it was intended to inform the subject of a test setup. The period during the calm video is considered the baseline period. The baseline threshold is yielded by averaging the readings taken during the baseline period. Each person's stress percentage is calculated by keeping the baseline threshold as a reference point.

4 Implementation and Result Analysis

The PSI consists of both hardware and software elements. The main functionalities of the overall system are sensing data with wearable bands and transmission of data to the ThingsSpeak IoT platform with the help of an Arduino microcontroller. In addition to data reception, its storage, analysis and transmission to other devices are also performed in the ThingSpeak platform.

The smart band and chest strap module were developed in the initial phase of the research study. An accelerometer module is added to the existing model to improve the efficiency of the system by reducing the chance of predicting an erroneous result. The activity profile of the person is obtained using an ankle band, which mainly consists of a triple-axis accelerometer module MPU6050, Arduino Nano microcontroller and an ESP8266 Wi-Fi module. MPU6050 module consists of an accelerometer and a gyroscope which will provide the activity data in terms of three-axis values and three gyro values. The axes and gyro data are fed to the Arduino microcontroller. Based on the axes and gyro values, the 'activity' variable will be set as 1, if any activity is indicated. Otherwise, the value will be 0. The activity data will be transmitted to the 'ThingSpeak' IoT platform using an ESP8266 Wi-Fi module. Figure 3 represents the GSR, HR, and activity data received in the ThingSpeak platform.

4.1 MATLAB Analysis

The data from the 'stress analysis' channel fields is retrieved for MATLAB analysis to evaluate the stress percentage and preprocessed to remove the missing data. The baseline period average is considered the threshold and finds the values above the threshold during stress video duration. If the activity is detected as 1, the data would not be considered stressed. The final output is stored in the 'stress monitoring' channel using the 'ThingSpeak Write' function. Stress percentage is calculated based on physiological parameters such as GSR and HR, and activity data and the output are stored in the output channel field. Finally, offline analysis is performed to analyze the stress percentage of all the participants, and the participant 'p14' has indicated high stress with a stress percentage of 38%, as shown in Fig. 4.

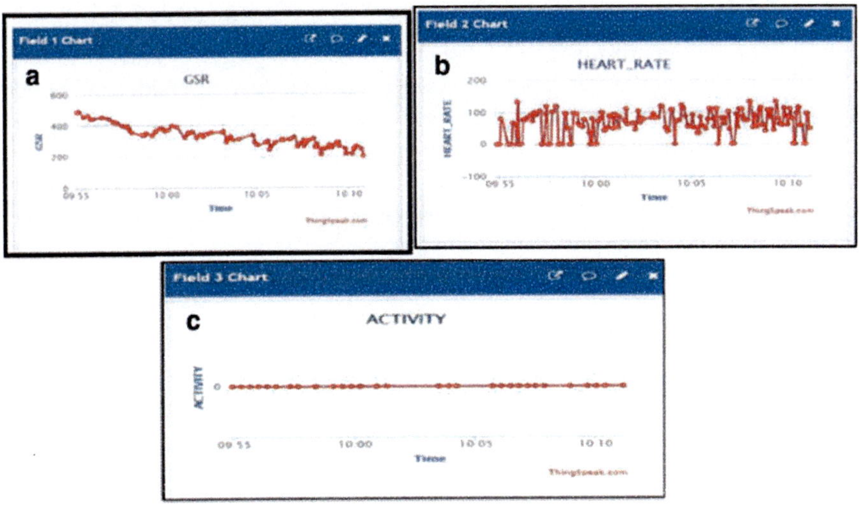

Fig. 3 Physiological data in ThingSpeak platform: **a** GSR, **b** heart rate, **c** activity data

Fig. 4 Stress percentage obtained for each participant

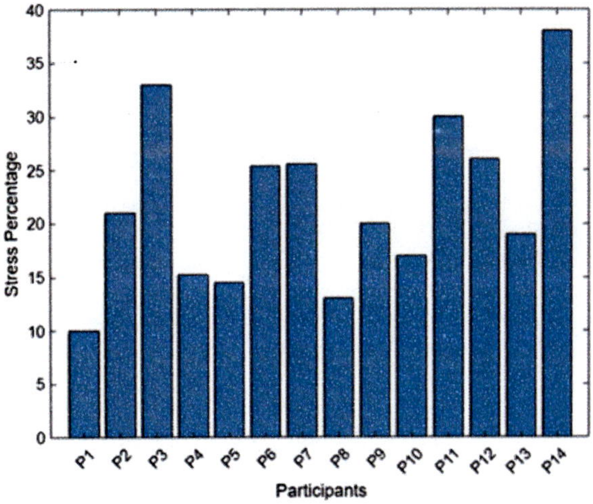

5 Conclusion and Future Enhancement

Applying emerging technologies like IoT and wearable sensor technologies can vastly enhance the healthcare sector. Such application aids the patients and doctors alike. Analysis of the mental health status and alert provided to the doctor or caretaker are the benefits of PSI. Since most of the COVID affected patients are treated at their homes, continuous monitoring of their physical and mental health is required. Since it is not feasible to access most mental health services during the COVID-19 pandemic, PSI can play a significant role in monitoring their stress. The proposed

IoT system to evaluate the stress percentage of a person assures to be a better health caretaker by providing regular feedback on his stress percentage.

The future direction of the work is to add a machine learning model and build an intelligent system for better prediction of a user's mental state. Based on the previous study [16], we could indicate the stress level of students based on their eye gaze measures. A multimodal system can improve the efficiency and prediction accuracy of a system. Since eye measures are good indicators of a person's mental state, incorporating eye measures into PSI can enhance predicting a person's mental state.

References

1. R.P. Rajkumar, COVID-19 and mental health: a review of the existing literature. Asian J. Psychiatry **52**, 102066 (2020)
2. S. Bhaskar, A. Rastogi, K.V. Menon, B. Kunheri, S. Balakrishnan, J. Howick, Call for action to address equity and justice divide during COVID-19. Front. Psych. **11**, 1411 (2020)
3. R. Rossi, V. Socci, D. Talevi, S. Mensi, C. Niolu, F. Pacitti, A. Di Lorenzo, COVID-19 pandemic and lockdown measures impact on mental health among the general population in Italy. Front. Psychiatry **11**, 790 (2020)
4. V. Jha, T.A. Dinesh, P. Nair, Are we ready for controlling community transmission of COVID 19 in India? Epidemiol. Int. (E-ISSN: 2455–7048) **5**(1), 10–13 (2020)
5. S.S. Shastri, P.C. Nair, D. Gupta, R.C. Nayar, R. Rao, A. Ram, Breast cancer diagnosis and prognosis using machine learning techniques, in *The International Symposium on Intelligent Systems Technologies and Applications* (Springer, Cham, 2017), pp. 327–344
6. D. Padmini Pragna, S. Dandur, M. Meenakzshi, C. Jyotsna, J. Amudha, Health alert system to detect oral cancer, in *International Conference on Inventive Communication and Computational Technologies* (IEEE, 2017), pp. 258–262
7. C. Jyotsna, J. Amudha, R. Rao, R. Nayar, Intelligent gaze tracking approach for trail making test. J. Intell. Fuzzy Syst. (2019)
8. Q.J. Campbell, D.F. Henderson, Occupational stress: preventing suffering, enhancing well-being. Int. J. Environ. Res. Public Health (2016)
9. B. Dorsemaine, J.-P. Gaulier, J.-P. Wary, N. Kheir, P. Urien, Internet of things: a definition and taxonomy, in *International Conference on Generation Mobile Applications, Services and Technologies* (IEEE, 2015), pp. 72–77
10. S. Uday, C. Jyotsna, J. Amudha, Detection of stress using wearable sensors in IoT platform, in *IEEE International Conference on Inventive Communication and Computational Technologies* (2018)
11. S.K. Panigrahy, S.K. Jena, A.K. Turuk, Study and analysis of human stress detection using galvanic skin response (GSR) sensor in wired and wireless environment. Res. J. Pharm. Technol. **10**, 545–550 (2017)
12. M. Parent, I. Albuquerque, A. Tiwari, R. Cassani, J.-F. Gagnon, D. Lafond, S. Tremblay, T.H. Falk, PASS: a multimodal database of physical activity and stress for mobile passive body/brain-computer interface research. Front. Neurosci. **14**, 1274 (2020)
13. F. Luay, T. Basmaji, O. Hassanin, A mobile mental health monitoring system: a smart glove, in *14th International Conference on Signal-Image Technology and Internet-Based Systems (SITIS)*, IEEE (2018)
14. N.C. Basjaruddin, F. Syahbarudin, E. Sutjiredjeki, Measurement device for stress level and vital sign based on sensor fusion. Healthc. Inform. Res. **27**(1), 11–18 (2021)

15. J. Sabrina, M. Shamim Kaiser, M. Mahmud, Towards artificial intelligence driven stress monitoring for mental wellbeing tracking during COVID-19, in *IEEE/WIC/ACM International Joint Conference on Web Intelligence and Intelligent Agent Technology (WI-IAT)*, IEEE (2020)
16. C. Jyotsna, J. Amudha, Eye gaze as an indicator for stress level analysis in students, in *International Conference on Advances in Computing, Communications and Informatics (ICACCI)* (IEEE, 2018), pp. 1588–1593
17. M. Hassanalieragh et al., Health monitoring and management using internet of things (IoT) sensing with cloud based processing: opportunities and challenges, in *International Conference on Service Computing* (IEEE, 2015), pp. 285–292
18. S.S. Machiraju, N. Konijeti, A. Batchu, N. Tata, Stress detection using adaptive threshold methodology, in *5th International Conference on Communication and Electronics Systems (ICCES)* (IEEE, 2020), pp. 889–894

Offloading Strategy of D2D Communication and Computing Resources Based on Shared Tasks

MingChu Li, Dengxu Li, Xiao Zheng, and Chuan Lin

Abstract This paper proposes a joint resource offloading strategy based on NOMA communication technology. In the scene, cellular network communication and D2D communication network communication coverage coexist. Cellular network users have social attributes. They can share computing resources between user devices through the trust index between users. The calculation task can also be offloaded to the edge server MEC for task calculation. D2D users need to forward tasks through the edge server MEC. We analyze the needs of two types of users for modeling and minimize the overall cost of the two types of users in this scenario through adjustments such as task division, trust decision-making, and power control. Finally, a low complexity DSG algorithm based on game theory Nash equilibrium is proposed to solve the target problem. The experimental results show that this method can reduce the cost of both and maintain a good user fairness experience index.

Keywords Mobile edge computing (MEC) · Non-orthogonal multiple access (NOMA) · Shared tasks · Device-to-device (D2D) · Game theory

1 Introduction

With the continuous increase in the order of magnitude of computing requirements of various users, scenes of intensive computing tasks in life are becoming more and more common. Although the central processing unit is also developing rapidly, mobile devices are still limited by battery energy and storage space. It is difficult for them to maintain the user experience and have strong computing power at the

M. Li (✉) · D. Li
School of Software, Dalian University of Technology, Dalian, China
e-mail: mingchul@dlut.edu.cn

X. Zheng
School of Computer Science and Technology, Shandong University of Technology, Zibo, China

C. Lin
School of Software, Northeastern University, Shengyang, China

A. K. Bashir et al. (eds.), *Proceedings of International Conference on Computing and Communication Networks*, Lecture Notes in Networks and Systems 394,
https://doi.org/10.1007/978-981-19-0604-6_27

same time. In the network hotspot area, the emergence of 5G provides the possibility of supporting high-speed and high traffic density transmission. Although 5G technology has brought us great innovations in transmission technology in the field of communication and reduced a lot of communication costs, the cost of communication between the cloud and users is still difficult to ignore. Therefore, researchers propose to use edge cloud computing technology to solve this problem. Using edge computing technology, it can unite various types of IoT devices to cooperate with mobile users. Through the transmission power control and reasonable task allocation on the edge cloud, all users can achieve the optimal balance strategy in the regional scenario. But we also need to consider many factors, such as the communication capabilities of the radio channel, the heterogeneous computing resources of the edge cloud, and the pairing of users and the edge cloud.

In the field of communication, the traditional 3G transmission technology has serious near-far effects. The 4G multiple access technology uses orthogonal frequency division multiplexing (OFDM)-based orthogonal frequency division multiple access technology and uses link adaptive coding technology. Compared with the traditional orthogonal multiple access technology (OMA), in scenes with near-far effects, wide coverage, and dense access points, due to its serial interference technology and power multiplexing technology, the use of non-orthogonal multiple access (NOMA) technology can bring obvious performance advantages.

At the same time, we also consider that the user scenario of this article is complex. In multiple areas of the cellular network, there are a large number of mobile device users who have an inconsistent number of computationally intensive tasks that need to be offloaded to the edge server MEC for calculation, or task sharing among trusted mobile users. At the same time, it is also faced with a complex network overlap situation, and multiple D2D communication users will choose to use the base station in the cellular network for data forwarding. Therefore, since D2D users and mobile device users may share an MEC base station, their communication will also be affected by the transmission power of the other party. The scene is shown in Fig. 1.

2 Our Contribution

Since the downlink time is very short during the entire communication process, this article considers describing the scenario in the uplink. The independent variables related to cellular mobile users are other trusted users who choose to offload computing tasks, the MEC base station, and the amount of offloaded tasks. The independent variables related to the D2D user are the MEC base station selected for task forwarding and the power for task forwarding. At the same time, our optimization goal is also subject to a series of constraints, such as restrictions on money and time costs for mobile users, restrictions on the mutual trust threshold between users and users, restrictions on the computing power of MEC base stations, and restrictions on the communication and transmission environment. Therefore, we believe that the problem is a complex, splittable, multi-user non-cooperative optimization problem.

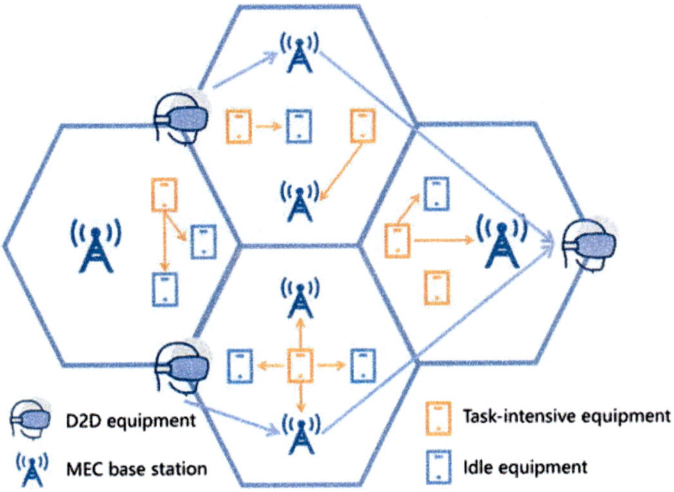

Fig. 1 Scene

In terms of solution ideas, we hope to reduce the complexity of the solution algorithm. Finally, we choose game theory with lower complexity as the theoretical basis and describe the competitive relationship between users as a non-cooperative load balancing game to solve the Nash equilibrium optimal solution of the problem. The main contributions of this article are as follows:

- Under the NOMA communication technology, the paper proposes a cooperative optimization solution to the communication needs of users of multiple types and needs in the same scenario.
- The trust index between users and users is proposed, and the historical connection situation, address book matching and other factors are used to measure whether users can delegate their own uninstall tasks to other users for calculation and processing.
- To solve this problem, we put forward the DSG algorithm based on the theory of game theory, joint power control, joint decision-making, and resource allocation, which minimized the overall user cost and increased the utilization rate of MEC and users. Fairness index.

3 Related Work

This section does the current research status of offloading strategies in power control and resource allocation. With the advent of the Internet of things and 5G era, massive amounts of terminal data are generally supported by edge computing. The ensuing problem of computing unloading faces many challenges. At present, more researches on the offloading strategy of MEC computing offloading mainly focus on the three themes of the offloading strategy based on energy consumption, the offloading strategy based on delay and the strategy of jointly adjusting energy consumption and delay. Literature [1] proposed a branch and bound method based on reconstruction linearization technology to obtain sub-optimal results to minimize energy consumption. Literature [2] proposed a closed-form optimal task segmentation strategy. Using convex optimization theory to obtain a closed form of computing resource allocation strategy, to minimize the weighting and waiting time of all mobile devices. Based on the DAG model [3, 4], the literature describes a fine-grained task offloading algorithm that minimizes the energy consumption of mobile devices under delay constraints. Literature [5] designs the optimal transmission precoding matrix and computing resource allocation and proposes an iterative algorithm based on a novel continuous convex approximation technique to simultaneously minimize the overall user energy consumption and meet the delay constraints.

At the same time, there are many resource allocations for complex user scenarios. For example, literature [6] studied the computing and traffic offloading used for content delivery and delay-sensitive task offloading services in a cache-assisted device-to-device multicast (D2MD) network. It proposes multicast-aware coding and cooperative caching schemes to improve the efficiency of content distribution and optimize the energy consumption of content delivery. Literature [7] studies the offloading of latency-aware computing of IoT devices with confidential settings and the joint optimization problem in scenarios where malicious eavesdroppers may cause interruptions.

With the in-depth research on the direction of communication resource allocation, the research on power control in the MEC system has gradually attracted academic research attention. Literature [8] proposed an online optimization strategy for computing task shunting of MEC server with sleep control scheme and proposed a Lyapunov optimization method based on decision-making in time slot blocks to optimize long-term problems. In [9], the edge cloud server optimizes the equipment transmission power and computing resource allocation ratio and designs a unique injective function from transmission power to computing resource allocation for each user. Literature [10] established a joint optimization model for task offloading and power allocation and proposed a centralized joint optimization algorithm for task offloading and power allocation. Literature [11] proposes a two-stage alternate method framework based on Lagrangian dual decomposition based on task calculation and task data transmission, using flow shop scheduling theory and greedy strategy for offloading decision-making and task scheduling. In terms of practical application, the current research includes monitoring applications and routing schemes related to big data and Internet of things industry [12–14].

Researchers pay more attention to user feedback and service quality for edge computing. Literature [15] studied the problem of maximizing the service provider's revenue under the condition of guaranteeing the user's service quality, constructing it as a semi-Markovian problem, and transforming it into a linear programming problem. And the literature [16, 17] considers various aspects of service quality.

4 System Model and Problem Modeling

This article simulates a complex scenario of 5G users with overlapping D2D communication networks and cellular communication networks. There are two types of users in the scene. The first type of cellular network mobile device users' goal is to minimize their own time and money costs. The second type of D2D network users establish a goal for power control: find the maximum transmission power of the sender when the upper limit of environmental interference is met. We assume that there are a total of N cellular network users $U = \{1, 2, \ldots, I\}$ in the scene, and each user randomly generates S_n computing tasks. D2D network users $D = \{1, 2, \ldots, Q\}$, and each D2D user randomly generates S_q transmission tasks. We believe that users are rational. When there are multiple choices, users will take rational behaviors within the limits to minimize their costs. At the same time, there are a total of J small base stations to provide services for users in the scene, which is expressed as $M = \{1, 2, \ldots, J\}$. The service provider MEC base station is heterogeneous in processing power, so its pricing is positively correlated with computing power.

4.1 Analysis of NOMA Transmission and Traditional Transmission

We assume that the channel transmission characteristics in the scene are implemented based on the Rayleigh probability density function. The paper considers that other conditional factors in each time slot T are static. The channel gain of the user's divisible task is:

$$h_1 < h_2 < \cdots < h_x < h_y < \cdots < h_I \tag{1}$$

The user's choice of base station is uncertain, $\omega_{x,j}$ is a 0/1 variable, indicating whether the two establish a connection. Assume that the user evenly distributes his tasks to two different MECs for offloading. In the traditional communication transmission mode, users can only queue up in order to transmit tasks, so the communication transmission rate is expressed as:

$$R_x = B\log\left(1 + \frac{P_i h_x}{N_0}\right) \tag{2}$$

$$T_{\text{trad}} = T_x + T_y = \frac{N}{2B\log\left(1 + \frac{P_i h_x}{N_0}\right)} + \frac{N}{2B\log\left(1 + \frac{P_i h_y}{N_0}\right)} \tag{3}$$

The sub-channels of the NOMA transmission base station are orthogonal and do not interfere with each other, so they can be shared by multiple users. However, it is also subject to interference among users. So its communication transmission rate is expressed as:

$$R_x = B\log\left(1 + \frac{P_i h_x}{P_i h_x + N_0}\right) \tag{4}$$

Since it is sent to multiple users at the same time in time slot T, its time is expressed as:

$$\begin{aligned} T_{\text{noma}} &= \max(T_x, T_y) = (1+\theta)T_y \\ &= \max\left(\frac{N}{2B\log_2\left(1+\frac{P_i h_x}{P_i h_x + N_0}\right)}, \frac{N}{2B\log_2\left(1+\frac{P_i h_x}{N_0}\right)}\right) \end{aligned} \tag{5}$$

Therefore, the difference between the delay in the NOMA transmission mode and the delay in the traditional mode is:

$$\begin{aligned} T_{\text{trad}} - T_{\text{noma}} &= (T_x + T_y) - \max(T_x, T_y) \\ &= \frac{N}{2B\log_2\left(1+\frac{P_i h_x}{N_0}\right)} + \frac{N}{2B\log_2\left(1+\frac{P_i h_y}{N_0}\right)} \\ &- \max\left(\frac{N}{2B\log_2\left(1+\frac{P_i h_x}{P_i h_x + N_0}\right)}, \frac{N}{2B\log_2\left(1+\frac{P_i h_x}{N_0}\right)}\right) \geq 0 \end{aligned} \tag{6}$$

This shows that the reduction of the delay cost brought by the NOMA transmission mode in the current scenario is significant. In terms of energy consumption, we also made a comparison under the same conditions:

$$E_{\text{trad}} - E_{\text{noma}} = 2TP_x - (1+\theta)TP_x = (1-\theta)TP_x \tag{7}$$

In the same way, when the user divides the task into multiple subtasks and transmits them to different MEC base stations, the transmission time is expressed as:

$$T_i = \max(T_1, T_2 \ldots T_J) \tag{8}$$

And its energy consumption is expressed as:

$$E_i = E_1 + E_2 + \cdots + E_J \tag{9}$$

4.2 Cellular Network Users with Social Attributes

In this section, we will numerically measure and describe the social attributes of cellular network users. Through our research, we found that in reality, users' social attributes are very universal. The connection between a mobile device and a mobile device usually represents the connection between the owners of the device, that is, the social relationship between people. The social relationship can reflect the trust relationship between people. Therefore, we believe that a device with a trust relationship is willing to provide the computing resources of its own device to serve the other party when one party has a task that needs to be calculated, and the other party has free computing resources. We use several comprehensive indicators to describe trustψ_i. The first is the intimacy index of the address book of mobile devices. We use

$$\psi_{a,b}^{\text{add}} = \begin{cases} 1 & U_a \text{ and } U_b \text{ exist in each others address book} \\ 0.5 & U_a \text{ or } U_b \text{ exist in each others address book} \\ 0 & \text{Both } U_a \text{ and } U_b \text{ have no others address book} \end{cases} \tag{10}$$

to describe whether there is a connection in the address book between device x and device y. The second indicator is the connection frequency indicator of the mobile device. We use

$$\psi_{a,b}^{\text{fre}} = \frac{F_{a,b}}{\sum_{i=0}^{N} F_{a,i}} \tag{11}$$

to describe the connection frequency of device x and device y in a certain period of time, including the number of times in a WIFI environment and the number of Bluetooth connection transmissions. Third, we use ψ_a^{fri} to describe the friendliness index. It is the product of the amount of computing resources received by user x from other mobile devices and the amount of computing resources from device y in a certain period of time in history, divided by the amount of computing resources provided to all devices by x.

$$\psi_a^{\text{fri}} = \frac{N_y^s}{\sum_{i=0}^{N} N_i^r} \times \frac{\sum_{i=0}^{N} N_i^s}{\sum_{i=0}^{N} N_i^r} \tag{12}$$

From these three ratios, we get a trust index of device x to device y.

$$\psi_i = \delta_1 \psi_{a,b}^{\text{add}} + \delta_2 \psi_{a,b}^{\text{fre}} + \delta_3 \psi_a^{\text{fri}} \tag{13}$$

4.3 Cellular Communication Network Users and D2D Communication Network Users

We use matrix $G(j) = (\omega_{1,j}^T, \omega_{2,j}^T, \ldots, \omega_{i,j}^T)$ to describe the base station user cluster established by MEC and each user. The tasks of cellular network users are separable. Therefore, in the time slot t, users can perform calculations locally, or they can transmit tasks to multiple MECs for offloading calculations. Therefore, the user's task volume is expressed as:

$$N_i = \sum_{j \in M} N_{i,j} + \sum_{k \in U \text{ and } k \neq i} N_{i,k} + N_{\text{loc}} \tag{14}$$

For users of the D2D communication network, we use

$$\mu_{q,g} = \begin{cases} 1 & D_q \text{ is connected to group } G_j \\ 0 & D_q \text{ is not connected to group } G_j \end{cases} \tag{15}$$

to represent the connection relationship between the D2D user and the cellular user base station cluster. The user's transmission task power and local calculation power are limited by the performance of the device itself and can only be used within the specified range:

$$\sum_{j \in M} P_{i,j} + \sum_{k \in U \text{ and } k \neq i} P_{i,k} + P_{\text{loc}} \leq P_i^{\max} \tag{16}$$

The users signal power is expressed as

$$S_{i,j} \triangleq P_i h_i \omega_{i,j} \tag{17}$$

The signal power of other D2D users in the same group is expressed as

$$O_{q,g} \triangleq \sum_{k \in D} P_q h_q \mu_{q,g} \tag{18}$$

and the signal power of cellular network users in the group is expressed as

$$I_{i,j} \triangleq \sum_{k=i+1}^{I} P_k h_k \omega_{k,j} \tag{19}$$

When the D2D user transmission network overlaps with the cellular user offloading network, the transmission power between users will affect each other. Therefore, the combined transmission rate of cellular network users is expressed as:

$$H_{i,j} \triangleq B\log_2\left(1+\frac{S_{i,j}}{I_{i,j}+O_{q,g(j\in g)}+N_0}\right)$$
$$= B\log_2\left(1+\frac{P_i h_i \omega_{i,j}}{\sum_{k=i+1}^{I} P_k h_k \omega_{k,j}+\sum_{k\in D} P_q h_q \mu_{q,g}+N_0}\right) \tag{20}$$

The transmission power of D2D communication network users is expressed as:

$$H_{q,g} \triangleq B\log_2\left(1+\frac{S_{q,g}}{O_{q,g}+N_0}\right)$$
$$= B\log_2\left(1+\frac{P_q h_q \mu_{q,g}}{\sum_{k\in D \text{ and } k\neq q} P_k h_k \mu_{k,g}+N_0}\right) \tag{21}$$

Therefore, the analysis shows that the transmission time of the cellular network user in the upload stage is expressed as

$$T_i^{\text{tran}} = \max\left\{\frac{N_{i,j}}{H_{i,j}} | j\in M\right\} \tag{22}$$

The computing tasks of cellular network users can choose to be executed locally, and the time cost is expressed as:

$$T_i^{\text{loc}} = \frac{N_i - \sum_{j\in M} N_{i,j}}{\gamma^{\text{loc}}} \tag{23}$$

Or the user can offload the task to the MEC base station for task calculation. The time cost is expressed as:

$$T_i^{\text{pro}}(j\in M, k\in U) = \max\left\{\frac{N_{i,j}}{\gamma_m}, \frac{N_{i,k}}{\gamma^{\text{loc}}}\right\} \tag{24}$$

The amount of data in the downlink transmission phase is generally the result data after processing, which is very small compared to the uploaded data, so this article ignores the time consumption of the downlink transmission phase. So it can be obtained that the time cost of a cellular network user is expressed as:

$$T_i = \max\left\{T^{\text{tran}} + T^{\text{pro}}, T^{\text{loc}}\right\} \leq T_{\max} \tag{25}$$

The user's task is generally time-sensitive, so it is constrained by the maximum delay of the user's task. From the transmission power, the energy consumption in the upload phase is expressed as:

$$E_i^{\text{tran}} = T_i^{\text{tran}} P_i \tag{26}$$

The locally calculated energy consumption is expressed as:

$$E_i^{\mathrm{loc}} = T_i^{\mathrm{loc}} p^{\mathrm{loc}} \tag{27}$$

At the same time, it should be noted that cellular users generally have mobility characteristics, so it is limited by energy consumption. Expressed as:

$$E_i = E_i^{\mathrm{tran}} + E_i^{\mathrm{loc}} \le E_i^{\max} \tag{28}$$

The MEC server provides users with services of different computing speeds, so the unit price per unit time of different base stations is different. According to the pricing of Amazon Web Services, it can be obtained that the monetary cost of the user when the task is uninstalled is affected by which server it is connected to and the size of the uninstalled task. So expressed as:

$$G_i = \sum_{j=1}^{j=J} \frac{N_{i,j}}{\gamma_m} \rho_m \le G_i^{\max} \tag{29}$$

The user will only pay a limited monetary cost for each task, so the user's monetary cost should be less than the maximum value of its monetary cost. For D2D communication network users, its transmission power will be much larger than that of cellular network users, but its size is also limited by the performance of its own equipment: $P_q \le P_q^{\max}$. Its transmission delay cost is also affected by the cellular network user group, expressed as:

$$\begin{aligned} T_q &= \frac{N_q}{H_q} \\ &= \frac{N_q}{B \log_2\left(1 + \frac{P_q h_q}{\sum_{k=q+1}^{Q} P_q h_q \mu_{q,g} + N_0}\right)} \end{aligned} \tag{30}$$

Since D2D users are generally large-scale immovable devices, their energy will not be limited by their own power, so the energy limit of D2D users is not considered. So its cost is expressed as: $C_q = T_q = \frac{N_q}{H_q}$. Considering that users, as independent individuals, hope to minimize their own costs, but the satisfaction of different types of users is generally related to the degree of importance they place on time and money costs. Therefore, the goal of this article is to minimize the total cost of users of all parties, but two weight values α and β are used to fine-tune the cost emphasis of different types of users. It is expressed as:

$$C_i = \alpha T_i + \beta G_i \tag{31}$$

4.4 Problem Modeling

Finally, in summary of the above analysis, we formulate the problem as an optimization problem that minimizes the cost of various users. The objective function and constraints of the problem are expressed as follows:

$$\text{Problem:} \quad \text{Min} \ \sum_{i \in U} C_i + \eta \sum_{q \in D} C_q$$

$$\text{Constraint:} \quad E_i^{\text{tran}} + E_i^{\text{loc}} \leq E_i^{\max} \qquad i \in U$$

$$\sum_{j \in M} P_{i,j} + \sum_{k \in U \text{ and } k \neq i} P_{i,k} + P_{\text{loc}} \leq P_i^{\max} \qquad i \in U$$

$$P_d \leq P_d^{\max} \qquad d \in D$$

$$\alpha + \beta = 1$$

$$\max\left\{T_i^{\text{tran}} + T_i^{\text{pro}}, T_i^{\text{loc}}\right\} \leq T_i^{\max} \qquad i \in U$$

$$\sum_{j=1}^{J} \frac{N_{i,j}}{\gamma_m} \rho_m \leq G_i^{\max}$$

$$\sum_{j \in M} N_{i,j} + \sum_{k \in U \text{ and } k \neq i} N_{i,k} + N_{\text{loc}} = N_i \qquad i \in U$$

$$\text{Variables:} \quad \mu_{q,g}, \ \mu_{i,j}, \ P_i, \ P_d, \ N_{i,j}, \ N_{i,k}$$

$$\{k, m, d | k \in U \quad m \in M \quad d \in D\}$$

From the objective function, we can see that our goal is to minimize the total cost of all users in the scene. The total cost is composed of the time and money costs of cellular network communication users and D2D network communication users. In the constraints on the algorithm, we first consider the constraints of the limited energy of the mobile users of the cellular network. And it is considered that the device power of the mobile user is used by the shared task part, the offload processing task part and the local task processing part. At the same time, the communication transmission power of the D2D communication device will also be restricted by the performance of the device. And the sum of cost weights should be 1. The user's time cost should be the maximum value of the time for tasks to be unloaded to the MEC equipment and processed and the local processing time, and this time cost is also constrained by the user time cost. Similarly, the user's money cost is related to the unit price of the MEC equipment uninstalled and the processing time, and it also receives the constraint of the user's maximum money cost. Since the users task needs to be fully completed in the end, the amount of tasks required by the user should be composed of three parts. The first part is the amount of tasks shared by the user to other users, and the second part is the amount of tasks that the user unloads to MEC. The third part is the amount of tasks completed by the device itself.

5 Solving Algorithm

Faced with this complex multi-objective optimization function, it is very complicated to directly solve it. In order to reduce the complexity of the algorithm and reduce the time cost of the algorithm, this section decomposes this optimization problem.

First, we use the switching algorithm to adjust the cluster grouping situation of D2D communication network users with the transmission time as an indicator and control the power output of the two types of users in each group, so that the overall cluster can achieve the optimal transmission time consumption. After determining the grouping of cellular network users, MEC mobile devices, and D2D communication network users, we evaluated the trust of cellular network users between the groups and matched the sharable users with idle computing resources. Then, according to the communication environment of these three types of participants and the existing computing resources, based on the theoretical basis of game theory, we adjust the user offloading tasks and the allocation of MEC computing resources, hoping to achieve the lowest total cost of the two types of users. And any unilateral offset change of users in the system cannot make the overall profit decrease in a steady state. The solution algorithm in this paper consists of two parts. Algorithm 1 is used to adjust the sum of transmission time and trust calculation, and Algorithm 2 is used to adjust the matching of users, as follows:

In Algorithm 1, the input is the parameter data of the user, D2D equipment, and MEC equipment, as well as the current communication environment parameters and the users task requirement data set. First, the algorithm will be initialized and all D2D devices will be allocated to the matching cluster of the first group of cellular users and MEC. Set the algorithm temperature to the highest temperature, and set the number of iterations to currently 1. The outermost loop is when the algorithm temperature is lower than the minimum temperature we set, which means that the algorithm has reached the approximate optimal solution. In this cycle, we use Formula 22 to obtain the total user transmission time T1 under the current allocation strategy. Then randomly assign D2D devices to the $G(r)$ user MEC matching cluster, and obtain the total user total transmission time T2 under the new allocation strategy. If this exchange can cause the total transmission time of users to drop, or the random number is less than the current exp value, then we will update the current exchange to packet G. And update l and the current algorithm temperature T.

In the second cycle, we judged the sharable users of cellular network users. Each user with idle computing resources will calculate the trust between itself and other cellular network users based on multiple indicators in Formula 13. At the same time, the trust of other users in it will be calculated. When the idle user's trust in the target user is greater than its own trust threshold, and the target user's trust in the idle user is greater than its own threshold, we record the sum of the trust in the user. After one round of the outer loop, the idle computing resources of user k are allocated to the user with the largest sum of trust in the record.

Algorithm 1 DSUG

Input: $U, D, M, Device\ parameters,\ Communication\ parameters,\ User\ task\ requirements\ parameters$
Output: G, U^{share}

Initialization: Put D into G(1), $T = T_{\max}, T_{\min}, l = 1$
while $T > T_{\min}$ **do**
  Use (22) to calculate $T_1 = \sum_{i \in U} T_i^{tran} + \sum_{q \in D} T_q$
  Randomly change a D2D temporarily assign it to G(r), and get T_2 by (22).
  if $T_1 - T_2 > 0$ or $random(0, 1) < Exp(\frac{T_1 - T_2}{T})$ **then**
    Update the last change to G
  end if
  $l{+}{+}, T = \frac{T_{\max}}{log(l)}$
end while
$U^{share} = [[]]\ and\ U^{tmp}.length = i$
for $U_k (k < I)$ **do**
  $U_k^{tmp} = new\ Map()$
  for $U_i (i <= I)$ **do**
    Use (13) to calculate $\psi_{i,k}$ and $\psi_{k,i}$.
    if $N_b = 0\ and\ psi_{i,k} > psi_a^{\min}\ and\ psi_{k,i} > psi_b^{\min}$ **then**
      $U_k^{tmp}[psi_{i,k} + psi_{k,i}] = U_i$
    end if
  end for
  $U^{share}[i].push(U_k^{tmp}[Max(U_k^{tmp})])$
end for
return

In Algorithm 2, the user MEC matching cluster, user task transmission time, user task requirement parameters, and equipment parameters need to be input. First, a stable user set needs to be initialized, and then all MEC devices are allocated to cluster one. Here we use the dichotomy to approximate the sum of the user's optimal cost. Outer loop control when all users have not entered a stable state, users need to always try to perform MEC exchange behavior. The inner loop controls that each user i will try to exchange the MEC it owns to other users. If the exchange brings a reduction in the total cost, then the current exchange user will be recorded, and the actual exchange mec user depends on the maximum cost reduction caused by the exchange behavior. If the user i fails to bring cost benefits through the exchange behavior in a round of the cycle, then the user will enter a personal stable state. Only when each user enters a personal stable state in the outer loop, will the final game optimal stable state be obtained. The cost at this time is that each user has no motivation to actively exchange behavior and will make the current cost of all users reach the smallest sum.

Algorithm 2 GAR

Input: $G, T, User\ task\ requirements\ parameters, Device\ parameters$

Output: $LastCost$

1: Initialization: $U_{sta} = \{\}$, Allocate all MEC to $G_{(1)}$. Use the dichotomy to find the C_i^{opt} that can complete the users task within the user cost $(0, C_i^{\max})$ range, and then get the sum of the optimal costs for all users $C_1 = \sum_{i \in U} C_i^{opt}$

2: **while** $U_{sta}.length < U.length$ **do**

3:   **for** $U_i(i \in G_{(g)})$ and $M_a(a \in G_{(g)})$ **do**

4:     $U_{sta} = \{\}$

5:     **for** $U_j(j \in G_{(g)}$ and $M_b(b \in G_{(g)})$ **do**

6:       $C_2 = \sum_{i \in U} C_i^{opt}$ Repeat operation 1 to get it, if $C_1 > C_2$, record the current exchange ER.

7:     **end for**

8:     **if** $\{ER\} \neq \emptyset$ **then**

9:       Find the exchange record of the maximum profit $ER_{\max} = Math.max(ER)$ and update it to G.

10:     **else**

11:       $U_{sta}.push(U_i)$

12:     **end if**

13:   **end for**

14: **end while**

15: Repeat operation 1 to get $C_{opt} = \sum_{i \in U} C_i^{opt}$

16: **return**

6 Experimental Results

This section compares and analyzes the experimental results obtained by the solution algorithm of the thesis. Generally speaking, the model and algorithm proposed in this paper can greatly improve the difficulty of user multi-task assignment in dense network scenarios. While ensuring the completion of user tasks, starting from the user as a whole, the cost of the user group is reduced. At the same time, in the algorithm evaluation method, we also consider the user's service quality and fairness index. The main comparison indicators of the experiment include the comparison of the total cost of users of different sizes, MEC of different sizes, and the fairness index of users of different sizes. The comparison algorithms are based on local computing LOC algorithm, random RAN algorithm, and exchange fade SWAP algorithm.

From Fig. 2, we can see that the DSG algorithm can still maintain a small sum of user costs even when the user scale changes. The DGT algorithm will reasonably and optimally allocate the user's tasks to the available MEC base stations. And when the user group is larger or the number of idle users increases, the DSG algorithm can make good use of these idle resources to help task-intensive users reduce their money costs. On the whole, compared with the best random algorithm, it can achieve a cost increase of nearly 10%. In these four algorithms, local computing can only share tasks between the user equipment itself and trusted users. However, local computing

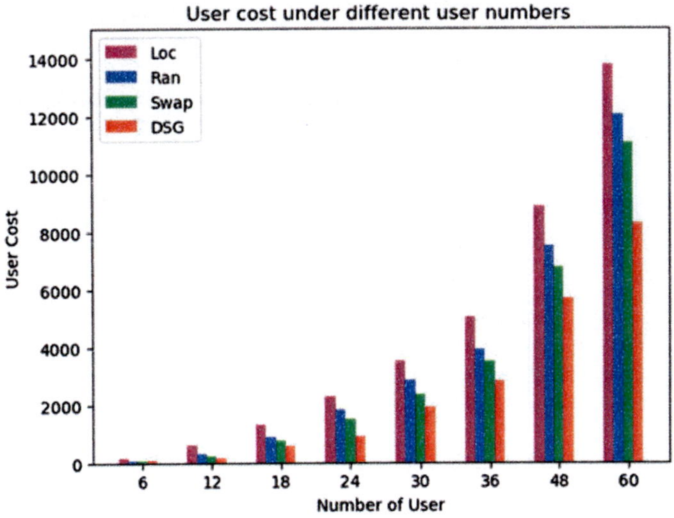

Fig. 2 ChangeUSER

is greatly constrained by the performance of the mobile device itself, which will cause a long time cost. However, due to the algorithm itself, the random algorithm cannot obtain the optimal solution in a complex user situation. Therefore, although the cost caused by it is lower than that of the local calculation, it is still higher than the switching algorithm and the DSG algorithm. The exchange algorithm can achieve very good results. When the user scale is small, it is more consistent with the performance of the DSG algorithm. But because the role of idle users is ignored, when the user scale becomes larger, the advantages of the DSG algorithm become more obvious.

Figure 3 shows the performance of the four algorithms on the user fairness experience index when the number of MECs changes. First of all, you can see that the purple line is the performance of the local algorithm. In the local algorithm, the user's offloading task is only affected by the computing power of its own device and the number of trusted users and computing power. Therefore, the number modification of MEC devices will not change the user's fair experience index. In the RAD algorithm, SWAP algorithm, and DSG algorithm, they will be more affected by changes in the number of MECs. First of all, in the early stage of the algorithm, due to the extremely small number of MECs, most users are in a situation where they cannot use MEC to offload tasks. Therefore, the fairness experienced by users is similar. With the increase in the number of MECs, the competition between users for MECs becomes greater and greater. There may be cases where some users have sufficient money costs, so they can use one or more MECs. However, some users cannot share MEC due to the limitation of money cost. At this time, the fairness experience index between users will be very poor. Although the supplier provides users with a service for calculating offloading, since the distribution can only be measured by monetary

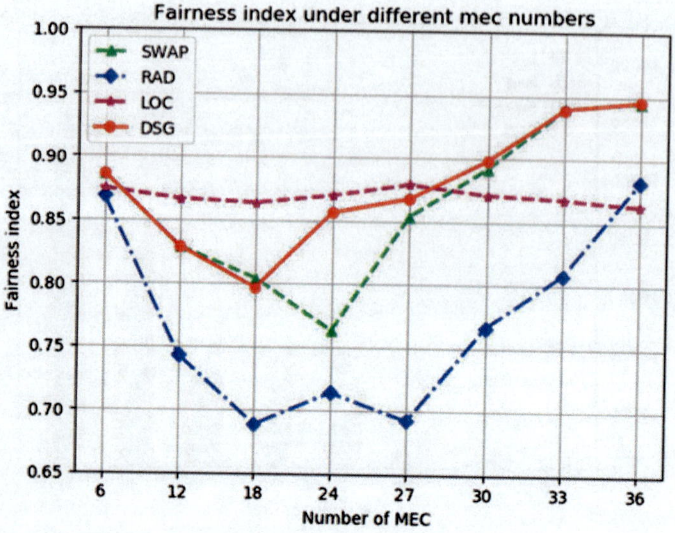

Fig. 3 FairChangeMEC

income, these situations cannot be taken into account. While the DGT algorithm reduces the cost, it will attribute everyone's cost to a reference value for adjustment and control and incorporate the fairness index into the influencing factor. This makes the DSG algorithm pay more attention to the whole in the distribution of MEC. When the number of MECs is saturated and the available computing resources exceed user needs, there is no longer a strong competitive relationship between users and users. At this time, the three algorithms will achieve a better fairness situation.

Figure 4 shows the performance of the four algorithms on the fairness experience index when the number of users changes. From a mathematical point of view, fairness index is the ratio of users' satisfaction with the use of resources to that of other users, so it is affected by two aspects. From the perspective of local algorithms, since the number of idle users is certain, and as the user scale grows, users with idle computing resources can match more users in the initial stage, resulting in an increase in the fairness index of small-scale users. After this, it has been in a downward situation, until after a certain degree of reduction, it gradually stabilized. The DSG algorithm, RAD algorithm, and SWAP algorithm will all be affected by the large scale of users. It can be seen that the DSG algorithm can achieve a good fairness index in resource allocation in combination with idle resources when the MEC has sufficient computing resources. In the middle section where competition is fierce, the DSG algorithm can also maintain a better fairness index. In the end, it is maintained at a relatively good fairness index. In general, compared with other algorithms, it can maintain a good fairness index in the entire user scale. However, the performance of RAD algorithm and SWAP algorithm is relatively poor under fierce competition. It performs better when the supply of computing resources is unbalanced with user needs.

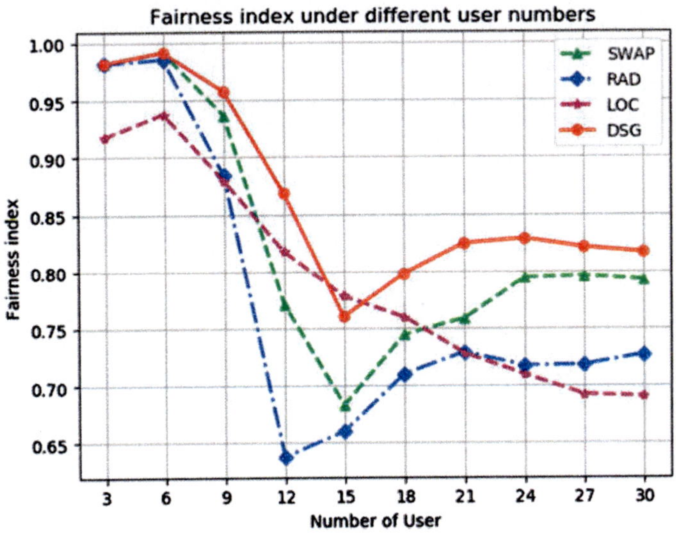

Fig. 4 FairChangeUSER

7 Conclusion

This article is oriented to mobile device users of multiple cellular networks and large data transmission users of D2D communication networks, and the optimization goal is the sum of their time consumption cost and the weight of money expenditure cost. In the scenario of this article, there is also a third type of mobile device users without demand. Their devices have certain computing capabilities and they have no requirements for computing tasks. Considering that in real life, there are many scenarios where such users exist. Therefore, we hope to be able to reasonably and safely use idle resources to complete the tasks of users who need them in the scene. Therefore, according to the characteristics of mobile devices, we propose a comprehensive description of the trust between devices through four indicators of intimacy in the address book, connection frequency, friendliness, and friendliness between devices. Matching idle users is performed by judging whether the trust degree between the user and the user meets the trust threshold of both parties. In the communication process, we choose to simulate the transmission mode of NOMA for task transmission.

In the scenario of this article, the composition of users is complicated, and the transmission power between users will affect the transmission time of themselves and other users. Therefore, we have jointly controlled the communication power of the two types of users. Since users have expectations of minimizing transmission time, we use this as a standard to allocate the access base station group of D2D communication users through the DSUG algorithm. Due to the need to minimize the time consumption cost and minimize the money expenditure, we use the GAR algorithm to allocate the access matching situation between the user and the MEC base station. Finally, by comparing with other algorithms, it is proved that the algorithm can play a role in reducing the total cost in the scene. In addition, the DSG algorithm also considers the user fairness experience index as an influencing factor for judging whether to connect or not in the process of computing resource allocation between the user and the MEC. Therefore, the DSG algorithm also has a better performance on the user fairness index than other algorithms.

For future work, we think we should pay attention to the changes brought by the update of communication technology and try to explore more effective algorithms.

Acknowledgements This work was support from the National Science Foundation of China (Grant No. 61877007), Fundamental Research Funds for the Central Universities (Grant No. DUT20GJ205), and the Shandong National Science Foundation of China (Grant No. ZR202103040468).

References

1. P. Zhao, H. Tian, C. Qin, G. Nie, Energy-saving offloading by jointly allocating radio and computational resources for mobile edge computing. IEEE Access **5**, 11255–11268 (2017)
2. J. Ren, G. Yu, Y. He, G.Y. Li, Collaborative cloud and edge computing for latency minimization. IEEE Trans. Veh. Technol. **68**(5), 5031–5044 (2019)
3. M. Deng, H. Tian, B. Fan, Fine-granularity based application offloading policy in cloud-enhanced small cell networks, in *2016 IEEE International Conference on Communications Workshops (ICC)* (2016), pp. 638–643
4. P. Zhao, H. Tian, B. Fan, Partial critical path based greedy offloading in small cell cloud, in *2016 IEEE 84th Vehicular Technology Conference (VTC-Fall)* (2016), pp. 1–5
5. S. Sardellitti, G. Scutari, S. Barbarossa, Joint optimization of radio and computational resources for multicell mobile-edge computing. IEEE Trans. Signal Inf. Process. Netw. **1**(2), 89–103 (2015)
6. D. Wang, Y. Lan, T. Zhao, Z. Yin, X. Wang, On the design of computation offloading in cache-aided D2D multicast networks. IEEE Access **6**, 63426–63441 (2018)
7. Y. Wu, J. Shi, K. Ni, L. Qian, W. Zhu, Z. Shi, L. Meng, Secrecy-based delay-aware computation offloading via mobile edge computing for internet of things. IEEE Internet Things J. **6**(3), 4201–4213 (2019)
8. S. Wang, X. Zhang, Z. Yan, W. Wenbo, Cooperative edge computing with sleep control under nonuniform traffic in mobile edge networks. IEEE Internet Things J. **6**(3), 4295–4306 (2019)
9. S. Barbarossa, S. Sardellitti, P. Di Lorenzo, Joint allocation of computation and communication resources in multiuser mobile cloud computing, in *2013 IEEE 14th Workshop on Signal Processing Advances in Wireless Communications (SPAWC)* (2013), pp. 26–30
10. J. Liu, X. Wei, J. Fan, Tolerable data transmission of mobile edge computing under internet of things. IEEE Access **7**, 71859–71871 (2019)
11. Z. Kuang, L. Li, J. Gao, L. Zhao, A. Liu, Partial offloading scheduling and power allocation for mobile edge computing systems. IEEE Internet Things J. **6**(4), 6774–6785 (2019)
12. M. Faheem, R.A. Butt, R. Ali, B. Raza, M.A. Ngadi, V.C. Gungor, CBI4.0: a cross-layer approach for big data gathering for active monitoring and maintenance in the manufacturing industry 4.0. J. Ind. Inf. Integr. **24**, 100236 (2021). [Online]. Available: https://www.sciencedirect.com/science/article/pii/S2452414X21000364
13. M. Faheem, G. Fizza, M.W. Ashraf, R.A. Butt, M.A. Ngadi, V.C. Gungor, Big data acquired by internet of things-enabled industrial multichannel wireless sensors networks for active monitoring and control in the smart grid industry 4.0. Data Brief **35**, 106854 (2021). [Online]. Available: https://www.sciencedirect.com/science/article/pii/S2352340921001384

14. M. Faheem, R.A. Butt, B. Raza, M.W. Ashraf, M.A. Ngadi, V.C. Gungor, A multi-channel distributed routing scheme for smart grid real-time critical event monitoring applications in the perspective of industry 4.0. Int. J. Ad Hoc Ubiquitous Comput. **32**(4), 236–256 (2019). [Online]. Available: https://doi.org/10.1504/ijahuc.2019.103264
15. D.T. Hoang, D. Niyato, P. Wang, Optimal admission control policy for mobile cloud computing hotspot with cloudlet, in *2012 IEEE Wireless Communications and Networking Conference (WCNC)* (2012), pp. 3145–3149
16. C. You, K. Huang, H. Chae, B.-H. Kim, Energy-efficient resource allocation for mobile-edge computation offloading. IEEE Trans. Wireless Commun. **16**(3), 1397–1411 (2017)
17. P.D. Lorenzo, S. Barbarossa, S. Sardellitti, Joint optimization of radio resources and code partitioning in mobile cloud computing. CoRR, vol. abs/1307.3835 (2013). [Online]. Available: http://arxiv.org/abs/1307.3835

Computation Offloading in Edge Computing Based on Deep Reinforcement Learning

MingChu Li, Ning Mao, Xiao Zheng, and Thippa Reddy Gadekallu

Abstract The development of 5G technology provides great convenience for edge computing. But it also means that the computing tasks generated by user equipment are more and more complex. These tasks are no longer simply to complete one target, but are often composed of multiple subtasks. In order to solve the edge computing problem of multiple subtasks, we propose a Task Mapping Algorithm based on Deep Reinforcement Learning (DRL-TMA). Firstly, we abstract the computation intensive task as a directed acyclic graph, and then propose a Graph Sequence Algorithm (GSA) to transform the DAG task into a specific topological sequence, and then determine the optimal offloading decision of all subtasks according to the sequence order. We model the offloading problem as a Markov Decision Process (MDP) to maximize the comprehensive profit. The experimental results show that the Task Mapping Algorithm based on Deep Reinforcement Learning has stronger decision-making ability and can obtain the approximate optimal comprehensive profit which proves the effectiveness of the algorithm.

Keywords Mobile edge computing (MEC) · Computation offloading · Deep reinforcement learning (DRL)

1 Introduction

With the development of 5G technology and Internet of things technology, more and more smart sensors are interconnected and have achieved good results [1–3]. And it also promotes the development of wireless sensor networks (WSN) [4]. As

M. Li (✉) · N. Mao
School of Software, Dalian University of Technology, Dalian, China
e-mail: mingchul@dlut.edu.cn

X. Zheng
School of Computer Science and Technology, Shandong University of Technology, Zibo, China

T. R. Gadekallu
Vellore Institute of Technology, Vellore, India

A. K. Bashir et al. (eds.), *Proceedings of International Conference on Computing and Communication Networks*, Lecture Notes in Networks and Systems 394,
https://doi.org/10.1007/978-981-19-0604-6_28

for mobile computing, it has changed from centralized cloud computing to edge computing. Mobile edge computing (MEC) refers to moving the network control and storage of mobile computing to the mobile edge devices with limited resources, so that centralized computing and low latency applications can be carried out. MEC can reduce network latency and energy consumption, and solve the key technical problems in 5G scenario.

At present, the complexity of mobile applications is getting higher and higher, and it presents diversified development, such as virtual reality, augmented reality, face recognition and so on [5, 6]. While bringing convenience to people, it is also a great testing of the computing power of mobile devices. However, due to the high complexity of the application, it will cause a great latency in relying on the device itself to complete the task, and aggravate the power consumption of the device. MEC sets computing resources to the edge of the network, which is closer to users than cloud computing [7, 8]. The advantage of MEC computing is that MEC can greatly shorten the computing latency and save the power of the device compared with the user device itself. Compared with cloud computing, because MEC is closer to the user, it can greatly shorten the transmission latency. How to decide whether the task is to be computed locally or offloaded to the edge is the core of the edge computing. The offloading decision problem in MEC environment can be transformed into an optimization problem, which needs to be solved in a given environment. However, due to the complexity of MEC environment, the final optimization problem is often NP-hard problem.

With the increasing complexity of applications, the computing tasks generated by mobile devices generally complete multiple functions, that is, a computing task is usually composed of multiple subtasks. These subtasks cooperate and depend on each other to form a whole computing intensive task. Therefore, it is easy to abstract a computing intensive task as a DAG, Now the goal is to get the optimal offloading decision of each subtask in DAG, so as to minimize the trade-off between the overall computing latency and energy consumption. However, with the increase of the number of subtasks, it is difficult to get the optimal offloading decision of all subtask [7]. Now there are many heuristic algorithms to solve this problem. For example, in [9], the edge offloading problem was abstracted as a 0–1 programming problem, and the actual problem was expressed as DAG structure, which was solved by particle swarm optimization algorithm. In [10] a heuristic algorithm for wireless sensing joint scheduling and computing offloading of multi-component applications was designed to minimize the computing latency. In [11], the scenario of multiple servers was considered. The problem was modeled as a nonlinear integer programming problem and solved by genetic algorithm.

With the research of depth learning, more and more fields have developed rapidly, especially in the medical field, such as disease prediction. In [12], the authors completed the prediction of heart disease through Deep Belief Network. Similarly, Deep Reinforcement Learning can realize the complex calculation with the help of neural network. Deep reinforcement learning has strong perception and decision-making ability [13]. Therefore, there are many applications of deep reinforcement learning in edge computing. In [14], authors considered the MEC computation offloading in wireless time-varying channel environment, which needed to solve the combinatorial optimization problem in the channel coherence time. In [15], authors proposed an adaptive DRL framework to complete the offloading decision in real time, a reinforcement learning algorithm based on DQN was proposed to solve the resource allocation problem in edge computing and improved the end-to-end average reliability. In [16], the vehicle was used as mobile edge server to provide computing services for nearby users, and a DQN algorithm was proposed to maximize the long-term utility of vehicle edge computing network. In [17], authors considered a scenario where multiple tasks arrived randomly. The goal was to minimize the long-term average computational cost of buffer latency. They used DDPG algorithm to learn the effective task offloading strategy of each mobile user independently. In [18], an online task scheduling optimization algorithm based on time difference learning was proposed, and the scheduling process was modeled as Markov Decision Process. In [19], this paper proposed a multi-task-related task offloading algorithm based on deep reinforcement learning, and used graph convolution network (GCN) to extract the depth information of DAG tasks.

In this paper, we design a Task Mapping Algorithm based on Deep Reinforcement Learning, which makes full use of the strong perception and decision-making ability of deep reinforcement learning to solve the offloading decision problem of DAG task. Our contributions are as follows:

- In order to transform DAG task into the data that can be input into the model. We propose a Graph Sequence Algorithm (GSA) to convert the task nodes in DAG task into a sequence of task vectors according to their priority without violating the dependency of DAG task.
- We formulate the offloading decision problem of DAG task into Markov Decision Process, define state, action and reward, and design Task Mapping Algorithm based on Deep Reinforcement Learning which maps task vector into offloading decision. This algorithm can get the approximate optimal offloading decision without any expert knowledge.
- The special actor network and critic network are designed to handle the topological sequence. The PPO deep reinforcement learning algorithm is adopted to train the model. The evaluation results show that the model is superior in performance.

The rest of the paper is organized as follows: In Sect. 2, we present the system model in detail. In Sect. 3, we established a mathematical model for practical problems. In Sect. 4, the details of Task Mapping Algorithm based on Deep Reinforcement Learning will be discussed in detail. Section 5 is the experimental part, we present the evaluation results here. Section 6 is the last section of this paper, which summarizes the full text.

2 System Model

In this paper, we set a base station in MEC environment to provide computing services for users, and the edge server in the base station can provide private computing and transmission resources for users [20], that is, the offloading services between users will not interfere with each other. The edge server will provide the same computing and transmission resources for each user. All computing and transmission resources of the edge server are equally divided. The user equipment can provide local computing, offload tasks and receive results, which are respectively completed by its local CPU and transmission component. The ability of edge server is to complete task calculation, receive task and return result.

In this paper, the dependency between subtasks in a computing intensive task is abstracted as a kind of computing dependency. Computing dependency is defined as: a subtask can start to execute if and only if all the subtasks it depends on complete the calculation. Thus, the computing dependency between subtasks can be compared to a directed edge in DAG. Through the constraint of computing dependency, a computing intensive task can be abstracted as a DAG, which is called DAG task. As shown in Fig. 1, we can get that the precondition for a subtask to start execution is that all its predecessor subtasks have been executed.

Like DAG in graph theory, we use $G = (N, E)$ to represent a computing intensive task, where N and E represent the set of subtasks and the set of computing dependencies, respectively, and n_i represents the ith subtask. If n_i has no predecessor, then it is called an entry subtask; If n_i has no successor, then it is called an exit subtask. A DAG task can have multiple entry subtasks and multiple exit subtasks. Subtask n_i should contain the following information:

- task data size d_i.
- the number of required CPU cycles c_i.
- result size b_i after calculation.

Figure 2 depicts the whole decision-making process of a computing intensive task. First, the user device generates an original computing intensive task. The computation intensive task is converted into a DAG task after passing through the parser. The DAG task passes through the core algorithm module (DRL-TMA) to get the optimal

Fig. 1 DAG task

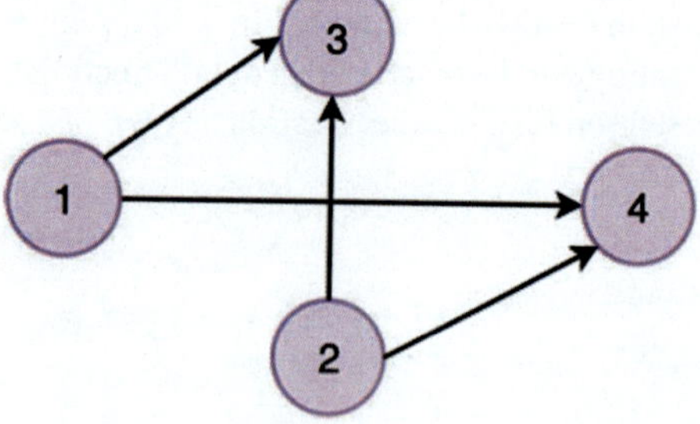

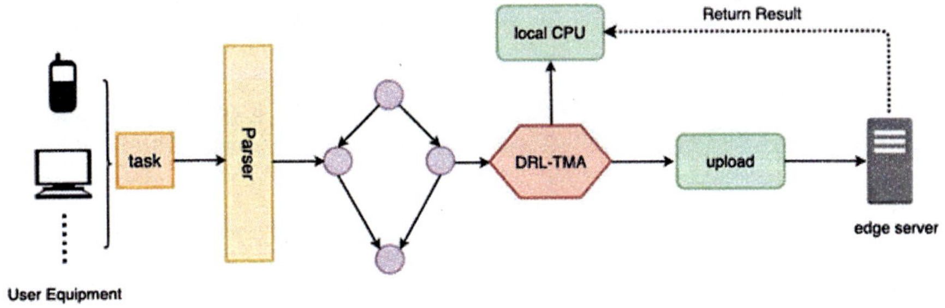

Fig. 2 System flow

offloading decision of each subtask. The subtask determined as local calculation will be left in the user equipment and completed by the local CPU; The subtask determined as edge computing will be uploaded by the user equipment, and the base station receives the task to the edge server. After computing, the base station returns the result.

We need to analyze the latency and energy consumption of subtask n_i. If n_i is determined to be local computing, the latency and energy consumption are only generated by the local CPU. So they can be expressed as:

$$\begin{cases} t_1^i = \dfrac{c_i}{v_1} \\ e_1^i = \sigma(v_1)^\tau t_1^i \end{cases} \quad (1)$$

where v_1 is the computing rate of user equipment, σ and τ are constants, and $\sigma(v_1)^\tau$ is the local calculation power.

If subtask n_i is decided to edge computing, the processing of n_i consists of three parts. First, it is uploaded to the edge server by the user device, and then it is processed in the edge server. After computing, the result is returned to the user device. The total latency is the sum of the above three parts, and the energy consumption is the sum of uploading n_i and receiving the result. The expressions of latency and energy consumption are as follows:

$$\begin{cases} t_s^l = \dfrac{d_i}{u} + \dfrac{c_i}{v_s} + \dfrac{b_i}{w} \\ e_s^l = p^u \dfrac{d_i}{u} + p^w \dfrac{b_i}{w} \end{cases} \quad (2)$$

where, u and w represent the upload rate and download rate of the user equipment respectively. v_s represents the computing rate of the edge server. p^u and p^w represent the upload power and download power of user equipment respectively.

3 Problem Formulation

This paper stipulates that the user equipment can only calculate one task at a time and upload one task at a time, and so can the edge server. If there are multiple subtasks to be processed at the same time, they need to wait in line. With this stipulation, we can express the following characteristics of subtask n_i:

- fm_{l}^i: the earliest completion time of local computing.
- fm_{u}^i: the earliest completion time of uploading.
- fm_{s}^i: the earliest completion time of edge computing.
- fm_{w}^i: the earliest completion time of result returning.

When subtask n_i is determined to be local computing, fm_{u}^i, fm_{s}^i and fm_{w}^i are meaningless and set to 0; When it is determined to edge computing, fm_{l}^i is meaningless and set to 0. Because the user equipment and the edge server can only process one subtask at the same time, we have to express available time when processing subtask n_i:

- am_{l}^i: the earliest available time that local CPU can start computing.
- am_{u}^i: the earliest available time that user equipment can upload subtask.
- am_{s}^i: the earliest available time that edge server can start computing.
- am_{w}^i: the earliest available time that edge server can start to return the result.

Then, according to the computing dependency, we can get the following expression:

$$\begin{cases} fm_{\mathrm{l}}^i = \max\left(am_{\mathrm{l}}^i, \max\limits_{j \in \mathrm{pre}(i)}\left(fm_{\mathrm{l}}^j, fm_{\mathrm{w}}^j\right)\right) + \dfrac{c_i}{v_{\mathrm{l}}} \\ fm_{\mathrm{u}}^i = \max\left(am_{\mathrm{u}}^i, \max\limits_{j \in \mathrm{pre}(i)}\left(fm_{\mathrm{l}}^j, fm_{\mathrm{u}}^j\right)\right) + \dfrac{d_i}{u} \\ fm_{\mathrm{s}}^i = \max\left(am_{\mathrm{s}}^i, \max\left(fm_{\mathrm{u}}^i, \max\limits_{j \in \mathrm{pre}(i)} fm_{\mathrm{s}}^j\right)\right) + \dfrac{c_i}{v_{\mathrm{s}}} \\ fm_{\mathrm{w}}^i = \max\left(am_{\mathrm{w}}^i, fm_{\mathrm{s}}^i\right) + \dfrac{b_i}{w} \end{cases} \tag{3}$$

$\mathrm{pre}(i)$ is the set of precursor subtasks of n_i. The earliest start time plus computing latency is the earliest completion time. First equation indicates that local computing can only be started when the local CPU is available and all the precursor subtasks of n_i have finished computing. Second equation indicates that user equipment can only be started uploading when all the precursor subtasks of n_i have been processed locally or uploaded. Third equation indicates that edge computing can only be started when the edge server is computable, subtask n_i completes the uploading, and all its precursor subtasks complete the calculation. Fourth equation indicates that result returning can only be started when the edge server can start to return and subtask n_i has completed the computing.

Set the start time of DAG task as B, then subtract start time from the last completion time of all export subtasks to get the overall computing latency. Therefore, we define the time and energy profit of a specific DAG task offloading decision as follows:

$$\begin{cases} T = \dfrac{T_{\mathrm{L}} - \max\left(\max\limits_{n_i \in \mathrm{exit}} \left(fm_{\mathrm{l}}^i, fm_{\mathrm{w}}^i\right) - B\right)}{T_{\mathrm{L}}} \\ E = \dfrac{E_{\mathrm{L}} - \left(\sum_{a_i=0} e_{\mathrm{l}}^i + \sum_{a_i=1} e_{\mathrm{s}}^i\right)}{E_{\mathrm{L}}} \end{cases} \tag{4}$$

T_{L} and E_{L} respectively represent the total latency and energy consumption when all subtasks in DAG task are processed locally. When the latency and energy consumption are smaller, the profit T and E are larger. The comprehensive profit of latency and energy consumption are defined as:

$$I = \alpha T + (1 - \alpha) E \tag{5}$$

where α is a constant in (0, 1), the goal of algorithm is to find the optimal decision of DAG task, so as to maximize the comprehensive profit I.

4 DRL Based Task Mapping Algorithm

Because the execution order of subtasks can't violate the computing dependency, the predecessor must make the decision before successor. Therefore, DAG task needs to be converted into a topological order, and the decision should be made according to the topological order. We propose Graph Sequence Algorithm (GSA) to transform DAG tasks into topological sequences composed of task vectors.

4.1 Graph Sequence Algorithm

If a subtask has a higher priority, it will be processed earlier without violating the computing dependency. We use **computing cost** to specify priority. The computing cost expression of subtask n_i is as follows:

$$\mathrm{cost}_i = \max\left(\alpha t_{\mathrm{l}}^i + (1 - \alpha) e_{\mathrm{l}}^i, \alpha t_{\mathrm{s}}^i + (1 - \alpha) e_{\mathrm{s}}^i\right) \tag{6}$$

That is to say, the computing cost of subtask n_i is the higher one of the two computation modes. Combined with the computing dependency, the priority of subtask n_i is obtained:

$$\text{pri}[i] = \begin{cases} \max\limits_{j \in suc(i)} (\text{pri}(j)) + \text{cost}_i & n_i \in \text{exit} \\ \text{cost}_i & n_i \notin \text{exit} \end{cases} \tag{7}$$

If subtask n_i is an export subtask, its priority is its own computing cost. On the contrary, priority is the highest priority among the its successor subtasks plus its own computing cost, and then the subtasks are sorted in descending order according to the priority. In this way, the priority of the predecessor subtask must be higher than that of the successor subtask. In addition, for subtasks located at the same level of DAG, whoever has a high computational cost has a high priority. According to the priority, the subtasks are sorted in descending order to get a topological sort. Starting from the following, in order not to cause ambiguity, we use the symbol n_i to represent the ith subtask in topological sorting, and the sequence length is represented by LEN.

In order to make DRL-TMA algorithm be able to process subtasks, we need to deal with the subtasks numerically. Because each subtask contains more than one information, so it is reasonable to abstract the subtask into a vector. Defining subtask n_i contains the following information:

1. Priority: $\text{pri}(i)$.
2. Local computing cost: $C_\text{L}^i = \alpha t_1^i + (1 - \alpha)e_1^i$.
3. Offload computing cost: $C_\text{S}^i = \alpha t_\text{s}^i + (1 - \alpha)e_\text{s}^i$.
4. Information of the precursor subtasks: $\boldsymbol{info}_{\boldsymbol{pre}}^i$.
5. Information of the successor subtasks: $\boldsymbol{info}_{\boldsymbol{suc}}^i$.

where $\boldsymbol{info}_{\boldsymbol{pre}}^i$ and $\boldsymbol{info}_{\boldsymbol{suc}}^i$ are vectors, the calculation method is as follows:

$$\begin{cases} \boldsymbol{info}_{\boldsymbol{pre}}^i = \left(\dfrac{\sum_{j \in \text{pre}(i)} \text{pri}(j)}{|\text{pre}(i)|}, \dfrac{\sum_{j \in \text{pre}(i)} C_\text{L}^j}{|\text{pre}(i)|}, \dfrac{\sum_{j \in \text{pre}(i)} C_\text{S}^j}{|\text{pre}(i)|} \right) \\ \boldsymbol{info}_{\boldsymbol{suc}}^i = \left(\dfrac{\sum_{j \in \text{suc}(i)} \text{pri}(j)}{|\text{suc}(i)|}, \dfrac{\sum_{j \in \text{suc}(i)} C_\text{L}^j}{|\text{suc}(i)|}, \dfrac{\sum_{j \in \text{suc}(i)} C_\text{S}^j}{|\text{suc}(i)|} \right) \end{cases} \tag{8}$$

the task vector of subtask n_i is obtained

$$\boldsymbol{vec}_i = \left(\text{pri}(i), C_\text{L}^i, C_\text{S}^i, \boldsymbol{info}_{\boldsymbol{pre}}^i, \boldsymbol{info}_{\boldsymbol{suc}}^i\right) \tag{9}$$

After each subtask is processed by Graph Sequence Algorithm, a sequence Seq composed of subtask vectors is obtained.

4.2 Markov Decision Process

Before introducing the DRL-TMA, we need to establish the Markov Decision Process. Because the edge server provides users with private computing and transmission

Algorithm 1 Graph sequence algorithm

Input: $t_1^i, t_s^i, e_1^i, e_s^i$ of subtask n_i
Output: Seq
1: **for** each task node n_i in DAG task **do**
2: calculate cost:

$$\begin{cases} C_L^i = \alpha t_1^i + (1-\alpha)e_1^i \\ C_S^i = \alpha t_s^i + (1-\alpha)e_s^i \\ \text{cost}_i = \max(C_L^i, C_S^i) \end{cases}$$

3: calculate priority:

$$\text{pri}[i] = \begin{cases} \max\limits_{j\in \text{suc}(i)} (\text{pri}\,(j)) + \text{cost}_i & n_i \in \text{exit} \\ \text{cost}_i & n_i \notin \text{exit} \end{cases}$$

4: calculate $\boldsymbol{info}_{pre}^i$ and $\boldsymbol{info}_{suc}^i$ according to (8)
5: then get:

$$\boldsymbol{vec}_i = \left(\text{pri}(i), C_L^i, C_S^i, \text{info}_{\text{pre}}^i, \text{info}_{\text{suc}}^i\right)$$

6: put $\boldsymbol{vec}_i$ into Seq
7: **end for**

resources, the network state can be regarded as static, so the Markov Decision Process is created from the perspective of DAG task.

- ***State space***: In this paper, In this paper, the state is defined as task vector Seq and historical offloading decision sequence his, so the state of subtask n_i is defined as:

$$A = \{Seq, his_i\}$$

- ***Action space***: Each subtask has only two choices: local computing and edge computing, so the action space of subtask n_i is

$$A = \{0, 1\}$$

  where 0 represents local computing and 1 represents edge computing.
- ***Reward***: Every time a subtask is executed, the current total latency and total energy consumption can be obtained. Therefore, when subtask n_i is executed, the increment of total latency and total energy consumption is represented by ΔT_i and ΔE_i, and the reward is defined as the opposite number of the sum of the two.

$$r_i = -(\Delta T_i + \Delta E_i)$$

Then we can get the cumulative reward: $R = \sum \gamma^i r_i$. It can be proved that when the cumulative reward R is maximized, the comprehensive profit I is also maximized.

4.3 Algorithm Design

The DRL-TMA is improved on the basis of sequence to sequence model. The algorithm takes the *Seq* as the input, and maps the subtask vector to its optimal offloading decision. As shown in Fig. 3, the network structure consists of an encoder and a decoder, both of which are composed of recurrent neural networks. According to the sequence *Seq*, each subtask vector is input into the encoder in turn for encoding. When all the subtask vectors in the sequence are encoded, the parameters of the last hidden layer of the encoder are used to initialize the decoder, and then decoding is started. Two processes are carried out in the output layer. The first process is to get the state value through a fully connected neural network, and the second process is to get the offloading decision through the *softmax* layer. It can be found that *Actor* network and *Critic* network share all network parameters except output layer. When subtask n_i is processed, state value is expressed as $V_\theta(S_i)$ and offloading decision is $\pi_\theta(a_i|S_i)$.

This paper uses *PPO* algorithm to complete the network training [21]. The overall optimization objective function is the combination of *Actor* network's objective function and *critic* network's objective function:

$$L(\theta) = E[L_{\mathrm{A}}(\theta) + L_{\mathrm{C}}(\theta)] \tag{10}$$

where $L_{\mathrm{A}}(\theta)$ and $L_{\mathrm{C}}(\theta)$ represent the objective function of actor network and critical network respectively. According to the definition of *PPO* algorithm and the gener-

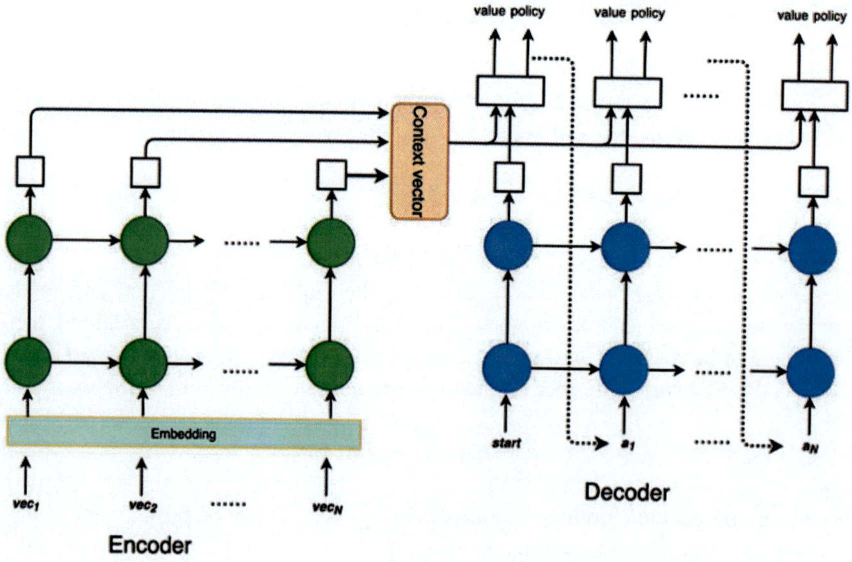

Fig. 3 Network structure

alized advantage estimation function gae, the expression of $L_A(\theta)$ can be obtained as follows:

$$L_A(\theta) = E[\min(pr_i(\theta)gae(\gamma, \phi), prune)] \tag{11}$$

where:

$$\begin{cases} pr_i(\theta) = \dfrac{\pi(a_i|Seq, his_i; \theta)}{\pi(a_i|Seq, his_i; \theta_{old})} \\ gae(\gamma, \phi) = \sum_j^{LEN-i+1} (\gamma\phi)^j (r_{i+j} + dis) \\ prune = clip(pr_i(\theta), 1-\epsilon, 1+\epsilon)gae(\gamma, \phi) \\ dis = \gamma V(S_{i+j+1}; \theta_{old}) - V(S_{i+j}; \theta_{old}) \end{cases} \tag{12}$$

the objective function of $Critic$ network is expressed in the form of mean square error:

$$L_C(\theta) = \left(\sum_j^{LEN-i+1} \gamma^j r_{i+j} - V(S_i; \theta) \right)^2 \tag{13}$$

training algorithm is shown in Algorithm 2.

Algorithm 2 DRL based task mapping algorithm

```
The parameters of strategy network a for training are initialized to θ
Use θ to initialize the policy network for sampling, θ_old = θ
for i = 1, 2, ... do
  for j = 1, 2, ..., M do
    Sampling strategy network is used to sample the whole network in the environment, and
    the resulting trajectory is cached as track_i
    According to (12), use track_i to calculate the value of gae function
    Use track_i to calculate Σ_j^{LEN-i+1} γ^j r_{i+j}
  end for
  Cache (track_i, gae, Σ_j^{LEN-i+1} γ^j r_{i+j})
  for epoch = 1, 2, ..., N do
    Based on the cache data, Adam function is used to optimize (10) and update the parameters
  end for
  Synchronizing the parameters of two policy networks
  Delete cache data
end for
```

5 Performance Evaluation

5.1 Simulation Environment

According to Ref. [22], a DAG generator is implemented, then the training set and testing set can be produced. DAG tasks in two datasets contain 12, 16, 20, 24, 28, 32, 36, 40 subtasks. In order to make the DAG task closer to the actual computing intensive task, the width and density of the graph are randomly generated. d_i and b_i are randomly generated in [20 kB, 800 kB], and c_i is randomly generated in [30 kB, 700 kB]. Under each category, there are 1000 DAG tasks in the training set and 500 DAG tasks in the test set. The encoder and decoder are composed of two layers of GRU, and the number of hidden layer neurons is 512, discount factor $\gamma = 0.99$, learning rate $\mathrm{lr} = 0.0005$. *Adam* is used as the training optimizer. The bandwidth of upload and download is set to 10 Mbps. The computing rate v_l and v_s are set to 1×10^6 B/s and 8×10^6 B/s respectively. p^u and p^w are set to 1.3 W and 1.2 W respectively. The balance factor α is set to 0.5.

In order to highlight the advantages of task mapping algorithm based on deep reinforcement learning, this paper compares it with the following algorithms:

- ***HEFT algorithm***: HEFT algorithm [23] is a heuristic static DAG scheduling algorithm, which is based on the earliest completion time.
- ***Round Robin (RR)***: The subtasks are assigned to the user equipment and the edge server in turn [24].
- ***Temporal-Difference learning (TD)***: Compared with Monte Carlo algorithm, this algorithm can solve reinforcement learning problems without using complete state sequence. The estimation is made by TD error in each time slot [18].
- ***Greedy policy***: If the comprehensive profit computed locally is greater than that computed in the server, it is determined to be local computing; otherwise, it is determined to be computed in server.

5.2 Simulation Results

The DRL-TMA is trained under the above experimental conditions. After the training, the performances are compared through the test data sets. The results are shown in Table 1. It can be concluded that DRL-TMA has the best performance on all test data sets. We get the optimal average comprehensive profit (shown in brackets) of test sets with 12, 16 and 20 task nodes. It can be found that the average comprehensive profit obtained by DRL-TMA is very close to the optimal solution.

In order to test the performance improvement of DRL-TMA in comprehensive profit with the increase of bandwidth, the training data set with 12 subtasks is selected, and the model is retrained under different bandwidth, and the test set with 12 subtasks is used to test. It can be seen from the results shown in Fig. 4 that with the increase of

Table 1 Comparison of average comprehensive profit

Nodes	DRL-TMA	HEFT	RR	TD	Greedy
12	0.621 (0.636)	0.480	0.345	0.528	0.492
16	0.659 (0.667)	0.522	0.389	0.553	0.320
20	0.623 (0.638)	0.433	0.312	0.541	0.338
24	0.665	0.557	0.325	0.565	0.462
28	0.677	0.541	0.351	0.553	0.346
32	0.646	0.518	0.398	0.549	0.326
36	0.671	0.531	0.338	0.531	0.429
40	0.650	0.490	0.434	0.569	0.489

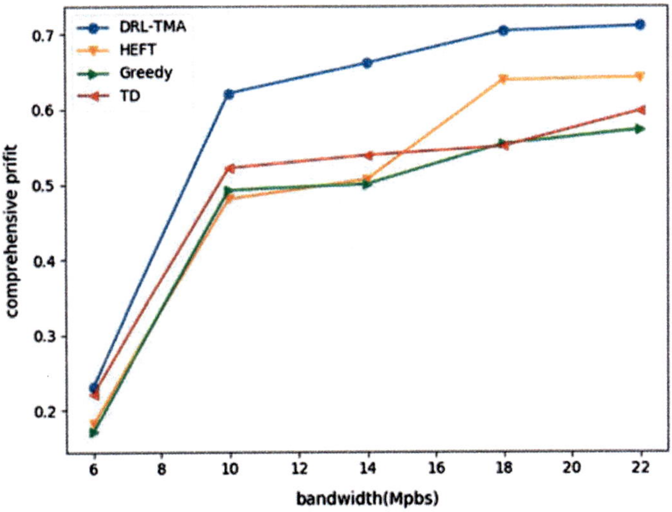

Fig. 4 Comparison of comprehensive profit of each algorithm with bandwidth increasing

bandwidth, the performance of the DRL-TMA is the best under the same conditions. As the network condition gets better and better, the average comprehensive profit of each algorithm is gradually increasing. However, when the network state is poor, DRL-TMA still has the highest average comprehensive profit.

6 Conclusion

In this paper, we design a Task Mapping Algorithm based on Deep Reinforcement Learning, which is used to solve the task offloading problem with dependent subtasks in MEC environment. Firstly, We model the user's computing task as DAG, and then the DAG task is transformed into task vector sequence task by Graph Sequence

Algorithm. Based on sequence to sequence model, we construct parameter shared actor network and critic network through GRU recurrent neural network. Finally, the model is trained by PPO deep reinforcement learning algorithm. By comparing the experimental results, we can conclude that the Task Mapping Algorithm based on Deep Reinforcement Learning proposed in this paper can achieve higher user comprehensive profit, which proves its effectiveness and feasibility. In the next work, we hope to design a more concise network structure to deal with DAG tasks. And want to change the network state to dynamic, that is, the bandwidth is time-varying, the computing power of the edge server is dynamic, and so on. These improvements will make the algorithm more applicable.

Acknowledgements This work was supported by the National Science Foundation of China (Grant No. 61877007), Fundamental Research Funds for the Central Universities (Grant No. DUT20GJ205), and the Shandong National Science Foundation of China (Grant No. ZR202103040468).

References

1. S.B.H. Shah, L. Wang, M.E. Haque, M.J. Islam, A. Carie, N. Kumar, Lifetime improvements of smart sensors maintenance protocol in prospect of IoT-based Rampal power plant, in *2020 16th International Conference on Mobility, Sensing and Networking (MSN)* (2020), pp. 260–267
2. S.B.H. Shah, Z. Chen, S.H. Ahmed, F. Yin, M. Faheem, S. Begum, Depth based routing protocol using smart clustered sensor nodes in underwater WSN, in *Proceedings of the 2nd International Conference on Future Networks and Distributed Systems*, Amman, Jordan, 26–27 June 2018, ed. by A. Abuarqoub, B. Adebisi, M. Hammoudeh, S. Murad, M. Arioua (ACM, 2018), pp. 53:1–53:7. [Online]. https://doi.org/10.1145/3231053.3231119
3. M. Faheem, G. Fizza, M.W. Ashraf, R.A. Butt, V.C. Gungor, Data acquired by internet of things-enabled industrial multichannel wireless sensors networks for active monitoring and control in the smart grid industry 4.0. Data Brief **35**(4), 106854 (2021)
4. S.B. Shah, C. Zhe, F. Yin, I.U. Khan, S. Begum, M. Faheem, F.A. Khan, 3D weighted centroid algorithm & RSSI ranging model strategy for node localization in WSN based on smart devices. Sustain. Cities Soc. **39**, 298–308 (2018). [Online]. Available: https://www.sciencedirect.com/science/article/pii/S2210670717312982
5. Y. Mao, C. You, J. Zhang, K. Huang, K.B. Letaief, A survey on mobile edge computing: the communication perspective. IEEE Commun. Surv. Tutor. **19**(4), 2322–2358 (2017)
6. T. Taleb, K. Samdanis, B. Mada, H. Flinck, S. Dutta, D. Sabella, On multi-access edge computing: a survey of the emerging 5G network edge cloud architecture and orchestration. IEEE Commun. Surv. Tutor. **19**(3), 1657–1681 (2017)
7. W. Shi, J. Cao, Q. Zhang, Y. Li, L. Xu, Edge computing: vision and challenges. IEEE Internet Things J. **3**(5), 637–646 (2016)
8. C. Tsai, J.J.P.C. Rodrigues, Metaheuristic scheduling for cloud: a survey. IEEE Syst. J. **8**(1), 279–291 (2014)
9. M. Deng, H. Tian, B. Fan, Fine-granularity based application offloading policy in cloud-enhanced small cell networks (2016), pp. 638–643
10. S.E. Mahmoodi, R.N. Uma, K.P. Subbalakshmi, Optimal joint scheduling and cloud offloading for mobile applications. IEEE Trans. Cloud Comput. **7**(2), 301–313 (2019)
11. Z. Cheng, P. Li, J. Wang, S. Guo, Just-in-time code offloading for wearable computing. IEEE Trans. Emerg. Top. Comput. **3**(1), 74–83 (2015)

12. S.A. Ali, B. Raza, A.K. Malik, A.R. Shahid, M. Faheem, H. Alquhayz, Y.J. Kumar, An optimally configured and improved deep belief network (OCI-DBN) approach for heart disease prediction based on Ruzzo–Tompa and stacked genetic algorithm. IEEE Access **8**, 65947–65958 (2020)
13. S.S. Mousavi, M. Schukat, E. Howley, Deep reinforcement learning: an overview. CoRR, vol. abs/1806.08894 (2018). [Online]. Available: http://arxiv.org/abs/1806.08894
14. L. Huang, S. Bi, Y.J.A. Zhang, Deep reinforcement learning for online computation offloading in wireless powered mobile-edge computing networks. IEEE Trans. Mob. Comput. **19**(11), 2581–2593 (2020)
15. T. Yang, Y. Hu, M.C. Gursoy, A. Schmeink, R. Mathar, Deep reinforcement learning based resource allocation in low latency edge computing networks (2018), pp. 1–5
16. Y. Liu, H. Yu, S. Xie, Y. Zhang, Deep reinforcement learning for offloading and resource allocation in vehicle edge computing and networks. IEEE Trans. Veh. Technol. **68**(11), 11158–11168 (2019)
17. Z. Chen, X. Wang, Decentralized computation offloading for multi-user mobile edge computing: a deep reinforcement learning approach. EURASIP J. Wirel. Commun. Netw. **2020**(1), 188 (2020)
18. Y. Zhang, Z. Zhou, Z. Shi, L. Meng, Z. Zhang, Online scheduling optimization for DAG-based requests through reinforcement learning in collaboration edge networks. IEEE Access **8**, 72985–72996 (2020)
19. Z. Tang, J. Lou, F. Zhang, W. Jia, Dependent task offloading for multiple jobs in edge computing (2020), pp. 1–9
20. J. Wang, J. Hu, G. Min, W. Zhan, Q. Ni, N. Georgalas, Computation offloading in multi-access edge computing using a deep sequential model based on reinforcement learning. IEEE Commun. Mag. **57**(5), 64–69 (2019)
21. J. Schulman, F. Wolski, P. Dhariwal, A. Radford, O. Klimov, Proximal policy optimization algorithms. EURASIP J. Wirel. Commun. Netw. (2017)
22. H. Arabnejad, J.G. Barbosa, List scheduling algorithm for heterogeneous systems by an optimistic cost table. IEEE Trans. Parallel Distrib. Syst. **25**(3), 682–694 (2014)
23. H. Topcuoglu, S. Hariri, M.-Y. Wu, Performance-effective and low-complexity task scheduling for heterogeneous computing. IEEE Trans. Parallel Distrib. Syst. **13**(3), 260–274 (2002)
24. G. Otsuru, Y. Sanada, User allocation with round-robin scheduling sequence for distributed antenna system, in *2019 IEEE 90th Vehicular Technology Conference (VTC2019-Fall)* (2019), pp. 1–5

Analysis of Machine Learning and Deep Learning in Cyber-Physical System Security

Ankita, Atef Zaguia, Shalli Rani, and Ali Kashif Bashir

Abstract Security and privacy are the major parameters needed to work upon in a variety of applications. The most widely used infrastructure known as cyber-physical system (CPS) is used for solving different types of challenges. In today's world of growing technology, decision-making should be very much fast. The Internet of Things or CPS brings changes into many industrial or manufacturing fields, thus taking many applications to the extreme level. But there exists the most important security issue in the network system where lots of personal or sensitive information resides and is at stake. To solve security and privacy matters different machine learning and deep learning approaches are being used. Our paper consists of various CPS layers along with the possible attacks on each layer and also the steps to detect attacks in CPS. Various machine learning (ML) and deep learning (DL) models are being discussed in the following paper to detect the attacks using datasets along with the accuracies.

Keywords ML · DL · CPS · Security

1 Introduction

A cyber-physical system (CPS) is a collaboration of a variety of elements, namely the physical elements, elements for communication, and storage purposes. CPS can be any of the Internet of Things (IoT), Internet of Vehicles (IoV), Internet of Medical

Ankita · S. Rani (✉)
Chitkara University Institute of Engineering and Technology, Chitkara University, Rajpura, Punjab 140401, India
e-mail: shalli.rani@chitkara.edu.in

A. Zaguia
Department of Computer Science, College of Computers and Information Technology, Taif University, P.O. BOX 11099, Taif 21944, Saudi Arabia
e-mail: zaguia.atef@tu.edu.sa

A. K. Bashir
Department of Computing and Mathematics, Manchester Metropolitan University, Manchester, UK
e-mail: dr.alikashif.b@ieee.org

A. K. Bashir et al. (eds.), *Proceedings of International Conference on Computing and Communication Networks*, Lecture Notes in Networks and Systems 394,
https://doi.org/10.1007/978-981-19-0604-6_30

Things (IoMT), Robotics, etc. There exist many CPS applications such as transportation, smart grids, healthcare domain, automatic pilot avionics, navigation, etc. According to the reports one of the organizations such as Cisco expecting the connectivity of billions of devices and smart objects which includes a lot of routers, meters, receivers, transmitters, servers, and also other smart objects [1]. In the world of technology, raw data comes with sensitive information giving rise to privacy and security issues. Protecting sensitive data from different types of attacks is the most challenging task for different organizations. As there is an increase in the number of Internet users, also there is an increase in the number of cyber-criminals as the main motive is to steal and destroy personal and confidential information [2]. As a result, the integrity of important data is at stake as malware attacks are very much common these days in terms of quality and quantity, malware threats are very costly. There are various security-providing techniques such as edge computing, fog computing among which machine learning (ML) and deep learning (DL) have gained a lot of attention in providing security and privacy and also able to deal with the real issues [3]. Different scenarios exist where the choice of machine learning and deep learning over classical techniques proves to be the best. Upgradation in some of the detection systems by adding additional components in both ML and DL affecting many different domains of cybersecurity [4].

Motivation: Our day-to-day life comes with a lot of tasks where the cyber-physical system plays a major role in dealing with that tasks. The CPS and IoT are integrated to give better results in many domains such as industry, healthcare, transportation, automotives, and other applications to the above level. Although having many advantages of CPS, still facing a large number of challenges due to the absence of smart tools as IoT generates a huge amount of data. As a result, skilled manpower is also not able to handle these types of issues. When the integration of security in ML, DL, and CPS along with skilled manpower results in tracking attacks in a very short period. With the arising of both ML and DL patterns are being analyzed in cybersecurity systems so that the network can be free from any attack. Also, ML and DL help the team to be more active in handling and preventing security threats, actively respond to those attacks in less time [5]. Machine learning and deep learning also make the efficient use of resources in many organizations. The time taken by the resources on daily basis tasks is also reduced by using both ML and DL enables the organizations to use their resources strategically. With the advancement in both ML and DL, cybersecurity is now inexpensive, simple, and efficient in dealing with attacks. Section 2 is having description of CPS Layers along with the attacks on each of the layer are discussed. Section 3 has the CPS Framework for attacks detection along with the steps. Section 4 will discuss the ML and DL for Cyber-Physical Systems. Section 5 has Results and Analysis part. Section 6 contains the challenges and limitations of CPS. Section 7 has the conclusion.

2 CPS Layers

The cyber-physical system operates on 3 C's such as communication, computation, and control. For example, from communication to computation where any cyber-attack can take place while transferring the data whether from device to device or from any software, at the time of computing the results, etc. [6], CPS operates. From computation to control various types of physical devices are connected including actuators, sensors, embedding systems, and many more physical devices where lots of physical attacks take place on any type of device connected. There exist three layers of CPS which include: (1) physical layer; (2) network layer; and (3) application layer. Each of these layers of CPS has its different function and each layer of CPS is vulnerable to attacks [7]. Figure 1 showing the IoT layers with the types of technologies being used at each layer with the types of devices supporting each layer, and the related applications at these three layers of CPS. Also, the following section discusses the various possible security and privacy threats in CPS for each of the three layers [8].

2.1 CPS Layers with Possible Attacks

The following section discusses each of the CPS layers along with all possible attacks on each layer.

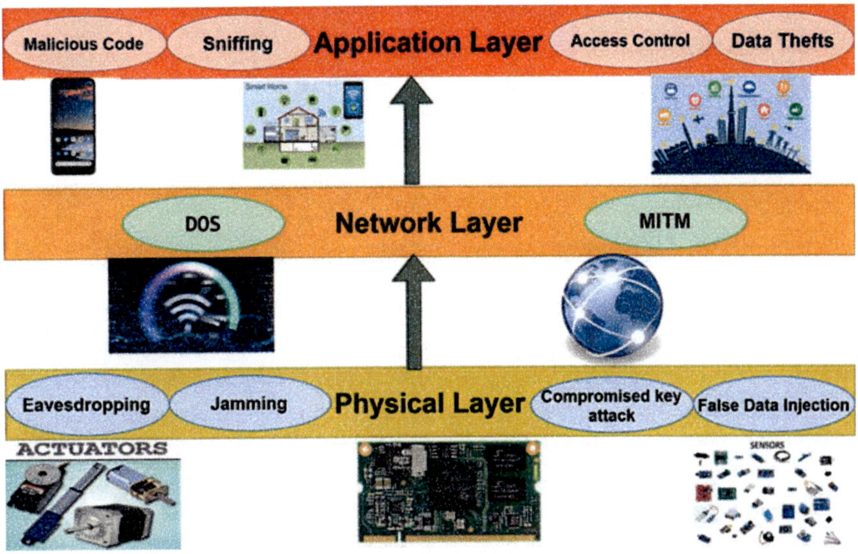

Fig. 1 CPS layers with possible attacks

Physical Layer: As Fig. 1 contains all the respective layers along with the attacks, the first is the physical layer. The major working function of cyber-physical systems is to make a link and establish a connection among communication channels. It works on the information obtained by the past data values then gives the feedback and works in the real-time environment. This layer has connected sensors, actuators, and embedded systems. Sensors are used for sensing physical phenomena. After obtaining the sensed data from the sensors, some actions can be performed on the physical environment with the help of actuators.

- **Eavesdropping**: Providing communication security in cyber-physical systems by securing wireless communication is the major aim of the CPS. Attacking the medium through which communication takes place gives rise to eavesdropping attacks [9].
- **Jamming**: The transceivers having low power are prone to jamming attacks. While using low power transceivers in cyber-physical systems the devices under CPS are vulnerable to jamming attacks [10].
- **Compromised Key Attack**: The use of a key in this type of attack is stolen by an attacker for gaining access in a secured transmission. But both sender and receiver are unaware of the key compromise attack [11].
- **False Data Injection**: After a key compromise attack, the attacker has a key that will decrypt the data and inject the false information thus manipulates the original data [11].

Network Layer: The network layer is used for the process of transferring the data or information from the physical to the application layer. This particular layer is also vulnerable to many attacks [12]. Some of them are as follows:

- **Denial of Service (DoS)**: This type of attack will target the server with lots and lots of unwanted requests so that huge traffic will be created making the server unable to respond to the genuine requests [11].
- **Man in the Middle (MITM)**: The communication model named publish–subscribe which is used by different subscribers and clients and act as a copy. The following model is very much helpful in differentiating the subscribing client from the publishing client [12].

Application Layer: After transferring the sensed data to the network layer, the network layer will pass the data to the application layer. This layer stands in support of the computation and allocation of the resources for the business-related works. There are many attacks on this respective layer. Some of them are as follows [11]:

- **Malicious Code**: The simplest way for an attacker to enter into and break the system. This can be done by the existing malicious or infected code sometimes which is vulnerable to attacks [12].
- **Sniffing**: In this type of attack, the attackers can make use of a sniffer application to steal sensitive or credential information by monitoring the traffic of the network. In case of weak security, the attackers can gain the sensitive information [13].

- **Access Control**: As per the name of the access control, it controls the access given to the user. Without permission, no one is allowed to gain access. Only the authentic one can get access [14]. All the applications contain private and personal data or information which are not for sharing.
- **Data Thefts**: As all the applications contain sensitive or personal data or information also moving of that data or information takes place from one source to another [13]. CPS applications have a lot of movable data and at the time of moving data from one place to another can arise a data theft attack.

3 CPS Framework for Detection of Attacks

Cyber-attacks these days became a serious problem or major threat to the population all around the world. The following section will discuss the process of detecting various attacks through the learning models [15]. Figure 2 discussing all the steps required for attack detection by using machine learning and deep learning models.

3.1 *Data Gathering*

The first and foremost step is to gather the data. The availability of raw data is high. But gathering important data or information from different sources is a very difficult task. Many types of datasets are available, but choosing the right one as per our requirement takes a lot of time [1].

3.2 *Data Preprocessing*

After the gathering of data from different resources as per requirement, the important data is then preprocessed. There exist many preprocessing techniques to achieve desired and efficient results [1]. In this case, the threat can be any flow of attack, but not the flow of packets as the flow is based on a set of protocols such as transmission control protocol (TCP) and user data paradigm (UDP).

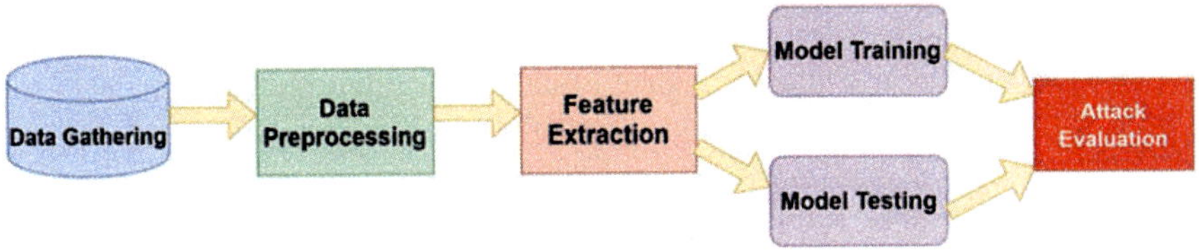

Fig. 2 Cyber attack detection using learning techniques

3.3 Feature Extraction

After collecting and preprocessing data, some of the features are being extracted for future analysis for the best results. The next step follows the feature extraction by building a model in which models training and testing can be done.

3.4 Model Testing and Training

In this following step, the model is trained by using some of the datasets according to the learning techniques such as machine learning and deep learning for the own evaluation process. There could be many different iterations needed to train the network according to the need. After training of the model, same or different number of iterations are used for the model testing as well to achieve the final output.

3.5 Attack Evaluation

The last step of the framework is the evaluation of different types of attacks such as intrusion detection, malware detection, and anomaly detection. The results related to the prediction of cyber-attacks give accurate results which predict the parameter accuracy [1].

4 ML and DL for CPS

As per the above discussion, the need for security arises for the researchers. The use of security analytics became important and gave priority to the signals and alerts. The use of these alerts and signals helps in finding a better solution to any problem in less time. Both ML and DL play an important role in providing security and privacy in cyber-physical systems. There exist various data science methods in which the large datasets of cyber-security companies can be analyzed and processed further for better results. By using DL and ML methods regression, classification, clustering, and dimensionality reduction-related problems, authentication-related problems can be solved easily. Traditional methods such as using mathematical models were not enough for providing security and privacy as they were unable to detect various security-related problems such as leakage of data, overfitting problems, detecting malware, intrusion detection, and many other cyber threats. The software has been built for security of various domains such as healthcare the system security from hackers, and attackers for accessing sensitive and personal information. Cyber-security threats can be detected by installing various machine learning and deep learning models for the healthcare network and the network activities can also be analyzed. Both machine and deep learning make the network aware of all the activities being done by attackers also able to determine the skeptical activities. Whenever any suspicious activity occurs for example anything that can be against the norms of the system, the anomaly can be detected in a network. By using ML, also, the cyber-security patterns can be analyzed and also makes the model able to learn the avoidance of attacks. Therefore, in the cyber-security of the networks, both ML and DL play an important role.

5 Results and Analysis

There exists the following datasets worked on different types of attacks used in various applications and achieved accuracy in Table 1.

In survey [3], DoS attack can be detected using dataset KDDCUP99 by using convolutional neural network (CNN) and achieved accuracy of 98.7%. In survey [9], detection of data falsification by creating their own network with 99.23% accuracy by using support vector machines (SVM). The following survey [13] attack DoS can be detected using dataset KDDCUP99 and achieved 98.3% accuracy using long short-term memory. In survey [15], analysis of network attacks can be done by using NSL-KDD and achieved accuracy of 80.4% using recurrent neural network. In the following survey [18], Black hole attack can be detected using DARPA dataset and achieved accuracy of 95.03% using k-nearest neighbor classifier.

6 Challenges and Limitations of CPS

As the introduction to cyber-physical systems gained a lot of attention in the growing technology world, also came up with a variety of issues and challenges that need to cope up with as many of them compromises with the security and privacy of CPS. The discussion to related issues and challenges are must. Some of them are as follows.

Table 1 Comparison of different attacks with accuracies

Ref. No	Dataset	ML/DL classifier	Type of attack	Accuracy (%)
[3]	KDDCUP99	CNN	DoS	98.7
[9]	Own network	SVM	Data falsification	99.23
[13]	KDDCUP99	LSTM	DoS	98.3
[15]	NSL-KDD	RNN	Network attack	80.4
	DARPA	k-nearest neighbor	Black hole	95.03

- **Consumption of Energy**: As there exist a variety of physical devices such as sensors and actuators consumes a lot of energy which is a major issue to work upon because to achieve stability these physical devices have limited energy. Therefore, a requirement of an efficient energy optimization protocol arises for balancing energy.
- **Intelligent Caching and Computing**: For intelligent caching and intelligent computing, the need for a variety of methods arises so that huge amount of data can be generated in a real-time environment.
- **Security Protocol**: To increase the security of a cyber-physical system, a secure channel with authentication is needed between physical devices such as sensors and actuators for tampering of data.
- **Trust Metrics**: For CPS component, trust metrics are designed and developed a large amount of data is being generated by the devices as there are chances of conflict between the reliability of a failure of one sensor and faulty sensor [1].

7 Conclusion

In the following paper, layers of the cyber-physical system have been discussed working on three C's that is communication, computation, and control along with the layers and each layer with all the possible attacks. A framework of CPS can be designed for the detection of attacks in which data can be gathered then it is to be preprocessed after that the main features can be extracted through which model can be trained and tested then evaluation of which type of attack can be evaluated. Both the machine learning and deep learning provides security for CPS system. Various ML and DL methods have been discussed along with the type of dataset used for detecting type of attack along with the accuracies achieved. The remaining ML and DL methods have been made using various datasets for detecting various types of other attacks for achieving better results is the further area of investigation. Also, the issues discussed above needs to be resolved.

References

1. Y. Ashibani, Q.H. Mahmoud, Cyber physical systems security: analysis, challenges and solutions. Comput. Secur. **68**, 81–97 (2017)
2. R. Alguliyev, Y. Imamverdiyev, L. Sukhostat, Cyber-physical systems and their security issues. Comput. Ind. **100**, 212–223 (2018). https://doi.org/10.1016/j.compind.2018.04.017
3. K. Sravanthi, M. Shamila, A.K. Tyagi, Cyber physical systems: the role of machine learning and cyber security in present and future. Comput. Rev. J. 66–80 (2019)
4. J.S. Raj, Machine learning based resourceful clustering with load optimization for wireless sensor networks. J. Ubiquitous Comput. Commun. Technol. (UCCT) **2**(01), 29–38 (2020)
5. T. Mehmood, H.B.M. Rais, Machine learning algorithms in context of intrusion detection, in *3rd International Conference on Computer and Information Sciences (ICCOINS)* (2016), pp. 369–373

6. L. Wang, M. Törngren, M. Onori, Current status and advancement of cyber-physical systems in manufacturing. J. Manuf. Syst. **37**, 517–527 (2015)
7. E.K. Wang, Y. Ye, X. Xu, S.M. Yiu, L.C.K. Hui, K.P. Chow, Security issues and challenges for cyber physical system, in *2010 IEEE/ACM International Conference on Green Computing and Communications & International Conference on Cyber, Physical and Social Computing*, Hangzhou (2010), pp. 733–738
8. M. Rungger, P. Tabuada, A notion of robustness for cyber physical systems. IEEE Trans. Autom. Control **61**(8), 2108–2123 (2016)
9. Y. Liu, P. Ning, M.K. Reiter, False data injection attacks against state estimation in electric power grids. ACM Trans. Inf. Syst. Secur. (TISSEC) **14**(1). Art. no. 13 (2011)
10. Y. Arjoune, F. Salahdine, M.S. Islam, E. Ghribi, N. Kaabouch, A novel jamming attacks detection approach based on machine learning for wireless communication, in *International Conference on Information Networking (ICOIN)* (2020), pp. 459–464
11. M. Hassan, M. Rehmani, J. Chen, Differential privacy techniques for cyber physical systems: a survey. IEEE Commun. Surv. Tutor. 1 (2019). https://doi.org/10.1109/comst.2019.2944748
12. Y. Zhou, F. Yu, J. Chen, Y. Kuo, Cyber-physical-social systems: a state-of-the-art survey, challenges and opportunities. IEEE Commun. Surv. Tutor. 1 (2019)
13. S. Parvin, F. Hussain, O. Hussain, T. Thein, J. Park, Multi-cyber framework for availability enhancement of cyber physical systems. Computing **95**(10–11), 927–948 (2012)
14. C. Neuman, Challenges in security for cyber-physical systems, in *Proceedings of DHS Workshop on Future Directions in Cyber-Physical Systems Security* (Newark, NJ, 2009), pp. 22–24
15. Y. Feng, H. Akiyama, L. Lu, Feature selection for machine learning-based early detection of distributed cyber attacks, in *16th IEEE International Conference on Dependable, Autonomic & Secure Computing* (2018), pp. 173–180

Visual Exploration in Glaucoma Patients Using Eye-Tracking Device

Sajitha Krishnan, J. Amudha, and Sushma Tejwani

Abstract Glaucoma is the second leading cause of blindness characterized by damage in the optic nerve and visual field loss. The disease does not show visible symptoms in the early stages, but it creates a scotoma in the peripheral vision and later impairs the central vision. The daily routines such as visual search, watching television, and reading can be screened to understand the visual decline in glaucoma. The visual field loss leads to the functional deficit and alters eye movements in performing day-to-day activities. The awareness of the disease tends to create a strategy of exploratory eye movements to compensate for the visual field loss. However, the association between exploratory eye gaze patterns and the severity of the disease is not well understood. The framework examines the performance of glaucoma subgroups concerning age and severity grade and whether exploratory eye movement patterns reflect in different tasks.

Keywords Visual field loss · Exploratory eye movements · Quality of life · Fusion map

1 Introduction

Eye diseases such as cataract, refractive errors, age-related macular degeneration, and glaucoma cause visual impairment or vision loss that affects day-to-day activities. According to the World Health Organization (WHO), there are 62 million visually impaired people, and among them, 11.2 million are affected by glaucoma in India

S. Krishnan (✉) · J. Amudha
Department of Computer Science and Engineering, Amrita School of Engineering, Amrita Vishwa Vidyapeetham, Bengaluru, India
e-mail: k_sajitha@blr.amrita.edu

J. Amudha
e-mail: j_amudha@blr.amrita.edu

S. Tejwani
Narayana Nethralaya, Bommasandra, Bengaluru, India

A. K. Bashir et al. (eds.), *Proceedings of International Conference on Computing and Communication Networks*, Lecture Notes in Networks and Systems 394,
https://doi.org/10.1007/978-981-19-0604-6_31

[1]. Glaucoma is a disease that affects gradual damage to the optic nerve. Three-fourth of reported glaucoma cases is primary open-angle glaucoma (POAG), a silent type of glaucoma that does not show visible symptoms in the early stages. The glaucoma risk factors are high internal eye pressure, family history, and prolonged use of corticosteroids [2]. Only, 1 in 5 Indians [3] go for a regular eye checkup, and out of that 84 percent does not follow the doctor's advice for their eye care [4].

Visual impairment changes visual sensory abilities and higher visual processing skills such as visual attention [5]. Visual attention is the ability to select relevant information and filter out irrelevant information. This ability requires the association of central and peripheral visual function in the interaction of real-world scenes [6]. The proposed method investigates different eye movement patterns. It analyzes them based on severity grade and age so that the glaucoma group can become aware of the disease to preserve the quality of life (QoL).

2 Related Works

A wide variety of research is presented in the clinical field to understand the association of eye movement patterns and the deficits of individuals while performing day-to-day activities [7, 8]. Researchers worldwide have found variations in eye movement patterns of glaucoma participants while performing visual exploration tasks such as reading, visual search, face recognition, watching TV and video, viewing images, driving, walking, and shopping to understand functional visual deficits of glaucoma patients. The research works investigated different tasks in their experiment using low-end, medium-end, and high-end eye trackers.

Visual search is used as a representative to understand the functional ability of glaucoma patients in performing everyday activities. Individuals with glaucoma do not show significant differences from normal people in the search performance in a controlled vision test. But, in real-world scenes, glaucoma patients show longer search duration than age-matched visually healthy participants. The glaucoma group shows more saccades per sec to compensate for the visual field loss. Studies show that only a few elder glaucoma patients and young glaucoma patients make compensatory patterns by making more saccades or head movements [9].

Restriction in the visual field is implied in glaucoma patients during free-viewing images and videos, which can be identified as signatures to discriminate between glaucoma and normal [10]. However, more research is needed to find the correlation between clinical measures and eye gaze parameters to understand the visual field loss in the early stages. A correlation exists between search performance and compensatory behavior in the form of more saccades in the on-road driving test and laboratory-based driving tests, but they fail to see the hazard in the driving scene [11].

Severe visual field loss affects mobility and finds difficulty in identifying faces. During any unfamiliar tasks, they use more saccade rates and fixation than controls to compensate for visual field loss [12]. The relation between scanning patterns and visual field loss needs to be explored to understand compensatory behavior. In addition, more research is required to understand the influence of severity in the visual field loss in eye movement behavior.

The controversy of compensatory eye movements enlightens that the performance variability depends on a specific task. Thus, there is an ambiguity in compensatory eye gaze patterns of glaucoma patients in the performance of different visual exploration tasks that are related to real-life scenarios. The proposed work investigated the performance of the glaucoma group in the visual exploration task. In addition to that it also investigated the influence of age and severity grade in their performance and the presence of exploratory gaze patterns to compensate for the visual field loss.

3 Materials and Methods

The proposed work is focused on analyzing the performance and eye movements of the glaucoma group in day-to-day tasks: simple dot task and visual search task. A non-invasive eye tracker Eye Tribe 60 Hz is connected to the laptop screen. The experimenter or person conducting an eye-tracking experiment explained the different tasks in English or their regional language to the participants.

The system first presented a simple dot task, in which the stimulus contains a white dot of size 12 pixels. There were 30 images in the task, and each image is displayed for 1.5 s. The task investigated the association between monocular (each eye separately) performance in the eye-tracking experiment and its clinical measures. The system subsequently checked the contribution of binocular vision (both eyes) in the performance during target-oriented approach such as visual search and how they compensated for their visual field loss. The visual search task included a set of 20 cartoon images with a trial time of 20 s. The target question was "Find the star in the image"? The participant responded by clicking the mouse button on the target or telling the experimenter the target's position.

The study group included participants with age group 30–70 years recruited from Narayana Nethralaya, Bengaluru. The participants underwent Humphrey field analyzer (HFA) perimetry test for the left eye and right eye separately to understand visual field loss by the clinicians. A total of 117 participants were recruited, which included 50 glaucoma participants and 48 normal participants. The data from 19 participants were excluded based on criteria signed by the ethics committee of the hospital. The participants were assigned as Sub_ID (where ID = 1, 2, 0.117 in the order of the data collection), and the records were separately marked as glaucoma and normal after the completion of clinical diagnosis.

The experimenter maintained a copy of the visual field report from HFA perimetry to find the correlation between clinical measures and eye gaze measures. The subgroups in glaucoma are identified based on severity grades (mild, moderate, and severe) [13]. The subgroups are also identified based on age group (younger glaucoma, mild-age glaucoma, and elder glaucoma) for study purposes. The higher severity grade is labeled on participants in case each eye has different severity grades.

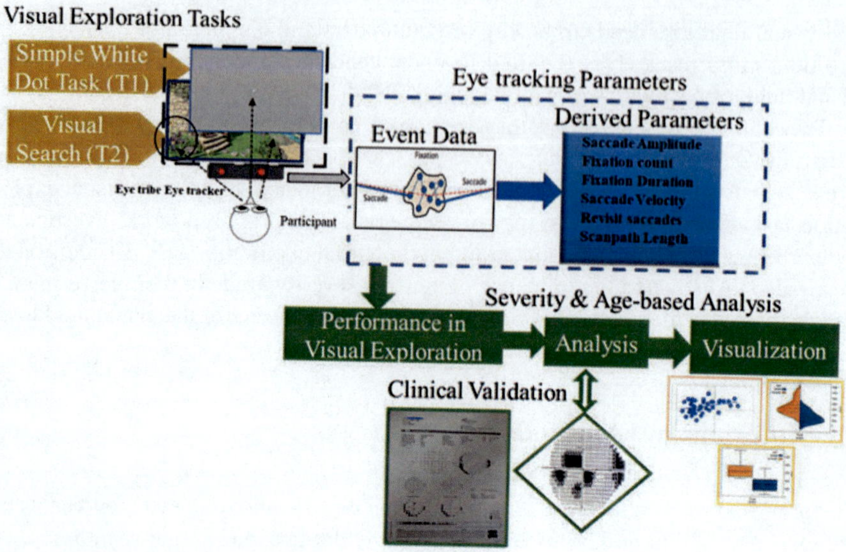

Fig. 1 Overview of system architecture

The proposed system focused on how compensatory eye movement reflects in different tasks. The proposed system included an OGAMA open-source software [14] that estimates fixations and saccades from the gaze samples. The customized software estimates comprehensive eye gaze measures [15]. Figure 1 shows the design of the system architecture.

The two eye-tracking experiments focused on the ability of participants to view monocularly and binocularly all parts of the screen and recorded their reaction time. The simple dot task estimated average miss, defined as the average of the miss/not seen on different target points. Following the monocular performance, participants viewed binocularly a goal-oriented activity that included visual search. Participants called out the target's position (here "star") and, if possible, click the mouse on the target. Each target point was marked as "seen/hit" if the number of fixations was within the bounding box of the target. Average reaction time is the performance measure defined as the average time taken by the participant/experimenter to click the mouse on the target on different images.

4 Analysis

4.1 Performance of Glaucoma Group Versus Normal Group

The performance measure of the normal and glaucoma group is compared using the two-sample t-test. Table 1 shows the summary of the statistical significance of normal

Table 1 Mean and standard deviation (in parenthesis) of age and performance measure of normal versus glaucoma group in the simple dot task

Measures	Normal ($n =$ 50)	Glaucoma ($n =$ 47)	p-value
Age	50.10(12.26)	53.04(12.03)	0.27
Average miss	0.29(0.30)	0.53(0.32)	**0.00014**
Average reaction time	5.0	7.0	**0.0076**

and glaucoma group in the sample dot task and visual search task. p-value less than equal to 0.05 is considered as statistically significant and marked in boldface in the tables. There is a significant difference in average miss between glaucoma and the normal group with p-value 0.00014. There is also a significant difference in average reaction time between normal and glaucoma group in visual search experiment with p-value 0.0076.

Severity-grade analysis shows that there is no significant difference in average miss among different severity grades (mild, moderate, and severe). In age-based analysis, there were 13 young glaucoma participants and 16 elder glaucoma participants. The mean value (standard deviation) of average miss is 0.403 (0.21) and 0.70 (0.36) of elder and younger glaucoma participants, respectively. There is a significant difference between young and elder glaucoma participants in average miss with p-value 0.018. But, no significant differences are seen in other subgroup comparisons in glaucoma.

In the visual search experiment, based on severity-based analysis, there were 11 severe glaucoma participants and 17 mild glaucoma participants. The mean value (standard deviation) of average reaction time is 11(3.0) and 6.5(3.4) of severe glaucoma and mild glaucoma participants, respectively. There is a significant difference between severe and mild glaucoma participants in average reaction time with p-value 0.04. Age-based analysis shows a significant difference between elder glaucoma and younger glaucoma with p-value $= 0.02$.

The glaucoma group took a longer response time to search the target than the normal group. Spearman correlation coefficient of age and reaction time is 0.629. When age increases, reaction time is longer for both the glaucoma group and the normal group. During the binocular performance, the higher severity grade worsens the performance of glaucoma participants.

4.2 *Eye Movement Measures of Glaucoma Group Versus Normal Group*

The eye movement measures are not estimated in the simple dot task since the trial time is short to obtain. Eye-tracking parameters are only estimated in the visual search experiment using customized software [15]. Table 2 shows that there is a significant difference between the glaucoma group and the normal group in fixation

Table 2 Mean and standard deviation (in parenthesis) of eye movement measures of normal versus glaucoma group

Measures	Normal (*n* = 41)	Glaucoma (n = 40)	*p*-value
Fixation count	6.30	5.48	**0.04**
Fixation count per sec	0.989	0.575	**0.0043**
Mean fixation duration	240.69	309.36	**0.04**
Fixation saccade ratio	256.10	185.06	0.08
Average saccade length	260.24	255.89	0.79
Fixation count until first click	6.84	3.95	**0.00063**

count, fixation count per sec, mean fixation duration, and fixation count until mouse click.

In severity-based analysis, there were 11 severe glaucoma participants and 17 mild glaucoma participants. There is no significant difference between the subgroups in eye gaze parameters. In age-based analysis, there were 10 elder glaucoma participants and 18 young glaucoma participants. The mean value (standard deviation) of mean fixation duration is 370.9(199.8) and 244(84.6) for elder glaucoma and young glaucoma participants, respectively. There is a significant difference between elder and young glaucoma participants in mean fixation duration with *p*-value 0.0128.

The Spearman correlation coefficient showed that the mean fixation duration (FD) is positively correlated to age with ρ-value 0.63. When age increases, glaucoma participants show greater restriction in their visual field that they spent time on greater fixation allocation.

5 Visualization

The single dot task performance of different participants was visualized onto a single image using gaze-fusion deep neural network (GFDM) model [16]. The hit/miss is overlaid on the fusion image of 30 images called the GFDM fusion map.

Table 3 illustrates the association between severity grade and the performance of participants. The dark spot in the visual field plot shows the damage in the quadrants of the visual field. The dark spot in the GFDM map shows that the participant missed seeing the target points in the location, and the red spot shows that the participant can see the points. The severe glaucoma participant, Sub_102 missed identifying almost all points on the screen.

Figure 2a is a combo chart of severe glaucoma participant Sub_45 in different visual search trials. The combo chart combined a line graph to represent hit/miss and bar graphs to understand the reaction time. The *X*-axis referred different images

Table 3 GFDM map generated from the simple dot task

Participant Id, class, study eye	Severity	Visual field plot	GFDM map
Sub_102, G, left eye	Severe		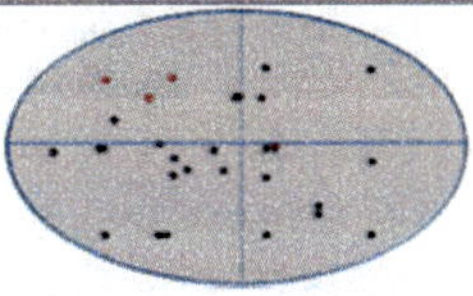

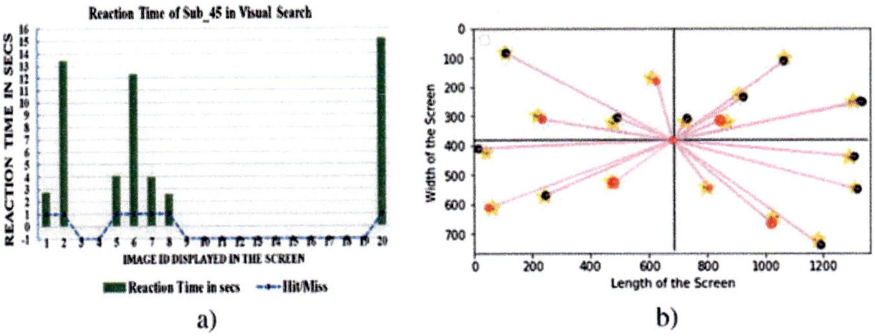

Fig. 2 Visualization illustrating reaction time and hit/miss in visual search of severe glaucoma participant Sub_45, **a** Combo chart, **b** GRTF map

displayed, and *Y*-axis referred the reaction time with −1 taken as "missed" and from +1 and so on considered as search duration. Figure 2b shows the average reaction time of the corresponding participant overly on the fusion of different images. The hit/miss, average reaction time, and position of different targets overlaid onto a single image, known as gaze-reaction time fusion map (GRTF map) created using the GFDM method. The fusion of targets in twenty images revealed the ability of different participants to identify the target and ignore the distractors. The red color referred to "seen," and the black color referred to "miss/not seen." The size of the circle linearly corresponded to the search duration. The targets toward the edge of the screen were almost missed or had taken longer reaction time by the participant. GRTF map implied the binocular performance which is different from the GFDM map.

6 Discussion

The proposed work focused on understanding exploratory gaze patterns and performance measures of different subgroups of glaucoma based on severity grade and age group. The analyses of different performance measures enlighten a significant difference between advanced stage and early-stage glaucoma.

Visual field loss restricts eye movement patterns and forced glaucoma participants to focus more on a certain region of the stimulus. Young glaucoma participants, aware of the disease, apply exploratory gaze patterns such as a greater number of fixations toward the region of visual field loss and reduction in fixation duration to improve the search performance. But, such exploratory gaze patterns were absent in the elder age group.

The visualization plots such as GFDM map and GRTF map imply that the glaucoma group has difficulty in viewing the bottom part of the screen. It also guides that the young participants look toward all parts of the screen than the elder glaucoma group. This also reveals that the age parameter creates a strategy of exploratory gaze patterns than the severity grade parameter.

The proposed work analyzed the subgroups of the glaucoma group based on independent variables: severity grade and age group. The study investigated the influence of each variable in the performance and exploratory eye movement patterns. In the analysis of each independent variable, the other independent variable is not controlled, which is taken as the limitation of the study.

7 Conclusion

The proposed work analyzed exploratory eye gaze patterns and investigated an association between exploratory eye gaze patterns and severity grade. In real-world scenes, when distractors are present, search duration is significantly different between advanced and early-stage glaucoma. Age also influenced the search performance of the glaucoma group. The elder subgroups in glaucoma showed lower fixation count and higher fixation duration which hinder the search tasks. Young glaucoma participants compensated visual field loss by making more fixations. GFDM map and GRTF map visualized the region where the glaucoma group faced search difficulty. The awareness of such exploratory gaze patterns helps the glaucoma group to improve day-to-day tasks by focusing toward the edge of their field of view by performing more fixations and head movements.

References

1. G.N. Rao, R.C. Khanna, S.M. Athota, V. Rajshekar, P.K. Rani, Integrated model of primary and secondary eye care for underserved rural areas: the L V Prasad Eye Institute experience. Indian J. Ophthalmol. **60**(5), 396–400 (2012)
2. C.W. McMonnies, Glaucoma history and risk factors. J. Optom. **10**(2), 71–78 (2017)
3. A. Azuara-Blanco, J. Burr, The rising cost of glaucoma drugs. Brit. J. Ophthalmol. **90**(2), 130–131 (2006)
4. S. Raj, L.P. Savla, F. Thattaruthody, N.G. Seth, S. Kaushik, S.S. Pandav, Predictors of visual impairment in primary and secondary glaucoma in a tertiary institute in North India. Eur. J. Ophthalmol. **30**(1), 175–180 (2020)

5. C. Owsley, G. McGwin, Jr Vision impairment and driving. Surv. Ophthalmol. **43**(6), 535–550 (1999)
6. E. David, J. Beitner, M.L. Võ, Effects of transient loss of vision on head and eye movements during visual search in a virtual environment. Brain Sci. **10**(11), 841 (2020)
7. J. Amudha, H. Nandakumar, A fuzzy based eye gaze point estimation approach to study the task behaviour in autism spectrum disorder. J. Intell. Fuzzy Syst. **35**(2), 1459–1469 (2018)
8. G. Gautam, G. Sumanth, K. Karthikeyan, S. Sundar, D. Venkataraman, Eye movement based electronic wheel chair for physically challenged persons. Int. J. Sci. Technol. Res. **3**, 206–212 (2014)
9. N.D. Smith, D.P. Crabb, D.F. Garway-Heath, An exploratory study of visual search performance in glaucoma. Ophthalmic Physiol. Opt. **31**(3), 225–232 (2011)
10. D.P. Crabb, N.D. Smith, H. Zhu, What's on TV? Detecting age-related neurodegenerative eye disease using eye movement scanpaths. Front. Aging Neurosci. **6**, 312 (2014)
11. R.P. Vega, P.M. van Leeuwen, E.R. Vélez, H.G. Lemij, J.C. de Winter, Obstacle avoidance, visual detection performance, and eye-scanning behavior of glaucoma patients in a driving simulator: a preliminary study. PloS one. **8**(10), e77294 (2013)
12. V.K. Gothwal, D.K. Bagga, H.L. Rao, S. Bharani, R. Sumalini, C.S. Garudadri, S. Senthil, S.P. Reddy, V. Pathak-Ray, A.K. Mandal, Is utility-based quality of life in adults affected by glaucoma? Invest. Ophthalmol. Vis. Sci. **55**(3), 1361–1369 (2014)
13. M.C. Sousa, L.G. Biteli, S. Dorairaj, J.S. Maslin, M.T. Leite, T.S. Prata, Suitability of the Visual Field Index according to glaucoma severity. J. Curr. Glaucoma Pract. **9**(3), 65 (2015)
14. A. Voßkühler, V. Nordmeier, L. Kuchinke, A.M. Jacobs, OGAMA (Open Gaze and Mouse Analyzer): open-source software designed to analyze eye and mouse movements in slideshow study designs. Behav. Res. Method. **40**(4), 1150–1162 (2008)
15. S. Krishnan, J. Amudha, S. Tejwani, Intelligent-based decision support system for diagnosing glaucoma in primary eyecare centers using an eye tracker. J. Intell. Fuzzy Syst. (Preprint), 1–8 (2021)
16. S. Krishnan, J. Amudha, S. Tejwani, Gaze fusion-deep neural network model for glaucoma detection, in *Symposium on Machine Learning and Metaheuristics Algorithms, and Applications* (2020), pp. 42–53

Impact of Deep Learning in the Analysis of Particulate Matter in the Air Pollution

Praveena Vasudevan and Chitra Ekambaram

Abstract Over the past few years, the perpetual threat of high concentrations of particulate matter (PM) particles in atmospheric air pollution urges serious health conditions to humans. The acquired inadequate PM air quality is not only affecting human health but also prompts ozone layer exhaustion, detriment of crops in agriculture, wildlife depletion, decelerating growth of plants and trees, etc. Thereby, it is required to equip contingency measures monitoring centres in all more polluted cities to monitor the atmospheric air quality and diminish the emissions as early as possible. In this article, the amount of particulate matter distribution in the air is monitored through IoT terminals, and the same has been tested with the training model developed with the help of machine learning using the dataset collected from the Indian government. The proposed multi-tier architecture had delivered fruitful results in terms of air quality maintenance and the analytical inference brought many unexplored insights that could potentially help in reducing the air pollution in the cities and other densely populated areas. The dataset used for training the machine learning algorithm has been taken from the open repository of the Indian government specifically belonging to the state, Tamil Nādu. The analysis of the developed system with the trained model showed higher efficiency when compared with the traditional algorithms.

Keywords Particulate matters · Deep learning · Machine learning · Multi-tier architecture · Air pollution

P. Vasudevan (✉) · C. Ekambaram
Department of ECE, SRM Institute of Science and Technology, Deemed to be University, Kattankulathur, Chennnai, India
e-mail: praveena@srmist.edu.in

C. Ekambaram
e-mail: chitrae@srmist.edu.in

A. K. Bashir et al. (eds.), *Proceedings of International Conference on Computing and Communication Networks*, Lecture Notes in Networks and Systems 394,
https://doi.org/10.1007/978-981-19-0604-6_32

1 Introduction

In recent years, the particulate matter scales have grown as a large planetary issue. The contemporary industrial capitalist society holds gas and air molecules and liquid droplets. Moreover, it leads to major causes of pathological disorders, lung cancer, viral infectivity and early deaths, etc. The air pollution has caused adverse effects, as the urbanization processes. The deep concentrated PM with sized 10 and 2.5 μm precipitate harmful health impacts [1]. Recent research has observed urban air pollution and its properties [2]. The properties of urban air pollution have temporal and spatial properties. The temporal factors will vary according to the time variations conversely the spatio-properties vary due to locations [3]. So, the spatiotemporal factors accordingly vary from different cities. As the immediate and long-term impression of suspended particulates acutely causes health risk issues to people, several countries have begun to manage their PM concentration scales. The requirement of PM concentration scales enables restoratives to minimize the harm and losses on the basis of rapid alert PM monitoring systems [4]. So, a large number of countries have established measuring stations for air pollution monitoring to detect atmospheric pollutants such as carbon monoxide (CO), nitrogen dioxide (NO_2), sulphur dioxide (SO_2), and particulate matter with the size of 10 and 2.5. As long as the PM contraction measures of PM10 and PM2.5 are highly correlated to public health, more epidemiological studies have been presented [5]. Whereas the greater concentrations of PM are toxic to people. Almost all countries have norms and criteria to maintain air quality standards to preserve public health. The WHO Air Quality Guidelines (AQG) and European Union have fixed some threshold range to contaminants which will not exceed the fixed range of air quality contingency measures. In India, the Central Pollution Control Board (CPCB) is the superior air pollution monitoring agency that monitors air pollution particulates among 731 CPCB stations [6]. In Korea, the government has taken contingency measures to protect from emissions according to high concentration atmospheric pollutants. The atmospheric contaminants emerge from serious health risks which lead to major dangerous causes over the upcoming decades too [7]. Subsequently, the weather-related factors to be considered to monitor PM in air. Those are the decline of intra-city mobility intensity (dIMI), multi-scale geographically weighted regression (MGWR), and geographically weighted regression (GWR) [8]. These geography and PM concentration variations are related according to the topographical factor, wind speed and temperature dependence of PM. Over the last few decades, a few template-based and computerized diverse approaches had been devised to recuperate air pollution levels in forecasting systems. The traditional PM monitor models were supported by the time-series-based models due to the spatiotemporal characteristics. An Autoregressive Integrated Moving Average (ARIMA) model with its alternatives was enabled from time-series-based models, and additionally, Holt Winter models also included. But the different kinds of models provided forecasting errors due to irregular variations of PM concentrations. Furthermore, the linear statistical models of ARIMA also do not provide air pollution countermeasures accurately [9]. So, the various older methods and machine learning approaches have been utilized in air pollution forecasting.

The rest of the article encompasses the following. Section II background works to monitor and analyse particulate matter distribution in air. Section III details the methodology with architecture and the mathematical modelling of the pollution information. The experimental setup and results are discussed in Section IV. Section V details the conclusion and future work.

2 Related Works

The previous investigations pursued precise effective prognostication models in atmospheric air pollution. Various machine learning methods have been utilized in air pollution forecasting [10]. The long short-term memory and gated recurrent unit methods (LSTM & GRU) using convolutional neural networks (CNN) have been used to predict PM concentrations in 39 stations in Seoul, South Korea. Hourly experimental results were obtained; the transcended hybrid models with LSTM and GRU method's calculating root mean square error (RMSE) and mean absolute error (MAE) values of PM10, PM 2.5 were observed. The hybrid models with CNN-GRU exploited better results than other conventional models [7]. In addition, the forecasting predictive models which are based on machine learning regression models can be experimented with Taiwan Air Quality Monitoring datasets. The performance measures can be employed with the errors measurement of root mean square error, mean absolute error, and mean square error [11]. Typically, the statistical methods with theoretical approaches have been utilized for predicting PM concentrations. Despite that, the conventional statistical models were not capable of identifying nonlinear patterns in complex times of air pollution forecasting. Since the hardware big data management has been developed, the reliable time-series forecasting approach of machine learning methods has been introduced to predict air pollution forecasting. The artificial neural networks (ANN)-based models delivered less features in finding long time-series connections due to its memory lacking cells. So, the recurrent neural network (RNN) models have been developed; in terms of consecutive time-series prediction, RNN can recall earlier time consecutive data due to its RNN layer features. The RNN with long-term time-series predictions can be obtained. Additionally, the tumultuous actions of air contaminants constitute significant challenges in screening three-dimensional movements around multifarious transitory domains among feasible air quality systems. A deep learning hybrid contextual long short-term memory model (CLSTM) is incorporated with convolutional neural network (CNN). The CLSTM model is utilized to predict total suspended particulate (TSP) in hourly basis observation measures due to the mapping scheme of the LSTM model. The CLSTM outwits various machine learning models. Those are multiple linear regression (MLR), M5 model tree, Volterra, and random forest (RF). These models can be performed to obtain hourly basis TSP observation. The LSTM algorithm with three-layered CNNs is employed to extract data features of the CLSTM model [12]. The severe control management of air quality monitoring is needed to avoid deleterious PM effects in an immense cityscape environment. So, the air quality monitoring over deployed sensors that established with the IOT platform could result in desirable solutions in air quality forecasting. The meticulous modelling of big data air quality compiled by IoT sensors thoroughly led to early detection by decision-making forecasting information. Additionally, the hybrid deep learning-based CNN-LSTM model has been employed to exploit the air quality monitoring scales from different locations through IoT air quality sensors [13]. The CNN-based encoder captures all spatiotemporal features based on data to facilitate accurate prediction. Whereas the convolutional long short-term memory (ConvLSTM) has been setting air pollution data into image sequences to automatically manage spatial and temporal features. The required sequences have interjected with the ConvLSTM model to predict the air quality for all over the city [14]. These spatial–temporal correlation analysis can be employed to view the parameters in between linear and nonlinear correlations by enabling mutual information (MI) parameter on to the analysis [15]. In addition to monitoring timely air quality, using image-based models under air quality monitoring has led to air pollution control. It is enabled with deep learning tools. The particular model evaluates the air quality scales from the scene images which can be captured by employed cameras. According to the image feature extraction, the deep learning image-based model can sort the air quality monitoring scales. Additionally, the model performance can be evaluated using high-quality outdoor air quality dataset. The comparison of experimental results has been observed from air quality control (AQC) net, deep residual network (ResNet) on air quality index (AQI) and space vector machine (SVM) [16]. But the persistent deciphering responses of PM2.5 and O_3 emission might be changed due to chemical reactions which cause changes in a predictive scale. The deep learning supported response surface model (DeepRSM) has delineated the reaction of O_3 and PM2.5 concentrations emission changes. The trained air quality dataset of deep RSM model is worked from brute-force simulations with the community multi-scale air quality (CMAQ) CTM to obtain the predictive scales of different emission changes [17]. Accordingly, the traffic volume and average driving speed data are contributing new databases in the research of air pollution. Furthermore, the recent investigations have sought to forecast models to be used in discrete locations [18]. The commercial particulates have technical challenges and limits in monitoring high PM concentration countermeasures with variations of data. Those constraints can be overcome by using smartphone-based digital holographic microscopy (S-DHM) systems with deep learning networks named Holo-Speckle Net. In this model, the deep autoencoder with regression layers can be highly trained using S-DHM recorded diverse PM concentrations of holographic speckle images. It is endorsed with enhanced hyperparameter optimization [19]. Consequently, the Data-driven, Open, Global (DOG) working paradigm to acquire automatic feature extraction using available data in order to enable pollution estimation. By using the DOG paradigm, the MapLUR model has been established. It can be employed by estimating pollution concentrations for particular locations using publicly available captured map images and satellite images. It provides the results of extracted features that are closely related to manually featured images to use land-use regression models [20].

3 Proposed Work

In this article, a multi-tier architecture for monitoring and analysing the particulate matter in the air has been developed. The proposal consists of three different tiers; the physical tier consists of the hardware IoT terminals that acts as the data source for the system. These nodes can collect information in different formats based on the type and behaviour of the data. These sensor nodes are directly connected to the internet through the IoT gateway and all the communication happens through the gateway. The second tier or the middle tier is the application tier. It acts as a bridge between the user dashboard and the physical tier. Main business logic of this application resides in this tier. The elements that are present in this tier includes, NoSQL database/data lake, IoT platform integration tool, Apache spark file system, storage component for the big data and finally the data processing engine powered by machine learning. In the machine learning component, we have modelled the convolution neural network (CNN) to fit for training the data and for analysis on the real-time data. The third tier or the bottom most tier is the dashboard region. This dashboard consists of the analytical and visualization tools that can be used to project the information/insights that are obtained from the sensor terminals.

The workflow of the proposed system is as follows: Initially, the data collected from the IoT terminals are fed into the system through the IoT gateway. The IoT terminals range from the simple gas detection sensor to high-end highly sensitive sensors. The IoT gateway serves as a mediator in collecting the required information from the hardware terminals and the same will be communicated to the data lake in the application tier. All the data collected will be dumped in the data lake, and it will be utilized by the system when there is a requirement.

Since the data received from the terminals will be not uniform and may not be clean, it has to undergo a phase called data cleaning. In the data cleaning process, the data received from the terminals are cleaned for null values, errors, missing values, etc. Immediately after the cleaning process, the data will be utilized for shaping up the data so as to make out the inferences. For instance, shaping may delete unwanted attributes, and at the same time, it may add derived attributes as well. The primary purpose of the shaping process is to assist the system when making an optimized decision. All these processes will happen with the help of the data store region that is available in the cloud environment.

Once the data are cleaned and shaped, it is fed into the machine learning component. The machine learning component consists of a pre-trained CNN model that is trained using the dataset of the Indian government. The prediction model is developed with the help of CNN, and once the raw data are sourced into the ML algorithm, it provides the insights on the data and also uses the same data to learn from the environment.

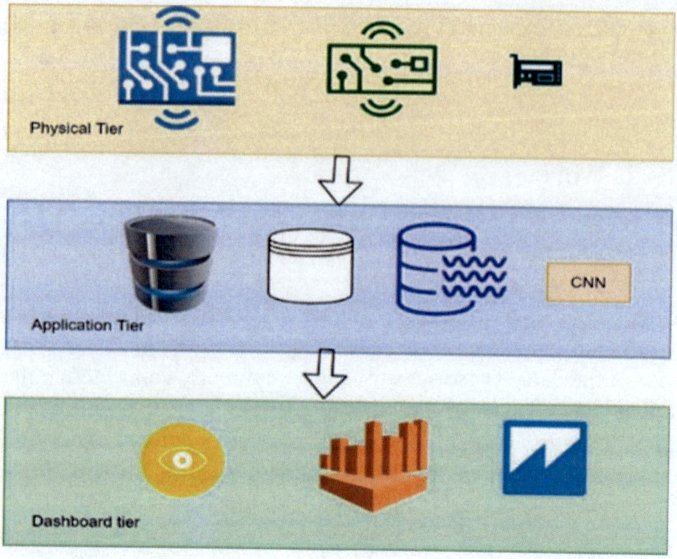

Fig. 1 Proposed multi-tier architecture

The dashboard environment is developed with frontend tools by integrating various visualization libraries. Based on the insights provided by the ML component, the data will be sourced into the visualization library, and the same will be transformed into charts for easier visualization. The outliers or the abnormal behaviour in the system can be easily found through this approach.

CNN has been integrated in application layer with the aid to make optimal decision from the data generated using IoT devices. Schematic diagram of the proposed is shown in Fig. 1.

4 Experimental Analysis and Results

The experimentation is carried out on Beijing air quality data []. From the experimental setup and the mathematical model, we bring the following inferences as the study of the particulate matter. So here, the proposed mechanism is compared with three machine learning mechanisms and the results of accuracy, root mean square error (RMSE), F-score. The machine learning techniques which are taken for the comparison are decision tree, random forest, and support vector machine. The metrics taken into account for experimentation are accuracy, RMSE, F-score. Table 1 represents the comparison of proposed CNN for measuring air quality.

Table 1 Comparison of proposed CNN with other machine learning algorithms

Performance metrics	Decision tree	Random forest	Support vector machine	Proposed CNN
Accuracy	89.31	91.23	91.34	94.34
F-score	85.3	87.3	89.1	92.3
RMSE	0.54	0.45	0.32	0.001

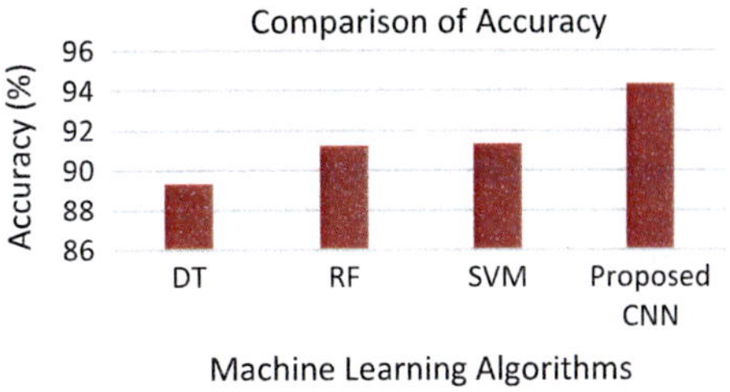

Fig. 2 Comparison of accuracy

4.1 *Comparison of Accuracy*

Accuracy is measured as the ratio of number of instances correctly classified to the total number of instances. It is represented in Eq. (1).

The accuracy of the proposed CNN is higher than other machine learning algorithms. The accuracy of proposed CNN is 5.6 times greater than decision tree. Similarly, the accuracy of SVM is 3.2 times lesser than accuracy of CNN. The diagrammatic representation of comparison of accuracy is shown in the Fig. 2.

$$\text{Accuracy} = \frac{\text{TP} + \text{TN}}{\text{TP} + \text{TN} + \text{FP} + \text{FN}} \tag{1}$$

4.2 *Comparison of F-Score*

F-score represents the harmonic mean of precision and recall. A good classifier will have good F-score. F-score. CNN achieves 8.2% greater F-score than decision tree and 5.7% greater F-score than random forest. The reason behind is that CNN involves multiple hidden layers where each layer learns necessary features for having better prediction. Equation (2) represents the calculation of F-score. The diagrammatic representation of F-Score is shown in Fig. 3.

$$\text{F} - \text{score} = \frac{2 * \text{precision} * \text{recall}}{\text{precision} + \text{recall}} \tag{2}$$

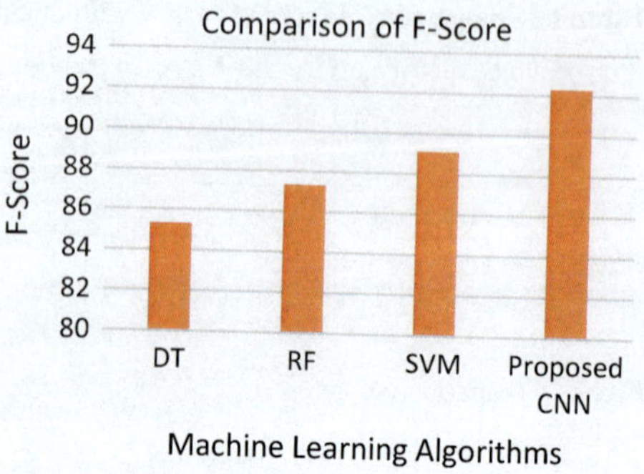

Fig. 3 Comparison of F-score

4.3 Comparison of RMSE

RMSE measures the difference between the actual and predicted values. Equation (3) represents the computation of RMSE. The RMSE of CNN is very less when comparing to decision tree, random forest, support vector machine. The RMSE of decision tree is 99.81 times greater than CNN. Similarly, the RMSE of SVM is 99.88 times greater than CNN. From Table 1, it is evident that CNN achieves minimum RMSE than other machine learning algorithms.

$$\text{RMSE} = \sqrt{\frac{\sum_{i=1}^{n} P_i - A_i}{n}} \tag{3}$$

where P_i represents the predicted value, and A_i represents the actual value.

5 Conclusion

The proposed CNN in the multi-tier architecture for measuring the air quality has been implemented in Python and the method achieves 94.34% accuracy. The proposed method achieves 5.7% greater F-score than random forest, which really proves that it is the best classifier for measuring the air quality. The future work focuses on improving the performance of CNN by learning per parameters.

References

1. R. Arul, R. Alroobaea, U. Tariq, A.H. Almulihi, F.S. Alharithi, U. Shoaib, IoT-enabled healthcare systems using block chain-dependent adaptable services, *Personal and Ubiquitous Computing* (2021), pp. 1–15

2. I. Manisalidis, E. Stavropoulou, A. Stavropoulos, E. Bezirtzoglou, Environmental and health impacts of air pollution: a review. Front. Public Health **8**, 14 (2020)
3. D. Balasubramaniam, C. Kanmanipappa, B. Shankarlal, M. Saravanan, Assessing the impact of lockdown in the US, Italy and France–what are the changes in air quality? *Energy Sources, Part A: Recovery, Utilization, and Environmental Effects* (2020), pp. 1–11
4. H. Fan, C. Zhao, Y. Yang, A comprehensive analysis of the spatio-temporal variation of urban air pollution in China during 2014–2018. Atmos. Environ. **220**, 117066 (2020)
5. R. Cao, B. Li, Z. Wang, Z.R. Peng, S. Tao, S. Lou, Using a distributed air sensor network to investigate the spatiotemporal patterns of PM2. 5 concentrations. Environ. Pollut. **264**, 114549 (2020)
6. J. Ngarambe, S.J. Joen, C.H. Han, G.Y. Yun, Exploring the relationship between particulate matter, CO, SO_2, NO_2, O_3 and urban heat island in Seoul, Korea. J. Hazard. Mater. **403**, 123615 (2021)
7. K. Shukla, P. Kumar, G.S. Mann, M. Khare, Mapping spatial distribution of particulate matter using Kriging and inverse distance weighting at supersites of megacity Delhi. Sustain. Cities Soc. **54**, 101997 (2020)
8. G. Yang, HwaMin Lee, G. Lee, A hybrid deep learning model to forecast particulate matter concentration levels in Seoul, South Korea. Atmosphere **11**(4), 348 (2020)
9. Z. Fan, Q. Zhan, C. Yang, H. Liu, M. Zhan, How did distribution patterns of particulate matter air pollution (PM2. 5 and PM10) change in China during the COVID-19 outbreak: a spatiotemporal investigation at Chinese city-level. Int. J. Environ. Res. Public Health **17**(17), 6274 (2020)
10. A. Dairi, F. Harrou, S. Khadraoui, Y. Sun, Integrated multiple directed attention-based deep learning for improved air pollution forecasting. IEEE Trans. Instrum. Meas. **70**, 1–15 (2021)
11. K. Kirwa, A.A. Szpiro, L. Sheppard, P.D. Sampson, M. Wang, J.P. Keller, M.T. Young, S.Y. Kim, T.V. Larson, J.D. Kaufman, Fine-scale air pollution models for epidemiologic research: insights from approaches developed in the multi-ethnic study of atherosclerosis and air pollution (MESA Air). Curr. Environ. Health Rep. 1–14 (2021)
12. K.S. Harishkumar, K.M. Yogesh, I. Gad, Forecasting air pollution particulate matter (PM2. 5) using machine learning regression models. Procedia Comput. Sci. **171**, 2057–2066 (2020)
13. E. Sharma, R.C. Deo, R. Prasad, A.V. Parisi, N. Raj, Deep air quality forecasts: suspended particulate matter modeling with convolutional neural and long short-term memory networks. IEEE Access **8**, 209503–209516 (2020)
14. S. Abirami, P. Chitra, R. Madhumitha, S. RagulKesavan, Hybrid spatio-temporal deep learning framework for particulate matter (PM 2.5) concentration forecasting, in *2020 International Conference on Innovative Trends in Information Technology (ICITIIT)* (IEEE, 2020), pp. 1–6
15. V.D. Le, T.C. Bui, S.K. Cha, Spatiotemporal deep learning model for citywide air pollution interpolation and prediction, in *2020 IEEE International Conference on Big Data and Smart Computing (BigComp)* (IEEE, 2020), pp.55–62
16. U. Pak, J. Ma, U. Ryu, K. Ryom, U. Juhyok, K. Pak, C. Pak, Deep learning-based PM2. 5 prediction considering the spatiotemporal correlations: a case study of Beijing, China. Sci. Total Environ. **699**, 133561 (2020)
17. Q. Liao, M. Zhu, L. Wu, X. Pan, X. Tang, Z. Wang, Deep learning for air quality forecasts: a review. Curr. Pollut. Rep. 1–11 (2020)
18. J. Xing, S. Zheng, D. Ding, J.T. Kelly, S. Wang, S. Li, T. Qin et al., Deep learning for prediction of the air quality response to emission changes. Environ. Sci. Technol. **54**(14), 8589–8600 (2020)
19. V.D. Le, T.C. Bui, S.K. Cha,Spatiotemporal deep learning model for citywide air pollution interpolation and prediction, in *2020 IEEE International Conference on Big Data and Smart Computing (BigComp)* (IEEE, 2020), pp. 55–62
20. J. Kim, T. Go, S.J. Lee, Volumetric monitoring of airborne particulate matter concentration using smartphone-based digital holographic microscopy and deep learning. J. Hazard. Mater. 126351 (2021)

Toward Machine Learning and IoT Environment-Enabled Smart Personal Health Monitoring Framework

V. Muthumanikandan, A. Bhuvaneswari, and R. Radhika

Abstract The healthcare industry is mostly "information rich;" however, sadly, not all the information square measure mined that is needed for locating hidden patterns and effective higher cognitive process. Progressed information handling methods square measure is used to find information in data and for clinical examination, fundamentally in heart condition forecast. This paper has investigated forecast structures for heart condition abuse a ton of assortment of enters credits. The contraption utilizes logical terms like sex, circulatory strain, and ideal cholesterol like thirteen credits to expect the likelihood of patient getting a heart ailment. The insights mining class procedures, in particular, multi-layer perceptron (MLP), random forest, support vector (Linear), Naïve Bayes, and AdaBoost are broke down on cardio-vascular disease dataset. The general exhibition of these methods is as analyzed, in light of precision. As in accordance with our results, precision of multi-layer perceptron (MLP), random forest, support vector (Linear), Naïve Bayes, and AdaBoost are 86.88%, ninety.16%, eighty three.60%, eighty five.24%, and ninety.16%, individually. Our assessment recommends that out of these three class models, neural networks predict heart condition with most extreme exactness.

Keywords Coronary heart illness · Decision tree · K-nearest neighbor · Machine learning · Naïve Bayes · Support vector machine

V. Muthumanikandan · A. Bhuvaneswari (✉)
Vellore Institute of Technology, Chennai, Tamil Nadu, India
e-mail: bhuvaneswari.a@vit.ac.in

V. Muthumanikandan
e-mail: muthumanikandan.v@vit.ac.in

R. Radhika
SRM Institute of Science and Technology, Kattankulathur, Chennai, Tamil Nadu, India
e-mail: radhikar2@srmist.edu.in

A. K. Bashir et al. (eds.), *Proceedings of International Conference on Computing and Communication Networks*, Lecture Notes in Networks and Systems 394,
https://doi.org/10.1007/978-981-19-0604-6_33

1 Introduction

In our everyday life, individuals are going through a daily practice and occupied timetable which prompts pressure and tension [1]. This outcomes in illness like heart condition [2], disease [3], and so on. The extended use of flexible developments and splendid devices has caused phenomenal impacts [4, 5]. Prosperity experts are logically taking advantage of the advantages these progressions bring, as such making an enormous improvement in clinical benefits in clinical settings [6, 7]. The test behind these sicknesses is its assumption [8]. Each individual has extremely amazing potential gains of urgent sign and squeezing factor level. Coming up next are the kind of heart sickness: Heart signifies "cardio." Thus, all heart illness worry to class of cardiovascular illness. The various types of heart ailment, for example, coronary heart illness, angina pectoris, congestive cardiovascular breakdown, cardiomyopathy, and congenital heart illness [9].

The strategies and computations will be directly used to make critical conclusions and deductions from the dataset. In this research article, the current innovation gives an effective method of checking the individual strength of people. The outcomes show that five of these eight calculations accomplished exactness of more than 90%. Utilizing these five calculations will give viable and exact expectation and ID of possible cases. In this paper, we propose a portable application-based model savvy home medical care framework for productive and powerful well-being observing for the older and crippled for their helpful and free living while at home. A segment of the proposed framework permits the patient to distantly transfer or catch fundamental well-being indications data during a time of a pandemic, for example, the continuous sickness for their primary care physician's helped determination. The remaining section of the paper is organized as follows. Section 2 presents the related work on smart health care using IoT and machine learning. Section 3 discusses the proposed system of patient through IoT. Section 4 discusses the experiments, and results demonstrate healthcare implementation and its working techniques. Section 6 concludes the paper with significant future directions.

2 Related Work

Prediction of heart Illness utilizing WEKA device with accuracy, they came to 86.3% for testing stage 87.3% for preparing stage [6]. Naïve Bayes, decision tree, and neural organization calculations and dissected using social network data [7]. There are a colossal amount of choices included. Thus, there is a necessity to downsize the amount of alternatives. This should be possible by include choice. On doing this, they say that time is diminished. They utilized choice tree and neural organizations with k-closest neighbor calculation, neural organization, and credulous Bayes and choice tree for heart sickness forecast [8]. They utilized information mining methods to identify the heart ailment hazard rate. Particle swarm optimization, artificial neural organization, genetic calculation for expectation. Cooperative grouping is a new and affordable strategy which coordinates affiliation rule mining and characterization to a model for expectation and accomplished great exactness [9].

A robotized framework in clinical conclusion would upgrade clinical consideration, and it can likewise diminish costs [10] with framework that can financially find the guidelines to foresee the danger level of patients dependent on the given boundary about their well-being. The principles can be focused on dependent on the client's prerequisite. The presentation of the framework is assessed as far as grouping precision, and the outcomes show that the framework has incredible potential in anticipating the heart disease hazard level all the more precisely." Heart illness prediction utilizing machine learning and data mining technique [11] with accuracy they arrived at J48 gives 56.76% which is higher than LMT calculation of precision 55.75%. Multi-Illness prediction utilizing data mining techniques using Naïve Bayes [12] accuracy they arrived at heart illness: 79% Diabetes: 77.6% Breast Cancer: 82.5%. An applied strategy to upgrade the forecast of heart illness is utilizing the information methods [13]. They utilized SVM in equal style. SVM gives higher and prudent exactness of 85 and 82.35%. SVM in equal style gives higher exactness than successive SVM. The congestion utilized decision tree classifier and support vector machine [14].

Heart illness prediction system evaluation utilizing C4.5 rules and partial tree. They utilized C4.5 rules and Naive Bayes calculation [15]. C4.5 gives higher precision than Naive Bayes. There are seven vital variables for heart sickness like smoking, actual idleness, sustenance, corpulence, cholesterol, diabetes, and hypertension [16]. Evidently, users lack in data privacy and the access control mechanisms available to avoid the risk of disclosure in IoT environment [17]. They likewise examined the measurements of heart ailment including stroke and cardiovascular ailment. The intermittent neural organization gives great exactness when contrasted with different calculations like CNN, Naïve Bayes, and SVM. Information entropy-based healthcare framework is proposed to identify using publicly available data. Shortest path fast rerouting technique was imposed during link failure and to ensure the recovery process. The accessing of data can be done using the proposed technique in the smart healthcare monitoring [18].

3 Proposed Machine Learning IoT Framework

Today, a few clinics oversee medical services data exploitation medical care data framework; in light of the fact that the framework contains huge amount of data want to remove stowed away information for making astute conclusion. The principle objective of this investigation is to make intelligent heart medical issue prediction system that gives recognizable proof of heart medical condition exploitation recorded heart data. To foster this method, clinical terms like sex, circulatory strain, and cholesterol like 13 information credits are utilized. The information mining grouping methods, viz., multi-facet perceptron (MLP), random forest, support vector

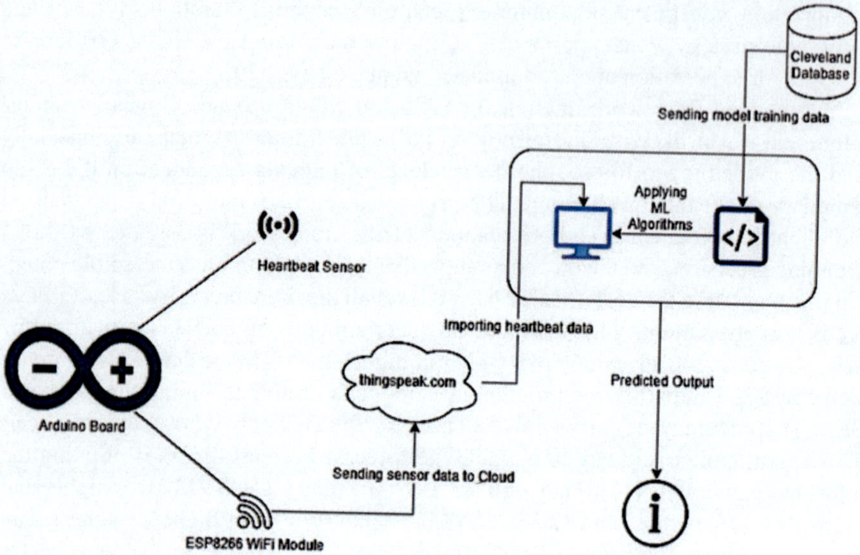

Fig. 1 Proposed architecture

(Linear), Naïve Bayes, and AdaBoost are utilized. The information source contains the openly accessible heart ailment dataset (Cleveland database). The Cleveland Heart Illness dataset comprises of 303 records. Irregular Forest Classifier—an arbitrary woods could be a meta-PC that coordinates with a measure of call tree classifiers on fluctuated sub-examples of the dataset and utilizations averaging to improve the prescient exactness and command over-fitting. AdaBoost Classifier—it is used to predict the accuracy. It also builds the strong classifier. The proposed architecture of patient monitoring system is shown in Fig. 1. Multi-facet Perceptron (MLP) Classifier—a multi-facet perceptron (MLP) could be a classification of feed forward counterfeit neural organization. MLP uses a directed learning method known as back engendering for coaching. It will recognize data that are not straightly severable. Backing Vector Machine (SVM) Classifier (rbf/direct). Innocent Bayes—in AI, gullible mathematician classifiers a group of simple "probabilistic classifiers" upheld applying Bayes' hypothesis with powerful (guileless) freedom presumptions between the decisions.

4 Experiments and Results

The heartbeat sensor is fixed to the patient's hand for reading pulse. This covers an IR sensor in it. Each siphoning we get beat from that sensor. This sensor yield is given to the Arduino through signal trim unit for improvement. NTC type thermistor is used as a temperature sensor. This temperature sensor yield shifts reliant upon

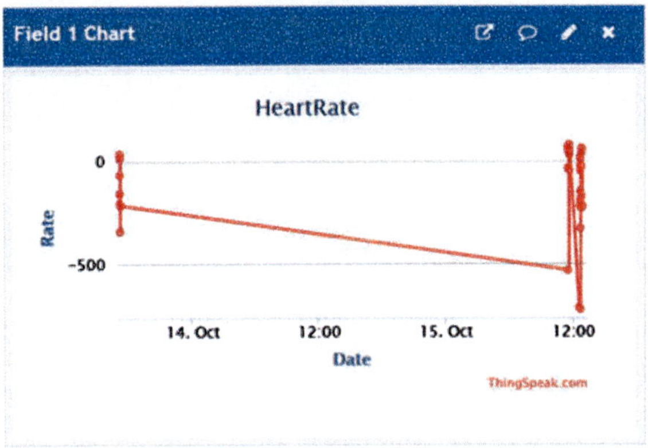

Fig. 2 ThingSpeak graph

the temperature; its resultant value is highly similar as given to Arduino. Clients' imperative signs information, for example, circulatory strain, pulse, weight, and blood glucose (BG) from sensor hubs to cell phones, while constant information handling was used to deal with the huge measure of consistently created sensor information. EEG sensor is a used to quantify the electrical movement of the heart. The electrical action can be graphed as an ECG yield as a simple examining model. The outcomes show that business forms of the proposed health monitoring sensors and the proposed constant information preparing are adequately effective to screen the fundamental signs information of diabetic patients. Moreover, AI put together grouping techniques were tried with respect to a health dataset and showed that a multilayer perceptron can give early expectation of disease given the client's sensor information. The yield is shown through values obtained in a specific time frame. The ThingsSpeak heart rate and data at real-time environment are shown in Fig. 2.

This framework is additionally intended to send a suggestion to patients on the utilization of specific drugs with input provided by the client. A calculation was created dependent on hyperspace analog context (HAC) for an inescapable climate, to address the requirement for decision for the client among various gadgets set in the shrewd home medical care framework. The specialized commitments of this paper are as per the following. Execution of IoT-based shrewd home medical services supports structure equipped for lessening superfluous weights on the clinics because of illness flare-up. The new framework which is likewise ready to give fundamental solaces utilizing the IoTs empowered home machines just, energizes patients with serious and basic conditions to use medical clinic offices. The correlations between the attributes are shown in Fig. 3.

Figure 3 this graph is showing correlation between sex age and target. Extraction and translation of patient's well-being information are studied and obtained from wearable, underlying, and versatile sensors for clinical judgments and remedies. The software and hardware used are as follows: Arduino board, wires, ESP2866 Wi-Fi module, finger detecting heartbeat module, IDE-Jupyter Notebook, Arduino IDE, ThingSpeak, machine learning library, scikit, NumPy, pandas Python packages. The feature extraction and dimensionality, correlation between the attributes are shown

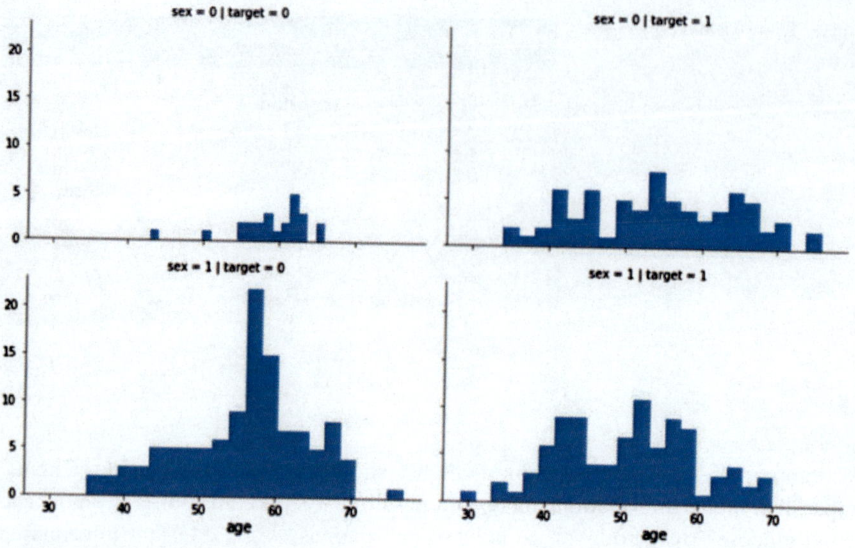

Fig. 3 Correlation between attributes

in Fig. 4. The correlation between the primary and important attributes is shown in Fig. 5. The planned versatile and Web application, once completely created, can be connected to existing Web spaces of medical clinics as an entry and can be dispatched as a new application for medical clinics without existing areas.

The results obtained using accuracy are shown in Fig. 6. It is additionally suggested that new components, for example, a physiological information catching gadget be consolidated into the current framework. A depiction of the persistently gathered information in Web worker (ThingSpeak) is represented. The level harmful gases (CO and CO_2) are estimated in ppm unit. Any clinical staff can undoubtedly screen

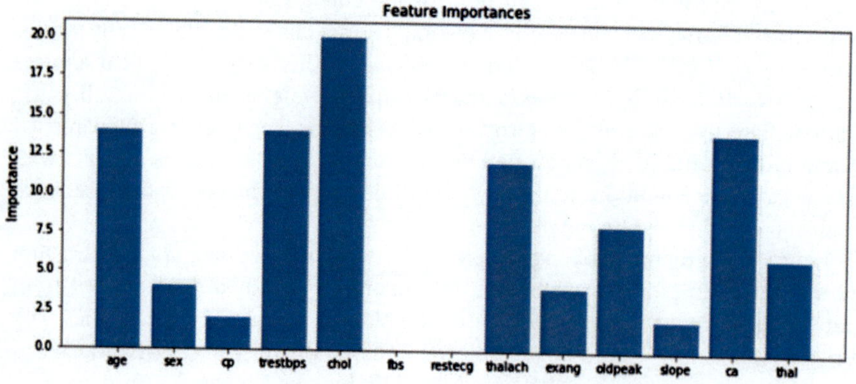

Fig. 4 Feature extraction of all options. "chol," "age," 'trestbps'

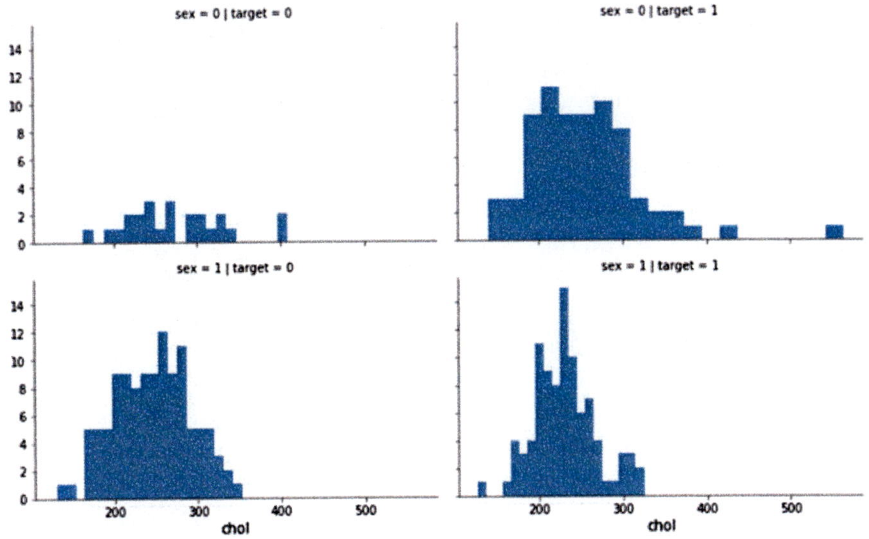

Fig. 5 Correlation between chol age and target

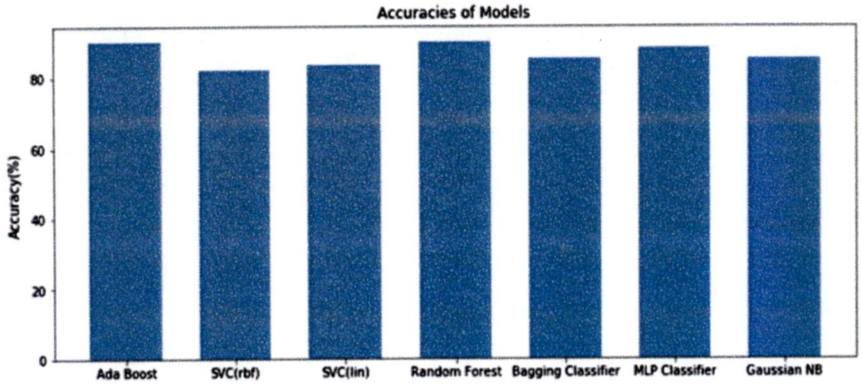

Fig. 6 Accuracy of machine learning classifiers

the particular patient just as the room condition where the patient is presently through a gadget which has Web access. The patient information here is secure in light of the fact that for getting to information, one requirement to go through a secret phrase ensured framework, i.e., just the true staff can screen the framework.

By examining information, a specialist can without much of a stretch settle on from a distant area. At the point, when the information crosses the level, then, at that point, the clinical staff can undoubtedly make vital strides for medical services. The accuracy of various machine learning algorithms is shown in Table 1.

Table 1 Machine learning algorithms accuracy

Method	Accuracy (%)
RFC [5]	90.16
ABC [6]	90.16
MLP [11]	85.24
SVC (Linear) [15]	83.60
SVC (rbf) [18]	81.96
Proposed machine learning framework	90.3

5 Conclusion and Future Work

In this paper, five information mining characterization methods were applied to be specific multi-layer perceptron(MLP), random forest, support vector (Linear), Naïve Bayes, and AdaBoost. The exactness accomplished for every classifiers is as following MLP = 85.24, RFC = 90.16, ABC = 90.16, SVC (linear) = 83.60, SVC (rbf) = 81.96, NB = 85.24. The error level of the created conspire is inside a specific breaking point (<5%) for each case. From results, it has been seen that AdaBoost classifier gives precise outcomes as contrast with different classifiers. This framework can be additionally extended. It can utilize greater amount of information ascribes like smoking and obesity. Different information mining procedures can likewise be utilized for predication, for example, bunching, time series, association rules. The text mining can be utilized to mine gigantic measure of unstructured information accessible in medical care industry dataset.

References

1. K. Purushottam, R. Saxena, Sharma, *Economical Heart Illness Prediction System* (2016), pp. 962–969
2. B. Brahmi, M.H. Shirvani, Prediction and diagnosis of heart illness by data mining techniques. J. Mult. Eng. Sci. Technol. **2**, 164168 (2015)
3. L. Sharan Monica, B. Sathees Kumar, Analysis of cardiovasular illness prediction using data mining techniques. Int. J. Mod. Comput. Sci. **4**, 55–58 (2016)
4. R. Radhika, A. Bhuvaneswari, G. Kalpana, An intelligent semanticification rules enabled user-specific healthcare framework using IoT and deep learning techniques. Wirel. Pers. Commun. (2021)
5. M. Sultana, A. Haider, Heart illness prediction using WEKA tool and 10-fold cross-validation, in *The Institute of Electrical and Electronics Engineers* (2017)
6. A. Bhuvaneswari, C. Valliyammai, Information entropy based event detection during disaster in cyber-social networks. J. Intell. Fuzzy Syst. **36**(5), 3981–3992 (2019)
7. J. Thomas, R. Theresa Princy, Human heart illness prediction system using data mining techniques (2016)
8. S. Bharti, S.N. Singh, Analytical study of heart illness prediction comparing with different algorithms (Amity University, Noida, India, 2015)
9. Purushottam, K. Saxena, R. Sharma, Economical heart illness prediction system using Decision tree (2015)

10. K. Gomathi, D. Shanmuga Priyaa, Multi illness prediction using data mining techniques. Int. J. Syst. Softw. Eng, 12–14 (2016)
11. M. Vanamoorthy, V. Chinnaiah, Congestion-free transient plane (CFTP) using bandwidth sharing during link failures in SDN. Comput. J. **63**(6), 832–843 (2020). https://doi.org/10.1093/comjnl/bxz137
12. S. Seema Shedole, K. Deepika, Predictive analytics to prevent and control chronic illness. https://www.researchgate.net/punlication/316530782. (2016)
13. P. Sharma, K. Saxena, R. Sharma, Heart illness prediction system evaluation using C4.5 rules and partial tree, in *Computational Intelligence in Data Mining*, vol. 2 (Springer, 2015), pp. 285–294
14. E.J. Benjamin et al., *Heart Illness and Stroke Statistics 2018 At-a-Glance* (2018)
15. A. Bhuvaneswari, C. Valliyammai, Social IoT enabled emergency event detection framework using geo tagged microblogs and crowdsourced photos, in *Emerging Technologies in Data Mining and Information Security, Advances in Intelligent Systems and Computing*, vol. 813, Chapter No. 13, ed. by A. Abraham, et al., (Springer Nature, Singapore, 2018), pp. 151–162
16. A. Kishore, A. Kumar, K. Singh, M. Punia, Y. Hambir, Heart attack prediction using deep learning (2018). (Department of Computer Engineering, Army Institute of Technology, Pune, Maharashtra Professor, Department of Computer Engineering, Army Institute of Technology, Pune, Maharashtra)
17. C. Valliyammai, A. Bhuvaneswari, Semantics-based sensitive topic diffusion detection framework towards privacy aware online social networks. Clust. Comput. **22**(1), 407–422 (2019)
18. V. Muthumanikandan, C. Valliammai, Link failure recovery using shortest path fast rerouting technique in SDN. Wirel. Pers. Commun. **97**, 2475–2495 (2017). https://doi.org/10.1007/s11277-017-4618-0

Exploration Study of Ensembled Object Detection Models and Hyperparameter Optimization

Jayesh Gupta, Arushi Sondhi, Jahnavi Seth, Tariq Hussain Sheikh, Moolchand Sharma, and Farzil Kidwai

Abstract Object identification models are becoming more accurate as processing capabilities improve. It is our goal to improve the accuracy of object recognition by the use of several ensembles of distinct state-of-the-art object detection models. The use of single architectures and models to handle object detection challenges has been demonstrated in prior studies; however, each model was later shown to have its own bias and variation. "Ensemble Learning" is currently being studied in recent research after the success of fundamental ensembled models like XGBoost. Ensemble learning in object detection is proposed to be expanded through this research by grouping different permutations of existing models to reduce individual bias and variance while improving metrics, accuracy, and gathering metrics that will aid in hyperparameter optimization for future research on object detection ensembles. It took us a while to find a top-performing ensemble for PASCAL VOC problems.

Keywords Ensemble learning · Computer vision · Deep learning · Object detection

1 Introduction

Artificial intelligence and machine learning include object detection. Current applications include robotics, verification, and automated systems. For a computer vision project to succeed, selecting and implementing an appropriate object detection model is one of the most difficult tasks [1]. There is no rule that says one model must outperform the rest of the pack. A distinct design and set of qualities make each model more effective at solving a specific problem than another. When it comes to machine learning challenges, ensembling is a new industry-standard approach that

J. Gupta · A. Sondhi · J. Seth · M. Sharma (✉) · F. Kidwai
Maharaja Agrasen Institute of Technology, Delhi 110086, India
e-mail: moolchand@mait.ac.in

T. H. Sheikh
Department of Computer Science, Government Degree College Poonch, Poonch, J&K, India

A. K. Bashir et al. (eds.), *Proceedings of International Conference on Computing and Communication Networks*, Lecture Notes in Networks and Systems 394,
https://doi.org/10.1007/978-981-19-0604-6_36

has increased accuracy. Gradient boosted ensembled decision tree models such as XGBoost have shown higher accuracy and efficacy compared to individual models [2]. Weak predictive models will be used by ensembling instead of a single generic predictor. Ensembling synthesizes the predictions and facts of several systems in order to arrive at an accurate appraisal of their capacities. A model's bias and variance can be improved by selecting the right model. Perfectly, accurate models would have minimum bias and low variance error, resulting in a powerful algorithm that can solve issues. The bias-variance standoff is a phenomenon that occurs when these two flaws complement each other, with one fault reducing or amplifying the other's reduction.

For the same task, we will combine conventional object identification models with a variety of hyperparameters to create a powerful learner. To characterize and study the performance of each model when it is applied to the same problem. A similar problem can be solved using any one of these models, but each one has a different variance and bias threshold. It is possible to develop a model that is more successful at dealing with the same issue while showing less bias and volatility than the separate models by merging their predictions. It is possible to acquire a better grasp of the standards metrics established for individual models by looking at a group of object detection models [3–5]. To construct a model that is superior in terms of outcomes and accuracy, we wish to employ our ensembled model. The challenge of object detection models is approached in a novel way in our study, as well as a novel strategy for employing standard object detection is presented. Following are the contributions that this research makes through its efforts:

1. To improve performance and accuracy, we show and expand on the use of numerous ensembled object identification models, which are made up of distinct state-of-the-art object detection models.
2. We compare individual object detection models as well as numerous ensembled object detection models in order to demonstrate the technique's merits and consequences.
3. We have gathered a lot of data to help us design a strategy for making model selection easier and figuring out how ensembling could help with difficult-to-detect items and certain classes.

Recognizing and classifying items are a fundamental component of computer vision, which involves identifying and classifying objects in images. There are many techniques to localize an object, such as using bounding boxes around the object or labeling each pixel in a frame that contains that object. Localization and classification of objects are combined into one task, which is known as object detection. As opposed to localization, classification involves determining the nature of an object by categorizing it with the type and likelihood that it belongs to that category [6]. Localization involves finding regions or sections of a picture that are likely to contain a subject or simply identifying regions of interest or high value in the image [7, 8]. One-stage and two-stage processes are the latest cutting-edge approaches. Among the most popular one-stage techniques is CenterNet, followed by YOLO and SSD, while faster R-CNN is an example of a two-stage technique that emphasizes detection

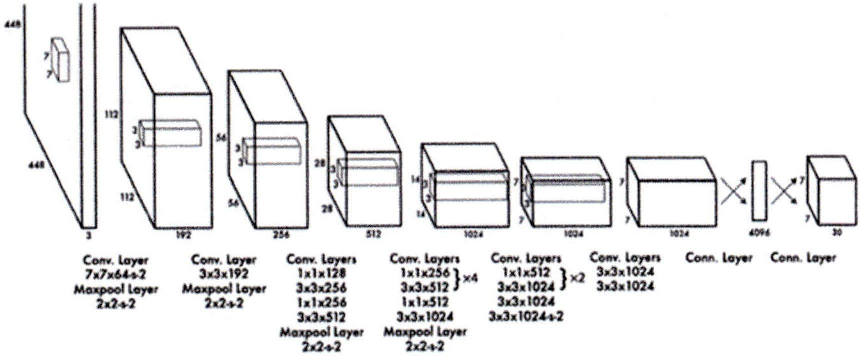

Fig. 1 Architecture of YOLO

precision. In this case, it is arranged as follows: As a result of this, Sect. 2 explores key ideas about the implementation of already known systems, as well as the use of YOLO, SSD, CenterNet, and faster region-based convolutional neural networks. According to the third section, the suggested protocol is described, along with the procedures for executing it. A discussion of the dataset is included in Sect. 4. System implementation is described in detail in section five, along with any conclusions that were drawn from it. Last, but not least, there is the report, together with some concluding remarks and references.

2 Existing Methods

2.1 *You Only Look Once (YOLO)*

Figure 1 illustrates how Yolo organizes the challenge of object recognition as a single regression problem, moving from picture pixels to bounding box coordinates and class probabilities. In order to determine what things are present and where they are located, you merely need to glance at a photograph. Single convolutional networks predict many bounding boxes and their class probabilities at the same time. Training is done on full images with careful optimization of recognition efficacy [9, 10].

2.2 *Single Shot Detector (SSD)*

Unified proposal area network (DPAN) is not present in single shot detector. This predicts the boundary boxes and groups using function mappings in a single pass. For default anchor boxes, it uses tiny convolutional filters to anticipate classes and

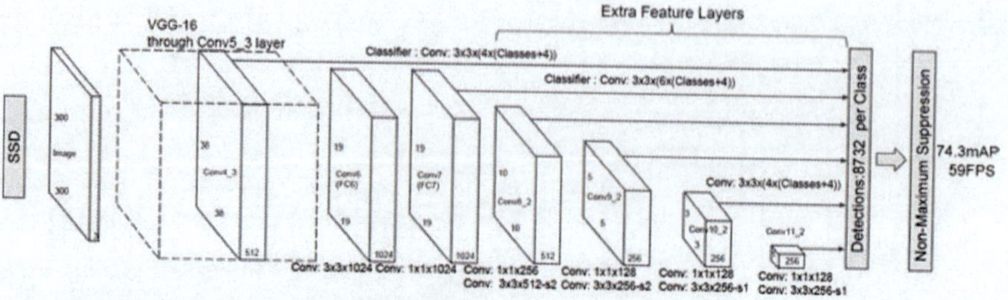

Fig. 2 Architecture of single shot detection (SSD)

offsets of objects as well as filtering to handle aspect ratio changes. A convolutional feed-forward network, shown in Fig. 2, is used to construct a set of bounding boxes that have a predetermined size [11]. A non-maximum suppression stage is then used to generate the final detections.

2.3 *Faster Region-Based Convolutional Neural Networks*

There are two parts to Faster R-CNN: An R-CNN detector that employs the regions provided by the first module is depicted in Fig. 3. For object detection, the entire network functions as a single, unified system. For example, an RPN takes an image as input and outputs an array of rectangular object recommendations, each with its own point value [12].

Fig. 3 Architecture of faster RCNN

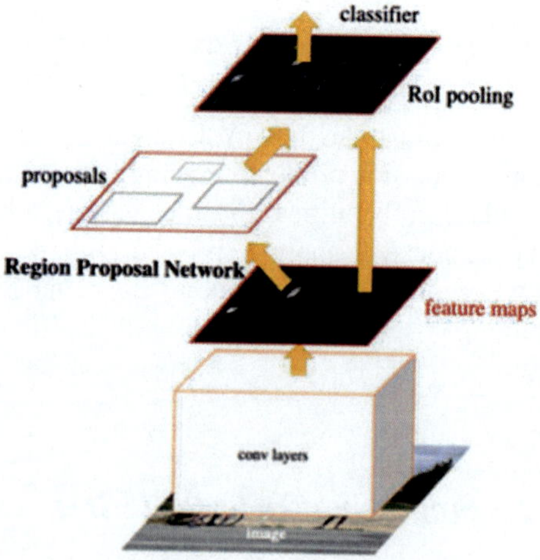

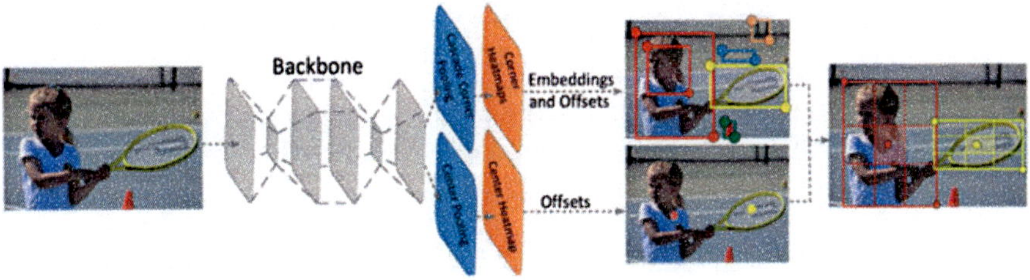

Fig. 4 Architecture of CenterNet

2.4 CenterNet

When an object is detected by CenterNet, it is detected as a triplet of key points rather than as a pair (see Fig. 4). [13] It employs two modules that enhance the information collected by the top left and bottom right corners, as well as providing additional identifying information in the central areas [14]. A predicted bounding box is said to have a high intersection over union with the ground-truth box if it has a high intersection over union with the ground-truth box, and vice versa.

2.5 Ensembling

By integrating independent detectors, assembling is a strategy for increasing detection performance. When it comes to deep learning studies on remote sensing, ensembling methodologies are rarely used. A wide range of machine learning problems benefit from ensemble approaches [15, 16]. It has been shown that combining several algorithms can reduce the bias or variance of a single detector. It is well known that there is a correlation between prediction bias and variance. A large model with a small dataset can yield them, as might data that are too diverse for the model to capture [17]. So, no single model can accurately represent the complete collection of data to be analyzed. Examples of this include algorithm bias-reducing stacking and variance-reducing bagging or test-time augmentation. It is true that the bulk of assembly-based methods require more computing (training and prediction) yet they are an effective way to progress technology. To combine weak learners, there are three types of meta-algorithms to consider: bagging, which combines weak learners that are homogeneous, learning them in parallel, then merging them using some kind of statistically based averaging technique [18, 19].

3 Proposed Method

We used a bagging-based compilation technique to create various combinations of four widely used object recognition models: CenterNet, YOLO, faster RCNN, and single shot detection (SSD). The PASCAL VOC dataset 2012 + 2007 was used as data, and each detector was trained on the entirety of the training and validation data sets. By training on the dataset, we created four learners, each of which learnt various bits and properties of the dataset and had its own bias and variance. Weaker predictions were blended in various ways to obtain a single strong prediction for a picture using weighted boxes. Following affirmative selection of the predictions, bounding box fusion is used to combine overlapping predictions. The weights of the various models in the ensemble were modified to generate various ensemble combinations, which we refer to as our ensembling activation parameters, as illustrated in Fig. 5.

3.1 Weighted Boxes Fusion

Object detection models can be combined using the weighted boxes Fusion approach. While NMS and soft-NMS algorithms discard some predictions, the WBF method constructs average bounding boxes based on confidence ratings of all proposed bounding boxes. This method improves the quality of the predicted rectangles by a significant amount. Neither non-maximum suppression nor soft-non-maximum suppression include all of the boxes [14, 20]. IoU with the ground truth is low in some cases, and the model forecasts all boxes erroneously. When compared to NMS, weighted boxes fusion has a better chance of getting you closer to the truth.

4 Research Approach

4.1 Dataset

In this study, we employed the PASCAL VOC dataset. It provides standardized picture datasets for object detection. When using this dataset, you can compare different methodologies or models based on industry standards. 11,530 images with 27,450 ROI-tagged objects make up the training set. The classes include: airplane, bicycle, boat, bottle, bus, vehicle, cat, chair, cow, dining table, dog, horse, motorbike, person, potted plant, sheep, train, and television, to name a few of them. Both PASCAL VOC 2007 and 2012 have 20 classes apiece; therefore, we integrated them for training. For validation, we used the PASCAL VOC 2007 test dataset with 4952 images [21].

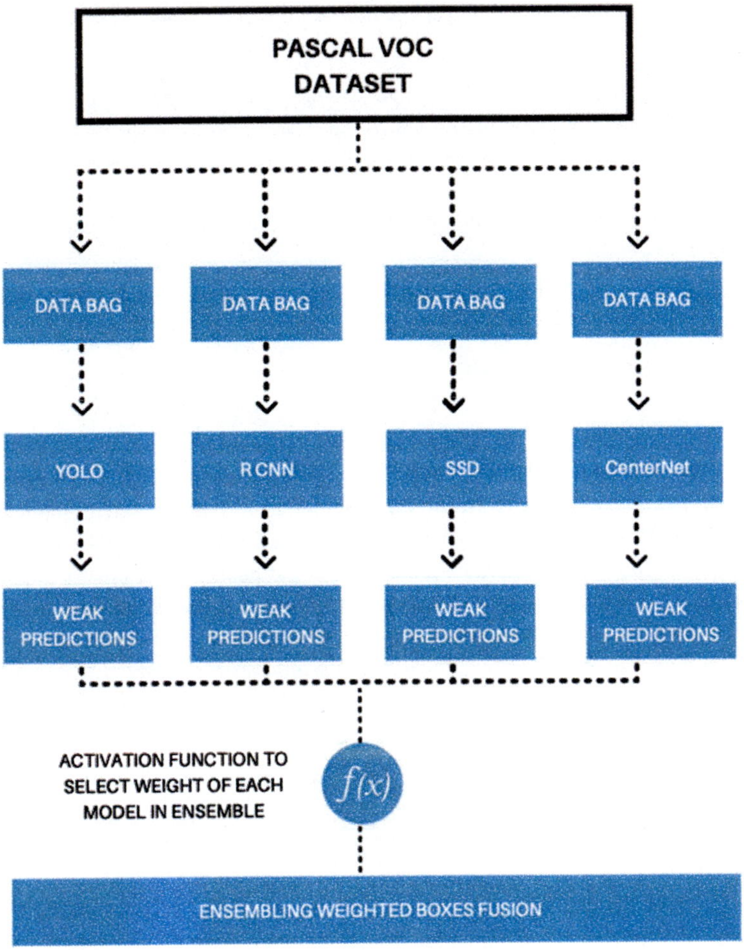

Fig. 5 Proposed ensembled object detection model architecture

4.2 Training

Training and validation datasets for the four models were obtained from the union of PASCAL VOC 2007, 2012, and 2013. It took roughly 200 epochs for each model to reach a preset industry-standard level of category and localization losses. In order to avoid overfitting, training was halted when the accuracy between models decreased between epochs. Written in Python using the Gluon CV module, the training software was developed using Gluon. A file for each image was created in XML format. After each session, the validity accuracy was examined.

4.3 Testing

Using 2007 testing PASCAL VOC data as input, the ultimate test accuracy was calculated. All 4952 images in the test dataset were predicted by each model. A separate text file was created for each image and model containing all the boxes, confidence scores, and labels. A preprocessing step followed the prediction, and then, each predicted file was evaluated and scored. Immediate action was taken after the forecasts were made.

4.4 Prediction Ensembling

It was decided to combine the model predictions for each image after taking note of the predictions made by each model for each photo individually. To create a single graphic, the forecasts from the four different models were combined. All of these predictions were then combined using weighted boxes fusion (WBF), and the resulting predictions were saved in separate files for each image. For the ensemble model [14], these forecasts were used. This was done to eliminate any mismatch between the picture size augmentations performed by each model. The coordinates for weighted boxes fusion were standardized. Each box was resized to its original size before being recorded. This preprocessing was done after the predictions were recorded.

- Box coordinates that exceeded the picture dimensions were constrained to the image dimensions by applying a max and min filter to each anticipated coordinate.
- To decrease false positives, boxes having a confidence value of less than 10% were removed.
- Boxes having an insignificant area or with inconsistencies in format, coordinates were removed.

5 Results and Discussion

5.1 Evaluation Metrics

Two separate metrics were recorded for each model. On the basis of the PASCAL VOC format, these metric computations were made; these two measurements are as follows:

1. Precision-recall curve for each class of the dataset
2. Average precision for each class and model.

Precision and recall were measured at various confidence score thresholds in order to calculate precision-recall. If an intersection over union (IoU) was more than or

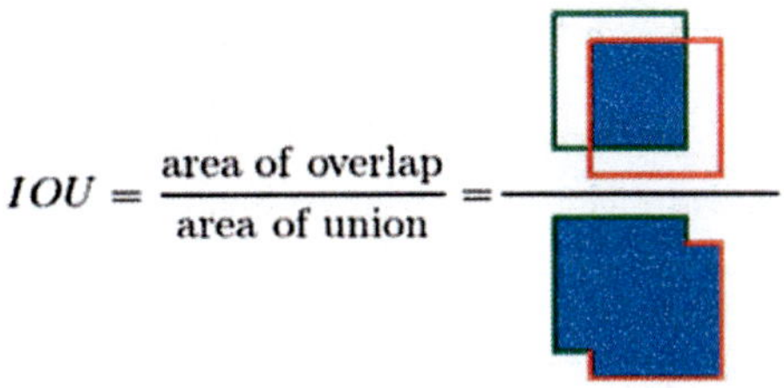

Fig. 6 Intersection over union illustration

equal to 50% in relation to the ground-truth box, a prediction was termed true positive (TP).

5.2 *Intersection Over Union (IOU)*

The Jaccard Index is used to calculate the overlap between two bounding boxes. A truth-bounding box for the truth Bp is required, however. With an IOU threshold, we can determine whether the detection is true (True positive) or not. It is calculated by multiplying the projected bounding box's union area by the ground truth's union area using Eq. (1):

$$\text{IOU} = \frac{\text{area}(B_\text{p} \cap B_\text{gt})}{\text{area}(B_\text{p} \cup B_\text{gt})} \tag{1}$$

Pictured here shown below, i.e., Figure 6 is a ground-truth box (green) and a detected box (blue) (red).

5.3 *Precision X Recall Curve*

For each object class, the precision × recall curve is a great approach to evaluate the output of an object detector. As recall rises, an object detector in a particular class is deemed healthy if accuracy remains high. This means the accuracy and recall will remain high even if the level of trust is changed. Alternatively, you might look for an object detector that only detects essential objects and does not detect objects of everyday life.

This means that a low object detector must detect more objects (leading to more false positives and lower precision) in order to extract all ground-truth items (high recall). Since accuracy decreases with increasing recall, the precision x recall curve begins with high precision values.

5.4 Average Precision

Object detectors can also be compared based on their area under the curve (AUC). A lot of the time, it is difficult to compare several detector curves in the same graph because they often cross each other. As a result, the average precision (AP) statistic can also be used to compare different detectors. Accuracy is calculated by averaging all recall values from zero to one over a period of time. For the first time since 2010, an issue with PASCAL's VOC has changed the way AP is calculated. Currently, all data points are used in the PASCAL VOC challenge for interpolation. In order to duplicate the default implementation of their existing submission, our research technique is in line with their current submission (interpolating all data points).

5.5 Recorded Metrics

Figure 7 and Table 1 demonstrate that the ensembled models outperformed all the individual models, resulting in an average increase in mean average precision of the top-performing individual model by 5–10%. Although the models were able to boost average precision in general, they also managed to increase it by a class-by-class basis. Due to the difficulty of localizing small objects such as bottles and potted plants, separate models were missing them at different times. By assembling, all models built on each other's mistakes, resulting in a considerably greater detection rate for such challenging classes. There was a minor rise across all models for classes where the individual model was able to predict more accurately and with greater ease.

To improve localization and reduce false positives and false negatives, ensemble models were used, resulting in better precision and recall. As can be observed in

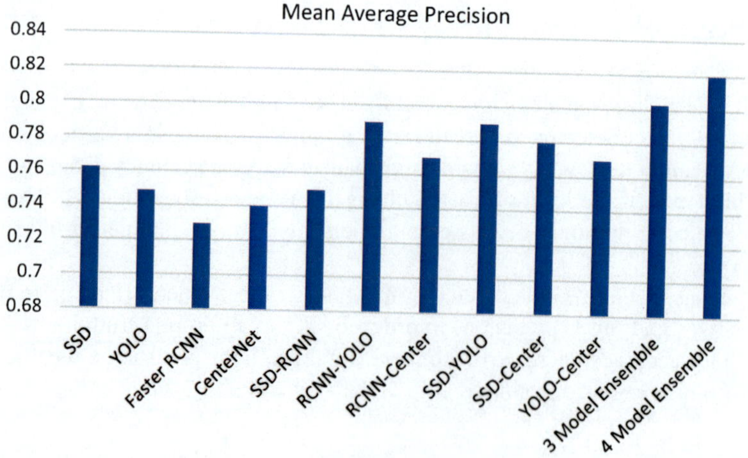

Fig. 7 Comparative plot of object detection model performance

Table 1 Mean average precisions for each model

Model	Mean average precision
SSD	0.7618
YOLO	0.7485
Faster RCNN	0.7294
CenterNet	0.74
SSD-RCNN	0.75
RCNN-YOLO	0.79
RCNN-center	0.77
SSD-YOLO	0.79
SSD-center	0.78
YOLO-center	0.77
3 Model ensemble (YOLO, faster RCNN, and SSD)	0.803
4 Model ensemble (faster RCNN, YOLO, SSD, and CenterNet)	**0.82**

Fig. 8, there were flaws in the models when compared to the ground truth. While RCNN and CenterNet's bounding box predictions were good in comparison with ground truth, they predicted the same class numerous times, resulting in more false positives. However, both the SSD and YOLO variants had problems with localization. All of these flaws reduced the particular model's precision and recall. However, false positives were removed in the ensemble detection, and object localization was improved compared to a single model. As a result, the ensemble model performed far better than any of the individual models. Comparing ensembled models, it was found that a model's number had a direct correlation with its mean average precision. These models can be deemed to be competitive with industry-standard models producing mAP exceeding 80%.

6 Conclusion

As a result of our research, we can confidently assert ensembled models outperform individual state-of-the-art object recognition approaches. There was a significant improvement in both accuracy and mean average precision using the suggested methods, as well as enhanced object localization, as well as a reduction in false positives and negatives. For all intents and purposes, ensembles outperformed individual object detection models.

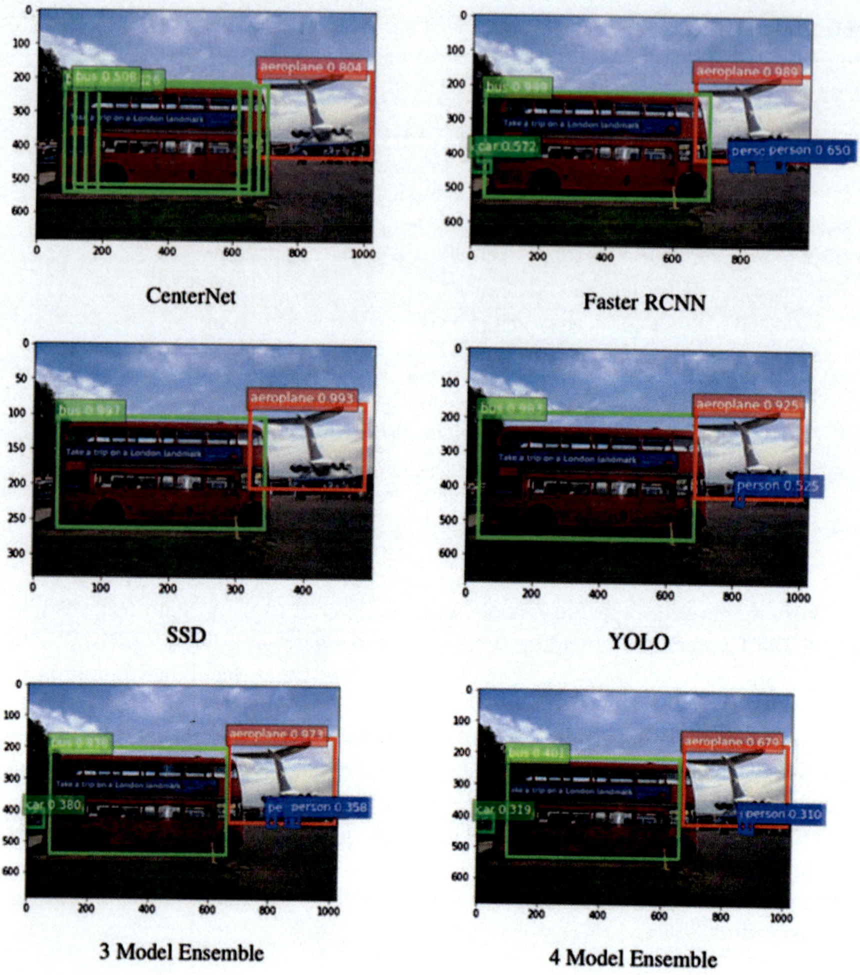

Fig. 8 Comparative predictions of different object detection models

7 Future Scope

There are a number of models, both in terms of architecture and backbones that can be explored as a weak learner in preparation for ensembling. It is possible that different combinations of hyperparameters might be examined to increase the performance of such ensemble models. To better understand how each model should be weighed in an ensemble model, more combinations and weights could be tried.

References

1. A. Groener, G. Chern, M. Pritt, A comparison of deep learning object detection models for satellite imagery, in *2019 IEEE Applied Imagery Pattern Recognition Workshop (AIPR)* (2019), pp. 1–10. https://doi.org/10.1109/AIPR47015.2019.9174593
2. T. Chen, C. Guestrin, XGBoost: a scalable tree boosting system, in *Proceedings of the 22nd ACM SIGKDD International Conference on Knowledge Discovery and Data Mining* (2016), pp. 785–794. https://doi.org/10.1145/2939672.2939785
3. R. Ray, S.R. Dash, Comparative study of the ensemble learning methods for classification of animals in the zoo, in *Smart Intelligent Computing and Applications*, vol. 159, ed. by S.C. Satapathy, V. Bhateja, J.R. Mohanty, S.K. Udgata (Singapore, Singapore, 2020), pp. 251–260. https://doi.org/10.1007/978-981-13-9282-5_23
4. X. Dong, Z. Yu, W. Cao, Y. Shi, Q. Ma, A survey on ensemble learning. Front. Comput. Sci. **14**(2), 241–258 (2020). https://doi.org/10.1007/s11704-019-8208-z
5. L. Rokach, Ensemble-based classifiers. Artif. Intell. Rev. **33**(1–2), 1–39 (2010). https://doi.org/10.1007/s10462-009-9124-7
6. J. Xu, W. Wang, H. Wang, J. Guo, Multi-model ensemble with rich spatial information for object detection. Pattern Recog. **99**, 107098 (2020). https://doi.org/10.1016/j.patcog.2019.107098
7. Z.-Q. Zhao, P. Zheng, S.-T. Xu, X. Wu, Object detection with deep learning: a review. IEEE Trans. Neural Netw. Learning Syst. **30**(11), 3212–3232 (2019). https://doi.org/10.1109/TNNLS.2018.2876865
8. Y. Wu et al., Rethinking classification and localization for object detection, in *2020 IEEE/CVF Conference on Computer Vision and Pattern Recognition (CVPR)* (Seattle, WA, USA, 2020), pp. 10183–10192. https://doi.org/10.1109/CVPR42600.2020.01020
9. J. Redmon, A. Farhadi, YOLOv3: an Incremental improvement. arXiv:1804.02767 [cs], Apr. 2018, [Online]. Available: http://arxiv.org/abs/1804.02767.
10. J. Redmon, S. Divvala, R. Girshick, A. Farhadi, You only look once: unified, real-time object detection, in *2016 IEEE Conference on Computer Vision and Pattern Recognition (CVPR)* (Las Vegas, NV, USA, 2016), pp. 779–788. https://doi.org/10.1109/CVPR.2016.91
11. W. Liu et al., SSD: single shot MultiBox detector. arXiv:1512.02325. [cs], vol. 9905, pp. 21–37 (2016). https://doi.org/10.1007/978-3-319-46448-0_2
12. S. Ren, K. He, R. Girshick, J. Sun, Faster R-CNN: towards real-time object detection with region proposal networks. IEEE Trans. Pattern Anal. Mach. Intell. **39**(6), 1137–1149 (2017). https://doi.org/10.1109/TPAMI.2016.2577031
13. K. Duan, S. Bai, L. Xie, H. Qi, Q. Huang, Q. Tian, CenterNet: keypoint triplets for object detection, in *2019 IEEE/CVF International Conference on Computer Vision (ICCV)* (Seoul, Korea (South), 2019), pp. 6568–6577. https://doi.org/10.1109/ICCV.2019.00667
14. R. Solovyev, W. Wang, T. Gabruseva, Weighted boxes fusion: ensembling boxes for object detection models. arXiv:1910.13302. [cs] (2020). [Online]. Available: http://arxiv.org/abs/1910.13302
15. P. Singh, Comparative study of individual and ensemble methods of classification for credit scoring, in *2017 International Conference on Inventive Computing and Informatics (ICICI), Coimbatore* (2017), pp. 968–972. https://doi.org/10.1109/ICICI.2017.8365282
16. Y. Ren, L. Zhang, P.N. Suganthan, Ensemble classification and regression-recent developments, applications and future directions [Review Article]. IEEE Comput. Intell. Mag. **11**(1), 41–53 (2016). https://doi.org/10.1109/MCI.2015.2471235
17. O. Sagi, L. Rokach, Ensemble learning: a survey. WIREs Data Mining Knowl. Discov. **8**(4) (2018). https://doi.org/10.1002/widm.1249
18. B. Ghojogh, M. Crowley, the theory behind overfitting, cross-validation, regularization, bagging, and boosting: tutorial. arXiv:1905.12787. [cs, stat], May 2019, [Online]. Available: http://arxiv.org/abs/1905.12787
19. S. González, S. García, J. Del Ser, L. Rokach, F. Herrera, A practical tutorial on bagging and boosting based ensembles for machine learning: algorithms, software tools, performance

study, practical perspectives and opportunities. Inf. Fusion **64**, 205–237 (2020). https://doi.org/10.1016/j.inffus.2020.07.007

20. N. Bodla, B. Singh, R. Chellappa, L.S. Davis, Soft-NMS—improving object detection with one line of code, in *2017 IEEE International Conference on Computer Vision (ICCV)* (Venice, 2017), pp. 5562–5570. https://doi.org/10.1109/ICCV.2017.593
21. J. Gupta, A. Sondhi, J. Seth, M. Sharma, F. Kidwai, A. Jain, EnSOTA: ensembled state of the art model for enhanced object detection, in *International Conference on Innovative Computing and Communications. Advances in Intelligent Systems and Computing*, vol. 1394, ed. by A. Khanna, D. Gupta, S. Bhattacharyya, A.E. Hassanien, S. Anand, A. Jaiswal (Springer, Singapore, 2022). https://doi.org/10.1007/978-981-16-3071-2_57

Evaluation of DSRC and LTE-V2x: Need for Next-Generation V2X Communication Systems

Zeeshan Hameed Mir and Fethi Filali

Abstract Predominantly, there are two wireless technologies for providing vehicle-to-everything (V2X) communications, i.e., ad hoc IEEE 802.11p and cellular LTE-V2x. This paper provides a qualitative and quantitative evaluation of both technologies. The qualitative evaluation aims to measure how these technologies meet the functional requirements of the V2X applications. To obtain a quantitative analysis, an extensive simulation-based performance evaluation is presented of the most common communication metrics such as latency, packet delivery ratio, and network load. The overall evaluation shows that the IEEE 802.11p provides superior support for safety-critical V2V applications in low-load scenarios. Similarly, LTE-V2x offers extensive support for services requiring infrastructure assistance and network coverage. This paper also provides a better understanding of the apparent need for evolutionary next-generation communication technologies to meet the future V2X application requirements

Keywords IEEE 802.11p · DSRC · LTE-V2X · Next-Generation V2X

1 Introduction

The vehicular communication technologies enable vehicles to connect to everything, i.e., vehicle-to-everything (V2X). The V2X consists of four types of communications, including vehicle-to-vehicle (V2V), vehicle-to-infrastructure (V2I), vehicle-to-network (V2N), and vehicle-to-pedestrian (V2P). The V2X applications vary from road safety and traffic efficiency to enhancing the in-vehicle experience and, more recently, paving the path to enhanced V2X (eV2X) use cases such as advanced driver-

Z. Hameed Mir (✉)
Faculty of Computer Information Science, Higher Colleges of Technology (HCT), PO Box 4114, Fujairah, United Arab Emirates
e-mail: zhameed@hct.ac.ae

F. Filali
Qatar Mobility Innovations Center (QMIC), Qatar University, Box 210531 Doha, Qatar
e-mail: filali@qmic.com

A. K. Bashir et al. (eds.), *Proceedings of International Conference on Computing and Communication Networks*, Lecture Notes in Networks and Systems 394,
https://doi.org/10.1007/978-981-19-0604-6_38

assistance systems (ADAS). These applications and use cases pose a diverse set of functional and performance requirements. These requirements must be met simultaneously to provide reliable active safety, improved traffic efficiency, and situational awareness.

The IEEE 802.11p [1] enabled direct short-range communication (DSRC) [2] and fourth generation/long-term evolution (4G/LTE) by third generation partnership project (3GPP) enabled LTE-V2x [3] are the two major V2X communication technologies. IEEE 802.11p is an altered version of IEEE 802.11a, primarily to support high-speed vehicular mobility, reduced response time, and extended range. The standard utilizes orthogonal frequency division multiple access (OFDM) at the physical (PHY) layer and the carrier sense multiple access with collision avoidance (CSMA/CA) protocol at the medium access control (MAC) layer with the enhanced distributed channel access (EDCA) to prioritize channel access according to the traffic categories. In 2015, the 3GPP initiated the V2x study item to explore the feasibility of LTE-based services. The standard defines two types of communications, direct communication over the PC5 interface and network communication using the Uu interface. Direct communication is based on device-to-device (D2D) functionality with two transmission modes. Transmission mode 3 requires infrastructure assistance to allocate resources, whereas transmission mode 4 allows resources allocation without the support of infrastructure, i.e., evolved Node B (eNodeB).

Several empirical and theoretical comparative studies of IEEE 802.11p and LTE-V2x are presented in the literature. However, the focus has been primarily on comparing IEEE 802.11p and LTE-V2x PC5 interface [4, 5]. The main contribution of this paper is to evaluate IEEE 802.11p and LTE-V2x, both qualitatively and quantitatively. Seven V2X application functional requirements were selected given in [6], to perform the qualitative evaluation. For the quantitative analysis, the emphasis is on assessing the scalability of the IEEE 802.11p and LTE-V2x Uu interface under high data traffic scenarios. The simulation results show that in high load and vehicle densities environments, the performance of IEEE 802.11p degrades considerably. Similarly, for the LTE-V2x, the cell load increases significantly as background and vehicular data traffic increases, requiring more radio resources.

The rest of this paper is organized as follows. Section 2 presents a comparison between IEEE 802.11p and LTE-V2x in terms of several functional requirements. Section 3 describes the IEEE 802.11p and LTE-V2x Uu simulation results in high load and vehicle density scenarios. Finally, the conclusion is given in Sect. 4.

2 Current V2X Communications Systems

In the following sub-sections, we review IEEE 802.11p and LTE-V2x technologies concerning seven functional requirements of V2X applications. Table 1 summarizes the qualitative comparison between the two technologies.

Table 1 Qualitative analysis of the V2X communication technologies

Classes	IEEE 802.11p	LTE-V2x
Operation in absence of network	✓ ✓ ✓	✓
Support of V2V	✓ ✓ ✓	✓
Support of safety-critical use cases	✓ ✓ ✓	✓
Support of V2I/I2V	✓ ✓	✓ ✓ ✓
Support of multimedia services	✓ ✓	✓ ✓ ✓
Network coverage	✓	✓ ✓ ✓
Advanced PHY	✓	✓ ✓ ✓

✓: Rarely meet requirements [Not suitable], ✓ ✓: Do meet requirements (but not in all conditions) [Somewhat suitable], ✓ ✓ ✓: Almost Always requirements (almost in all conditions) [Suitable]

2.1 IEEE 802.11p for Standalone Deployment

Operation in the absence of network: The IEEE 802.11p-equipped vehicles can communicate uncoordinated in an ad hoc fashion without relying on the presence of a network or the intervention from a centralized network entity.

Support of V2V: Most awareness and warning use cases require an immediate broadcast of notification messages to all vehicles in the surrounding area. Therefore, a direct IEEE 802.11p-based V2V link will be necessary for communication among the nearby vehicles to meet the latency requirements.

Support of safety-critical use cases: With seven channels dedicated licensed spectrum, the IEEE 802.11p standard has been designed explicitly for safety-critical application and use cases keeping their low latency and security requirements in consideration.

Support of V2I/I2V: Typically, at intersections, the beyond-line-of-sight communication is achieved by the V2I/I2V mode of information transmission. However, this requires installing the roadside units (RSUs) to relay messages, thus enhancing the coverage and the probability of message delivery. While RSUs can be deployed using the existing roadside infrastructure such as traffic signs and traffic lights, city-wide deployment is considered inevitable but not imminent.

Support of multimedia services: Several of the infotainment and convenience applications require higher data rates and Internet access. Especially, if multimedia contents have to be distributed through the IEEE 802.11p link, the RSUs must connect to a fixed network or the Internet access via a gateway.

Network coverage: The network coverage is very challenging due to the high 5.9GHz frequency band, thus making the standard less attractive for infrastructure deployment [7]. Moreover, comparatively shorter communication ranges, especially high-speed scenarios, are the leading reasons behind smaller link duration and sporadic connectivity.

Advanced PHY: Vehicular propagation channels have frequency and time-selective fading due to multipath/delay spread and mobility, respectively [8]. The vehicular channel tends to exhibit a more significant delay spread and a higher Doppler spread that is inversely proportional to channel coherence bandwidth and time, respectively [9]. The standard employs the OFDM technique to gain spectral efficiency. However, high mobility aggravates the impact of Doppler's spread on OFDM, causing a higher packet error rate and lower channel capacity. Moreover, sparse pilot design as defined in IEEE 802.11p standard, i.e., the adjacent pilot subcarriers space is more than 2 MHz (larger than the coherence bandwidth) coupled with longer packet transmission time (longer than coherence time), results in obsolete channel estimate. Thus, making it challenging to adapt transmissions according to the current channel conditions [10]. From the system performance point-of-view, these factors have a significant impact on the reliability of the data transmission [8, 10]. Other relevant issues include channel congestion in high-density vehicle deployment and asymmetric links due to higher relative speed. Both of these issues result in lower throughput and higher end-to-end latency.

2.2 *Cellular LTE-V2x for Infrastructure-Assisted Deployment*

Operation in the absence of network: Infrastructure-less V2X operation over 4G/LTE relies on cellular uplink technology and wireless access network infrastructure deployment. A message transmitted by an information source needs to traverses to the base station (i.e., eNodeB) first. Also, possibly involve core network elements (i.e., Evolved Packet Core or EPC) and other specialized network entities (e.g., GeoServer [11]) before it is forwarded toward all potential information sinks. Alternatively, mechanisms defined for transmission mode 4 can be used in the absence of a mobile cellular network.

Support of V2V: While the IEEE 802.11p-based V2V communication is distributed in nature, the centralized architecture of the 4G/LTE cellular network lacks native V2V communication support [12] due to the infrastructure assistance required to transmit information to the vehicles. Transmission mode 3 can be used to support V2V communication. Still, it requires intervention from the network to allocate resources for vehicles requesting a direct connection.

Support of safety-critical use cases: Although cellular networks are capable of low-latency transmissions, their functions and processes are not explicitly designed to meet safety-critical use cases and stringent latency requirements. This is particularly true when the uplink and down channels carry voice and data simultaneously from other user equipment (UEs), i.e., background traffic (BT). The unicast downlink transmissions would result in excessive delays and demand high data bandwidth depending on the number of vehicles and the offered network load by other UEs in the cell. The use of a technique like Evolved Multimedia Broadcast Multicast Services (eMBMS) (also LTE broadcast) can mitigate downlink delay and capacity issues [13]; however, eMBMS protocol deployment is well underway.

Support of V2I/I2V For most non-safety-related use cases involving the infrastructure, i.e., V2I or I2V communication, 4G/LTE is the wireless access technology of choice since it best meets the data rate, longer-range, latency, and mobility requirements.

Support of multimedia services: The multimedia content transmission requires data rates of several hundreds of kbps or even Mbps, depending on the type and quality of data requested for download or streaming. The 4G/LTE technology is well suited for high-speed multimedia download since it offers the required performance and bandwidth.

Network coverage: The 4G/LTE networks operate at frequencies between 700 MHz and 2600 MHz, offering wide coverage ranges. The cell coverage range varies from few meters across to a cell with several kilometers of radii with comparatively lower rates delivering superior coverage.

Advanced PHY: The LTE cellular network employs several techniques to satisfy two leading requirements, i.e., high transmission rate and spectral efficiency. To achieve this, LTE utilizes OFDMA and single carrier-frequency division multiple access (SC-FDMA) schemes for downlink and uplink transmissions, respectively. For LTE downlink, higher reliability and data rates are attained by combining OFDMA with the multiple input multiple output (MIMO) technique. Whereas, SC-FDMA for LTE uplink transmission offsets associated loss of efficiency caused by high peak-to-average power ratio (PARP) [14]. Additionally, methods like adaptive modulation and coding, and frequency selective scheduling improve the system throughput performance.

3 Performance Evaluation of IEEE 802.11p and LTE-V2x Uu

The simulation environment represents an urban scenario, combining network and road traffic simulators. During the simulation, the traffic load is increased by increasing the number of vehicles or the beaconing frequency of the Cooperative Awareness Messages (CAMs). The details regarding the simulation tool and setup are given in [15]. Figure 1a shows the impact of beaconing frequency on delay and packet delivery ratio (PDR). As the beaconing frequency increases, the delay increases, and PDR decreases. For example, 10 Hz, only 20% of transmissions reached destinations with 100ms or lesser delays. In contrast, over 60% of beacons were received with 500ms

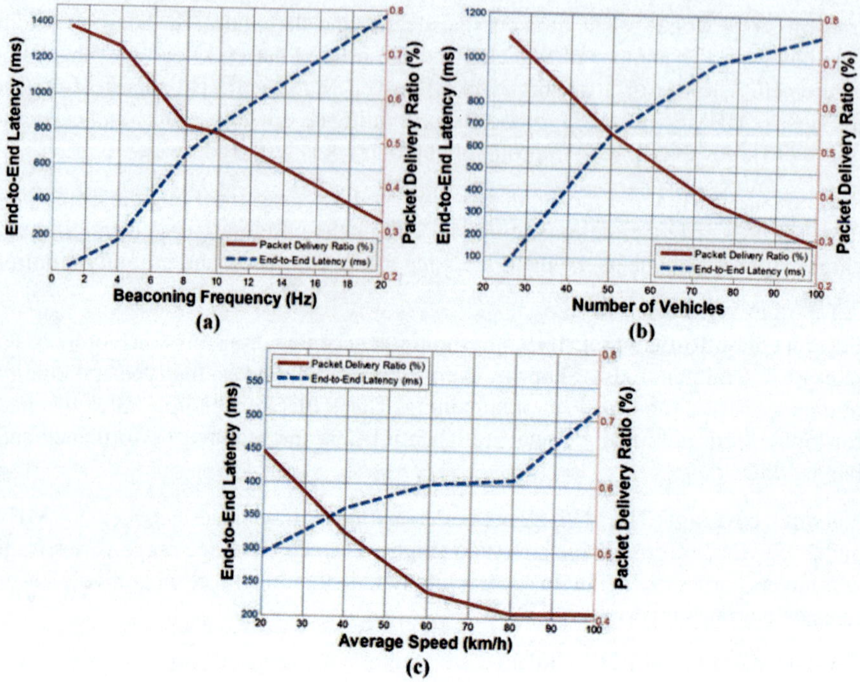

Fig. 1 IEEE 802.11p performance evaluation. Impact of **a** Beaconing frequency (Hz), **b** number of vehicles and **c** average speed (km/h) on end-to-end delay (ms) and packet delivery ratio (%)

or above delays. An excessive number of beacon transmissions without the decentralized congestion control (DCC) mechanism resulted in more packet drops and collisions, thus lower PDR as the beaconing frequency increases.

Figure 1b shows how the increase in vehicles in the region affects the IEEE 802.11p performance in delay and PDR. The beaconing frequency remains unchanged 8 Hz. Only the vehicular density in the simulation area is increased. The net effect is the same, i.e., the more the vehicles transmit beacons simultaneously, the fewer the beacons reaching the destination successfully, while retransmissions, medium access contention, and queue waiting contributed significantly toward higher delays. Mobility is the hallmark of vehicular networks where limited coverage, rapid topological changes, and environmental conditions often bring an adversary to communication performances. Figure 1c shows the delay and PDR as the average speed increases. Faster moving vehicles tend to contend for the shared channel for longer, thus incurring more delays. For instance, at 60 km/h speed, over 80% of beacons were transmitted with lower than 100 ms delay, whereas at 80 km/h, only 60% of beacons were delivered with 100ms or lower delays. Vehicles moving more rapidly also attained lower PDR because of frequent network fragmentation at higher speeds.

To investigate how vehicular traffic impacts the cellular network, we implemented a realistic reference scenario consisted of a macro-cellular LTE network with 48 cells

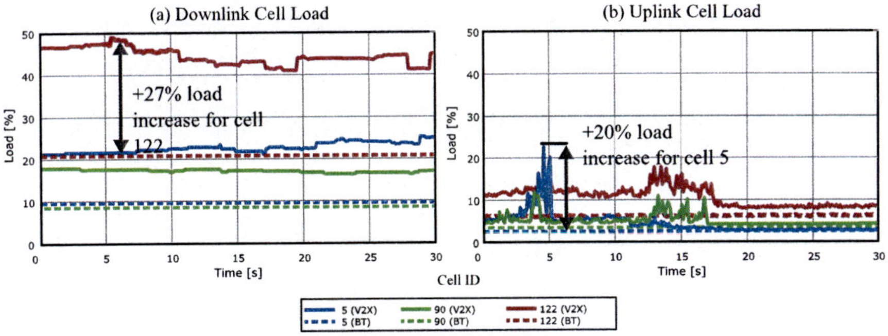

Fig. 2 LTE-V2x Uu performance evaluation. **a** Downlink and **b** uplink cell load of the exemplary cells 5, 90, and 122 in the simulation scenario when serving only the background traffic (BT) (shown as dashed lines) and after adding vehicular (V2X) traffic (shown as solid lines)

[15]. Each cell operated at 800 Mhz with both static background and mobile vehicular subscribers generating background traffic (BT) and vehicular (V2X) traffic. At first, the network load of all cells inside the scenario was evaluated considering only the background subscribers, i.e., BT traffic, and second, after adding vehicular traffic, i.e., V2X traffic. Moreover, the up- and downlink cell load was analyzed in detail for the simulated reference scenario. The cell load is defined as the ratio between the percentage of occupied and available resources. The results for the up- and downlink cell load of three selected cells covering the simulation scenario are shown in Fig. 2. The dotted lines indicate the cell load, which served only the background subscribers. For the first simulation setup without the V2X traffic, the maximum loaded cell 122 has a downlink load of 21% while cell 90 and cell 5 have a load of 9% and 10%, respectively. All cells are less loaded in uplink based on the asymmetric uplink–downlink traffic ratio. Since the background subscribers are static, and the throughput request is based on a constant bit rate model for up- and downlink, the cell loads do not vary over time.

After adding the V2X traffic, a heavy load increase for all LTE cells inside the scenario can be observed. For cell 122, the load increases from 21 to 48% at time 5.6 s, while on the other side, the maximum gain for the uplink cell load can be observed at time 13.5 s, showing an increase from 6.2 to 17.8%. Figure 2a shows that the load increases much more for the downlink than for the uplink after adding vehicular subscribers. This is because one CAM message transmitted over a unicast uplink is broadcasted in multiple cells leading to increased downlink traffic. Furthermore, for eMBMS, a robust modulation and coding scheme (MSC) is chosen to ensure a high probability for successful decoding. A vehicle with a downlink signal-to-noise ratio (SINR) greater or equal to 5dB can decode the messages for this simulation.

For the uplink transmission, the network load strongly depends on the time-dependent uplink SINR of each user when transmitting a CAM message. Based on the current link condition, the uplink cell load can vary enormously within a few time steps, as shown in Fig. 2b. For cell 5, the load increases from 6.1% at time 2.7 s to 22.6% at time 4.6 s and decreases again to 5% at time 5.3 s. The system-level evaluation points out that up- and downlink load for a realistic scenario when applying eMBMS performs differently. The downlink cell load increases for all cells significantly while the temporal development of the cell load remains almost constant. However, the results for the uplink demonstrate a slight load increase for the uplink cell load but a high load fluctuation over time. These scenarios demand a more robust MSC with a lower code rate. Consequently, it requires more radio resources for the same amount of data transmission.

4 Conclusion

In this paper, we have reviewed and evaluated the IEEE 802.11p and LTE-V2x. The IEEE 802.11p is a comprehensive technology capable of supporting safety and time-critical applications. However, the standard has extreme scalability and reliability issues in the presence of high traffic load, vehicle densities, and speed environment. Utilizing cellular technologies to facilitate V2I and V2N, which involve mobile network infrastructure and access to cloud-enabled services, makes LTE-V2x a desirable alternative technology to provide V2X communications. However, the presence of a higher number of background subscribers and vehicular traffic demands higher data rates. There is a requirement for an evolutionary path to next-generation to meet the higher throughput, reduced latency, better reliability, and extended range of the future V2X applications requirement. The primary design goal of IEEE 802.11bd is to support $2\times$, i.e., double the throughput, relative speed, and range while remaining interoperable with other existing technologies. Similarly, the emerging 5G NR C-V2X introduces several advanced features to achieve twofold spectral efficiency and higher capacity while reducing the transmission latency. Future work will identify the critical functional and performance requirement of the V2X applications and analyze to which extent different communication technologies support them.

References

1. IEEE Std 802.11-2016, *IEEE Standard for Information technology–Telecommunications and information exchange between systems Local and metropolitan area networks— Part 11: Wireless LAN Medium Access Control (MAC) and Physical Layer (PHY) Specifications* (2016)
2. J.B. Kenney, Dedicated short-range communications (DSRC) standards in the United States. IEEE **99**(7), 1162–1182 (2011)
3. 3GPP TR 22.885, *Study on LTE Support for Vehicle to Everything (V2X) Services*, (Release 14), 3GPP Technical Specification Group Radio Access Network, v1.0.0 (2015)
4. T. Shimizu, H. Lu, J. Kenney, S. Nakamura, *Comparison of DSRC and LTE-V2X PC5 Mode 4 Performance in High Vehicle Density Scenarios,* 26th ITS World Congress (2019)
5. K.Z. Ghafoor, M. Guizani, L. Kong, H.S. Maghdid, K.F. Jasim, Enabling efficient coexistence of DSRC and C-V2X in vehicular networks. IEEE Wireless Commun. **27**(2), 134–140 (2020)

6. A. Filippi, *Wireless Connectivity in Automotive*, CWTe 2016 Research Retreat, Centre for Wireless Technology Eindhoven (CWTe) (2016)
7. 5GAA, *The Case for Cellular V2X for Safety and Cooperative Driving*, White Paper (2016)
8. A.F. Molisch, F. Tufvesson, J. Karedal, C.F. Mecklenbrauker, A survey on vehicle-to-vehicle propagation channels. IEEE Wireless Commun. **16**(6), 12–22 (2009)
9. L. Liang, H. Peng, G.Y. Li, X. Shen, Vehicular communications: a physical layer perspective. IEEE Trans. Veh. Technol. **66**(12), 10647–10659 (2017)
10. J. Li, M. Wódczak, X. Wu, T.R. Hsing, Vehicular networks and applications: challenges, requirements and service opportunities, in *2012 ICNC* (2012)
11. A. Festag, M. Wiecker, N. Zahariev, Safety and traffic efficiency applications for GeoMessaging over cellular networks, in *19th ITS World Congress* (2012)
12. G. Araniti, C. Campolo, M. Condoluci, A. Iera, A. Molinaro, LTE for vehicular networking: a survey. IEEE Commun. Mag. **51**(5), 148–157 (2013)
13. Z.H. Mir, F. Filali, LTE and IEEE 802.11p for vehicular networking: a performance evaluation, in *EURASIP JWCN*, vol 89, pp 1–15 (2014)
14. S.H. Sun, J.I. Hu, Y. Peng, X.M. Pan, L. Zhao, J.Y. Fang, Support for vehicle-to-everything services based on LTE. IEEE Wireless Commun. **23**(3), 4–8 (2016)
15. N. Dreyer, A. Moeller, J. Baumgarten, Z.H. Mir, T. Kuerner, F. Filali, On building realistic reference scenarios for IEEE 802.11p/LTE-based vehicular network evaluations, in *2018 IEEE 87th Vehicular Technology Conference (VTC Spring)* (2018)

A Novel Secure Data Processing Mechanism in IoT Using Deep Learning with Ontology

Auxilia Michael, K Raja, Kannan Kaliyan, and Rajakumar Arul

Abstract Security has become the biggest concern in recent technologies like the Internet of things (IoT) as it involves heterogeneous users and resources sharing sensitive data through cloud. These heterogeneous users and service providers have multifarious privacy and security requirements and need a common mechanism to share the same across the heterogeneous environments. Ontology proves to be an efficient means to handle heterogeneous environments for the following reasons: It provides simple means to share domain knowledge among entities, it is simple to reuse domain knowledge, and it facilitates convenient means to manage and manipulate domain entities and interrelationships among entities. The performance of ontologies is determined by their reasoning ability. In IoT, many devices are involved, and hence, multiple ontologies are to be involved. Still, most of the works mainly focus on only single ontology for reasoning. They lack considering reasoning involving multiple ontologies. This article proposed a deep learning method for associating various ontology rule bases and thus learning new inference rules and thereby providing efficient security in IoT applications. To verify the usefulness of the proposed work, it is realized in healthcare application and proves to achieve better security.

Keywords Ontology · IoT applications · Heterogeneous environment · Cloud · Domain knowledge

A. Michael (✉)
Department of Artificial Intelligence and Data Science, Sri Manakula Vinayagar Engineering College, Puducherry, India
e-mail: auxiliaaids@smvec.ac.in

K. Raja
Department of CSE, SRM Institute of Science and Technology, Ramapuram, Chennai, India
e-mail: rajak1@srmist.edu.in

K. Kaliyan
Department of CSE, Sree Vidyanikethan Engineering College, Tirupati, India
e-mail: kannan.k@vidyanikethan.edu

R. Arul
Department of CSE, Vellore Institute of Science and Technology, Chennai, India
e-mail: rajakumar.arul@vit.ac.in

A. K. Bashir et al. (eds.), *Proceedings of International Conference on Computing and Communication Networks*, Lecture Notes in Networks and Systems 394,
https://doi.org/10.1007/978-981-19-0604-6_39

1 Introduction

The commonly useful connection between the Internet of things (IoT) and artificial intelligence (AI) is empowering troublesome advancements in wearables and implantable biomedical gadgets for healthcare; shrewd observation and checking applications like the utilization of a self-governing drone for calamity management and salvage tasks [1]. The combination of AI and IoT empowers the frameworks to be prescient, prescriptive and self-sufficient. This combination of AI and IoT is advancing the idea of arising applications from being helped to expanded and at last to self-sufficient knowledge. This continuum affects all enterprises going from assembling, retail, medical care, media transmission, and transportation and so on. IoT sensors will permit the assortment of a tremendous measure of information, though AI can assist with determining insight for conceiving more brilliant applications for a more intelligent world [2]. In addition, the arising 5G technology gives an establishment to understanding the maximum capacity of AI-controlled IoT. The enormous availability offered by 5G alongside super-low latency capacity will open up roads for bracing applications across all verticals.

The emerging era of artificial intelligence and Internet of things comprises of three major parts: (i) brilliant gadgets, (ii) savvy systems of systems and (iii) end-to-end analytics. Innumerable complications may arise while deploying such innovative, sensitive and data centric environments incorporating various strategies and algorithms to gratify quality of service requisites such as latency, transfer speed and stall; it is mandatory to design mechanisms for protecting IoT data and suggest secured support for clients realizing the applications; it is necessary to choose models capable of balancing both huge volume of data and fast reactive IoT data utilizing Intelligent edge platform. Additionally, from the application point of view, it is still mandatory to plan a compliant and deep processing of IoT data by making use of an integrated learning mechanisms, and also, it is necessary to find the concepts collaboratively to achieve high intelligence about the multifarious environments and users.

Ontology is used for describing and organizing domain-specific knowledge [3]. Construction of ontology, its life cycle and development are the main focus in its research area. Knowledge acquisition, knowledge representation and reasoning are crucial tasks. Ontology is used for the following reasons: 1) it provides a simple way to share domain knowledge among entities; 2) it makes it simple to reuse domain knowledge; and 3) it provides a convenient way to manage and manipulate interrelationship among the domain entities. Reasoning involves the process of combining ontology and rules for expressing explicit and implicit knowledge in a domain. The various semantic interrelationships in an ontology can be reduced into a problem of subsumption resulting in the extraction of implicit knowledge, and new inference rules can be derived from the existing rules. Also, it can result in new properties which are necessary for reasoning but need not be explicitly specified in an ontology. This is possible in a single ontology. Since IoT can involve multivarious devices and users, it is necessary to deal with multiple ontologies. Hence, a mechanism is needed to realize multiple ontology interrelationships; thereby, many new semantic interrelationships can be mined and new rules are to be inferred from multiple ontology rule bases [4].

Therefore, multiple ontology-based reasoning methods employing deep learning are proposed in this paper. This method normalizes values of the arity of parameters in the inference rule database and hence resulting in the reduction of setting parameters manually and evading the setting of some unreasonable parameters in the reasoning process. Recurrent neural network is then utilized to realize the semantic interrelations among the multiple ontologies, and thereby discovering new rules and also elaborate the rule base for knowledge reasoning.

The rest of this article is ordered in the following manner. Section 2 elaborates various works proposed by the scientific community about the use of deep learning dealing with multiple ontologies. Section 3 explains the proposed work handling multiple ontologies with the help of recurrent neural networks. Section 4 exhibits case study of healthcare IoT application and experimental analysis and results. Finally, in Sect. 5, conclusion and future work are proposed.

2 Related Work

In this section, a detailed description of the solutions that have already been proposed by the scientific community for securing the IoT 1 have been presented. It also includes the other works for health care data by the scientific community. Ontologies are mainly supportive for specifying the conceptualization and interrelations of a domain of knowledge formally [4] from which specific domain objects (e.g., users and resources) are defined as instances of this conceptualization.

Ontology is a prescribed portrayal of a model comprising of collective concepts [4]. Ontology can be useful information retrieval [5], artificial intelligence, NLP and so on. Various general ontology library systems, viz WordNet, DBpedia, Cyc, etc., and a wide range of domain ontology library systems have been developed to satisfy requirements from both industry and academia.

Prasad et al. [6] have devised a method based on Bayesian, while Doan et al. [7] have coined a method based on probability distribution in the mapping process. Heuristic rules are used to map ontologies in a work done by Ehrig et al. [8]. Gruber [9] introduced an ontology-based classification method employing the decision tree classier method for multi-source classification. But the research challenge is how to suitably pick the- appropriate mapping method according to their characteristics when more than two ontologies are involved. As an answer, a wide range of selection methods has been proposed at this stage. State-of-the-art approaches for distribution of weights include: approaches focusing on convict sets [10], approaches focusing on logical hierarchy [11], approaches focusing on trilateral fuzzy figures [12], approaches focusing on decision making relying upon entropy weight.

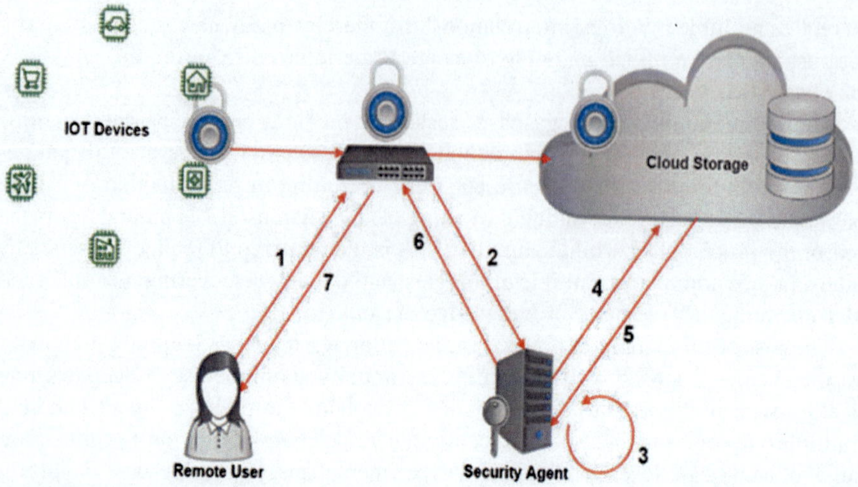

Fig. 1 End-to-end security for the healthcare IoT

3 Proposed Work

In this section, the deep learning-enabled ontology approach has been proposed to secure data in IoT applications. The basic IoT concept involved multifarious users, devices, and sensors. Due to these reasons, multiple ontologies must be involved. The reasoning is a vital task in any ontology-based approach. But the problem is that reasoning in a single ontology consumes more time. Thus, a machine learning concept is being introduced to reason multiple ontologies. The proposed model is given in Fig. 1. The scientific community has offered many solutions to deal with the speeding up of the reasoning process. A deep learning technique called long short-term memory (LSTM) is used to create a new large semantic network. This network is created based on the similarity of concepts among the ontologies. The concepts of ontologies are named C_1, C_2, …, C_n. The starting node of the network is represented as Cs, and the terminating node is named as Ce.

4 Experimental Analysis and Results

To understand the proposed work and its efficacy, a case study of the healthcare IoT application is considered. The architectural view of the healthcare IoT is given in Fig. 2. The following is the workflow of the model.

1. The remote user makes an access request.
2. It is propagated to the security agent.
3. After evaluating the request, it requests the essential information to be evaluated from the cloud storage.

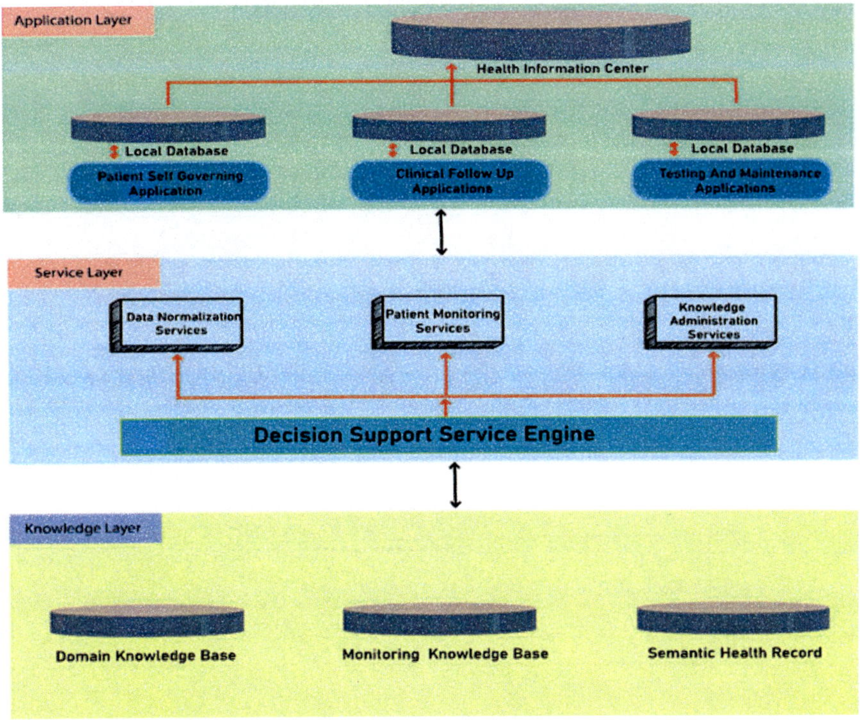

Fig. 2 Layered architecture of ontological approach to HealthCare IoT

4. Security agent sends get information message to cloud storage.
5. Then the information is sent to the agent from cloud storage.
6. Checks whether the user is authorized.
7. If it is authorized, then the access is granted else denied.

Figure 3 shows the patient monitoring ontology which involves the personal information of the patient like condition and disease type and whether the system report is normal, risk and alert must be given or not and can determine whether the patient needs immediate care by taking him/her to the hospital or it is enough to take him to a general practitioner. Based on these three important healthcare-related multiple ontologies, the deep learning model proposed in Section III is constructed to find novel inference rules.

The accuracy of the proposed work is given in Fig. 4, which shows that the model has high level of accuracy when trained with different trained data sets.

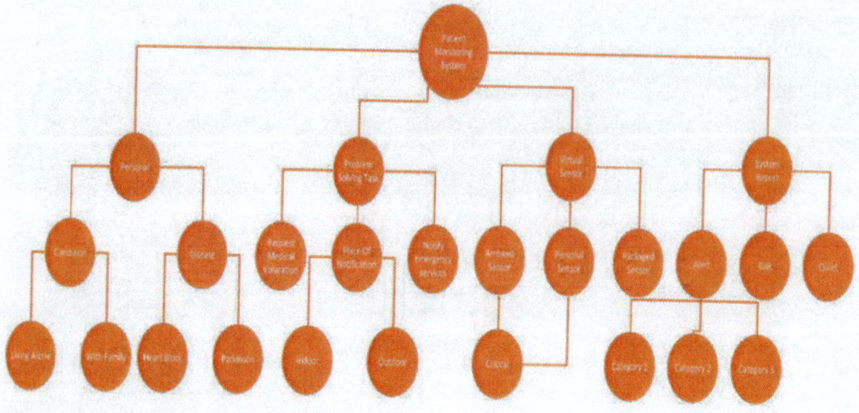

Fig. 3 Patient monitoring ontology

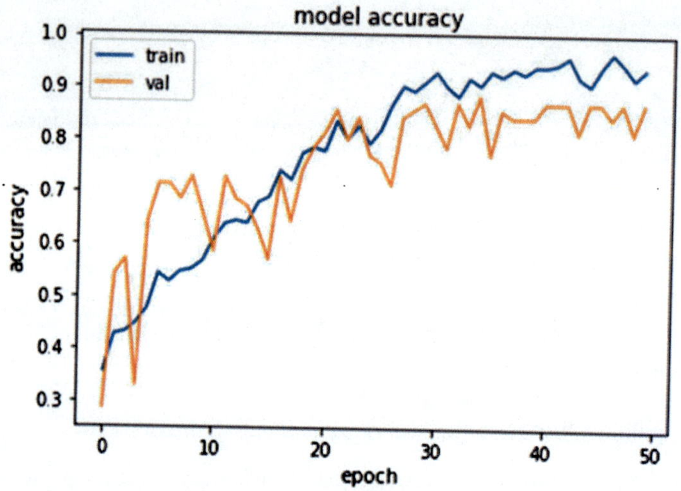

Fig. 4 Accuracy of the deep learning RNN Model

5 Conclusion

A deep learning-based technique for realizing new inference rules discovery automatically using combining multi-domain ontology rule bases for IoT applications has been proposed. Primarily, a semantic network is framed using the concept pair triples from the multiple ontologies. Later, new inference rules were discovered using recurrent neural networks, and thus, rule bases of multiple ontologies are associated together, thereby expanding rule bases for effective reasoning. In this experimental analysis, the important elements of healthcare system are efficiently modeled using ontologies, and using the deep learning methods, the inference rule bases are expanded to hold newly derived inferences, and healthcare services can be exploited by the users more effectively. Consequently, the improved understanding of securing IoT applications is realized through the proposed method.

References

1. T. Qiu, X. Liu, K. Li, Q. Hu, A.K. Sangaiah, N. Chen, Community aware data propagation with small-world feature for Internet of Vehicles. IEEE Commun. Mag. **56**(1), 86–91 (2018)
2. J. Wang, Z. Zhang, B. Li, S. Lee, R.S. Sherratt, An enhanced fall detection system for elderly person monitoring using consumer home networks. IEEE Trans. Consum. Electron. **60**(1), 23–29 (2014)
3. Z. Song, A.A. Cárdenas, R. Masuoka, Semantic middleware for the Interanet of Things, in *Proceedings of IEEE Internet Things (IoT)* (2010), pp.1–8
4. D. Zeng, Y. Dai, F. Li, S. Sherratt, J. Wang, Adversarial learning for distant supervised relation extraction. Comput. Mater. Continua **55**(1), 121–136 (2018)
5. R. Studer, V.R. Benjamins, D. Fensel, Knowledge engineering: principles and methods. Data Knowl. Eng. **25**(1), 161–197 (1998)
6. Z. Ma, F. Zhang, L. Yan, *Fuzzy semantic Web ontology mapping* (In Fuzzy Knowledge Management for the Semantic Web, Berlin, Germany, Springer, 2014), pp. 157–180
7. A.H. Doan, J. Madhavan, P. Domingos, Learning to map between ontologies on the semantic Web, in *Proceedings of 11th International Conference World Wide Web* (2002), pp. 662–673
8. T.R. Gruber, A translation approach to portable ontology specifications. Knowl. Acquis. **5**(2), 199–220 (1993)
9. M.A. Helou, M. Palmonari, M. Jarrar, Effectiveness of automatic translations for cross-lingual ontology mapping. J. Artif. Intell. Res. **55**(1), 165–208 (2016)
10. L.-Y. Zhang, J.-D. Ren, X.-W. Li, OIM-SM: A method for ontology integration based on semantic mapping. J. Intell. Fuzzy Syst. **32**(3), 1983–1995 (2017)
11. Z. Wang, R. Bie, M. Zhou, Hybrid ontology matching for solving the heterogeneous problem of the IoT, in *Proceedings of IEEE 11th International Conference Trust, Security, Privacy Computer Communication (TrustCom)* (Liverpool, U.K., 2012), pp. 1799–1804
12. R. Wang, L. Wang, L. Liu, G. Chen, Q. Wang, Combination of the improved method for ontology mapping. Phys. Procedia **25**(22), 2167–2172 (2012)

An End-to-End System for Text Extraction in Indian Identity Cards

Arjun S Kedlaya and J. Amudha

Abstract The process of customer verification has become a mandatory procedure to follow for a large number of sales and financial organizations. Traditionally, it has required collecting the required data from the uploaded images of identity cards and then manually entering them into a system which makes it a resource-intensive and error-prone task. Automation of this entire process is extremely beneficial for these organizations. The paper proposes an end-to-end system for preprocessing, text detection, and text recognition of identity cards by identifying the suitable methods and model architectures for each of the components in the system. The end-to-end system presented in the paper gives an accuracy of 93%. This paper is useful for researchers who want to work on improving the automation of text extraction from images, optical character recognition, and other similar use-cases

Keywords ID card · Text detection · Text recognition · Text extraction · Image processing · OCR · Computer vision

1 Introduction

Identity card verification is the process of an organization authenticating its clients and assessing their suitability, along with the potential risks of bad intentions toward the business relationship [1]. The automation of this entire process aids in reducing the time required for identity card verification and mitigating the security risks as confidential information of the customer is involved. All these factors make the automation of the verification process using computer vision a worthwile objective for any organization to fulfill.

A. S. Kedlaya (✉) · J. Amudha
Department of Computer Science and Engineering, Amrita School of Engineering, Amrita Vishwa Vidyapeetham, Bengaluru, India
e-mail: arjunskedlaya@gmail.com

J. Amudha
e-mail: j_amudha@blr.amrita.edu

A. K. Bashir et al. (eds.), *Proceedings of International Conference on Computing and Communication Networks*, Lecture Notes in Networks and Systems 394,
https://doi.org/10.1007/978-981-19-0604-6_40

The paper proposes an end-to-end system for text extraction from identity cards which consists of four stages digital image processing, text detection, text extraction, and application deployment. The first stage consists of image sizing, image deskewing, and image thresholding. For the next two stages, the paper aims to find the most suitable model architecture by testing them on the test dataset of identity card images and calculate the model's performance and accuracy. Finally, the best performing model is integrated with a web or mobile application.

Applications of text recognition and its challenges have been addressed in [15]. Many of the solutions proposed for this use-case make identity-card-specific assumptions such as using template matching methods for text extraction, using the location of the face or emblem to deskew the image. The main advantage of the end-to-end system proposed in the paper is that it is a generic and identity-card-independent solution. It can also be easily used for similar use-cases or transformed into an identity-card-specific solution by making assumptions about the image data.

This paper is divided into five sections as follows: In Sect. 2, an overview of related work, models, and datasets that was used is mentioned. In Sect. 3, the test dataset used in the paper to evaluate the models is described. Section 4 summarizes the system design of the end-to-end text extraction system and the performance of the MobileNet model. Section 5 tabulates the performance of each model against the test dataset.

2 Related Works

The solutions proposed in some of the other papers for text extraction have been described, and the models that have been used to extract text and the methods have been mentioned.

Salunkhe et al., 2017, have implemented a multilingual end-to-end framework where stroke width transform (SWT) has been used for text detection and Tesseract has been used for text recognition to extract text from multilingual signage boards [2].

Vailente Romero et al., 2016, have implemented an approach to recognize text in generic identity cards where the image taken is processed in the cloud server. They have used Hough transform rectification, MSER text detection, and Tesseract for the text recognition stage [3].

Sathish et al., 2016, have implemented a road sign detection and recognition system where shape of the road sign is detected by drawing contours and Hough transform method [4]. Template matching is done after using SIFT to detect the road sign, and this system has achieved an accuracy of 90% on the test dataset.

Shrivastava et al., 2019, have implemented a deep learning model using CNN for low-level feature extraction, AON module for high-level feature extraction, and Bi-LSTM to predict character sequence, and the proposed model has been able to achieve an accuracy of 91.4% on Synth90K dataset [14].

The various text detection and recognition models that have been used on the test dataset are listed below.

- Tesseract : Tesseract 4.0 has been used in this paper for text detection and recognition purposes. Tesseract 4 consists of a neural net (LSTM)-based OCR engine which is focused on line recognition [5].
- Stroke Width Transform and Maximally Stable Extremal Regions (SWT): Özgen et al., 2018, have developed a text detection algorithm where maximally stable extremal regions to acquire text region candidates which is reduced in quantity by using geometric and stroke width properties [6]. Candidate regions are joined to obtain final text groups.
- EAST Text Detection (EAST) : Zhou et al., 2017, have presented a text detection algorithm to predict words or text lines of quadrilateral shapes in full images and arbitrary orientations with a single neural network with intermediate steps like candidate aggregation and word partitioning eliminated [7].
- Pixel Link: Deng et al., 2018, have developed a text detection algorithm that is based on instance segmentation algorithm [8]. Text regions are segmented by aggregating pixels within the same instance. Bounding boxes are then extracted directly from the segmentation result without location regression.
- Text Spotter: Gupta et al., 2016, have developed an architecture to detect and recognize text in images using a fully convolutional regression network (FCRN) which can perform text detection and bounding box regression at multiple scales in an image [9].
- Convolutional Recurrent Neural Networks (CRNNs): Shi et al., 2015, have developed a framework with convolutional recurrent neural network (CRNN) and connectionist temporal classification (CTC) function which combines feature extraction, sequence modeling, and transcription [10]. It can handle sequences of unknown lengths and generates an effective and smaller model.
- MobileNet: Howard et al., 2017, have developed a deep learning neural network architecture which is efficient and lightweight and uses convolutions that are separable depth-wise. It uses two hyper-parameters that determine model size and balance between latency and accuracy of the model [11].

The models have been trained and tested on a variety of benchmark datasets for evaluating their performance. Models [8], [9], [10], and [11] have been tested result and methodology with ICDAR-2013 dataset, and models [7, 8] have been tested with ICDAR-2015 and MSRA-TD500. Models [10, 11] performance has been evaluated on Street View Text Database which contains outdoor street image of business signboards of varying resolution.

3 Test Dataset

A test dataset of Indian identity cards has been used to evaluate the models and compare their performances to determine the best models for text detection and text recognition phase. The models used in this project are pre-trained and have been directly used on the test dataset to evaluate their accuracy and performance. The test

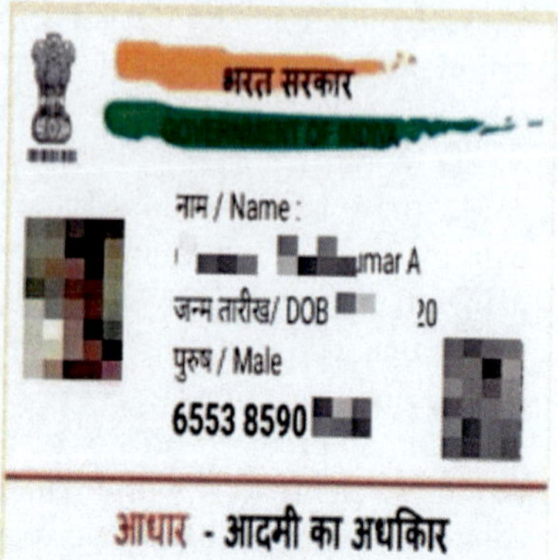

Fig. 1 Example of a good quality image from the test dataset on which the model performance is evaluated

Fig. 2 Example of a bad quality image from the test dataset which is highly pixellated

dataset contains ten images of Indian identity cards which have been separated into two equal categories:

- Good Quality Images: Centered, aligned, unskewed, and light to moderately pixellated images. Refer Fig. 1 for example.
- Bad Quality Images: Off-centered, unaligned, skewed, and highly pixellated images. Refer Fig. 2 for example.

Strict accuracy and partial accuracy are calculated for all the models. Strict accuracy follows binary system of scoring where unless the bounding box is perfectly drawn or all the letters in the field are recognized correctly the corresponding text detection and recognition models will not achieve a score of 1 for that particular field. Partial accuracy is where it is possible to get a score in between 0 and 1 provided that at least 50% of the letters are recognized or the bounding boxes cover at least 50% of the letters in the word.

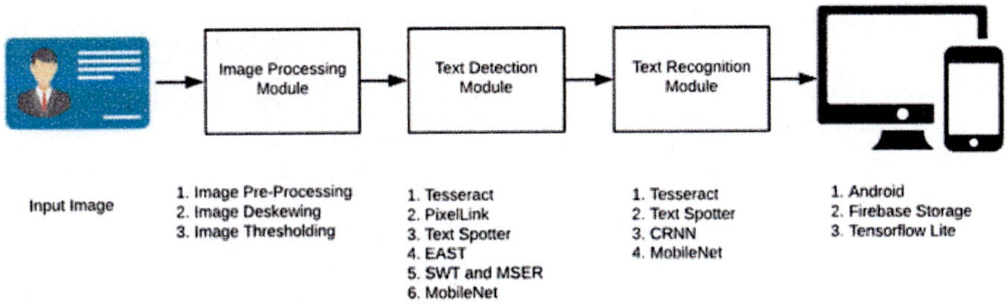

Fig. 3 System design of the end-to-end model presented in the paper

4 System Design

The end-to-end model presented in the paper takes the image of the identity card as an input and gives out the text recognized in the image as a string for output. The end-to-end model presented in the paper has four main stages. See also Fig. 3. Detailed explanation of each stage is provided below.

4.1 Image Processing

In the image processing module, the raw image given by the user is sized to the correct dimensions, deskewed, converted to greyscale image and given as the output to the next module. It has three stages:

1. Image Preprocessing: The images are transformed into a standard size of 250*250 pixels of height and width. Intensity values of images are scaled down to 8 bits to reduce the computational complexity of the input image.
2. Image Deskewing: The image deskewing module makes an attempt to detect and correct the skewness of the image. It uses the following methods:

   (a) Canny Edge Detector: The canny edge detector is an multistage algorithm used to detect wide range of edges in image[12]. This mathematical operator can also be used to find the text lines in the image. A Gaussian filter is used to smooth the image to remove noises, and then final coordinates of the text lines in the image are obtained through double thresholding, finding intensity gradients, local maxima, and through hysteresis.

   (b) Hough Transform: Hough transform is a technique to extract features of objects that fall in a certain shape class by a procedure of voting [13]. Hough transform is used to find the skew of the image, and it uses text lines detected by the Canny edge detector to determine the skew angle. The technique determines the peaks in the image based on the angle, distances, and Hough transform accumulator array. The deviation of each of the peaks from 45° angle is found and segregated into num_peaks bins. The probable skew angle

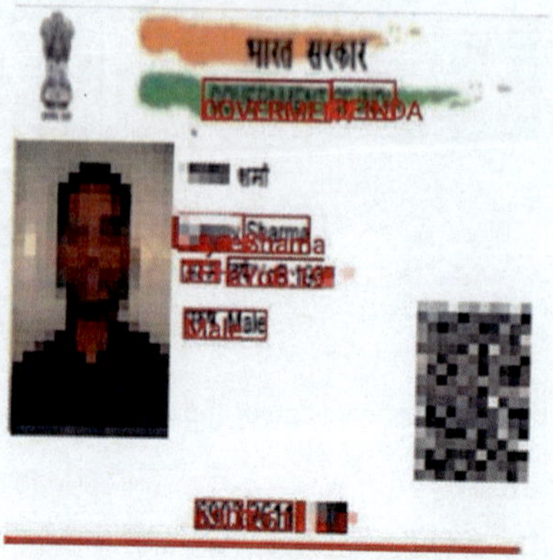

Fig. 4 MobileNet bounding box and text recognition output on an image from test dataset

of the image is computed by taking the average of the deviation angle values in the bin with the most number of values.

3. Image Thresholding: The image obtained after deskewing is converted to grayscale using adaptive binary thresholding as it reduces the complexity and computation efforts required for the text detection and recognition stage.

4.2 *Text Detection*

The processed image from the image processing module is used as input and given to a text detection model for detecting location of text regions in the image. The six models tried for the text detection task are Tesseract, SWT, EAST, TextSpotter, PixelLink, and MobileNet and their performance evaluated against the test dataset (Fig. 4).

Out of the models tried in the paper, MobileNet has achieved the highest accuracy and can even detect the boundaries of text regions correctly in bad quality images. Refer Table 1 for each model's performance for text detection task. MobileNet, when tested on the sample dataset, was able to draw non-overlapping bounding boxes and also was able to distinguish text regions from non-text regions of an identity card like the emblem, face, and QR code, unlike some of the text detection models tried in this paper. It gives a strict accuracy of 100% for both good and bad quality images in the dataset. Apart from the superior performance in terms of accuracy, the MobileNet model also offers other advantages that make it a suitable model to implement for this use-case. It is a lightweight deep neural network that has been designed to perform well even on low computing power devices like mobiles.

Table 1 Text detection performance of the six models tried in the paper

Text detection accuracy		
Models	Mean strict accuracy	Mean partial accuracy
Tesseract	0.75	0.75
SWT	0.48	0.53
EAST	0.33	0.53
Text Spotter	0.45	0.63
Pixel_ Link	0.60	0.66
MobileNet	1	1

4.3 *Text Recognition*

The text recognition module tries to recognize the text embedded in text regions. The final output from the text recognition module is a string consisting of all the text recognized in the image. The text recognition models that have been tried in the paper are Tesseract, TextSpotter, CRNN, and MobileNet models.

Out of the four text recognition models that have been tried, MobileNet model has achieved the highest accuracy. Refer Table 2 for each model's performance for text recognition task. It gives a strict accuracy of 100% for good quality images and 85% for bad quality images. It generates less noise than the other text recognition models which makes text extraction easier. MobileNet model is also very robust and has a good accuracy even with bad quality images. However, in a few of the bad quality images, it fails to recognize all of the letters in lengthy words.

4.4 *Android/Web App Deployment*

The app is used as an user interface for taking input image from the user, and the text returned by the text recognition module is displayed for the user on the screen.

Table 2 Text recognition performance of the four models tried in the paper

Text recognition accuracy		
Models	Mean strict accuracy	Mean partial accuracy
Tesseract	0.40	0.49
Text Spotter	0.20	0.34
CRNN	0.45	0.68
MobileNet	0.93	0.99

A real-time text extraction android application was developed using the MobileNet model as it was the best text detection and recognition model when tested on the test dataset. The MobileNet model is integrated with the android app using TensorFlow Lite which is a machine learning framework for integrating a TensorFlow model with smartphones and edge devices. The android app uses the smart device's camera for taking input image, and the user gets the recognized text from the image using the toast functionality. It is a real-time text recognition app, and hence, the results can be previewed live. Optionally, a web application can also be built using the Flask Framework and Python.

5 Results of Various Models

The performances of various text detection and recognition models have been evaluated on the test dataset. MobileNet model is the best text detection and text recognition model as it has a much better accuracy than all the other models tested in the paper.

The image deskewing module which is used has some limitations as it fails when the images are highly pixellated or contain background noise in the image, which makes it difficult to determine the skew angle of the image. However, it works correctly with images of good quality and images with less background noise.

In testing phase, the pre-trained models are evaluated against the images in the test dataset and their performance evaluated. For text detection task, the MobileNet model has a mean strict accuracy of 100%. It has segmented the four pertinent fields of the identity card perfectly and even performs text detection of bad quality images correctly. MobileNet model has a mean strict accuracy of 93% for the text recognition task when its accuracy was measured with the test dataset. It recognizes most of the fields correctly, but it fails to recognize some characters in the name field which can have very long words.

The performance of the MobileNet model is much better than the other models tested in the paper as it has an overall strict accuracy of 93% on the test dataset. Tesseract is the best alternative for text detection task, while CRNN with CTC is the best alternative for text recognition task. However, both of the alternatives mentioned have their own set of disadvantages which will significantly bring down the performance when compared with the MobileNet model.

Since the MobileNet model has a superior performance compared to the other models in the paper, it is integrated with the application using TensorFlow Lite Machine Learning Framework as it shows the best performance for the text extraction from identity card use-case.

6 Conclusion

An end-to-end system for text extraction from Indian identity cards has been presented in the paper. Although the test dataset is small, MobileNet is superior in performance when compared with other models. So, MobileNet model gives the best accuracy for text detection and recognition purposes. The system described in the paper can be repurposed for similar use-cases or extended to implement additional functionalities. The end-to-end model described in the paper is robust, accurate, and a viable solution for fulfilling the objective.

Future enhancements can include testing some of the latest text detection and recognition modules and comparing their accuracy with the models described in the paper. Assumptions about the data can be made like using the location of emblem or a person's face for image deskewing operation in a specific use-case. Efforts can be made to increase the size of the dataset through manual collection of data, increasing the size of the data by applying transformations on the original image and by implementing a synthetic identity card generator.

References

1. S. Mathur, What is KYC? Why is it important? (2019, June 18). Retrieved 20 April 2020, from https://veri5digital.com/veri5-blog/what-is-kyc-why-is-it-important/
2. P. Salunkhe, S. Bhaskaran, A. Joseph, D. Gupta, Recognition of multilingual text from signage boards (2017), pp. 977-982. https://doi.org/10.1109/ICACCI.2017.8125968
3. R. Valiente Romero, M. Sadaike, J. Gutiérrez Menéndez, D. Soriano, G. Bressan, W.V. Ruggiero, A process for text recognition of generic identification documents over cloud computing (2016)
4. P. Sathish, D. Bharathi, Automatic road sign detection and recognition based on SIFT feature matching algorithm (2016). https://doi.org/10.1007/978-81-322-2674-1_39
5. Tesseract 4.0 with LSTM. Retrieved August 29, 2020. From https://tesseract-ocr.github.io/tessdoc/4.0-with-LSTM.html
6. A.C. Özgen, M. Fasounaki, H. Ekenel, Text detection in natural and computer-generated images (2018), pp. 1–4. https://doi.org/10.1109/SIU.2018.8404600
7. X. Zhou, C. Yao, H. Wen, Y. Wang, S. Zhou, W. He, J. Liang, EAST: An Efficient and Accurate Scene Text De-tector (2017), pp. 2642–2651. https://doi.org/10.1109/CVPR.2017.283
8. D. Deng, H. Liu, X. Li, D. Cai, PixelLink: Detecting scene text via instance segmentation (2018)
9. A. Gupta, A. Vedaldi, A. Zisserman, Synthetic data for text localisation in natural images (2016), pp. 2315–2324. https://doi.org/10.1109/CVPR.2016.254
10. B. Shi, X. Bai, C. Yao, An end-to-end trainable neural network for image-based sequence recognition and its application to scene text recognition. IEEE Trans. Pattern Anal. Mach. Intell. (2015). https://doi.org/10.1109/TPAMI.2016.2646371
11. A. Howard, M. Zhu, B. Chen, D. Kalenichenko, W. Wang, T. Weyand, M. Andreetto, H. Adam, *MobileNets: Efficient Convolutional Neural Networks for Mobile Vision Applications* (2017)
12. S, Akshay, A, Rahul, B.M.Y. Brahma, Traffic sign board detection using Canny edge detection for compressed images. J. Adv. Res. Dyn. Control Syst. **10**, 474–477
13. N. Parameswaran, E. Achan, S. Shree, R. Manjusha, Road detection by boundary extraction technique and hough transform (2019). https://doi.org/10.1007/978-3-030-00665-5_165

14. A. Shrivastava, J. Amudha, D. Gupta, K. Sharma, Deep learning model for text recognition in images, in *2019 10th International Conference on Computing, Communication and Networking Technologies (ICCCNT)* (2019), pp. 1–6. https://doi.org/10.1109/ICCCNT45670.2019.8944593.
15. M. Nevetha, B. Arumugam, Applications of text detection and its challenges (2015), pp. 712–721. https://doi.org/10.1145/2791405.2791555

Static Malware Analysis Using Machine and Deep Learning

Himanshu Kumar Singh, Jyoti Prakash Singh, and Anand Shanker Tewari

Abstract In the era of digital advancement and innovation, malware (malicious software) still poses major threats to users' privacy and leads to many security breaches. Due to the exponential rise in malware attacks, malware analysis and detection continue to be a hot research topic. Malware analysis plays a vital role in the malware detection process. Currently, the detection process adopts the malware signatures (static analysis) and behavior patterns (dynamic analysis) that have been proven time-consuming and less effective in identifying unknown malware in real time. Recent malware uses abstraction, packing, encryption, polymorphic, and other cryptic methods to hide and change the malware behavior and its signature which makes the detection process complex. Most of the new malware is the variants of existing malware, where machine learning techniques are effective in identifying such malware. However, the traditional machine learning technique is time-consuming because it requires substantial feature engineering and learning. By using the state-of-the-art learning technique such as deep learning, compel the learning process faster. By utilizing the high-level machine learning techniques, the training stage can be completely avoided. In this paper, first, we analyze the old-style MLAs and profound learning models for malware detection using publicly available datasets. Second, we analyze the deep learning models to examine the accuracy over the traditional machine learning technique. Third, our major commitment is in proposing an efficient and accurate model which combines the capabilities of the machine and deep learning technique which detect the zero-day malware efficiently. Our model shows that our proposed method outflanks traditional MLAs and deep learning models.

Keywords Malware · Static analysis · Dynamic analysis · Machine learning · Deep learning

H. K. Singh · J. P. Singh · A. S. Tewari (✉)
National Institute of Technology Patna, Patna, India
e-mail: anand@nitp.ac.in

H. K. Singh
e-mail: himanshus.pg19.cs@nitp.ac.in

J. P. Singh
e-mail: jps@nitp.ac.in

A. K. Bashir et al. (eds.), *Proceedings of International Conference on Computing and Communication Networks*, Lecture Notes in Networks and Systems 394,
https://doi.org/10.1007/978-981-19-0604-6_41

1 Inroduction

In this digital world, where modern technologies like 5G, Internet of Things (IoT), and artificial intelligence (AI) lead to the advancement and innovation of digital society. However, privacy and security breaches pose major challenges as cyber-criminals attack the users computer and networks for stealing sensitive data, spy on the infected system, or take control of the system for self-gain [1]. The attackers use malware (malicious software) to gain access to the target system. Malware is a software, code, or program which performs malicious actions. The term malware is used to generalize any form of malicious software and code. It can get different names based on behavior and purpose like virus, Trojan, adware, worm, and spyware. Malware analysis is used to understand the behavior of malware and also helps in the detection process. Currently, the analysis of the malware process is signature-based or behavior-based, but these are proven to be time-consuming as well less effective in identifying unknown malware in real time. This paper aims to propose a novel architecture that combines the concept of machine learning and deep learning which effectively detects zero-day malware.

1.1 Research Background

When the very first computer virus appeared in 1988–89, antivirus software were designed to detect only the known viruses by searching the virus definition databases which is updated time to time; this method is called signature-based detection. But the challenges with this approach are virus variants use different types of obfuscation which hides the viruses signature. Hence, signature-based method are less efficient in terms of detecting the zero-day attack [2]. Signature-based analysis needs domain-level knowledge to reverse engineer the malware using static and dynamic malware analysis techniques. These techniques are used to identify the important features of the malware which helps in signature-based detection. These methods take larger time to reverse engineer the malware; during that time, hackers might take many valuable information. It is also a resource-extensive method.

Many potential researchers have identified that hackers use obfuscation methods to against signature-based detection. To tackle this problem, software are used to manually unpack the file and analyze the APIs calls. But this process is resource-intensive. In [3], author presented a model which automatically extract the APIs call and analyze the binary in four-step. In step 1, unpacking of malware. In step 2, disassembling of binary. In step 3, extraction of APIs call, and in step 4, APIs call mapping and feature analysis. This work was further enhanced in [4] by adding a extra step using machine learning. SVM is used with n-gram feature extraction from both goodware and malware binary with tenfold cross validations. In [5], author proposed a hybrid model which combines support vector machine (SVM) and maximum-relevance minimum redundancy filter (MRMRF) with API calls feature for enhanced malware detection. With the increase in malware variants due to obfuscation, recently many potential researcher are improving the malware detection methods [6]. This forms the motivation of this research.

2 Related Work

Machine learning algorithms works on feature engineering, selection, and representations. The set of features of different class is used to train the model in order to create a plane of goodware and malwares. This plane helps to classify the malwares and goodwares. Both feature selection and engineering requires domain level knowledge. Various features can be obtained by static and dynamic analysis explained in Sect. 3 of this paper.

The problem with classical machine learning-based malware detection system is that they rely on the feature engineering, learning, and feature representation [7–9] and once an attacker have the knowledge about the features used in model, the malware detector can be easily bypassed [10].

To be accurate, machine learning algorithms requires variety of data. The publicly available data for malware analysis is very less due to privacy and security concerns, and each available data has their own limitations. Many researchers prepare their own datasets and preparing their own dataset by using data science explained in [11] for research is a daunting task. These are the major limitations for developing a machine learning-based malware detection system that can be used in real time.

Nowadays, deep learning models, an improved model of neural networks better performed compared to machine learning models in many of the task in the field of natural language processing, robotics, and others [12]. In training phase, it tries to grab high-level representation of features in hidden layers with the capability to learn from mistakes. These are [7–9, 13–20] are the few research studies which uses the application of deep learning models for malware analysis.

3 Methods for Malware Analysis

3.1 Static Analysis

In static analysis, executables are analyzed without actually executing them. It is the very first and less risky process and does not require any safe environment or sandbox for analyzing them. Static analysis involves the analysis of the internal structure of the program. It involves various steps: (a) Determining the file type of the malware: It helps in identifying the malware's target operating system and architecture. (b) Fingerprinting the malware: By fingerprinting means generating the hash value based on its file content. It helps in identifying whether this particular malware is identified before by searching in multi-anti-virus databases like VirusTotal. (c) Extracting Strings: Executable strings can be extracted using the string utility tool available in the linux system. Extracted strings can give clues about the program functionality and indicators associated with a suspect binary. (d) Determining file obfuscation: Obfuscation is a method used by the malware authors to hide the inner working of the binary. Packers and cryptors are obfuscation methods used by the malware authors.

3.2 Dynamic Analysis

Dynamic analysis is the way toward extricating data from malware while it is running. Not at all like the restricted view, the static analysis gives of the malware being broke down, powerful examination offers a more top to bottom view into the malware's capacities since it is gathering data while the malware is executing its capacities and orders. To lead dynamic malware analysis, two things are required: malware test environment and dynamic analysis tools.

A malware test environment is a framework where malware is executed with the end goal of examination. It should comprise of a working framework that the malware is composed for and should have most, if not all, of the conditions the malware needs to execute appropriately.

The dynamic analysis tool, otherwise called framework checking apparatuses, is the one observing the malware test environment for any progressions made by the malware to the objective framework. A portion of the progressions that are observed and recorded remember changes for the document framework, adjustments in setup documents, and whatever other important changes that are set off by the malware's execution. The powerful investigation devices likewise screen inbound and outbound organization correspondences and any working framework assets utilized by the malware. With these tools, the investigator can comprehend what the malware is attempting to never really target framework.

A completely executed malware test climate with the fitting powerful investigation instruments is otherwise called a malware sandbox. A malware sandbox is a place where an examiner can run and notice a malware's conduct. A malware sandbox can be a solitary framework or an organization of frameworks planned exclusively to break down malware during runtime.

4 PE File Format

The Windows PE document is the record sort of Windows working frameworks beginning in Windows NT and Windows 95. It is called Portable Executable because

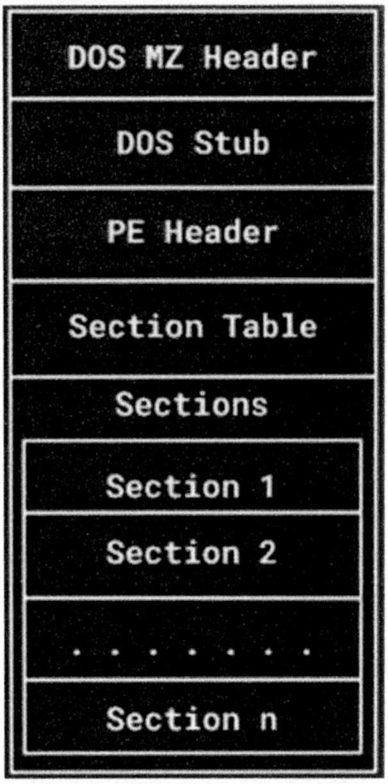

Fig. 1 PE file format

Microsoft's vision was to utilize a similar document design in future kinds of Windows, making the PE document basic to all Windows stages paying little mind to what central processing unit (CPU) they support.

The Windows PE document design is gotten from the Common Object File Format (COFF) that was utilized in Virtual Address extension (VAX) frameworks running the Virtual Memory System (VMS) working framework created by Digital Equipment Company (DEC), which was procured by Compaq in 1998 and converged with HP in 2002. The majority of the first Windows NT improvement group came from DEC (Fig. 1). The PE File design comprise of the accompanying:

- DOS MZ Header
- DOS Stub
- PE Header
- Section Table
- Sections

5 Dataset Description

The dataset is obtained from the publicly available dataset from IEEE Dataport. It contains information from around 48K malware and goodware. The dataset is gotten by exploiting the openly accessible reports from malware administration. It is a free online assistance that does a static and dynamic examination on submitted records utilizing the Cuckoo sandbox, which are then available in an HTML report.

To ensure the credibility of the dataset, we turned to two other online archives: National Software Reference Library (NSRL) and VirusShare.com, these give meta-data (for example MD5 hash) in regards to known goodware and malware samples, separately. As NSRL contains an assortment of advanced marks of known, traceable software applications, if an example is available in this assortment, we are more sure it is without a doubt goodware. Then again, VirusShare.com is a vault of malware tests, henceforth an example present in this storehouse gives us higher certainty it is malware.

When the information validness is affirmed, we began the extraction process, where online information is saved locally or in a central database for additional examination. This method is called scraping, and it is done by using Python scraping library. Concerning the NSRL repository, data was given in textual format, which drove us to utilize Pandas, a Python data analysis library, to extract and dissect the information. The data extracted from the PE samples are visualized in three different datasets:

1. PE_Import dataset contains the top 1000 imported function information extracted from the import section of the PE sample.
   It has 1002 columns in which 1000 columns are the features, one column is for the hash, and one column for the label.
2. PE_Section_Header contains the information of the section header of .text, .data, .code, and code section of the PE sample.
   It has 6 columns in which 4 columns are the features, one column is for the hash, and one column for the label.
3. PE_Raw_Image contains the raw PE byte stream re-scaled to 32×32 grayscale images of PE sample.
   It has 1026 columns in which 1024 columns are the pixel value, one column is for the hash, and one column for the label.

6 Model Implementation

Proposed model uses the combination of both machine and deep learning as shown in the Fig. 2 based on static analysis. It uses deep learning for the feature extraction process and classical machine learning model for the classification process. For the PE_Section_Header, we used fully connected artificial neural network, for the PE_Import, we also used fully connected neural network, and for Raw_PE_Image, we have used convolution neural network.

For the fully connected neural network, we have used adam optimizer and binary cross entropy as loss function and ReLU as the activation function.

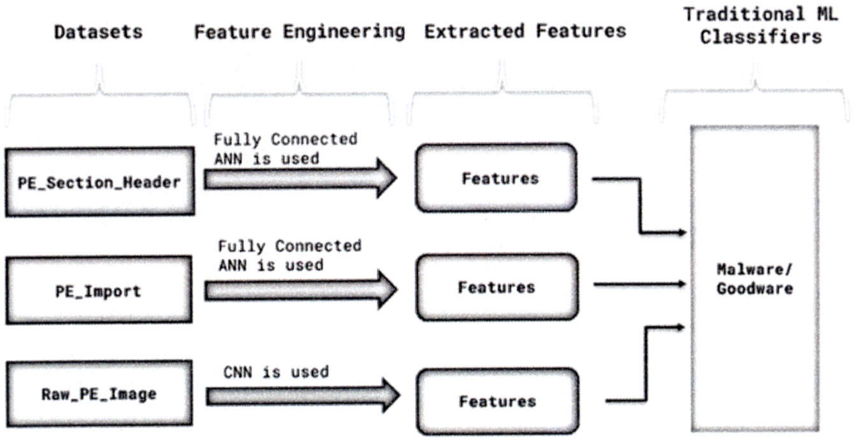

Fig. 2 Proposed model

6.1 Performance Evaluation

We have performed various experiments based on the number of features of the datasets. We evaluate the optimal number of features for our model for that we initially used 50 features out of 1000 from the PE Import, 50 features out of 1024 from the Raw PE Image, and 4 features from the PE Section Header. For the selection process, we tuned the second-last layer of the neural network as per our requirements and later stored these intermediate values in a new .csv file using which we create a more informative and efficient datasets. Later, these new datasets are given to machine learning models for the classification process. The experiment results are in Table 1.

The best result we got when we selected 100 features from the Import dataset and 100 from Image dataset and 4 from the Section dataset. We have also performed our experiment using various machine learning algorithm and also deep learning model and compared our model which can be seen in Table 2.

Table 1 Features analysis result

S. No.	Number of features from import	Number of features from image	Number of features from section	Accuracy
1	50	50	4	97.22
2	50	100	4	97.86
3	100	50	4	97.86
4	100	100	4	98.91

Table 2 Experiment result

Model	Classifier	Accuracy
Voting method based on Machine Learning	Ensemble Voting	97.66
Voting method based on Deep Learning	–	95.99
[21]	CNN 2 layer + LSTM	98.8
[22]	RNN	96.01
[23]	–	98.4
[24]	–	97.4
Proposed model	KNN	96.95
Proposed model	Random Forest	98.91
Proposed model	SVM	98.64

7 Conclusion

In this work, we analyzed how ML and DL techniques fit into the scope of malware detection and how could the chosen dataset influence the results of the classifier. We analyzed, trained, and validated multiple models to better understand how laboratory conditions vary from real-world conditions. We compared our model with others models which is based on machine learning, deep learning, and combinations on both, but doing so our model provided us with results as high as 98.91%.

We have also concluded that the model combined with ML and DL both gives better and promising results. Having a solid knowledge of the effects of temporal consistency in the task of malware detection, we improved our base model for better results. This was done by using the DL feature extraction approach to provide the ability to extract information regarding malware classes and by adding more features to the model.

The task we set ourselves to achieve was not without its difficulties, but all in all, we believe our work shows that the path to malware detection via machine learning and deep learning is feasible, not only theoretically, as related work as shown, but also with practical implications.

8 Future Work

In the above work, we used static analysis for the feature analysis and selection; in future, we want to incorporate the dynamic analysis for the feature engineering doing so give us a new prospect toward the datasets and also obtain new feature which can increase the accuracy of the model.

References

1. G. McGraw, G. Morrisett, Attacking malicious code: a report to the Infosec Research Council. IEEE Softw. **17**(5), 33–41 (2000)
2. B. Li, K. Roundy, C. Gates, Y. Vorobeychik, Large-scale identification of malicious singleton files, in *Proceedings of 7th ACM conference on data and application security and privacy, New York, NY, USA: ACM*, Mar 2017, pp. 227–238
3. M. Alazab, S. Venkataraman, P. Watters, Towards understanding malware behaviour by the extraction of API calls, in *Proceedings of 2nd Cybercrime Trustworthy Computing Workshop*, July 2010, pp. 52-59
4. M. Tang, M. Alazab, Y. Luo, Big data for cybersecurity: vulnerability disclosure trends and dependencies. IEEE Trans. Big Data (to be published)
5. S. Huda, J. Abawajy, M. Alazab, M. Abdollalihian, R. Islam, J. Yearwood, Hybrids of support vector machine wrapper and filter based framework for malware detection. Fut. Gener. Comput. Syst. **55**, 376–390 (2016)
6. E. Raff, J. Sylvester, C. Nicholas, Learning the PE header, malware detection with minimal domain knowledge, in *Proceedings of 10th ACM Workshop Artificial intelligence and security* (ACM, New York, Nov 2017), pp. 121–132
7. E. Raff, J. Barker, J. Sylvester, R. Brandon, B. Catanzaro, C. Nicholas, Malware detection by eating a whole exe (2017). [Online]. Available:https://arxiv.org/abs/1710.09435
8. M. Krcál, O. Švec, M.Bálek, O. Jašek, Deep convolutional malware classifiers can Learn from raw executables and labels only (2018).[Online]. Available: https://openreview.net/forum?id=HkHrmM1PM
9. M. Rhode, P. Burnap, K. Jones, Early-stage malware prediction using recurrent neural networks. Comput. Secur. **77**, 578–594 (2018). (Aug.)
10. H.S. Anderson, A. Kharkar, B. Filar, P. Roth, *Evading Machine Learning malware Detection* (Black Hat, New York, 2017)
11. R. Verma, Security analytics: adapting data science for security challenges, in *Proceedings of 4th ACM International Workshop on security and privacy analytics* (ACM, New York, Mar 2018), pp. 40–41
12. Y. LeCun, Y. Bengio, G. Hinton, Deep learning. Nature **521**(7553), 436–444 (2015)
13. A.F. Agarap, F.J.H. Pepito, Towards building an intelligent anti-malware system: a deep learning approach using support vector machine (SVM) for malware classification.v [Online] (2017). Available:https://arxiv.org/abs/1801.00318
14. E. Rezende, G. Ruppert, T. Carvalho, A. Theophilo, F. Ramos, P. de Geus, Malicious software classification using VGG16 deep neural network's bottleneck features, in *Information Technology-New Generations* (Springer, Cham, 2018), pp. 51–59
15. J. Saxe, K. Berlin, Deep neural network based malware detection using two dimensional binary program features, in *Proceedings of International Conference on Malicious and Unwanted Software (Malware)*, Oct 2015, pp. 11–20
16. S. Tobiyama, Y. Yamaguchi, H. Shimada, T. Ikuse, T. Yagi, Malware detection with deep neural network using process behavior, in *Proceedings of IEEE 40th Annual Computer Software and Applications Conference (COMPSAC)*, vol. 2, June 2016, pp. 577–582
17. W. Huang, J. W. Stokes, Mtnet: a multi-task neural network for dynamic malware classification, in *Proceedings of International Conference on Detection of Intrusions and Malware, and Vulnerability Assessment* (Springer, Cham, July 2016), pp. 399–418
18. R. Pascanu, J. W. Stokes, H. Sanossian, M. Marinescu, A. Thomas, Malware classification with recurrent networks, in *Proceedings of IEEE International Conference on Acoustics, Speech, and Signal Processing (ICASSP)*, Apr 2015, pp. 1916–1920

19. T. Shibahara, T. Yagi, M. Akiyama, D. Chiba, T. Yada, Efficient dynamic malware analysis based on network behavior using deep learning,' in *Proceedings of IEEE Global Communications Conference (GLOBECOM)*, Dec 2016, pp. 1–7
20. B. Kolosnjaji, A. Zarras, G. Webster, C. Eckert, Deep learning for classification of malware system call sequences, in *Proceedings of Australasian Joint Conference on Artificial Intelligence* (Springer, Cham, Dec 2016), pp. 137–149
21. R. Vinayakumar, M. Alazab, K.P. Soman, P. Poornachandran, S. Venkatraman, Robust intelligent malware detection using deep learning. IEEE Access **7**, 46717–46738 (2019). https://doi.org/10.1109/ACCESS.2019.2906934
22. M. Rhode, P. Burnap, K. Jones, Early stage malware prediction using recurrent neural networks. Available: arXiv:1708.03513
23. S.Z.M. Shaid, M.A. Maarof, Malware behaviour visualization. J. Teknol. **70**(5), 25–33 (2014). (Sep.)
24. H. Zhou, Malware detection with neural network using combined features, *Presented at the Australasian Joint Conference on Artificial Intelligence, Beijing, China, Aug. 2018)*

An Elaborative Approach for the Histopathological Classification of the Breast Cancer using Residual Neural Networks

P. S. S. Madhulika and Nalini Sampath

Abstract Breast cancer is the most conventional type of cancer that takes place predominantly in women. Whenever a transition is made in DNA, then it is confirmed that it is caused by cancer. Nowadays, the rate at which this disease is spreading is highly exponential. Currently, it has become a number one morbidity that can happen to women. Tumor in the human body can be detected in early stages using mammography, ultrasound, biopsy, etc., by performing necessary procedures and can be prevented in the early stages. Breast cancer detection in early stages is very crucial, otherwise there is chance of spreading of cancer cells all over the body. The proposed solution is used to identify whether the tumor present in the body is benign which means harmless or malignant. In this paper, the histopathological images are used as the dataset for the classification and detection of breast cancer. The images can be easily downloaded from the public domain. Transfer learning techniques are used to train the model, and later, a machine learning classifier is applied to achieve better results. The proposed model adapts usage of the residual neural network along with the linear support vector machine classifier. Linear SVM is used for binary class classification as it improves the performance of the model by rewarding with substantial accuracy. The final accuracy achieved by implementing this model is 96.92%.

Keywords Breast cancer · Linear SVM · Histopathological images · Residual neural networks

P. S. S. Madhulika (✉) · N. Sampath
Department of Computer Science and Engineering, Amrita School of Engineering, Amrita Vishwa Vidyapeetham, Bengaluru, India
e-mail: madhulika008@gmail.com

N. Sampath
e-mail: s_nalini@blr.amrita.edu

A. K. Bashir et al. (eds.), *Proceedings of International Conference on Computing and Communication Networks*, Lecture Notes in Networks and Systems 394,
https://doi.org/10.1007/978-981-19-0604-6_42

Table 1 Classification of dataset images

Categories	Number of images
Benign	2480
Malignant	5429
Total	7909

1 Introduction

Cancer is one of the most life-threatening conditions that has an enormous impact on the health condition of an individual. Typically, breast cancer can be seen mostly in the females. If the symptoms are noticed in the primitive stages, then it will become easy to treat the patients. Every year, many ladies pass away due to this breast cancer condition as tumor is located. By taking the assistance of the advanced technology, many deep learning and transfer learning approaches are being used to detect any type of malignancy. Performing classification manually becomes very difficult as it requires loads of the data. If a woman crosses the age 20, then they will likely have the chance to get attacked with the breast cancer. The health census taken from the World Health Organization gives us an information that 2.1 million women may die each year because of the breast cancer [1]. The women living in the rural areas must be educated and should be made aware of the equipment and services that are available to cure easily [2]. Since this is a severe issue, every woman should get tested once in every year [3]. The classification of breast cancer using the histopathological images dataset, i.e., Break His dataset which consists of 7909 images [4]. This database has been built in collaboration with the P&D Laboratory–Pathological Anatomy and Cytopathology, Parana, Brazil. All the images are sorted according to two categories such as benign and malignant, then the new rearranged dataset is used. The image classification of the dataset is shown in Table 1.

After the data preprocessing, apply the pretrained ResNet50 model along with the linear SVM classifier [5]. Linear SVM algorithm is used to improve the performance of the model and deliver better results. A detailed analysis of the Break His dataset which consists of histopathological classified breast cancer images implemented using the residual neural networks model along with the linear support vector machine classifier. Residual neural networks use a pretrained model value as an input and perform our model using the Break His dataset. Since the dataset consists of 7909 images and it is considered to be very large, and it requires an exceptionally rapid classifier that can cope up with the huge network model.

2 Related Work

A study by Zou et al. [6] focused breast cancer histopathological image classification model which was made using the CNN Inception V1 model. The images are divided into two categories such as benign and malignant. The aspect ratio used in this model is SPP, and a new ratio is introduced which is known as special GAP which is the full form of global average pooling is used. The outcome accomplished from the special GAP is much better when compared to SPP. The Inception V1 convolutional neural network is employed to perform the classification technique between the malignant and benign neoplasmic images. The final results that are obtained after the execution of the model show that the using the aspect ratio as special global average pooling achieved best results. This approach is proposed by de Matos et al. [7] perform biopsies that are considered as the best quality surgical procedures for breast cancer detection. This is performed usually using the computer-aided diagnosis (CAD) systems, reduction of the diagnostic time, and diminishing the viewer disparity. The improvement in computation system has made the systematic model very real. Normally, the datasets obtained from the histopathological classification are very insufficient and consist of a smaller number of images and cannot be used for processing deep learning methods as they require huge data. In this paper, texture-based filters are used to achieve maximum results. The experimental results show that using the large dataset like Break His justifies the proposed new texture CNN model. A paper by Ghosh et al. [8] explores an automatic segmentation method for breast images taken from the ultrasound dataset. A small dataset is taken, and preprocessing is performed on them before the application of the CNN model. Feature extraction is executed on the set of images to improve the stability of the model, thereby it decreases the appearance of errors. Segmentation decides whether the model that is constructed is able to with stand all the other parameters or not. In this case, usage of segmentation leads to yield better results. A work done by Tan et al. [9] uses the dataset which consists of the mammogram images. All the images are split into three types such as benign, malignant, and normal. Since the dataset is very small, the system can easily process the images and implement the model very quickly and can perform diagnosis to each and every person. First, the data preprocessing is done on the dataset so that they become easily adaptable with the system. By extracting all the features of the dataset using the feature extraction technique, it will be able to find the best matching image. In conclusion, the BCDCNN model which has been successfully developed is tested using 320 mammogram images. The benefits of this method are quick diagnosis time and large accuracy.

An article by Nguyen et al. [10] performs data augmentation on the ICIAR dataset, and more images are added to improve the performance of the model. Data augmentation should be performed during the testing, and the images are rotated in the ninety-degree angle anti-clockwise direction. Generative adversarial networks can also be used instead of performing the augmentation to obtain better results.

3 Methodology

This paper proposes a system that is used for the binary classification of the histopathological images using the ResNet model along with the linear SVM classifier. Figure 1 describes the architecture of the ResNet model.

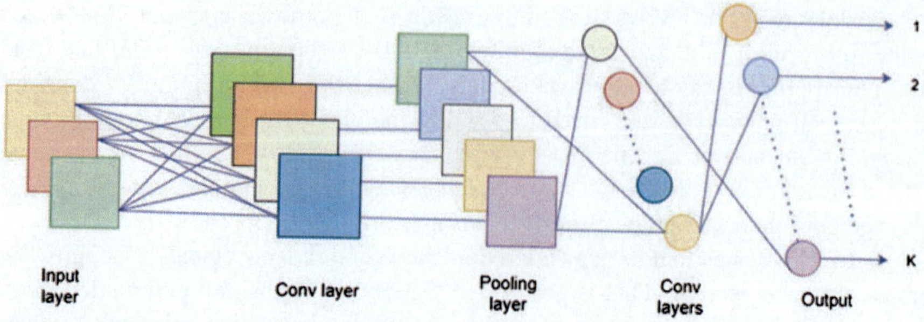

Fig. 1 Architecture diagram of the ResNet model

First, the raw Break His dataset must be taken, and it has to be rearranged so that it consists of one set of images for training the model and another set of images for the validation of the model. The dataset has benign and malignant images. All the images are distributed according to the system. The next step to be performed is the data preprocessing. It consists of five phases such as data mounting, data resizing, data shuffling, data rescaling, and label encoding. For data mounting, the platform used for the implementation of the model is Google Colab. It gives free access to RAM and GPU. Zip the dataset into a folder and upload it into the Google Drive. Then, unzip the data into Colab using suitable commands. In case of data rescaling, all the images which are of the size 700 $\times$ 460 pixels are converted into 224 $\times$ 224 pixels. For data shuffling, an uploaded dataset folder is divided into training set and validation set. The training set consists of the benign and malignant folders. The validation dataset also consists of benign and malignant folder. In the rescaling, the data will be multiplied with a value before the implementation of the model. The label encoding consists of the value 0 is assigned to benign. The value 1 is assigned to malignant. Figure 2 depicts the component-module diagram of the system:

The residual network constructed below consists of input layer, convolution layer, pooling layer, and output layer. The input layer consists of the pixel matrices of all the images that are present in the dataset. Then, the output of the first layer is then passed on to the second one. It administers various mathematical functions on the input matrices and extracts several features that are required into feature maps. Then, the pooling layer recapitulates all the feature maps that are extracted and lessen their proportions, and it results in the reduction of the training parameters.

4 The Proposed System

Data augmentation is performed to enlarge the dataset, and this can be done using the image data generator. With the help of larger number of images, the model will become more agile as it is training on a numerous amount of examples. Each image consists of large pixel values. The pixel value starts from 0 to 255. This type of

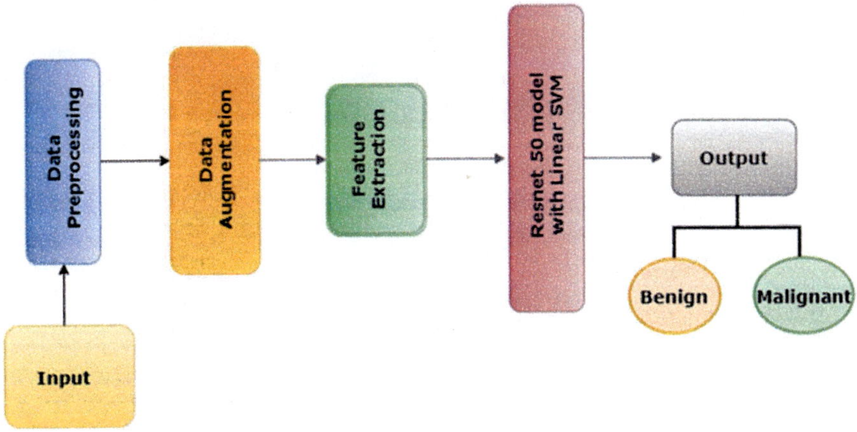

Fig. 2 Component-module diagram

huge values cannot be handled by the network, and the training set does not fit into the model perfectly. Hence to avoid all these issues, divide each value with 255 to convert all the coefficient values between 0 and 1. The decimal values are small when compared to numerical digits, and this type of approach is preferred to prevent any fitting issues in future. The class mode should be set to its default type which is categorical because of the implementation of the binary class classification. A sample plot is plotted using the matplotlib library to understand the dataset. Figure 3 shows the sample image representation.

In the subplot function, values are given as 1,3 which means that the plot is constructed in a single row and three columns which automatically results in the three plots. Each figure size should be given as 10×10. The flatten () command is used to convert the list of images into a one-dimensional array. ResNet is used to build a deeper network compared to normal CNN and simultaneously find optimized number of layers to negate the vanishing gradient problem. Linear SVM algorithm is the fastest machine learning algorithm that can solve the binary and multi-class

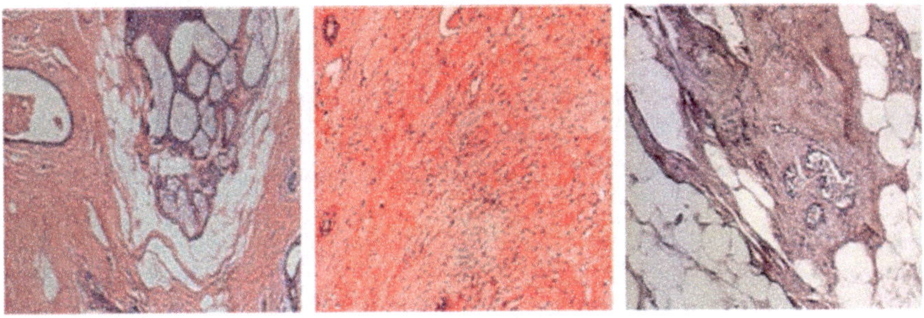

Fig. 3 Sample representation of images

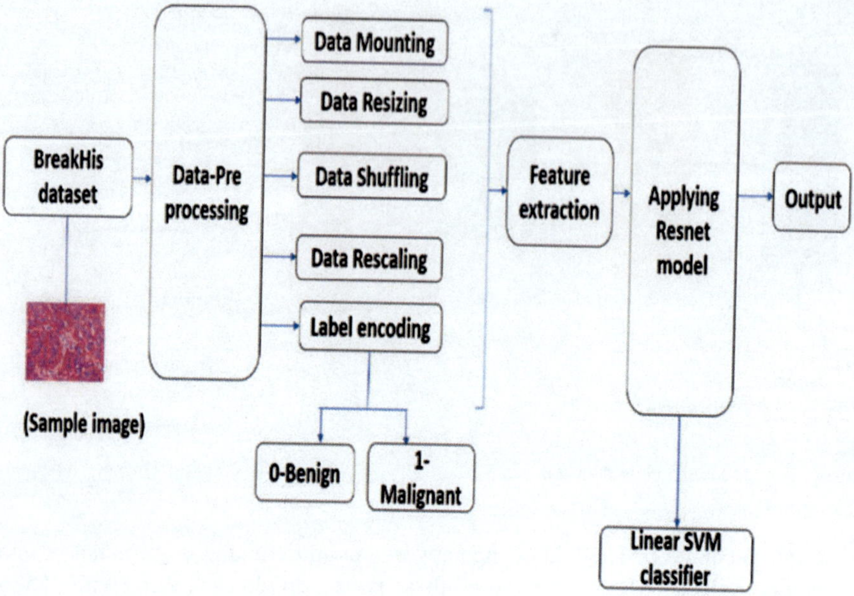

Fig. 4 Flowchart of the system

classifications. Building the required ResNet model using the linear SVM classifier. Figure 4 shows the various stages present in the flowchart of the system:

The input shape consists of the image height and width as the parameters, and the number three represents the RGB representation of the images. The ImageNet pretrained model is taken as an input to the ResNet-50 model. Average pooling is performed on the feature maps. The base model is named as model finally. The dense layer consists of 1024 neurons as input, and l2 regularizer is taken as one of the parameters to prevent the chances of overfitting. Batch normalization and dropout layers are added to build a stable model which do not collapse due to simple errors. The activation function value is taken as ReLU, since it impels all neurons to work at the same time. Other functions like sigmoid or Softmax can also be used. The SVM loss function should be defined which consists of a mathematical combination of the input weights and the loss values. The hinge loss and the regularization loss values are calculated above, and the value of categorical hinge function must be returned. The model is implemented using the fit function, and the epochs are used to count the number of steps in which the code is executed. The epoch steps can be calculated by dividing the total number of the training images with the batch size.

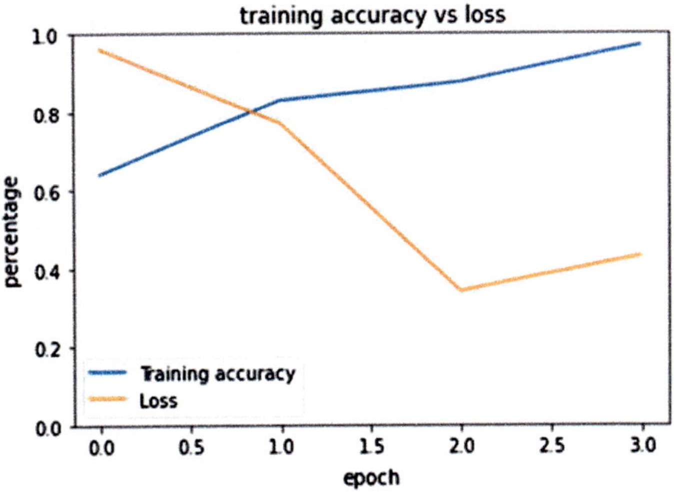

Fig. 5 Graph between training accuracy and loss

Table 2 Representation of various accuracies of the model

Observation	Loss value	Accuracy (%)
1	0.96	64.02
2	0.77	82.91
3	0.43	87.56
4	0.34	96.92

5 Results

The batch size and the epochs are assigned 160 and 5 values, respectively. The training accuracy obtained is 96.9%, and the validation accuracy is 62.9%. Figure 5 depicts the values of accuracy and loss in the form of a graph.

The training loss obtained is 0.34. The accuracies are described using Table 2.

Figure 6 is used to represent the graph between the training and validation accuracies:

6 Conclusion

In this paper, an effort is made to perform the binary classification of the histopathological images from the Break His dataset. The model constructed is called as residual neural networks model which comes under transfer learning algorithms. The proposed model involves data preprocessing, data augmentation, feature extraction, and application of ResNet architecture. The enlargement of the data must be

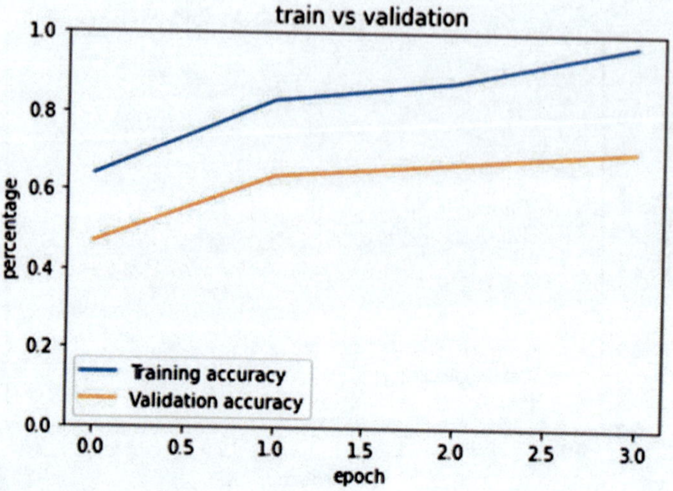

Fig. 6 Graph between training and validation accuracy

performed to improve the quality of the model. The model which is used is the ResNet-50 which is pretrained on the ImageNet dataset. After building the model, the hyperparameter optimization should be performed to analyze the best results. In this model, if the batch size is increased continuously, then the accuracy of the model decreases which results in the poor performance. Hence, the batch size should be appropriate, and the accuracy achieved alone using the residual neural network model is 92.55%. This can be further improved by the application of the linear SVM classifier. The accuracy achieved by the model along with the linear SVM classifier comes out to be 96.9%.

7 Future Work

In future, this work can be extended to other transfer learning approaches such as VGGNet and Inception V3 models. Various datasets can also be used to compare the performances of the models. An ensemble model can also be developed from the models.

References

1. WHO cancer information (on line). Available: https://www.who.int/news-room/fact-sheets/detail/cancer
2. N. Tripathi et al., Barriers for early detection of cancer amongst Indian rural women. South Asian J. Cancer **3**(2),122–127 (2014). https://doi.org/10.4103/2278-330X.130449

3. American Cancer Society Guidelines for the Early Detection of Cancer (on line). Available: https://www.cancer.org/healthy/find-cancer-early/american-cancer-society-guidelines-for-the-early-detection-of-cancer.html
4. F. Spanhol, L.S. Oliveira, C. Petitjean, L. Heutte, A dataset for breast cancer histopathological image classification. IEEE Trans. Biomed. Eng. (TBME) **63**(7), 1455–1462 (2016)
5. H. Basly, W. Ouarda, F.E. Sayadi, B. Ouni, A.M. Alimi, CNN-SVM learning approach based human activity recognition, in *Image and Signal Processing. ICISP 2020. Lecture Notes in Computer Science*, vol. 12119, ed. by El Moataz A., Mammass D., Mansouri A., Nouboud F. (eds) (Springer, Cham, 2020). https://doi.org/10.1007/978-3-030-51935-3_29
6. W. Zou, H. Lu, K. Yan, M. Ye, Breast cancer histopathological image classification using deep learning, in *2019 10th International Conference on Information Technology in Medicine and Education (ITME)* (Qingdao, China, 2019), pp. 53–57. https://doi.org/10.1109/ITME.2019.00023
7. J. de Matos, A. de Souza Britto, L.E.S. de Oliveira, A.L. Koerich,Texture CNN for histopathological image classification, in *2019 IEEE 32nd International Symposium on Computer-Based Medical Systems (CBMS)* (Cordoba, Spain, 2019), pp. 580–583. doi: https://doi.org/10.1109/CBMS.2019.00120
8. D. Ghosh, A. Kumar, P. Ghosal, T. Chowdhury, A. Sadhu, D. Nandi, Breast lesion segmentation in ultrasound images using deep convolutional neural networks, in *2020 IEEE Calcutta Conference (CALCON)* (Kolkata, India, 2020), pp. 318–322. doi: https://doi.org/10.1109/CALCON49167.2020.9106568
9. Y.J. Tan, K.S. Sim, F.F. Ting, Breast cancer detection using convolutional neural networks for mammogram imaging system, in *2017 International Conference on Robotics, Automation and Sciences (ICORAS)* (Melaka, 2017), pp. 1–5. https://doi.org/10.1109/ICORAS.2017.8308076
10. C.P. Nguyen, A. Hoang Vo, B.T. Nguyen, Breast cancer histology image classification using deep learning, in *2019 19th International Symposium on Communications and Information Technologies (ISCIT)* (Ho Chi Minh City, Vietnam, 2019), pp. 366–370. doi: https://doi.org/10.1109/ISCIT.2019.8905196
11. P. Mohapatra, B. Panda, S. Swain, G6270058719 cancer.8 (2019)
12. Z. Xiang, Z. Ting, F. Weiyan, L. Cong,Breast cancer diagnosis from histopathological image based on deep learning, in *2019 Chinese Control And Decision Conference (CCDC)* (Nanchang, China, 2019), pp. 4616–4619. doi: https://doi.org/10.1109/CCDC.2019.8833431
13. H.M. Ahmad, S. Ghuffar, K. Khurshid, Classification of breast cancer histology images using transfer learning, in *2019 16th International Bhurban Conference on Applied Sciences and Technology (IBCAST)* (Islamabad, Pakistan, 2019), pp. 328–332. https://doi.org/10.1109/IBCAST.2019.8667221
14. S. Lee, M. Amgad, M. Masoud, R. Subramanian, D. Gutman, L. Cooper, An ensemble-based active learning for breast cancer classification, in *2019 IEEE International Conference on Bioinformatics and Biomedicine (BIBM)* (San Diego, CA, USA, 2019), pp. 2549–2553. https://doi.org/10.1109/BIBM47256.2019.8983317
15. R. Singh, T. Ahmed, A. Kumar, A.K. Singh, A.K. Pandey, S.K. Singh, Imbalanced breast cancer classification using transfer learning, in *IEEE/ACM Transactions on Computational Biology and Bioinformatics*. https://doi.org/10.1109/TCBB.2020.2980831
16. F. Siddiqui, S. Gupta, S. Dubey, S. Murtuza, A. Jain, Classification and diagnosis of invasive ductal carcinoma using deep learning, in *2020 10th International Conference on Cloud Computing, Data Science & Engineering (Confluence)* (Noida, India, 2020), pp. 242–247. https://doi.org/10.1109/Confluence47617.2020.9058077
17. Z. Hameed, S. Zahia, B. Garcia- Zapirain, J. Javier Aguirre, A. María Vanegas, Breast cancer histopathology image classification using an ensemble of deep learning models. Sensors **20**, 4373 (2020)
18. P.T. Nguyen, T.T. Nguyen, N.C. Nguyen, T.T. Le,Multiclass breast cancer classification using convolutional neural network, in *2019 International Symposium on Electrical and Electronics Engineering (ISEE)* (Ho Chi Minh, Vietnam, 2019) pp. 130–134. doi: https://doi.org/10.1109/ISEE2.2019.8920916

19. S. Singh, R. Kumar, Histopathological image analysis for breast cancer detection using cubic SVM, in *2020 7th International Conference on Signal Processing and Integrated Networks (SPIN)* (Noida, India, 2020), pp. 498–503. https://doi.org/10.1109/SPIN48934.2020.9071218
20. S. Asare, F. You, O. Tettey, Efficient, ultra-facile breast cancer histopathological images classification approach utilizing deep learning optimizers. Int. J. Comput. Appl. **177**, 1–9 (2020). https://doi.org/10.5120/ijca2020919875
21. F. Spanhol, P. Cavalin, L.S. Oliveira, C. Petitjean, L. Heutte, Deep features for breast cancer histopathological image classification, in *2017 IEEE International Conference on Systems, Man, and Cybernetics (IEEE SMC 2017)* (Banff, Canada, 2017)
22. F. Spanhol, L.S. Oliveira, C. Petitjean, L. Heutte, Breast cancer histopathological image classification using convolutional neural network, in *International Joint Conference on Neural Networks (IJCNN 2016)* (Vancouver, Canada, 2016)
23. C. Thallam, A. Peruboyina, S.S.T. Raju, N. Sampath, Early stage lung cancer prediction using various machine learning techniques, in *2020 4th International Conference on Electronics, Communication and Aerospace Technology (ICECA)* (Coimbatore, India, 2020), pp. 1285–1292. https://doi.org/10.1109/ICECA49313.2020.9297576
24. N. Gouda, J. Amudha, Skin cancer classification using ResNet, in *2020 IEEE 5th International Conference on Computing Communication and Automation (ICCCA)* (Greater Noida, India, 2020), pp. 536–541. https://doi.org/10.1109/ICCCA49541.2020.9250855
25. P. Tyagi, T. Singh, R. Nayar, S. Kumar, Performance comparison and analysis of medical image segmentation techniques, in *2018 IEEE International Conference on Current Trends in Advanced Computing (ICCTAC)* (Bangalore, India, 2018), pp. 1–6. https://doi.org/10.1109/ICCTAC.2018.8370392
26. S.S. Shastri, P.C. Nair, D. Gupta, R.C. Nayar, R. Rao, A. Ram, Breast cancer diagnosis and prognosis using machine learning techniques, in *Intelligent Systems Technologies and Applications. ISTA 2017. Advances in Intelligent Systems and Computing*, vol. 683, ed. by S. Thampi, S. Mitra, J. Mukhopadhyay, K.C. Li, A. James, S. Berretti (Springer, Cham, 2018). https://doi.org/10.1007/978-3-319-68385-0_28
27. S. Tamuly, C. Jyotsna, J. Amudha, Deep learning model for image classification, in *Computational Vision and Bio-Inspired Computing. ICCVBIC 2019. Advances in Intelligent Systems and Computing*, vol. 1108, ed. by S. Smys, J. Tavares, V. Balas, A. Iliyasu (Springer, Cham, 2020). https://doi.org/10.1007/978-3-030-37218-7_36
28. T. Singh, S. Karanchery, Universal image segmentation technique for cancer detection in medical images, in *2019 10th International Conference on Computing, Communication and Networking Technologies (ICCCNT)* (2019)

Optimization of Textual Index Construction Using Compressed Parallel Wavelet Tree

Arun Kumar Yadav, Sonam Gupta, Divakar Yadav, and Bharti Shukla

Abstract Nowadays, large number of information are flowing through Internet that requires better storage and optimization technique to retrieve relevant information. In the past, a lot of researches have been carried out in the field of index optimization for reducing storage space, construction time and retrieval time. In recent times, researchers have used various succinct data structures for indexing the textual data. In this paper, we propose two algorithms for compressing the textual index that reduces the index construction time. In the baseline approach (Algorithm 1), we apply traditional wavelet tree with LZW compression technique (TWL) to construct the index. In the second approach (Algorithm 2), we propose LZW compression with parallel wavelet tree (PWTL) to construct the textual index. The algorithms are evaluated on I3, I5 and I7 processor with multiple sizes of index. The results show that proposed algorithms outperform as compared to traditional wavelet tree in terms of index construction time.

Keywords Compression · Parallel wavelet tree · Complexity · File size · Textual data

1 Introduction

Optimization of information is always a challenging issue, in terms of retrieving relevant information and storing information in structured form. Indexing plays an important role to optimize information in terms of retrieval and storage. In the past,

A. K. Yadav · D. Yadav
National Institute of Technology, Hamirpur, HP, India
e-mail: ayadav@nith.ac.in

D. Yadav
e-mail: dsy99@rediffmail.com

S. Gupta (✉) · B. Shukla
Ajay Kumar Garg Engineering College, Ghaziabad, India
e-mail: guptasonam@akgec.ac.in

A. K. Bashir et al. (eds.), *Proceedings of International Conference on Computing and Communication Networks*, Lecture Notes in Networks and Systems 394,
https://doi.org/10.1007/978-981-19-0604-6_43

many researches have been carried out to reduce index construction/retrieval time and storage optimization [1]. Also, researchers used many data structures for constructing the inverted index such as: B-tree, B+ tree, Hash tree, R-tree, R* tree and wavelet tree [2]. In recent years, the wavelet tree becomes popular, and versatile data structure is used to store the data in sequence format, solve range quartile query, range insertion query and range next value query. Most of the search engines use indexing approaches to store the data from web and retrieve it in optimal time. Due to increase of the exponential growth of data, optimal indexing approach is required. The optimal indexing approach works in two dimensions, firstly takes minimum space stored and minimum time to retrieve relevant data [3]. In the past, researcher applies multiple indexing approaches along with compression technique to reduce storage space of constructed index. In recent years, a succinct data structure (wavelet tree) is used for indexing and retrieval. Wavelet tree is succinct data structure and is used for less space to stored data and take minimum time for search data in database [4]. Its main applications are demonstrating of sequence, reorder of element and grid of point, etc. It works as binary trees which contain left child and right child. Left node is denoted by 0, and right node is denoted by 1. Wavelet tree applied string operation, graph inversion, document retrieval, etc. Basic tool required to construct wavelet tree is select and rank operation.

LZW is a popular compression technique, used for data compression. The formulation of this compression is to reduce the cost of coding. LZW codes are the simple method of text coding; it reduces the redundancy of LZW text. The LZW is having the one-pass compression method. It depends upon the efficiency. It is significantly faster but more complex to construct such a wavelet tree [5].

In this paper, we proposed two algorithms of index construction with variants of wavelet tree and LZW compression technique. For this, we apply the LZW compression scheme in traditional wavelet tree and a parallel wavelet tree. Also, we calculated the indexed constructed time of compressed wavelet tree.

Remaining paper organizes as follows. It consists of several sections. Section 2 consists of literature review which is categorized into three subsections to meet our objectives. Section 3 consists of the proposed methodology of compressed traditional wavelet tee and compressed parallel wavelet tree. Section 4 consists of results and experiments of proposed methodology. Section 5 consists of the conclusion and future scope of our proposed model.

2 Literature Review

In recent times, researchers used various data structures for index optimization, i.e., index construction time, storage optimization and reducing retrieval time. This section describes the application of variants of wavelet tree for index optimization.

In the paper [1], authors deployed the concept of self-indexing of wavelet tree. They joined the wavelet tree with char-based Huffman code and word-based Huffman code. In the paper [6], authors proposed textual indexing to access the index data quickly. In the paper [7], researchers proposed compressed textual indexing on variable query length. In the paper [8], the authors proposed wavelet tree-based algorithm for retrieving a piece of information. It focused on solving range quintile queries and range insertion queries, etc. using wavelet tree. In the paper [2], authors discuss the various data structures such as Tree, B-Tree and wavelet tree for index construction and evaluate the performance. They commented that Hash tree scored less than compared to others it scored up to level 5. The B-tree is scored 4 up to the level, and wavelet tree scored up to 3 levels up to 5. Again, in the paper [9], optimized text retrieval system is used to optimize textual indexing. In the paper [10], authors proposed wavelet tree-based geographical index to search spatial queries. Also, they proposed the concept of minimum bounded rectangle (MBR) to search geographical documents.

In the paper [11], authors proposed compression technique to reduce the redundancy of the database and increase the storage and retrieval efficiency. They used LZW compression technique and achieved redundancy gain ration 3–7. In the papers [12, 13], authors proposed forward moving and frequently used sentence compression using LZW compression. It is an improved algorithm for forward moving, and frequently used entries are optimized. LZW string table can store the prefix value. It implemented the LZW algorithm for the chained lists.

In the paper [14], the researches worked on the optimization of LZW codes. They comprise LZW codes and its optimization technique using modified compression technique based on optimality of LZW code (MOLZW). In the paper [15], the author performed LZW compression pattern matching on different dataset/format like txt, doc, and Java or in C format. They focused on the multiple patterns matching of codes for the perfect compression of codes. Again, in the paper [16], authors focused on LZW compression and color-coding technique for stenography. They used LZW-based color-coding approach to reduce the size of data and wrap the confidential data into the email address. In the paper [17], the authors proposed the select/rank operation of wavelet tree in practical implementation. Also, they proposed the empirical entropy concept in wavelet tree.

In the paper [18], authors proposed secured secure wavelet matrix to search data from bioinformatics databases. They commented that proposed algorithm work in logarithmic time $|\Sigma|$, where Σ is the number of unique characters. Also, in the paper [19], authors proposed the index constriction using wavelet tree in poly-logarithmic depth. In the papers [20, 21], the authors proposed parallel wavelet tree algorithm for index construction. They used level-by-level parallelism. The proposed algorithm takes $O(n)$ work and $O(\log n)$ depth. In this paper, they proposed two algorithms. The first algorithm takes $O(n)$ time, and working space is $O(n \log \sigma + \sigma \lg \sigma)$ bits. Second algorithm construction has $O(\log n)$ time, and working space is $O(n \log\sigma + p\, \sigma \lg n/\lg \sigma)$ bits, and p is the number of cores. In the paper [22], authors worked on the existing sequential algorithms constructing the rank/select structures and stored the data on node of wavelet tree. They imposed the concept of wavelet packets for texture indexing. It presented a wavelet packets solution on the synthetic data, which covered the real dataset of different sizes of images and texture classes. It divided the dataset used by wavelet packet in the proportion of 3:7 while testing and training.

As discussed in this section, a lot of researches have been done on index optimization using various data structures. The research has also moved to compression of index by applying compression technique along with data structures used in indexing. In this paper, we proposed LZW compression on traditional wavelet tree and parallel wavelet tree to optimize the index construction time.

3 Proposed Methodology

This section describes the LZW compression on wavelet tree and parallel wavelet tree for optimizing the textual index construction time. We define two algorithms for our proposed method, i.e., wavelet tree with LZW (WTL) and compressed parallel wavelet tree with LZW (PWTL). Both algorithms compress the textual index and reduce the index construction time as compared to traditional wavelet tree. The algorithms are evaluated on I3, I5 and I7 processors.

Figure 1 shows the working of proposed method of construction of compressed

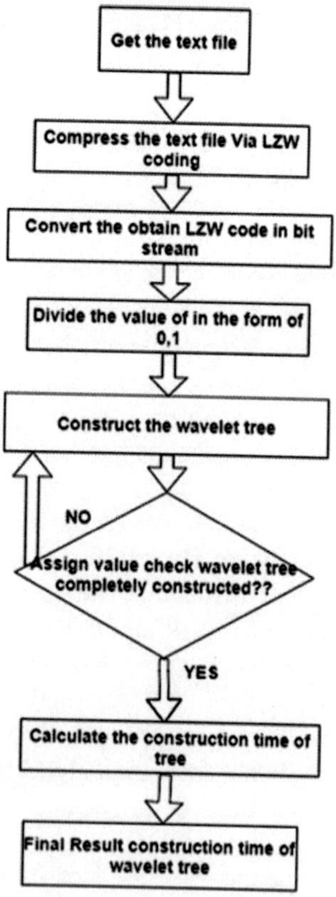

Fig. 1 Proposed methodology of wavelet tree

wavelet tree. The detailed explanation of algorithms is shown in Sects. 3.1 and 3.2.

3.1 *Wavelet Tree with LZW (WTL)*

To construct compressed wavelet tree with LZW, the textual data is mapped with LZW code. The construction of traditional wavelet tree is start from root node. At root node, textual data grasps the location at 0 and 1 according to indexing. Left-side array contains 0 value, and right-side array contains 1 value. Repeat the same step till wavelet tree is created.

To construct WTL, we proposed Algorithm 1. This algorithm uses abbreviations such as: 'C' is code, 'c' is character, where '||' is concatenation, 'Sn' is a new symbol, 'Sp' is a previous symbol, 'S' is the symbol of output, dictionary 'd', 'n' is length of new symbol.

Algorithm 1: Wavelet Tree with LZW(WTL)

Step 1	:	Apply the LZW algorithm on the textual data
		{
		I. Declare LZW code, Output's', Character c'.
		II. Set the sp'=s'.
		}

Step 2	:	Get the next C,
		{
		I. If Variable is not present in d
		II. Then, s = S p' || c'
		III. Else
		IV. Assign, s = sc' where sc' is the symbol of C in the dictionary Output s'
		V. Assign sn = Sp' || C'
		VI. Add sn to dictionary.
		}

Step 3	:	If length of the new symbol is greater than the dictionary,
		{
		I. Sb = Pb (n-1)…. Where Sb= Secret bit.
		II. If n = Dictionary symbol and exists
		III. Sb = Pb (n-1)….. Where Pb= Parity bit.
		IV. Set Sp'=s and continue to Step2
		}

Step 4	:	With the help of formula, create a first wavelet tree,
		{
		Mid = Start + (end–start) / 2.
		}

Step 5	:	Assign the value to nodes 0, 1 in wavelet tree.
Step 6	:	The nodes on the left side of wavelet tree represented with the 0, and nodes on the right side of wavelet tree with 1.
Step 7	:	Again, create a wavelet tree for left array and right array.
Step 8	:	Continue S2 to S7 step until we got minimum value at left side and maximum value at right side. Also, we say entire tree will split in a single array.
Step 9	:	Calculated the constructed tree time in milliseconds.
Step 10	:	Show the Construction time of traditional wavelet tree with LZW

3.2 *Parallel Wavelet Tree with LZW (PWTL)*

To construct PWTL, we used two WTL parallelly. The construction of both WTL starts from root node. At root node, textual data grasps the location at 0 and 1 according to indexing. Left-side array contains 0 value, and right-side array contains 1 value. Repeat the same step till wavelet tree is created parallelly.

To construct PWTL, we proposed Algorithm 2. This algorithm uses abbreviations such as: 'C' is code, 'c' is character, where '||' is concatenation, 'Sn' is a new symbol, 'Sp' is a previous symbol, 'S' is the symbol of output, dictionary 'd', 'n' is length of new symbol.

Algorithm 2: Parallel Wavelet Tree with LZW(PWTL)

Step 1	:	Apply the LZW algorithm on the textual data
		{
		III. Declare LZW code, Output's', Character c'.
		IV. Set the sp'=s'.
		}

Step 2	:	Get the next C,
		{
		VII. If Variable is not present in d
		VIII. Then, s = S p' || c'
		IX. Else
		X. Assign, s = sc' where sc' is the symbol of C in the dictionary Output s'
		XI. Assign sn = Sp' || C'
		XII. Add sn to dictionary.
		}

Step 3	:	If length of the new symbol is greater than the dictionary,
		{
		V. Sb = Pb (n-1)…. Where Sb= Secret bit.
		VI. If n = Dictionary symbol and exists
		VII. Sb = Pb (n-1)….. Where Pb= Parity bit.
		VIII. Set Sp'=s and continue to Step2
		}

Step 4	:	Parallel wavelet tree we require two traditional wavelet tree at the same.
Step 5	:	Time With the help of formula, create a first wavelet tree, { Mid = Start + (end–start) / 2. }
Step 6	:	Assign the value of nodes 0,1 in wavelet tree. We show the nodes on the left side of wavelet tree with the 0, nodes on the Right side of wavelet tree with 1.
Step 7	:	Again, create a wavelet tree for left array and right array.
Step 8	:	Continue S4 to S6 step until we got minimum value at left side and maximum value at right side. Also, we say entire tree will split in a single array.
Step 9	:	For another wavelet tree, continue step S4 to S7. One tree store on left side and another tree store on right side. Finally, we get last tree.
Step 10	:	Calculate, the constructed tree time in milliseconds of parallel wavelet tree.
Step 11	:	Show the Construction time of Parallel wavelet tree with LZW

4 Result and Analysis

This section describes the implementation and evaluation of both the proposed algorithms (TWL and PWTL). The algorithms are implemented in Java programming language and evaluated on I-3(CPU Intel® core™ processor, RAM 8 GB), I-5(Intel® core™ processor. Size of RAM is 8GB), I-7(I-7-9500 CPU is Intel® core™ processor, RAM 8GB) using NetBeansIDE-8.2 IDE.

Table 1 shows the comparison between TWL and PWTL on I3 processor. This experiment takes four different variants of file size and evaluates construction time on I3 processor. Table 1 shows that LZW compression on parallel wavelet tree (PWTL) required less time to construct than wavelet tree with LZW compression (TWL) on I3 processor.

Table 2 shows the comparison between TWL and PWTL on I3 processor. This experiment takes four different variants of file size and evaluates construction time on I3 processor. Table 1 shows that LZW compression on parallel wavelet tree (PWTL) required less time to construct than wavelet tree with LZW compression (TWL) on I5 processor.

Table 3 shows the comparison between TWL and PWTL on I3 processor. This experiment takes four different variants of file size and evaluates construction time on I3 processor. Table 1 shows that LZW compression on parallel wavelet tree (PWTL) required less time to construct than wavelet tree with LZW compression (TWL) on I7 processor.

Figure 2 shows the comparison of index construction time using WT, WTL and PWTL on I3, I5 and I7 processors. Results show that PWTL outperforms as compared

Table 1 Comparison of WTL and PWTL on I3 processor (different file size)

S. No.	File size	Wavelet tree with LZW (WTL) (s)	Parallel wavelet tree with LZW (PWTL) (s)
1	28 KB	15	4
2	237 KB	29	7
3	1 MB	25	6
4	1.20 GB	35	8

Table 2 Comparison of WTL and PWTL on I5 processor (different file size)

S. No.	File size	Wavelet tree with LZW (WTL) (s)	Parallel wavelet tree with LZW (PWTL) (s)
1	28 KB	16	5
2	237 KB	4	5
3	1 MB	12	6
4	1.20 GB	44	7

Table 3 Comparison of WTL and PWTL on I7 processor (different file size)

S. No.	File size	Wavelet tree with LZW (WTL) (s)	Parallel wavelet tree with LZW (PWTL) (s)
1	28 KB	3	5
2	237 KB	9	6
3	1 MB	5	5
4	1.20 GB	7	5

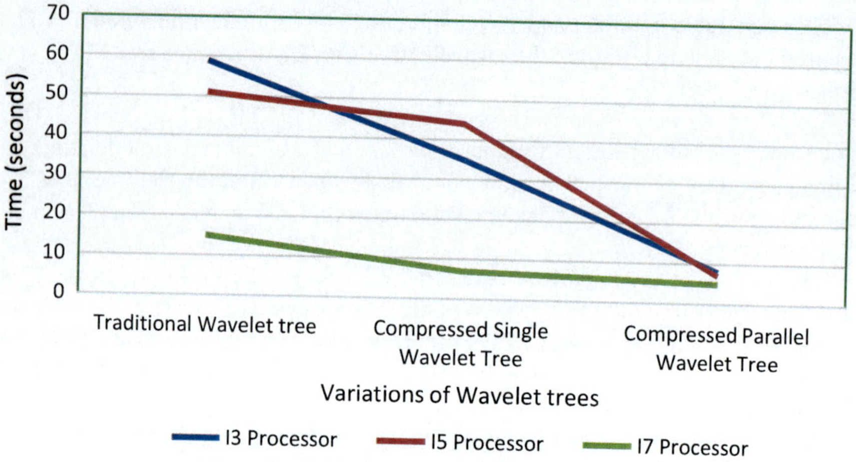

Fig. 2 Comparison of construction time between WT, TWTL and PWTL the file size 1.20 GB

to WT and WTL. To make fare evaluation, the construction time is evaluated on fixed length file size (1.20 GB).

5 Conclusion and Future Work

In this paper, we proposed WTL and PWTL algorithms to compress the index that causes to reduce the index construction time. The proposed scheme has been evaluated on I3, I5 and I7 processors with variant size of files. The basic advantage of this approach is to reduce the size of index and finally reduce the index construction time. Another advantage is that when we looking for a word in this tree, we get it easily by traversing the parallel wavelet tree. The parallel wavelet tree construction algorithm works efficiently during parallelism because it depends on linearly on the depth of small words. In the future, we will extend the work for the multithreaded tree that may provide the better outcomes. The ideas of researchers will extend to constructing the LZW compressed wavelet trees of arbitrary shape, multinary wavelet trees, as well as wavelet matrices

Acknowledgements Council of Science and Technology, Lucknow, Uttar Pradesh, supported this research via Project Sanction letter number CST/D-3330.

References

1. N.R. Brisaboa, Y. Cillero, A. Fariña, S. Ladra, O. Pedreira, A new approach for document indexingusing wavelet trees, in *Proceedings—International Workshop on Database and Expert Systems Applications*, DEXA (2007). https://doi.org/10.1109/DEXA.2007.118
2. A.K. Yadav, D. Yadav, R. Prasad, Efficient textual web retrieval using wavelet tree. Int. J. Inform. Retrieval Res. **6**(4), 16–29 (2016). https://doi.org/10.4018/ijirr.2016100102
3. C.K. Jha, M.H. Kolekar, Empirical mode decomposition and wavelet transform based ECG data compression scheme. IRBM **42**(1), 65–72 (2021). ISSN 1959-0318. https://doi.org/10.1016/j.irbm.2020.05.008
4. S. Gupta, et al. Technologies in health care domain: a systematic review. IJEC **16**(1), 33–44 (2020). https://doi.org/10.4018/IJeC.2020010103
5. N. Katiyar, et al. Wavelet tree ensembles with machine learning and its classification. J. Phys. Conf. Ser. 1 (1998). https://doi.org/10.1088/1742-6596/1998/1/012001
6. R. Grossi, A. Gupta, J. Scott Vitter, *High-Order Entropy-Compressed Text Indexes* (n.d.). http://www.ittc.ku.edu/~jsv/Papers/GGV03.entropy.pdf
7. A. Yadav, D. Yadav, Wavelet tree based hybrid geo-textual indexing technique for geographical search. Indian J. Sci. Technol. **8**(33), 1–7 (2015). https://doi.org/10.17485/ijst/2015/v8i33/72962
8. T. Gagie, G. Navarro, S.J. Puglisi, New algorithms on wavelet trees and applications to information retrieval. Theoret. Comput. Sci. **426–427**, 25–41 (2012). https://doi.org/10.1016/j.tcs.2011.12.002
9. A.K. Yadav, D. Yadav, D. Rai, Efficient methods to generate inverted indexes for ir. Adv. Intell. Syst. Comput. **435**, 431–440 (2016). https://doi.org/10.1007/978-81-322-2757-1_43
10. A.K. Yadav, D. Yadav, Wavelet tree based dual indexing technique for geographical search. Int. Arab J. Inform. Technol. **16**(4), 624–632 (2019)
11. G. Lakhani, Reducing coding redundancy in LZW. Inform. Sci. **176**, 1417–1434 (2004). https://www.researchgate.net/publication/220310804
12. F. Zhang, Z. Li, M.C.L. Wen, X. Jia, C. Chen, Implementation and optimization of LZW compression algorithm based on bridge vibration data. Procedia Eng., **15**, 1570–1574 (2011). https://doi.org/10.1016/j.proeng.2011.08.292
13. Z. Wang, H. Yang, T. Cheng, C. Cheng, A high performance reversible data hiding scheme for LZW codes. J. Syst. Softw. **86**, 2771–2778 (2013). https://doi.org/10.1016/j.jss.2013.06.024
14. U. Nandi, J.K. Mandal, Modified compression techniques based on optimality of LZW Code (MOLZW). Procedia Technol. **10**, 949–956 (2013). https://doi.org/10.1016/j.protcy.2013.12.442
15. P. Gawrychowski, Simple and efficient LZW-compressed multiple pattern matching. J. Discr. Algorithms **25**, 34–41 (2014). https://doi.org/10.1016/j.jda.2013.10.004
16. A. Malik, G. Sikka, H.K. Verma, A high capacity text steganography scheme based on LZW compression and color coding. Eng. Sci. Technol. Int. J. **20**(1), 72–79 (2017). https://doi.org/10.1016/j.jestch.2016.06.005
17. R. Grossi, J.S. Vitter, B. Xu, Wavelet trees: from theory to practice, in *Proceedings—1st International Conference on Data Compression, Communication, and Processing*, CCP 2011, pp. 210–221 (2011). https://doi.org/10.1109/CCP.2011.16
18. H. Sudo, M. Jimbo, K. Nuida, K. Shimizu, Secure wavelet matrix: Alphabet-friendly privacy-preserving string search for bioinformatics. IEEE/ACM Trans. Comput. Biol. Bioinf. **16**(5), 1675–1684 (2019). https://doi.org/10.1109/TCBB.2018.2814039

19. J. Shun, *Parallel Wavelet Tree Construction. Data Compression Conference Proceedings*, 2015-July, 63– 72 (2015). https://doi.org/10.1109/DCC.2015.7
20. J. Fuentes-, E. Elejalde, L. Ferres, D. Seco, Parallel construction of wavelet trees on multicore architectures. Knowl. Inf. Syst. **51**(3), 1043–1066 (2017). https://doi.org/10.1007/s10115-016-1000-6
21. J. Shun, Improved parallel construction of wavelet trees and rank/select structures. Inf. Comput. **273**, 104516 (2020). https://doi.org/10.1016/j.ic.2020.104516
22. A. Vidal, J.F. Silva, C. Busso, Discriminative features for texture retrieval using wavelet packets. IEEE Access **7**, 148882–148896 (2019). https://doi.org/10.1109/ACCESS.2019.2947006

Robust Approach for Detecting Face Mask Using Deep Learning and Its Comparative Analysis

Abhijeet Singh, Amandeep Kaur, and Sonali Vyas

Abstract The world is fighting against the novel coronavirus, and a lot of people have lost their lives with the scenario getting bad to worse and worst. This infection is communicated from one individual to another while wheezing or talking as drops. To prevent Covid-19, wearing masks is very beneficial. In this paper, an existing model, 'DenseNet201', is being modified to efficiently track the persons who are wearing a mask or not. This system uses a convolutional neural network (CNN) and computer vision to limit the risk of Covid-19 and make sure nobody violates the rule. The dataset used in the process contains two classes, namely 'with mask' and 'without a mask'. Data pre-processing and splitting take place before the model training; then comparative analysis has been made in between the modified versions of the three transfer learning models, viz. DenseNet201, InceptionResnetV2, and ResNet101V2 to validate the modified model's efficiency. Results suggest that the revised version of DenseNet201 is very effective and can detect the events where face masks are not used at all or in an improper manner, with an accuracy of 98.90%. Various other metrics for performance are also being evaluated and reported in the paper. This model can work with images and videos/CCTV cameras using the help of OpenCV, TensorFlow, and Keras.

Keywords Computer vision · Convolutional neural network · Deep learning · Transfer learning · Covid-19

1 Introduction

The Covid-19 pandemic has been a few each day's lives which have created a troubling circumstance. This pandemic is affirmed as a global pandemic by World Health

A. Singh (✉) · A. Kaur
Guru Tegh Bahadur Institute of Technology, New Delhi, India
e-mail: abhivirdi05@gmail.com

S. Vyas
University of Petroleum and Energy Studies, Dehradun, India

A. K. Bashir et al. (eds.), *Proceedings of International Conference on Computing and Communication Networks*, Lecture Notes in Networks and Systems 394,
https://doi.org/10.1007/978-981-19-0604-6_44

Fig. 1 Algorithm workflow

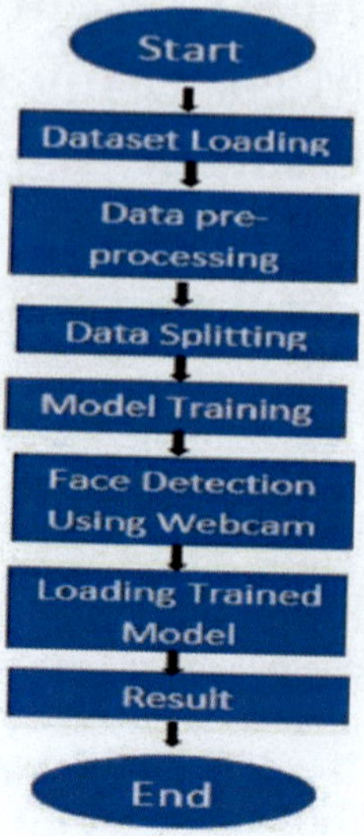

Organization because it affects the lives and businesses of billions around the globe [1]. Amidst the ongoing crisis, no pharmaceutical solution has been found to restrict the virus from spreading. As we roll back to the state of normalcy, there remains a comprehensiveness of how the future will unfold. While the research continues, as per studies till date, to curb the viral transmission of novel coronavirus, physical distancing and wearing a mask essentially helps [2]. However, adherence to the strict implementation of wearing masks by the people might not be possible manually. Here, technology could play a notable role. Artificial intelligence and its subfields could make things more apparent. With the influx of computer vision and deep learning approaches, its efficacy in classification and recognition through image processing helps in various sectors. In this paper, the dataset used has two categories of classes, one 'with mask' and another 'without a mask'. For class detection, a modified transfer learning model is proposed that takes an image of a person as input. It then categorizes the person in one of the two classes depending on whether the person is wearing a mask or not. The decision is made by extracting the predominant features using machine learning and deep learning tactics, along with the help of libraries such as TensorFlow, Keras, Sklearn, OpenCV to avoid the spread of Covid-19. Figure 1 depicts the approach mentioned in the paper.

1.1 Convolutional Neural Network

A convolutional neural network (CNN) is a kind of deep neural network that takes an input image and extracts the significant feature from it with the help of the CNN layers and gets the relevant output as result [3]. Convolutional neural network is used

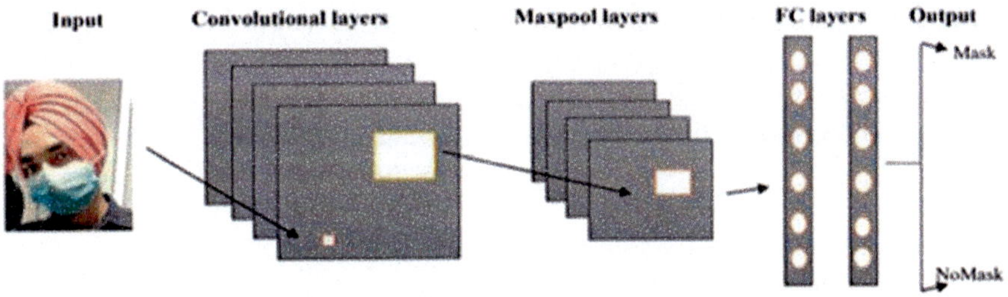

Fig. 2 Convolutional neural network layers

very efficiently in tasks like image classification, image processing, object detection, pattern detection. Figure 2 describes a convolutional neural network that contains several layers as first they take person's image as input then apply Conv2D layers, Maxpool layer, flatten layer to extract the dominant features from the image to get good accuracy and at last presenting the output layer.

1.2 Computer Vision

Computer vision is a field of integrative that tells a computer how to gain knowledge about the features from images and videos to acquire a good accuracy in a deep learning model. Also, computer vision has various techniques like object tracking, object detection, semantic segmentation, etc.

2 Literature Survey

In object detection, detecting the face of a person is a bit difficult task for a computer. Nevertheless, this issue gets resolved by the traditional object detection model Viola et al. [4], as this algorithm helps in recognizing the face of a person. This algorithm proves to be very efficient in real-time face detection. Oumina et al. [5] proposed a system to identify the two categories, people with mask and without a mask, by combining the MobileNetV2 pre-trained model with a support vector machine (SVM). The algorithm achieved a classification rate of 97.1%.

Qin and Li [6] proposed an algorithm to detect the face mask-wearing conditions. They proposed algorithm in four main steps, i.e., image pre-processing, face detection and crop, image super-resolution, and face mask-wearing categories, and the three face mask-wearing categories detailed as face mask-wearing, incorrect face mask-wearing, and no face mask-wearing. They used SRCNet to classify the images into these categories and achieved an accuracy of 98.70%.

Abidin and Harjoko [7] introduced an approach using a neural network to classify facial expression with the help of the Japanese Female Facial Expression (JAFFE). The proposed algorithm achieves a classification rate of 89.20%. Hussain et al. [8], mentioned in their paper that they have applied convolutional neural network for image emotion classification and recognition with the help of Haar cascade, Keras, and OpenCV. They proposed their approach in three steps: First, the algorithm detects the face, recognizes the expression, and classifies the emotions. Then, they test their model on a webcam which gives very satisfactory results. Mata et al. [9] deployed a system to see masked and unmasked images/videos using supervised learning algorithm tactics for computer vision on a small dataset of around 2000 images and achieved an accuracy of about 60% only and tested their model at different images.

Das et al. [10] have introduced a practical approach using the sequential convolutional neural network classifier to predict the wearing of masks or not. They implemented this approach in two different datasets and can also identify a face mask while the person is in motion. The first dataset contains 1376 masked and unmasked images that achieves an accuracy of 95.77%, and another dataset contains 853 images in which they attain an accuracy of 94.58%. Finally, Kalaiselvi et al. [11] have implemented a model using a convolutional neural network (CNN) and library such as TensorFlow, Keras, OpenCV to identify if the person is wearing a mask or not. The model obtained a recognition rate of 82.47% on a training set and 83.75% on a testing set.

3 Dataset

The underneath Table 1 represents the following dataset (https://github.com/chandrikadeb7/FaceMask-Detection/tree/master/dataset) consists of 2165 images that contain visuals of people with masks and 1930 visuals of people without masks which means there are a total of 4095 images. Therefore, the dataset is classified into two classes; one with people wearing a mask, and the second is people not wearing a mask. In the dataset, some visuals are tilted, slant and some have fingers on faces which are counted in without mask category only. Figure 3 represents the images of people with a mask from the dataset and Fig. 4 represents the images of people without a mask.

Table 1 Dataset description

Description	No. of images
People with mask	2165
People without mask	1930

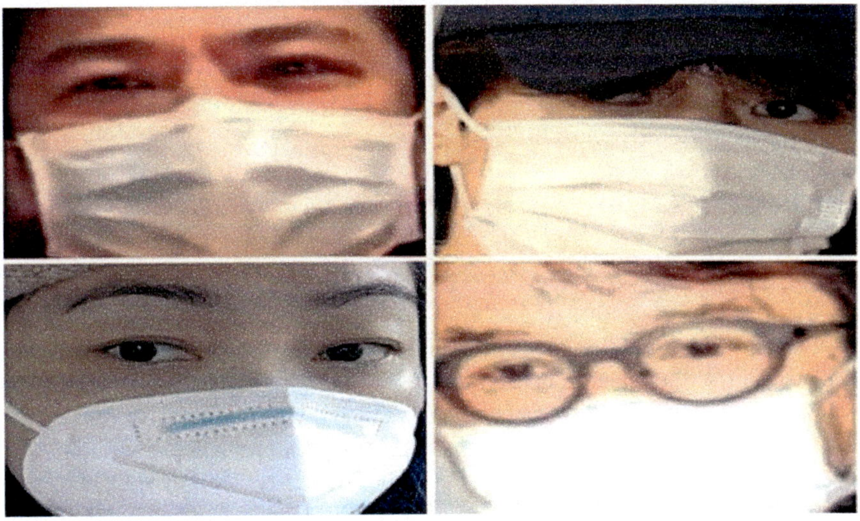

Fig. 3 Person with a mask

Fig. 4 Person without a mask

4 Proposed Revised Model

The proposed approach aims only to detect the person wearing a mask or not in images/videos, with the help of computer vision, OpenCV, Keras, TensorFlow libraries.

4.1 Data Importing and Pre-processing

The first step is to load the respective dataset. After loading, data pre-processing takes place. First, the image is resized into 162 × 162 dimensions. Then, the image is converted into the array form as per the model requirements, and at the last Label Binarizer techniques are applied on the labels so that they return the labels in binary format.

4.2 Splitting of the Dataset

After data pre-processing, the dataset is divided into the training and testing set. The test size is taken as 0.2, which means 80% of images are splitting as the training set. While the rest 20% is the testing set.

4.3 Model Building and Training

For building the model, data augmentation is applied. It is like a regularization technique that makes slight adjustments to the images by rotating, cropping, flipping, padding, scaling, etc. It also helps to prevent overfitting because it artificially increases the size of training data, reducing the effect of overfitting. Further, the data is trained on the DenseNet201 pre-trained model, and it has 201 deep layers. It is trained on a large amount from the ImageNet database.

To make the model more effective flatten layer, the dense layer with 200 neurons and having activation function as 'ReLU' is added in the model layers. After this, dropout layer is also added, which drops 0.5 neurons, and then the last dense layer is added with 2 neurons with activation function as SoftMax. After creating the model, the next step is model fitting in which the batch size is taken as 120, and the optimizer is 'Adam' with a learning rate of 0.001, the loss function used as 'categorical crossentropy'. The model is trained over ten epochs. Now the modified DenseNet201 model is ready to detect whether a person is wearing a mask or not from images as well as in videos.

4.4 *Face Detection Using Webcam*

However, once the model is trained, then the detection of the face is done with the help of the Haar cascade features. Once it visualizes the face, it converts the RGB image to a gray scale image by 'cv2.cvtColor'. After this, resizing of the image is done into 162×162. Further, the image is converted into the array format. At last, the image is adequate as per the model requirements.

4.5 *Loading the Trained Classifier*

Once the classifier is trained, then the trained classifier is loaded, and now the classifier is ready to predict whether the people are wearing a mask and or not. As the dataset [12] has two labels so with mask class is specified as with white color (255, 255, 255) and without mask class specified as orange color (0, 255, 255). Finally, with the help of a webcam, the model predicts the likelihood of each class, i.e., with a mask and without a mask.

5 Evaluation Metrics

Evaluation metrics perform a very key role in the field of machine learning. It estimates the model performance with the help of precision, recall, confusion matrix, classification accuracy.

5.1 *Precision*

Precision tells the percentage of the real prediction made will be correct [12]. Equation (1) is taking the TP in the numerator and adding the TP and FP in the denominator in order to get precision.

$$\text{Precision} = \frac{\text{TP}}{\text{TP} + \text{FP}} \tag{1}$$

5.2 *Recall*

It tells what percentage of actual positive values are successfully classified via classifier [12]. Equation (2) is taking the TP in the numerator and adding the TP and FN in the denominator in order to get the result of recall

$$\text{Recall} = \frac{\text{TP}}{\text{TP} + \text{FN}} \tag{2}$$

where TP, FP, and FN stand for the true positive, false positive, and false negative.

5.3 *F1-Score*

It the harmonic mean of precision and recall computed from the number of mispronunciations detected by both the computer and human evaluator [13]. Equation (3) is calculating the F1-score by multiplying precision and recall with 2 and dividing by the sum of precision and recall to acquire the F1-score.

$$\text{F1-score} = \frac{2 \times \text{Precision} \times \text{Recall}}{\text{Precision} + \text{Recall}} \tag{3}$$

6 Discussions

The transfer learning models are very effective in various approaches, and it carries the potential of detecting a face mask in an image/videos with many faces. Three models are applied in this approach DenseNet201, InceptionResNetV2, and ResNet101V2 on the respective dataset.

Based on comparative analysis, Table 2 represents that DenseNet201 and ResNet101V2 have higher validation accuracy, i.e., 98.90 and 96.45%, respectively, as compared to InceptionResNetV2 which obtains an accuracy of 89.92%. But DenseNet201 is performing very well among the other models.

Table 2 Comparison of model on the basis of validation accuracy

CNN model	Accuracy (%)
DenseNet201	98.90
InceptionRessNetV2	89.92
ResNet101V2	96.45

Table 3 Comparison of model on the basis of evaluation metrics

CNN model	Precision (%)	Recall (%)	F1-score (%)
DenseNet201	98.45	99.22	98.83
InceptionResNetV2	95.34	74.35	83.55
ResNet101V2	99.42	91.07	95.06

Table 3, DenseNet201 has the highest recall and F1-score which is 99.22 and 98.83% in contrast to other transfer learning models. Whereas ResNet101V2 is also showing the highest precision from others, i.e., 99.42%. However, Inception-ResnNetV2 has not given a much stronger result. Moreover, based on the overall observation, DenseNet201 gives the finer performance as compared to other models toward the face mask detection approach.

7 Results

In this paper, the data is trained on DenseNet201, InceptionResNetV2, ResNet101V2 transfer learning models. Among all, DenseNet201 has been employed, which gives 98.90% validation accuracy throughout the training process as compared to other transfer learning models in classifying persons with or without masks which we have discussed in Sect. 5 on the basis of evaluation metrics. In Fig. 5, the underneath graph portrays the highest true positive rate with the lowest false positive rate using the modified DenseNet201 model.

Figure 6 depicts the accuracy result of the trained DenseNet201 model, and in Fig. 7, the graph portrays the accuracy and loss of validation data and training data over only ten epochs, and the graph also manifests that as the epochs are increasing, the accuracy gets better and the loss gets decreased over the increasing epochs which means the model is performing well.

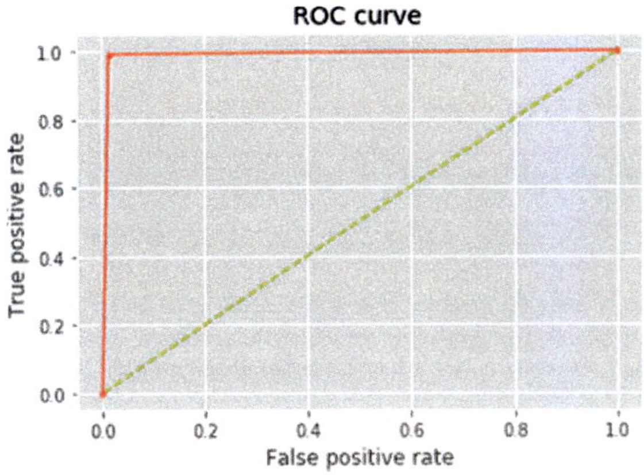

Fig. 5 Receiver operating characteristics curve (ROC) of the DenseNet201 classifier

	precision	recall	f1-score	support
with_mask	0.99	0.99	0.99	433
without_mask	0.98	0.99	0.99	386
accuracy			0.99	819
macro avg	0.99	0.99	0.99	819
weighted avg	0.99	0.99	0.99	819

Fig. 6 Screenshot of classification report of trained model

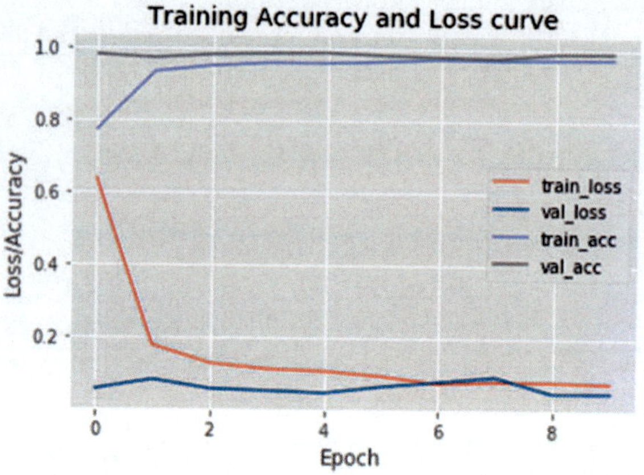

Fig. 7 Graph of training accuracy and loss

Based on the model's performance, Fig. 8 shows the result related to the person putting hand on their face instead of wearing a mask; which is also recognized as without mask, whereas Fig. 9 represents the results related to the person with the

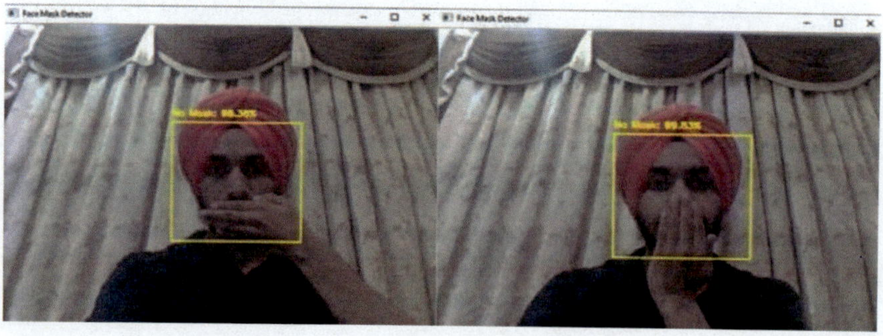

Fig. 8 Result of person while putting hand on face

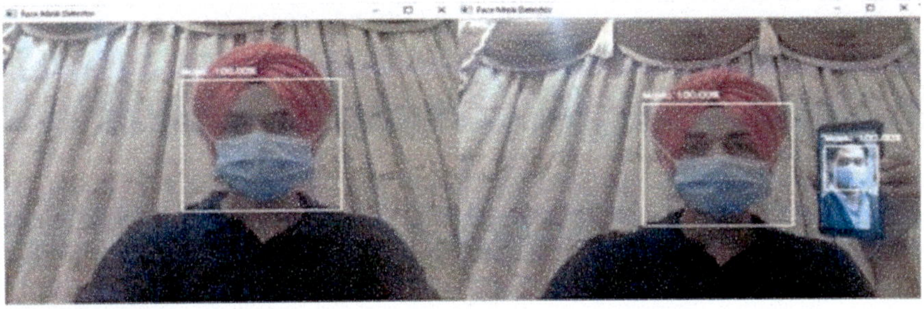

Fig. 9 Results with a mask

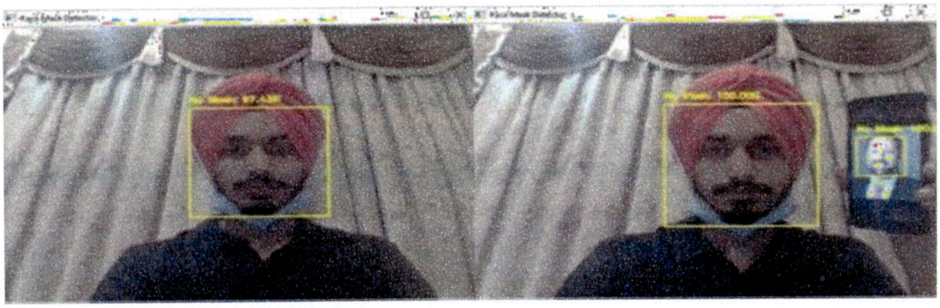

Fig. 10 Result of person without a mask

mask. Figure 10 shows predictions when a person is without a mask. These predictions are done using cv2.VideoCapture(0), which helps in capturing the videos using the webcam.

8 Managerial and Social Implications

This deployed approach conceivably adds to public places such as markets, shopping complexes, educational institutes to provide help to the government officials to ensure that nobody is violating the rules; moreover, it can also aid to stop the viral transmission of the Covid-19, individuals to meet their ends by going toward their respective works to earn money like earlier which also help country to uplift their economy.

9 Conclusion

This paper uses an approach toward the face mask detection system using computer vision and a deep learning tactic. Different pre-trained models like DenseNet201,

InceptionResNetV2, ResNet101V2 are modified and used to detect a person with or without a mask. But based on the overall result, DenseNet201 gives an excellent performance in classifying the two different categories. The DenseNet201 modified model is tested on real-time video analysis to identify a mask and is also capable to detect the face mask when the person is in motion. As the proposed algorithm attains 98.90% validation accuracy and recall of 99.22% followed by precision and F1-score, i.e., 98.45 and 98.83%, respectively, on the given dataset (4096 images). In the future, we will further try to add some new feature extraction techniques like surf, kaze, color histogram, with the help of a machine learning algorithm that also detects the type of face mask the person is wearing like a cloth mask, surgical, etc.

Acknowledgements This is to acknowledge that I express my special thanks of gratitude to Dr. Amandeep Kaur for her valuable guidance and constant support in the completion of my paper on robust face mask detection using deep learning.

References

1. https://www.who.int/director-general/speeches/detail/who-directorgeneralsopeningremarks-at-the-media-briefing-on-covid-19---11-march-2020
2. D.K. Chu, E.A. Akl, S. Duda, K. Solo, S. Yaacoub, H.J. Schünemann, A. El-harakeh, A. Bognanni, T. Lotfi, M. Loeb, A. Hajizadeh, Physical distancing, face masks, and eye protection to prevent person-to-person transmission of SARS-CoV-2 and COVID19: a systematic review and meta-analysis. The Lancet **395**(10242), 1–15 (2020)
3. S. Albawi, T.A. Mohammed, S. Al-Zawi, Understanding of a convolutional neural network, in *2017 International Conference on Engineering and Technology (ICET)*. IEEE (2017), pp. 1–6
4. P. Viola, M. Jones, Rapid object detection using a boosted cascade of simple features, in *Proceedings of the 2001 IEEE Computer Society Conference on Computer Vision and Pattern Recognition*. CVPR 2001, vol. 1, pp. 1–11. IEEE (2001, December)
5. A. Oumina, N. El Makhfi, M. Hamdi, Control The COVID-19 pandemic: face mask detection using transfer learning, in *2020 IEEE 2nd International Conference on Electronics, Control, Optimization and Computer Science (ICECOCS)*, pp. 1–5. IEEE (2020, December)
6. B. Qin, D. Li, Identifying facemask-wearing condition using image superresolution with classification network to prevent COVID-19. Sensors **20**(18), 1–23 (2020)
7. Z. Abidin, A. Harjoko, A neural network based facial expression recognition using fisherface. Int. J. Comput. Appl. **59**(3), 1–6 (2012)
8. S.A. Hussain, A.S.A. Al Balushi, A real time face emotion classification and recognition using deep learning model. J. Phys. Conf. Ser. **1432**(1), 1–14 (2020). 012087
9. B.U. Mata, Face mask detection using convolutional neural network. J. Nat. Rem. **21**(1), 14–19 (2021)
10. A. Das, M. Wasif Ansari, R. Basak, Covid-19 face mask detection using TensorFlow, Keras and OpenCV, in *2020 IEEE 17th India Council International Conference (INDICON)*, 2020, pp. 1–5. https://doi.org/10.1109/INDICON49873.2020.9342585
11. G. Kalaiselvi, R. Sonali, M. Sowmiya, Real time face mask detection for Covid-19 pandemic using supervised learning of convolutional neural network. J. Homepage. www. ijrpr.com ISSN 2582:7421
12. S.M. Shafi, R. Rather, Precision and recall of five search engines for retrieval of scholarly information in the field of biotechnology

13. H. Huang et al., Maximum F1-score discriminative training criterion for automatic mispronunciation detection. IEEE/ACM Trans. Audio Speech and Lang. Process. **23**(4), 1–11 (2015)

Utilizing Deep Belief Network for Ensuring Privacy-Preserved Access Control of Data

Puneet Goswami and Suman Madan

Abstract Deep belief network is an unsupervised deep learning methodology of neural networks and has gained increasing attention during recent years. In contrast to conventional learning methods, deep learning is powerful methodology that is greatly influencing cloud environments for model training purposes where collaborative learning between different data sources exists. It is utilized for wide application domains varying from health care to pattern or speech recognition. However, for the healthcare data domains, the privacy and security issues are the major concerns due to the presence of sensitive information regarding patient's medical data during collaborative learning phase. This paper aims at presenting an effective privacy-preserved access control solution, called PPAC, for the collaborative healthcare data. The overarching aim of improving privacy of health data is managed during three proposed phases, namely user-registration, control setup, and validation. Encryption and hashing functions are utilized for controlled access in collaborative cloud framework. In addition, the deep belief network algorithm of deep learning is used for the classification of cloud user as either genuine or attacker. This paper further elaborated the performance analysis of PPAC to reveal its effectiveness on performance metrics, such as precision, recall, and F-measure. The experiment results reveal that the proposed PPAC method provided best precision, recall, and F-measure of 0.9427, 0.945, and 0.9438, respectively.

Keywords Deep learning · Deep belief network · Access control · Privacy · Privacy preservation

P. Goswami
SRM University, Sonepat, India

S. Madan (✉)
Jagan Institute of Management Studies, Sec-5, Rohini, Delhi, India
e-mail: madan.suman@gmail.com

A. K. Bashir et al. (eds.), *Proceedings of International Conference on Computing and Communication Networks*, Lecture Notes in Networks and Systems 394,
https://doi.org/10.1007/978-981-19-0604-6_45

1 Introduction

One approach of machine learning that has pulled in a ton of consideration in recent research is deep learning (DL) or deep neural network (DNN) due to its unsupervised behavior especially for the cloud environments. The major domains using deep learning applications in cloud environment include health indicators, big data analytics, pattern or speech recognition, intrusion detection, etc. [1]. Cloud computing service-oriented architecture, distributed computing, and utility computing environment actually motivated the deep learning involvement in training models for collaborative learning. Cloud, pay-as-you-use environment liberates the cloud customers from maintenance expenses by leasing the remote resources instead of buying hardware and thus increases the number of organizations and individuals accessing the cloud [2]. This in turn facilitated collaborative learning on large datasets wherein data is provided from different sources, for example, patients data, laboratories data, or doctor's data. Although the cloud computing has numerous advantages, the security and the privacy still remain a major concern. Generally, the data is controlled by the trusted administrator through the trusted servers that are stored by the data owner, whereas the data is stored on the remote cloud servers in the public cloud storage, which are controlled by the semi-trusted cloud service provider. In the public cloud storage, the cloud servers cannot be trusted in the access control of the privacy-preserved secure data as they are not in the data owner's trusted domains. Thus, the secure and privacy-preserved access control in the public cloud storage has become a major challenge [3].

Most of the cloud storage system uses the server-dominated access control, such as certificate-based authentication [4] and password-based authentication [5]. This method trusted the cloud provider for the protection of sensitive data. The drawback of the server-dominated access control is the policy of the owner for controlling the user that lets the cloud providers as well as their employees to read the document of the customer who may not be authorized to access sensitive data of customers, such as deducing type of illness from health data of patients [6]. Additionally, the cloud providers exaggerated the resource consumption and make the customers to pay without any verifiable records as the system does not have the details regarding the usage of the resource [7]. The existing method failed to satisfy the requirement of the data owners to access the files on the cloud servers on their own hands and to sustain the data confidentiality against the malicious users. Although the ciphertext-policy attribute-based encryption (CP-ABE) methods access permission to the users in controlling the cloud data, the single user cannot maintain the secret information [8–11]. The existing methods failed to access the policy in a flexible and fine-grained way. Each user is labeled by the attribute set in which the responsibility of the user is expressed in a single attribute for the fine-grained access control [12]. The privacy of the users can be increased by enhancing the privacy technologies that can offer additional level of protection rather than relying only on laws and policies. In order to address the privacy concerns of users, several approaches like information manipulation, privacy and context awareness, access control, and data anonymization. As, the medical data is inevitable, there is a necessity to preserve the data.

The main contribution of this work is to ensure the privacy preservation access control of the healthcare-related data in cloud using DNN technique, specifically deep belief network (DBN). In the cloud, the access control is done using encryption, hashing, and so on. The steps involved in the access control mechanism are user registration, control setup, and validation. Finally, the anomalous behavior is detected using the deep belief network for the classification of the attacker and genuine user in the cloud.

The rest of the paper is organized as: Sect. 2 describes the motivation from existing privacy preservation and access control methods and the various research challenges related to healthcare data, Sect. 3 discusses system model for cloud, Sect. 4 explains proposed DBN-based privacy preservation method, Sect. 5 shows the result and discussion of proposed DBN-based privacy preservation method, and Sect. 6 concludes the paper.

2 Motivation

2.1 *Related Works*

In the literature, the major motivation for this work is provided from the literature related to privacy-preserved access control in cloud environment.

Trejo et al. [13] developed a DNS anomaly detection visual platform for the traffic anomaly detection. This method interpreted the traffic continuously and tested on the synthetic attacks successfully. In this work, the effectiveness of some techniques used to limit the effects of DDoS attacks targeting DNS servers is analyzed and also a set of measures to timely detect potential DDoS attacks against this service, such as amplification, and reflection attacks, among others were proposed. However, this method failed to reduce the latency of the system.

Chen et al. [14] designed hybrid supervised and unsupervised machine learning methods for anomaly detection. The data classification method (DCM) enables a self-learning capability that eliminates the requirement of prior knowledge of abnormal network behaviors and therefore can potentially detect unforeseen anomalies. They introduced a self-taught mechanism that transfers the patterns learned by the DCM to a supervised data regression and classification module (DRCM). The DRCM, whose complexity is mainly related to the scale of the applied supervised learning model, can potentially facilitate more scalable and time-efficient online anomaly detection by avoiding excessively traversing the original dataset. Although this method had higher anomaly detection rate along with lower false positive rate, it failed to detect the anomaly for a greater number of attributes and for diverse failure scenarios.

Xue et al. [12] modeled a combined data owner-side and cloud-side access control. This method reduced the overhead using the probabilistic check in the resource consumption. The introduction of this method provided a practical solution to the EDoS attack. A practical threat model which provides the higher security was used. However, this method had high computational complexity which became a major flaw to the model.

Armbrust et al. [3] developed an attribute-based encryption (ABE) for securing the access control of the data. This method had different attribute sets for gaining permission for accessing the data. Although this method had efficient computational overhead and storage, it was not implemented in real-time applications.

Harn and Ren [5] developed a generalized digital certificate (GDC) which provides user authentication and key agreement. The GDC is a simpler method than the X.509 public-key digital certification. This protocol maintains the enable authentication technique and secure privacy of the digital certificate.

2.2 Challenges

The major challenges as per the present privacy-preserved access control techniques for cloud computing applications are depicted below:

- The challenges like secure interoperability, semantic heterogeneity, and policy evolution management should be addressed by the policy of the cloud [15, 16].
- The attribute-based encryption method secured the access control of the data, but the implementation of the system in real time was an important challenge [3].
- Trejo et al. [13] exhibited the traffic anomaly detection using the DNS anomaly detection visual platform interpreted the traffic continuously but the challenge lies in the implementation of the system with low latency.
- Although the anomaly detection based on hybrid supervised and unsupervised machine learning method provided higher detection rate, it failed to anomaly for diverse failure scenarios [14].

3 System Model for Cloud

This section explains the system model for cloud for which the PPAC method is proposed. Three stakeholders who are accessing the collaborative dataset include: data server, cloud user, and cloud data owner. The cloud environment offers several benefits to both the users and providers, varying from data processing services to storage services. Since cloud data is outsourced to several third-party users, the privacy becomes very much significant issue. The cloud server services that are offered to its users include data management system, data storage agents, query-mapper and handler for request and thus arise the need of maintaining privacy of data by the data provider before passing it to the data storage agents. Figure 1 shows

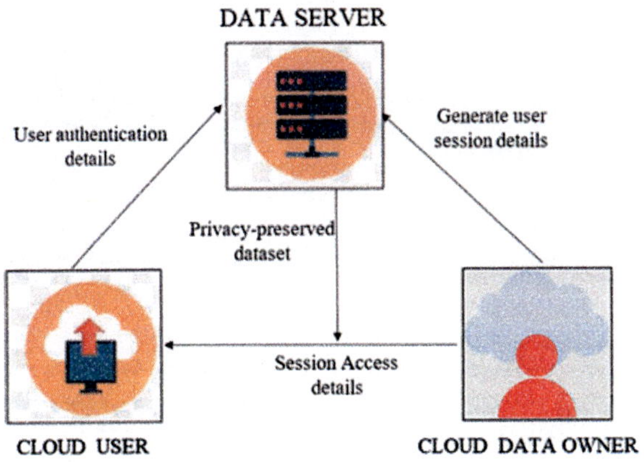

Fig. 1 System model showing users with their roles

system model that reflects the communication between three stakeholders involved proposed privacy-preserved access control process.

4 Proposed Privacy-Preserved Access Control (PPAC) Method

In this paper, an access control of the privacy preservation data is developed for controlling the access of the cloud by the user and the detection of the anomalous behavior. The block diagram of proposed PPAC method is explained in Fig. 2. The proposed PPAC method gives access of DBN-based dataset only to genuine or authorized users. The collaborative health data collected from patients, laboratories, doctors, etc., is inputted to proposed model. The DBN layer added to proposed model performs classification using the features extracted from the data packets of the user and then validates that user accessing the dataset is genuine or attacker. The proposed PPAC process involves three phases: registration phase, control setup and validation phase, and DBN-based anomaly detection phase. The process involved in every phase is explained in subsections below.

4.1 Registration Phase of PPAC Method

In the registration phase, the server and the user are registered with the cloud. Initially, the user ID and password is generated in the cloud. The ID and password of the user is then stored in the data server. Then, the stored user ID and password in the data server is shared to the cloud data owner. The shared user ID and password is then received and stored by the data owner of the cloud. After storing the information,

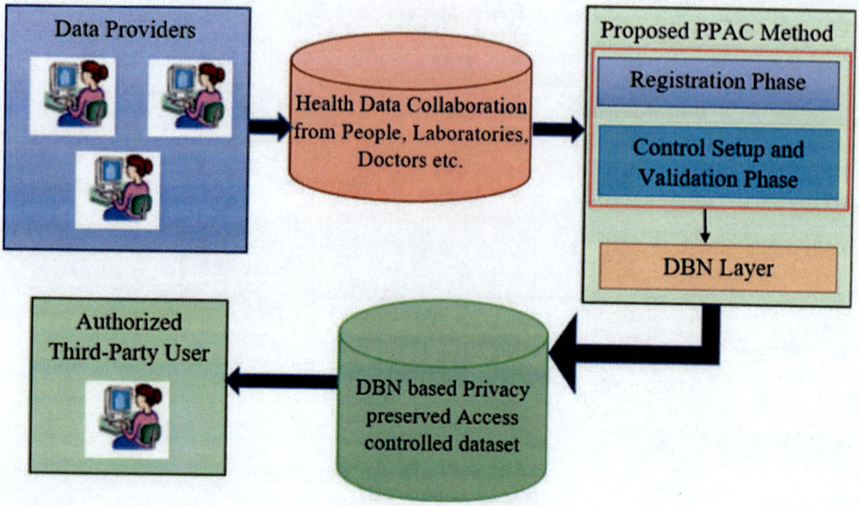

Fig. 2 Block diagram of proposed DBN-based PPAC method

the data owner generates a user session password, which is then shared to the data server. The shared user session password is then received and stored by the cloud user. If the user session password in the cloud user and the data server is same, then the user is registered in the cloud. Once the user is registered, the server ID and server password is generated in the data server, which is shared to the data server of the cloud. The data server stores the server ID and password. After storing the server ID and password, the cloud data owner generates server session password. Then, the server session password is shared to the data server. If the server session password in both data server and cloud data owner is same, then the server is registered in the cloud. Figure 3 shows the registration phase of the access control privacy preservation

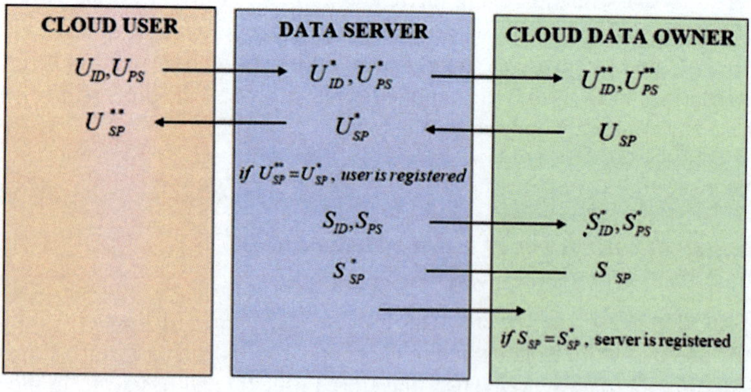

Fig. 3 System process during registration phase

Table 1 Comparison of different privacy techniques

Symbol	Description	Symbol	Description
U_{ID}	user identification	U_{PS}	user password
U^*_{ID}	user identification stored in the data server	U^*_{PS}	user password stored in the data server
U^{**}_{ID}	user identification received and stored by the cloud data owner	U^{**}_{PS}	user password received and stored by the cloud data owner
U_{SP}	user section password	U^*_{SP}	user section password stored in the data server
U^{**}_{SP}	user section password received and stored in the cloud server	S_{ID}	server identification generated in data server
S_{PS}	server password generated in data server	S^*_{ID}	server identification stored in the cloud data owner
S^*_{PS}	Server password stored in the cloud data owner	S_{SP}	server session password generated in the cloud data owner
S^*_{SP}	server session password stored in the data server		

method. The notations used in registration phase process Fig. 3 are explained in Table 1.

4.2 *Control Setup and Validation Phase of PPAC Method*

The registration phase is followed by control setup and validation phase. In the control setup and validation, there are two levels of verification. Initially, the private key of the user is generated in the data server. The private key of the user is generated by the modulation of the hashing function and the parameter n. The user private key is expressed as,

$$K_U = h(U^*_{ID}//r) \bmod n \tag{1}$$

where K_U is the private key of the user, r and n are the random numbers. The K_U generated in the data server is then shared to both cloud data owner and cloud user. After storing the user private key, the cloud user generates the control message m_1. The control message is generated by the encryption of the EX-OR between the user ID and the concatenation of K^*_U with U^{**}_{SP}. The control message is given as,

$$m_1 = E(U_{ID} \oplus (K^*_U//U^{**}_{SP})) \tag{2}$$

where K^*_U is the private key of the user stored in cloud user and cloud data owner. $\oplus$ is the EX-OR function. The control message from the cloud server is shared to the data server. The control message stored in data server is denoted as, $\tilde{m}_1$. If the control message in data server is same as that of cloud user, first level is verified. For second-level verification, the server private key is generated by the cloud data owner. The server private key is given as,

$$K_S = h(S_{ID}^* \oplus g//E(S_{PS})) \tag{3}$$

where g is the random number. The server key generated by the data owner of the cloud is stored by the data server as, K_S^*. The control message m_2 generated in the data server is given as,

$$m_2 = h(A//r) \otimes (S_{ID} \oplus K_S^*) \tag{4}$$

where $A = 8x^4 - 8x^2 + 1, x = h(K_U//r) \oplus S_{ID}$. The control messages $\tilde{m}_1$ and m_2 in the data server are shared to the cloud data owner and stored as, $\tilde{\tilde{m}}_1$ and $\tilde{m}_2$.

$$\tilde{\tilde{m}}_1 = E(U_{ID}^{**} \oplus (K_U^*//U_{SP})) \tag{5}$$

$$\tilde{m}_2 = h(A//r) \otimes (S_{ID}^* \oplus K_S^*) \tag{6}$$

If $\tilde{m}_1 = \tilde{\tilde{m}}_1$ and $m_2 = \tilde{m}_2$, then the second level is verified. After the access control, the data is communicated to the user through data packets. Figure 4 shows

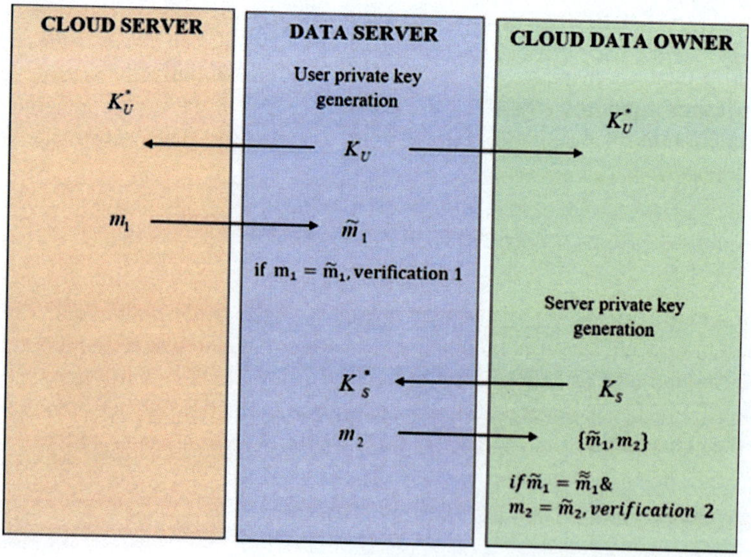

Fig. 4 System process during control setup and validation phase

the process of control setup and validation in access control privacy preservation method.

Once the user is verified, the data packets are communicated from the user end such that the anomalous user is detected using the data packets.

4.3 DBN-Based Anomaly Detection Phase of PPAC Method

This phase helps in detecting the anomaly behavior and determines whether the user is genuine authorized user or not. Initially, the features are extracted from the data packets and the extracted features include KDD feature duration, dst_bytes, src_bytes, protocol type, wrong fragments, flag, and logged-in. The features extracted in this initial process are described in Table 2. The extracted features are then given to the DBN classifier for the classification of user into anomaly and genuine user.

Architecture of DBN:

The DBN layer consists of two RBM layers and a MLP layer. Each layer of DBN is interconnected with neurons. The input features are multiplied with input neurons in RBM1 to form the input of RBM2. The hidden weights of the RBM2 are multiplied with the input of the RBM2 layer to form the input of MLP layer. Finally, the features are classified into the output of MLP layer to determine whether the user is anomaly or genuine. For the classification, we are using the deep belief network. As the DBN is pre trained, it is efficient in classification. Initializing the weights of all layers helps in better optimization during the pre-training process. The greedy learning algorithm is fast, efficient and learns one layer at a time. The training process is carried out sequentially from the bottom-most layer. Fine tuning is done in this method which helps to differentiate between different classes better. The optimal value can be obtained by adjusting the weights during the fine-tuning process. This makes the classification model more accurate. Figure 5 shows the architecture of DBN. The input to the DBN classifier is represented as, p, which is the features extracted from the user.

Table 2 Features extracted from connection

Feature name	Description	Feature type
Duration	Connection length in number of seconds	Continuous
dst_bytes	Data bytes number from destination to source	Continuous
src_bytes	Data bytes number from source to destination	Continuous
Protocol type	Protocol used (TCP, UDP etc.)	Discrete
Wrong fragments	Number of wrong fragments	Continuous
Flag	Status of connection: error or normal	Discrete
Logged-in	Successfully login: "1" else "0"	Discrete

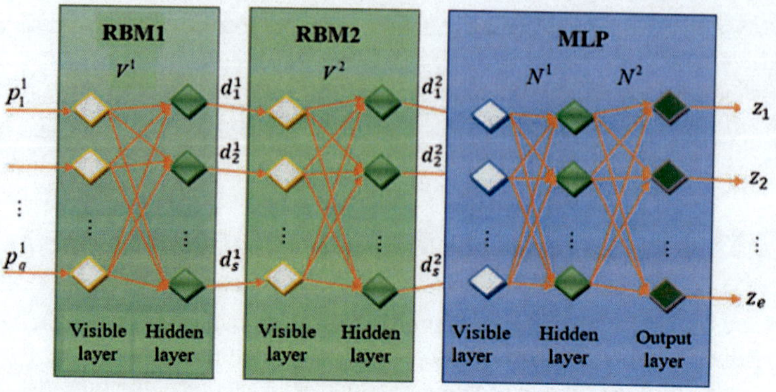

Fig. 5 Architecture of DBN

The input to the visible layer of RBM1 is the feature vector. In the input layer of RBM1, the visible and hidden neurons are represented as,

$$b^1 = \{b_1^1, b_2^1, \ldots, b_v^1, \ldots, b_u^1\};\ 1 \le v \le u \tag{7}$$

$$d^1 = \{d_1^1, d_2^1, \ldots, d_w^1, \ldots, d_a^1\};\ 1 \le w \le a \tag{8}$$

where the *v*th visible neuron and *w*th hidden neuron are denoted as, b_v^1 and d_w^1. The RBM-1 layer consists of *a* hidden neurons and *u* visible neurons. The weights and biases of the hidden and visible layer in RBM-1 are given as, G^1 and B^1. In RBM 1, the weights are initialized as,

$$V^1 = \{V_{vw}^1\};\ 1 \le v \le u;\ 1 \le w \le s \tag{9}$$

where the weight between the *w*th hidden layer and *v*th visible layer is given as, V_{vw}^1 and the dimension of the weight in the RBM-1 is given as, $[u \times a]$. The input to the visible layer of RBM-1 is the output from the hidden layer of RBM-1, which is represented as,

$$d_w^1 = \gamma\left[G_w^1 + \sum_v p_v^1 V_{vw}^1\right] \tag{10}$$

where the activation function is given as, γ and the feature vector is expressed as, p_v^1. The output of RBM-1 is given as,

$$d^1 = \{d_w^1\};\ 1 \le w \le a \tag{11}$$

The input to the RBM-2 layer is the output that is generated from the hidden layer in RBM-1. The weights generated in RBM-2 are represented as,

$$V^2 = \left\{V^2_{ww}\right\};\ 1 \le w \le a \tag{12}$$

where the weight between wth hidden neuron in the RBM-2 and the wth visible neuron in RBM-1 is given as, V^2_{ww}. The output of the RBM-2 is represented as,

$$d^2_w = \gamma\left[B^2_w + \sum_v b^2_v V^2_{ww}\right]\forall b^2_v = d^1_w \tag{13}$$

where the dimension of the weight vector generated in RBM-2 is given as, $[a \times a]$. The bias with respect to wth hidden neuron is given as, B^2_v. The output generated from the hidden layer of the RBM-2 is given as the input of the MLP layer, which is represented as,

$$d^2 = \left\{d^2_w\right\};\ 1 \le w \le a \tag{14}$$

In the MLP layer, the input is expressed as,

$$x = \{x_1, x_2, \ldots, x_w, \ldots, x_a\} = \left\{d^2_w\right\};\ 1 \le w \le a \tag{15}$$

where the neurons in the input layer are represented as, a. The hidden layer of MLP with C hidden neurons is given as,

$$y = \{y_1, y_2, \ldots, y_i, \ldots, y_C\};\ 1 \le i \le C \tag{16}$$

Let us assume the bias of the ith hidden neuron as, R_i. The output from the MLP layer is given as,

$$Z = \{z_1, z_2, \ldots, z_l, \ldots, z_e\};\ 1 \le l \le e \tag{17}$$

where the neurons in the output layer of the MLP are given as, e. The weight vectors N^1 in the MLP layer are the weight that links the hidden and input layers, whereas N^2 is the weight that links the output and hidden layers. The weight vector N^1 is given as,

$$N^1 = \left\{N^1_{wi}\right\};\ 1 \le w \le a;\ 1 \le i \le C \tag{18}$$

where the size of N_1 is $[a \times C]$ and the weight between the ith hidden neuron and the wth input neuron is given as, W^1_{wi}. The computed output is given as,

$$P_i = \left[\sum_{w=1}^{a} N_{wi}^{1} * x_w \right] R_i \forall x_w = d_w^2 \tag{19}$$

where the hidden neuron bias is given as, R_i. The weights between the output and the hidden layer are, N^P, and it is given as,

$$N^P = \{N_{il}^P\};\ 1 \le i \le C;\ 1 \le l \le e \tag{20}$$

The final output from the MLP layer is represented as,

$$Z_l = \sum_{i=1}^{C} N_{il}^P * P_i \tag{21}$$

The output from the DBN classifier demonstrates that the user is either genuine or attacker using the features extracted from the data packets of the user.

5 Results and Discussion

This section depicts the results and discussion of the developed DBN-based privacy preservation of access control method to evaluate effectiveness of performance with existing approaches using Cleveland dataset. The performance of the developed model is evaluated in terms of precision, recall, and F-measure. The experimentation of proposed PPAC method is performed in Java software that runs in PC with Windows 10 OS. The dataset employed for the experimentation is Cleveland from heart disease dataset from UCI machine learning repository [17]. The Cleveland dataset contains 76 attributes and 303 instances; however, 14 attributes out of them are mostly used for research. The attributes include sex, age, type of pain, blood pressure range, cholesterol, and so on. Age is described in years, and sex is indicated by “1” and “0”. The integer “1” indicates the male, and “0” indicates the female.

5.1 Evaluation Metrics

The performance of developed DBN-based privacy-preserving access control model is evaluated based on the metrics, such as precision, recall, and F-measure.

(a) F-measure: It is a measure, which calculates the weighted harmonic average of recall and precision.

$$\text{F-measure}, f = \frac{2 + mn}{m + n} * 100 \tag{22}$$

where m depicts the precision, and n depicts the recall.

(b) Precision: Precision depicts the ratio of true positive rate (TPR) and the sum of the true positive and false positive rate (FPR).

$$\text{Precision}, m = \frac{s_1}{S_1 + S_2} \tag{23}$$

where S_1 indicates the TPR and S_2 indicates the FPR.

(c) Recall: Precision depicts the ratio of true positive rate (TPR) and the sum of the true positive and false negative rate (FNR).

$$\text{Rec all}, n = \frac{S_1}{S_1 + S_3} \tag{24}$$

where S_1 indicates the TPR and where S_3 indicates the FNR.

5.2 *Comparative Methods and Analysis*

This section depicts the existing techniques utilized for the comparative analysis in order to evaluate the effectiveness of developed model. The existing approaches employed used in this work are: neural network (NN) [18], support vector machine (SVM) [19], and naïve Bayes (NB) [20]. The comparison of developed DBN model is performed based on precision, recall, and F-measure.

Analysis of evaluation metrics with number of features = 4

This section depicts the comparative analysis of performance metrics with number of features = 4 by varying the training data. Figure 6a shows the comparative analysis of developed DBN model based on accuracy by varying the training data. When the training data is 90%, the precision value obtained by the developed DBN model is 0.942, and the existing approaches, like NN, SVM, and NB, achieved the precision value of 0.846, 0.882, and 0.898, correspondingly.

The performance improvement achieved by the developed DBN model with existing approaches like NN, SVM, and NB based on precision is 10.17, 6.37, and 4.66%. Figure 6b shows the comparative analysis based on recall value achieved by the developed model through varying the training data. The recall value achieved by the developed model is 0.936 with training data = 80%, whereas the existing

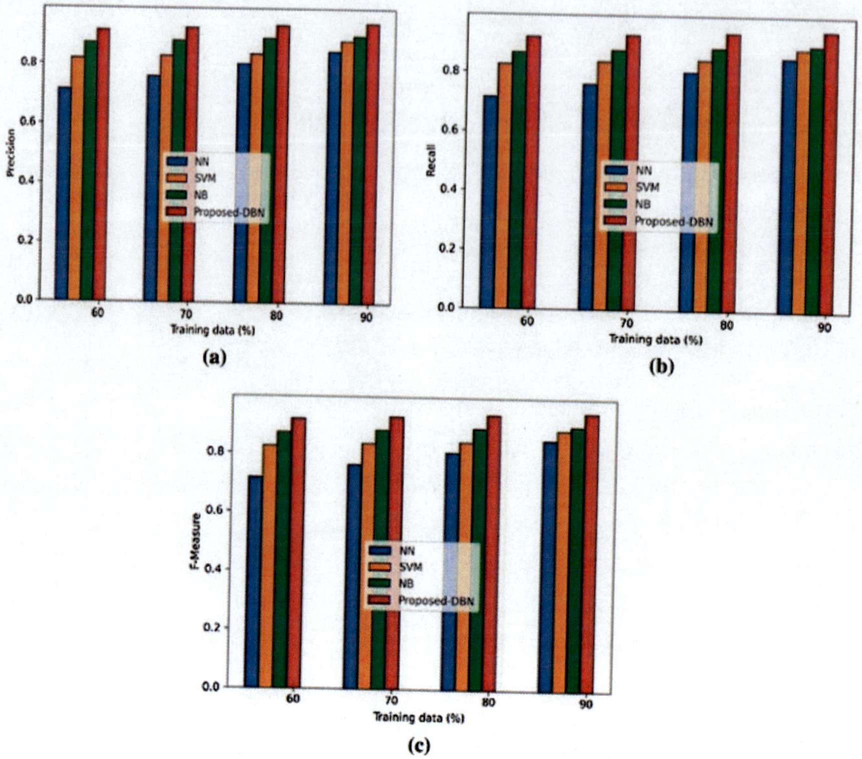

Fig. 6 Comparative analysis when number of features = 4, **a** Precision, **b** Recall, **c** F-measure with respect to the training percentage

approaches, such as NN, SVM, and NB achieved the recall value is 0.804, 0.843, and 0.884 with training data = 80%. The performance improvement of the developed DBN model with existing approaches like NN, SVM, and NB based on recall value is 14.10, 9.87, and 5.46%. Figure 6c depicts the comparative analysis of developed DBN model based on F-measure by changing the training data. The developed DBN model achieved the F-measure value is 0.924, and the existing approaches like NN, SVM, and NB achieved the F-measure value of 0.759, 0.831, and 0.877 while the training data is 70%. The performance improvement attained by the developed DBN model with existing approaches is 17.91, 10.06, and 5.14%.

Analysis of evaluation metrics with number of features = 8

This section depicts the comparative analysis of performance metrics with number of features = 8 by varying the training data. Figure 7a shows the comparative analysis based on precision value achieved by the developed model through varying the training data.

The precision value achieved by the developed model is 0.942 with training data = 90%, whereas the existing approaches, such as NN, SVM, and NB, achieved the

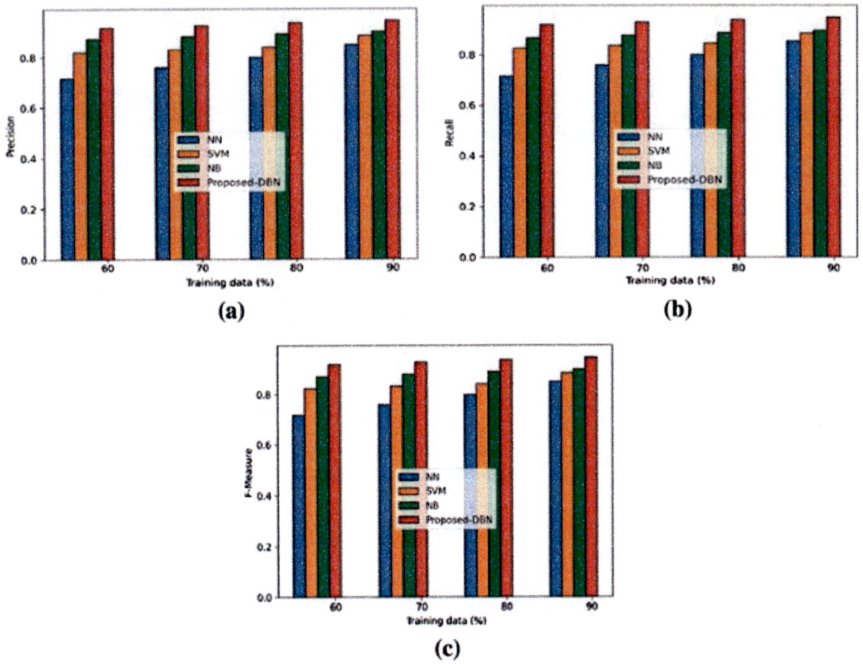

Fig. 7 Comparative analysis when number of features = 8, **a** Precision, **b** Recall, **c** F-measure with respect to the training percentage

precision value is 0.846, 0.882, and 0.898 with training data = 90%. The performance improvement of the developed DBN model with existing approaches like NN, SVM, and NB based on recall value is 10.20, 6.39, and 4.64%. Figure 7b depicts the comparative analysis of developed DBN model based on recall by changing the training data. The developed DBN model achieved the recall value is 0.917, and the existing approaches like NN, SVM, and NB achieved the recall value of 0.715, 0.825, and 0.865 while the training data is 50%. The performance improvement attained by the developed DBN model with existing approaches is 22.01, 10.08, and 5.72%. Figure 7c shows the comparative analysis of developed DBN model based on F-measure by varying the training data. When the training data is 70%, the F-measure value obtained by the developed DBN model is 0.925, and the existing approaches like NN, SVM, and NB achieved the F-measure value of 0.758, 0.831, and 0.877, correspondingly. The performance improvement achieved by the developed DBN model with existing approaches like NN, SVM, and NB based on F-measure is 18.02, 10.14, and 5.12%.

Table 3 illustrates the comparative discussion of the developed DBN method. Based on the number of features = 4, the precision value achieved by the developed model is 0.9427, and the precision value achieved by the existing techniques such as NN, SVM, and NB method is 0.8467, 0.8826, and 0.8987, respectively, while the training data is 90%. When the training data is 90%, the recall and F-measure value

Table 3 Comparative discussion

		NN	SVM	NB	Developed PPAC
Number of features = 4	Precision	0.8467	0.8826	0.8987	**0.9427**
	Recall	0.8525	0.8826	0.8944	**0.945**
	F-measure	0.8496	0.8826	0.8965	**0.9438**
Number of features = 8	Precision	0.8462	0.8821	0.8986	**0.9424**
	Recall	0.852	0.8821	0.8944	**0.9454**
	F-measure	0.8491	0.8821	0.8965	**0.9439**

obtained by the developed DBN model are 0.945 and 0.9438, whereas the recall value achieved by the existing approaches like NN, SVM, and NB is 0.8525, 0.8826, and 0.8944 and the F-measure value achieved by the existing approaches are 0.8496, 0.8826, and 0.8965, correspondingly.

6 Conclusion

In this paper, an access control model is developed for privacy preservation of the medical data. Three significant entities, like cloud user, cloud data owner, and the data server, are developed for cloud storage mechanism. The steps in the access control mechanism include user registration, control setup, and validation. The access control of the cloud is done through the functions, like encryption, hashing, and so on. Finally, the anomalous behavior is detected using DBN, which classifies the user in the cloud as genuine user and attacker. The unauthorized access of the sensitive medical data that is stored in the cloud is protected using this method. The analysis of the proposed DBN-based privacy preservation methods is done based on the performance measures, such as precision, recall, and F-measure. The proposed method provided a maximal precision, recall, and F-measure of 0.9430, 0.9454, and 0.9442, respectively, when compared to other existing methods. The future enhancement can be done by including more advanced classifiers for the classification.

References

1. W. Liu, Z. Wang, X. Liu, N. Zeng, Y. Liu, F.E. Alsaadi, A survey of deep neural network architectures and their applications. Neurocomputing **234**, 11–26 (2017)
2. M. Armbrust, A. Fox, R. Griffith, A.D. Joseph, R. Katz, A. Konwinski, G. Lee, D. Patterson, A. Rabkin, I. Stoica, A view of cloud computing. Commun. ACM **53**(4), 50–58 (2010)
3. Y. Xue, K. Xue, N. Gai, J. Hong, D.S. Wei, P. Hong, An attribute-based controlled collaborative access control scheme for public cloud storage. IEEE Trans. Inform. Forens. Secur. **14**(11), 2927–2942 (2019)

4. H.M. Sun, Y.H. Chen, Y.-H. Lin, oPass: a user authentication protocol resistant to password stealing and password reuse attacks. IEEE Trans. Inform. Forens. Secur. **7**(2), 651–663 (2012)
5. L. Harn, J. Ren, Generalized digital certificate for user authentication and key establishment for secure communications. IEEE Trans. Wirel. Commun. **10**(7), 2372–2379 (2011)
6. S. Madan, P. Goswami, k-DDD measure and mapreduce based anonymity model for secured privacy-preserving big data publishing. Int. J. Uncertain. Fuzziness Knowl.-Based Syst. **27**(2), 177–199 (2019). https://doi.org/10.1142/S0218488519500089
7. C. Chen, P. Maniatis, A. Perrig, A. Vasudevan, V. Sekar, Towards verifiable resource accounting for outsourced computation. ACM SIGPLAN Notices **48**(7), 167–178 (2013)
8. B. Waters, Ciphertext-policy attribute-based encryption: an expressive, efficient, and provably secure realization, in *Public Key Cryptography—PKC 2011* (Springer, Berlin, 2011), pp. 53–70
9. S. Boopalan, K. Ramkumar, N. Ananthi, P. Goswami, S. Madan, Implementing Ciphertext policy encryption in cloud platform for patients' health information based on the attributes, in *Computational Methods and Data Engineering. Advances in Intelligent Systems and Computing*, vol 1227, ed. by V. Singh, V. Asari, S. Kumar, R. Patel (Springer, Singapore, 2021). https://doi.org/10.1007/978-981-15-6876-3_44
10. S. Madan, P. Goswami, A privacy preservation model for big data in Map-reduced framework based on k-anonymization and Swarm-based algorithms. IJIEI **8**(1), 38–53 (2020). https://doi.org/10.1504/IJIEI.2020.105433
11. S. Madan, P. Goswami, Nature inspired computational intelligence implementation for privacy preservation in MapReduce framework. IJIIDS **13**(2/3/4), 191–207 (2020). https://doi.org/10.1504/IJIIDS.2020.109455
12. K. Xue, W. Chen, W. Li, J. Hong, P. Hong, Combining data owner-side and cloud-side access control for encrypted cloud storage. IEEE Trans. Inform. Forens. Secur. **13**(8), 2062–2074 (2018)
13. L.A. Trejo, V. Ferman, M.A. Medina-Perez, F.M.A. Giacinti, R. Monroy, J.E. Ramirez-Marquez, DNS-ADVP: a machine learning anomaly detection and visual platform to protect top-level domain name servers against DDoS attacks. IEEE Access **7**, 116358–116369 (2019)
14. X. Chen, B. Li, R. Proietti, Z. Zhu, S.J.B. Yoo, Self-taught anomaly detection with hybrid unsupervised/supervised machine learning in optical networks. J. Lightw. Technol. **37**(7), 1742–1749 (2019)
15. H. Takabi, J.B. Joshi, G.J. Ahn, Security and privacy challenges in cloud computing environments. IEEE Secur. Priv. **8**(6), 24–31 (2010)
16. S. Madan, P. Goswami, Adaptive privacy preservation approach for big data publishing in cloud using k-anonymization. Recent Adv. Comput. Sci. Commun. **14**(8) (2021). https://doi.org/10.2174/2666255813999200630114256
17. UCI machine Learning Repository: Heart Disease Dataset. https://archive.ics.uci.edu/ml/datasets/heart+disease. Accessed on May 2020
18. J. Yuan, S. Yu, Privacy preserving back-propagation neural network learning made practical with cloud computing. IEEE Trans. Parallel Distrib. Syst. **25**(1), 212–221 (2014)
19. Y. Rahulamathavan, R.C.W. Phan, S. Veluru, K. Cumanan, M. Rajarajan, Privacy-preserving multi-class support vector machine for outsourcing the data classification in cloud. IEEE Trans. Dependable Secure Comput. **11**(5), 467–479 (2014)
20. X. Liu, R. Lu, J. Ma, L. Chen, B. Qin, Privacy-preserving patient-centric clinical decision support system on Naïve Bayesian classification. IEEE J. Biomed. Health Inform. **20**(2), 655–668 (2016)

Wild OCR: Deep Learning Architecture for Text Recognition in Images

J. Amudha, Manmohan Singh Thakur, Anupriya Shrivastava, Shubham Gupta, Deepa Gupta, and Kshitij Sharma

Abstract With the emergence of the Industry 4.0 known as "smart factory", visual intelligence plays a vital role in various automation-related applications. Entrenched texts in images give plenty of information, which can be utilized in many applications. Various methods and technologies are existing in the realm of text extraction from the images. But due to diversity in arrangement of texts, scene background, shape, size, space, style variations, it is still a challenging task. In the past years, many researchers have been independently analyzing and solving issues related to detection and recognition of texts embedded in natural Images. This paper focuses on an end-to-end deep learning model for text detection and recognition in vehicle number plates. The proposed model's objective is to read text information in challenging car number plates. The model has been compared across state-of-the-art models to understand its advantages and limitations in cases of horizontal, perspective texts along with Indian/foreign car number plates.

Keywords Detection · Recognition · Deep learning · CNN

1 Introduction

Industry 4.0 is the prevailing trend of automation and data exchange in engineering technologies. The fourth generation (nowadays) is employing a combination of Internet of Things and cyberphysical security system for converting the traditional machines to self-aware and self-learning machines so that their total performance

J. Amudha · M. S. Thakur · A. Shrivastava (✉) · S. Gupta · D. Gupta
Department of Computer Science and Engineering, Amrita School of Engineering, Amrita Vishwa Vidyapeetham, Bengaluru, India
e-mail: 22.anupriya@gmail.com

K. Sharma
Paralaxiom Pvt. Ltd., Bangalore, India
e-mail: kshitij.sharma@paralaxiom.com

A. K. Bashir et al. (eds.), *Proceedings of International Conference on Computing and Communication Networks*, Lecture Notes in Networks and Systems 394,
https://doi.org/10.1007/978-981-19-0604-6_46

Fig. 1 Samples of Foreign/Indian car number plates

can get more improved, and also it can manage interactions efficiently [1]. This work proposes a deep learning pipelined architecture to solve the problem. The end-to-end scene text-recognition system, deep neural network, detects the regions of texts in an image followed by a deep neural network module to recognize the textual content. There are methodologies where deep neural network (for both detection and recognition steps) is trained for end-to-end recognition of texts in a semi-supervised way. The architecture suggested here follows a deep learning pipelined model of utilizing CNN architecture for detection and LSTM architecture for recognition. Moreover, the primary focus is on irregular texts like little blurred, oriented, or perspective style as shown in Fig. 1 (first, second images are little blurred, third image is oriented with dim light, and fourth image is perspective) along with regular texts.

For text detection, a convolution neural network (CNN) model is leveraged for extracting feature maps from a natural image and for text recognition, sequential prediction is performed using an LSTM architecture . The CNN-based model [2] as a detection module, based on instance segmentation method and the LSTM-based deep learning model [3] as a recognition module based on an arbitrary oriented network [4], which provides word-based accuracy. Here, an end-to-end text-recognition model has been implemented using deep learning technology, which comprises text detection and recognition modules for the automation processes of industries with high accuracy. In this implementation, a pre-trained model has been taken as a detection module and recognition module has been designed and trained for optimized parameters and hyperparameters. Initially, text regions get detected and further cropped by a cropping algorithm. Then the cropped images are given to the recognition module where texts get predicted. Here the models are implemented in Python with TensorFlow framework. Images are considered in .jpg format. The use case is taken as car/vehicle number plate recognition systems. The model's performance is compared on several benchmarks along with use case. Further, this paper comprises the following sections: Sect. 2 shows the literature survey (background works) for various text detection and recognition models. Section 3 defines the system architecture along with the layer's description. Section 4 elaborates the implementation and result analysis, and Sect. 5 presents conclusions and future works of the system.

2 Related Work

Reading texts in scene images is a very challenging topic in the applications of computer vision. In this section, a short introduction is given for background research in the area of text detection and recognition methods based on deep learning networks only. The literature survey looks at two aspects of model, one with single (same module for detection along with recognition) deep neural networks and second as combined (detection and recognition are separate modules) deep neural networks. Deng et al. [2] have focused on the detection of texts in natural images. They have used instance segmentation rather than traditional semantic segmentation for separating very closed text instances. The model requires very fewer training iterations and less training data. Cheng et al. [4] have presented an end-to-end trainable network called as arbitrary-oriented network (AON), which uses images along with word-level annotations for the training. The model focuses on extraction of deep features from text images, in horizontal and vertical directions. This model utilizes LSTM architecture for producing word-level accuracy. Liu et al. [5] have described a system which detects and recognizes the texts from image simultaneously. Uniqueness of this work is Region of Interest (RoI) rotate. They have trained both the tasks detection and recognition in a combined way, so that both can get benefited with each other. Ma et al. [6] have described an end-to-end detection system based on rotation approach. The key point of this paper is Rotation Region Proposal Networks (RRPNs), which generate inclined proposals with the information of orientation angle of text. Liao et al. [7] have described the single-shot-oriented detector (SSD) named Textboxes++. Basic idea behind this is: object detection algorithm single-shot-oriented detector (SSD). They have proposed superior designs for adjusting single-shot-oriented detector (SSD) networks for detecting capably oriented texts in scene images. Model uses a word bounding box with quadrilateral through a single neural network. Tang et al. [8] have described the text detection mechanism based on stroke feature transform using super pixels concepts and region classification using deep learning concepts. The first module is responsible for extracting candidate character regions and the second module is responsible for identifying character regions. Shi et al. [9] have proposed a model which comprises text-rectification and text-recognition network. This model is trained end-to-end with only images and ground truth annotations. It comprises an encoder and decoder module following a traditional sequence-to-sequence model. Bartz et al. [10] have proposed the model based on a semi-supervised neural network, which is end-to-end optimized. It has a single deep neural network, which checks the line of texts, words, and single character independently of each other. They have experimented on both character recognition along with word recognition. Shrivastava et al. [3] focuses on word-based recognition systems. The model uses CNN architecture which is based on the AON model [4]. Here, model has been trained on Amazon web services (AWS) platform using synth90k dataset [11]. Use cases considered are: sign boards, consumer products, vehicle number plates, some random images, etc. The best result was obtained at 390,000 iterations. Models [12–14] are based on recognition of different languages texts.The models discussed in the literature survey either use separate deep learning modules for detection and recognition or end-to-end deep learning modules for detection and recognition. Models [5, 7, 15] are based on end-to-end recognition of texts in images. Models [2, 6, 8] are representing individual detection modules. Models [3, 4, 9] represent individual recognition modules. Our model is extended work of [3]. Here, we have implemented an end-to-end recognition model mainly for vehicle number plates as use case, along with some standard datasets of challenging texts.

Along with standard detection and recognition datasets, car number plate dataset has been considered, which have two categories: standard and self-created. Standard car dataset includes 100 images of foreign cars and self-created dataset (captured by normal phone camera in day time) includes 100 images of Indian car number plates. All the car images are of size 300 × 300 in jpg/png format. Samples of car images are shown in Fig. 1 (first, second images represent foreign car dataset, and third, fourth images represent Indian car dataset).

3 Implementation

The implementation of the model is performed on system with configuration: Intel core i7-4770 processor, NVIDIA GeForce GTX TITAN Black GPU, 8 CPU @3.40GHz CPU, 32GB RAM. The project is encoded using Python language. Base OS is Ubuntu 16.0 (a Linux-based system). This end-to-end recognition model comprises a two-step method, one step for detecting texts and another one for recognizing text in scene text images. This system has two trainable modules, each for detection and recognition as shown in Fig. 2. The detection system is further integrated with the cropping module to provide cropped text images as an input to recognition module.

1. Detection module: Initially, input images are pre-processed, where it is resized to 1280 × 720 pixels. After pre-processing, images are directed for ground truth calculation of text regions using checkpoints. Then, it is passed to the detection module which is a pre-trained model [2]. Training loss is considered for text/non-text prediction, and link prediction is given below:

   a. Pixel prediction:

$$L\text{pixel} = \frac{1}{(1+r)^s} WL\text{pixel.ce} \tag{1}$$

Fig. 2 Block diagram of text-recognition model

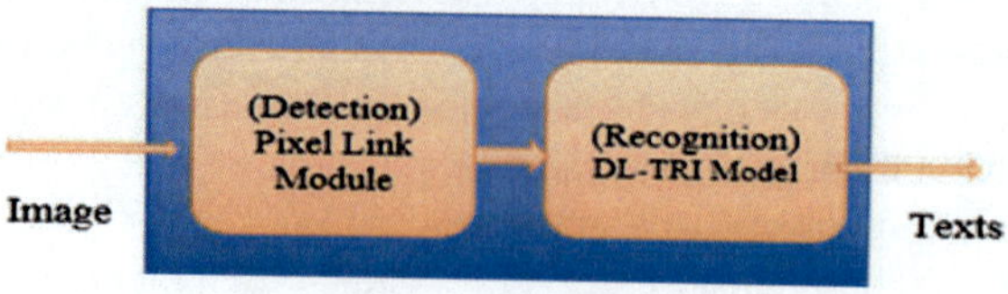

where Lpixel.ce = matrix of cross-entropy loss of text/non-text prediction
W = weight matrix
S = Instance area
r = negative positive ratio.

b. Link Prediction:

$$L_{\text{link}} = \frac{L(\text{link positive})}{r\text{sum}(W\text{positive link})} + \frac{L(\text{link negative})}{r\text{sum}(W\text{negative link})} \tag{2}$$

where L(linkpositive)= loss on positive links
L(linknegative)= loss on negative links
W(positivelink)= weight of positive links
W(negativelink)= weight of negative links
$r$sum= reduced sum.
The text regions are masked, which generates three pieces of information: text detected stored in compressed format, text file with x, y coordinates visualization folder (includes bounding boxes of texts) followed by an IOU (intersection over union) which is used to check the detection accuracy.

2. Cropping module: This is an intermediary stage where cropping algorithm has been used to crop the text regions. Cropping module first checks whether bounding boxes are present in the image or not. If bounding boxes are present, then it crops the texts regions and saves them in a specified folder. The relative path of each and every cropped image gets saved in TAGS File. This TAG file is further used for recognition modules.
3. Recognition module: The TAG file provided by the cropping module is preprocessed first; then it is given to decoder, LSTM (long short-term memory) which predicts the character sequence. Synth90k dataset of size 8 million cropped images has been used to train this model [12]. Training loss function L is given below:

$$L = \ln P(Y|I, \theta) \tag{3}$$

This is the loss function during the training of the whole network, where I is the image, θ is the combined parameters used in the whole network and Y is the ground truth of Yth character.

The recognition module is tested based on word accuracy instead of character accuracy. Word accuracy of recognition is given as:

$$\text{Word-Accuracy} = \frac{\text{Correctly-recognized-word}}{\text{Total-words}} \tag{4}$$

The Wild OCR model has been tested on various real-time use cases with good accuracy.

4 Result Analysis

In this section, performance of Wild OCR is discussed. The performance measures used are IOU (intersection over union) for detection and OCR accuracy for recognition (word level). The use case considered for testing the Wild OCR system is car number plates as shown in Fig. 3. Here, first two images are of foreign car number plates, and last two images are of Indian car number plates. The test case has been chosen to exhibit the systems performance in case of variation in fonts, styles, size, etc.

(a) Test Cases
(b) Performance of End-to-End Module: Intermediate results are shown in Fig. 4 (cropped text images) for given test cases.

Table 1 shows that the model satisfies the recognition process very well for horizontal/perspective/little blurred/dim light texts. This model is suitable for all the test cases. Broadly, we have worked on vehicle number plate datasets, and it performs very effectively. However, some challenging cases like very blurred texts, symbols, texts in different languages other than English are not recognized well. For fancy texts (different font style texts), an example is shown in failed case segment of Table 1, where the model has interpreted different letters for the whole text.

We have tested and validated Wild OCR model for datasets: ICDAR 2015 and SVT along with car number plates. Table 2 shows the testing results of our implemented model on various datasets. Here, 2245 images for IC15, 1187 images for SVT, 100 images for foreign cars and for Indian cars also 100 images are taken as input without changing the original image size. Our model performs very well on various datasets which will be very useful for industrial use cases.

Fig. 3 Samples of datasets used as test cases

Fig. 4 Intermediate results: cropped images of test cases

Table 1 Recognition results on different car number plates

1. Number plate (SI-819AK)		
Inputs	Recognition	Challenges
Image 1–2	SI 819AK (true)	Little blurry
2. Number plate (N0-450AM)		
Inputs	Recognition	Challenges
Image 3–4	NO 450AM (true)	Perspective
3. Number plate (KA 51 MG 9078)		
Inputs	Recognition	Challenges
Image 5–6–7–8	KA 51 MG 9078 (true)	Dim light text
4. Failed case-number plate (UP 32 CX 8055)		
Inputs	Recognition	Challenges
Image 9	BOSS (false)	Different font style

Table 2 Accuracy comparison on different datasets

S. No.	Datasets	Recognition (accuracy%)	Challenges
1.	ICDAR 2015	53	Perspective
2.	SVT	78	Different font style
3.	Foreign Car number plates	56	Perspective, blurred
4.	Indian Car number plates	70	Perspective, blurred

5 Limitations and Future Work

An end-to-end Wild OCR model has been discussed which utilizes a combination of CNN and LSTM deep learning architecture. The system is able to detect and recognize texts in horizontal/perspective/oriented style, making it suitable for various industrial use cases. The text detection is currently restricted to English language and will be extended to other languages. The model has to explore the limitations (e.g., license plate recognition: two-line number plates, fancy number plates, etc.), and it should be tested and validated for other use cases, e.g., sign boards, consumer goods, random images, etc., in the future.

Acknowledgements This project is part of the consultancy work with Paralaxiom Pvt. Ltd.

References

1. S. Vaidya, P. Ambadb, S. Bhosle, Industry 4.0—a glimpse, in *2nd International Conference on Materials Manufacturing and Design Engineering, ScienceDirect, Procedia Manufacturing*, vol. 20, pp. 233–238 (2018)
2. D. Deng, H. Liu, X. Li, D. Cai, *Pixel Link: Detecting Scene Text via Instance Segmentation*. arXiv:1801.01315v1 [cs.CV], 4 Jan 2018
3. A. Shrivastava, J. Amudha, D. Gupta and K. Sharma, Deep learning model for text recognition in images, in *2019 10th International Conference on Computing, Communication and Networking Technologies (ICCCNT)*, Kanpur, India (2019), pp. 1-6
4. Z. Cheng, Y. Xu, F. Bai, Y. Niu, S. Pu, S. Zhou, *AON: Towards Arbitrarily- Oriented Text-Recognition*. arXiv:1711.04226v2 [cs.CV] 22 Mar 2018
5. X. Liu, D. Liang, S. Yan, D. Chen, Y. Qiao, J. Yan, *FOTS: Fast Oriented Text Spotting with a Unified Network*. arXiv:1801.01671v2 [cs.CV] 15 Jan 2018
6. J. Ma, W. Shao, H. Ye, L. Wang, H. Wang, Y. Zheng, X. Xue, *Arbitrary-Oriented Scene Text Detection via Rotation Proposals*. arXiv:1703.01086v3 [cs.CV] 15 Mar 2018
7. M. Liao, B. Shi, X. Bai, Textboxes++: a single-shot oriented scene text detector. IEEE Trans. Image Proces. **27**, 3676–3690 (2018). arXiv:1801.02765v3
8. Y. Tang, X. Wu, Scene text detection using super pixel-based stroke feature transform and deep learning based region classification. IEEE Trans. Multimed. **20**(9), 2276–2288 (2018)
9. B. Shi, M. Yang, X. Wang, P. Lyu, C. Yao, X. Bai, ASTER: an attentional scene text recognizer with flexible rectification. IEEE Trans. Pattern Anal. Mach. Intel. **41**(9), 2035–2048 (2019)
10. C. Bartz, H. Yang, C. Meinel, *STN-OCR: A single Neural Network for Text Detection and Text Recognition*. arXiv:1707.08831v1 [cs.CV] 27 Jul 2017
11. M. Jaderberg, K. Simonyan, A. Vedaldi, A. Zisserman, *Synthetic Data and Artificial Neural Networks for Natural Scene Text Recognition* (2014). arXiv:1406.2227 [cs.CV]
12. P. Salunkhe, S. Bhaskaran, J. Amudha and Deepa Gupta, Recognition of multilingual text from signage boards, in *6th Internal Conference on Advances in Computing Communications and Informatics (ICACCI'17)* (2017)
13. R Anil, K Manjusha, S.S. Kumar and K.P. Soman, Convolutional neural networks for the recognition of Malayalam characters, in *Proceedings of the 3rd International Conference on Frontiers of Intelligent Computing: Theory and Applications (FICTA). Advances in Intelligent Systems and Computing*, vol. 328 (2014)
14. S. Bhaskaran, G. Paul, D. Gupta, J. Amudha, *Indian language identification for short text, in Advances in Intelligent Systems and Computing* (Springer, Berlin, 2021), pp. 47–58
15. P. He, W. Huang, T. He, Q. Zhu, Y. Qiao, X. Li, Single shot text detector with regional attention, in *IEEE International Conference on Computer Vision(ICCV)* (2017). eISSN: 2380-7504

ScamBlk: A Voice Recognition-Based Natural Language Processing Approach for the Detection of Telecommunication Fraud

Manoj Nandakumar, Ramanathan Nachiappan, Akhil Krishnan Sunil, João C. Neves, Hugo Pedro Proença, and Mithileysh Sathiyanarayanan

Abstract Telecom fraud has consistently caused extreme economic losses to telecom clients all around the world for years. Scammers use a wide variety of sophisticated techniques, making them indistinguishable from a harmless caller to access the victims' details or money. A voice recognition-based natural language processing-infused machine learning approach, ScamBlk, is proposed which uses real-world call content derived from the call audio. The audio from the call is used as input and is transcribed into the textual form which is further pre-processed and fed to the machine learning model. The machine learning model uses an ensemble approach deploying the bagging of a sequential model long-term short-term memory network and a linear model of support vector machine to classify the scam phrases in the phone conversation. The machine learning model is trained using a custom dataset with scam phrases obtained from various online sources. The ensemble model is found to be superior to other machine learning approaches obtaining an accuracy of 97.08%, allowing us to conclude that the ScamBlk approach is efficacious in detecting potential scam calls.

Keywords Call · Ensemble learning · Machine learning · Natural language processing · Spam · Transcribing · Voice spam

M. Nandakumar · R. Nachiappan · A. K. Sunil
Department of Computer Science and Engineering, SRM Institute of Science and Technology, Ramapuram, Chennai, India

J. C. Neves · H. P. Proença
IT: Instituto de Telecomunicações, University of Beira Interior, Covilhã, Portugal

M. Sathiyanarayanan (✉)
MIT Square Services Private Limited, London, UK
e-mail: mithileysh@mitsquare.com

A. K. Bashir et al. (eds.), *Proceedings of International Conference on Computing and Communication Networks*, Lecture Notes in Networks and Systems 394,
https://doi.org/10.1007/978-981-19-0604-6_47

1 Introduction

The number of smartphones and the users for smartphones is rising meteorically every passing day. Unfortunately, with this increase in the number of smartphone users, there is also an increase in the number of malicious fraudulent or scam calls that can potentially steal and misuse essential data like credit/debit card information, essential sites login details social security number, etc.

The scam calls have evolved to the point to which they are not distinguishable from legitimate calls. As a result, the number of victims of such fraud has steadily increased. Moreover, the contents of the call, i.e., the way the scammer converses and the terms they use, sound very similar to how a real would sound. Hence, it's difficult for people who are unaware of identifying such calls as a scam. Therefore, there is a dire need for effective solutions to help detect and prevent these scam calls, which threaten the fundamental human right of privacy.

Motivation. Conventional telecommunication scam identification techniques rely on naïve strategies based on call data records for extracting features such as call duration, frequency and phone number location. However, the scammer may exploit loopholes by changing his phone numbers and devising other techniques to make the call data record look innocuous. Therefore, to effectively combat telecommunication scams, there arises a need for the usage of the content of the phone calls, i.e., the phrases and specific keywords uttered by the scammer in the duration of the phone call. This content-based approach for telecommunication fraud enables better accuracy in results and helps devise a fool-proof system compared to traditional techniques that the scammer may subvert. In particular, the audio data from voice calls are an accurate source for deriving the content of the voice calls, which may be further used to detect fraudulent calls as fraudulent. Considering this, machine learning algorithms can be exploited to detect the aforementioned scam phrases when a plethora of call data is available. Thereby, a voice recognition-based approach that uses natural language processing (NLP) and machine learning techniques may prove to be effective in detecting scam calls.

Contribution

- ScamBlk is a voice recognition-based framework for telecommunication scam detection that employs NLP to detect telecommunication scams.
- The model is trained using a custom-made dataset with scam phrases scraped from various sources from the World Wide Web.
- The model uses the voice data of phone calls, pre-processes the audio and transcribes the audio into textual form.
- The textual data is subject to pre-processing by the usage of tokenization and vectorization and is fed into an ensemble machine learning model.
- The machine learning models used are the long short-term memory networks (LSTMs) and the support vector machine (SVM).
- The scam content is flagged with respect to a threshold set, and the user is alerted if the threshold is breached.

- The model is deployed as a service using the Flask API for ease of use for the user.

Organization. The remainder of the paper is structured as follows; the second section highlights the relevant research associated with the topic of voice scam detection. The third section illustrates the system architecture and proposed methodology of the system. The fourth section deals with the implementation and the result analysis for the proposed approach. The fifth section concludes the paper and is followed by the references.

2 Related Works

Zhao et al. [1] propose a methodology to detect fraudulent calls by the usage of the description of phone calls from news reports and posts on social media; machine learning algorithms like TF-IDF, logistic regression and neural networks validation are used to train a model which extracts the keywords related to the scam from the text by use of detection of patterns in the text. Huichen et al. [2] propose an application-based fraudulent call detection system; call-based features were used to train a machine learning model using vanilla neural networks and logistic regression with parameters based on the features to evaluate the reputation factor for a phone call. Ali et al. [3] have put forward a semantic-based approach to detect telecommunication fraud based on social engineering attacks using scam signatures; the scam signatures are automatically defined based on clustering utilizing K-means clustering. Bordjiba et al. [4] describe a data-driven approach for telephony threat analysis to detect fraudulent phone calls by complaints received about the phone number. Xing et al. [5] propose a method to detect fraudulent phone calls with the aid of deep learning. A classifier is developed using the call detail records. Javed et al. [6] devise a novel strategy to combat fraudulent calls by using a framework "N-Combat" where caller reputation is computed using a variety of factors like the call duration, received feedback and the participants of the communication. Yuhong et al. [7] build a framework DeMalC that uses call data records and a variety of other features associated with the call to detect fraudulent calls in a machine learning-infused approach. Mine et al. [8] provide an approach to classify fraudulent phone calls utilizing clustering with the usage of call data records which contain a total set of 97 features. PCA algorithm is used to reduce the dimensionality of the features. Balduzzi et al. [9] propose a methodology to detect telecommunication fraud by developing a tool to collect fraudulent calls and text messages. The unsolicited messages and calls are further analyzed using transcription using a tool and subject to a hierarchical clustering algorithm. Arafat et al. [10] introduce a methodology to detect a special kind of telecommunication fraud; ensemble machine learning has been used to detect this type of fraud. The machine learning algorithms AdaBoost, XGBoost and random forest algorithm are used for evaluating the dataset. Similar methods for fraudulent call detection are furnished in [11–15].

3 Proposed System Architecture

Figure 1 elucidates the proposed system for the detection of fraudulent phone calls. The system comprises six phases: user input, audio chunking, audio pre-processing, transcribing of call data, text pre-processing and prediction.

Firstly, the data is input as an audio file of the .wav format, which one can easily connect using the endpoint of the flask API. Flask is a Python framework that allows developers to control how users access data completely. It handles HTTP requests from users. This audio file is in essence, the audio recording of the phone call. The voice data is then subject to chunking, i.e., splitting the audio into separate segments based on an interval in the audio source. This is done to enhance the speech features. The chunking is done concerning the pauses in the call after every phrase is uttered.

Secondly, the audio is pre-processed by subjecting to tokenization, which converts the raw wav format file of the audio to signal data compatible to be predicted by the Wav2Vec2 model in the transcription phase. In the transcription phase, the audio contents are transcribed, i.e., converted to a textual form that facilitates the process of natural language processing.

The model training is performed using a custom-made dataset compiled from various online resources like social media articles, news clippings, scam call recordings and magazine reports of the different phrases used in a scam call (https://github.com/ManojN22/voiceBlk-dataset). The text extracted is then subject to pre-processing, where stemming and stop word removal are done. The stemming is done to remove the prefixes and suffixes of a word and reduce it to its base form, while the stop word removal is done to remove various stop words like and, for, the, etc. so that the emphasis is given to the important words present in the textual data.

Before the data is fed into the machine learning model the two different machine learning models are used, the LSTM and SVM models require different pre-processing methods. The LSTM model is a custom artificial neural network and the words from the text are subject to tokenization, which allows the conversion of complex paragraphs into fixed-size features that can be input to the NLP model

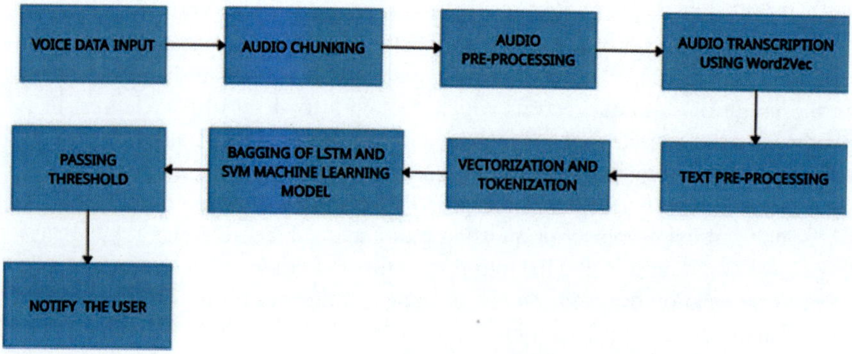

Fig. 1 System architecture design for the proposed ScamBlk model

and also padding is done which adds zeros to the end of the sequence of tokenized texts to make the sample size the same. The textual data is subject to vectorization regarding the SVM model, which converts a set of raw documents into a TF-IDF feature matrix. The data is then fed into the machine learning model, an ensemble of the LSTM and SVM machine learning models. The LSTM model is RNN with embedding layer, spatial_dropout1d layer, LSTM layer and dense layer. The SVM model is a linear classification model. Before feeding the data feature selection, we choose the features in our data that have the greatest impact on the target variable, i.e., the best predictors for the target variable are selected. Both models bring a different aspect from the conversation's content and produce a more accurate prediction. A threshold of 0.5 is used for discriminating between a scam and real content in the call, such that if the content of the call exceeds the threshold value, the user is notified of the call being a scam phone call, which is subsequently sent using the Flask API.

4 Implementation and Result Analysis

The implementation of the proposed spam detection coupled with audio transcription was implemented in Python using Jupyter notebook, using a computer with an i7 4.0 Ghz Intel CPU, 16 GB RAM and 10 GB of disk drive storage. The Sklearn machine learning library was used for the implementation of ensemble models. The dataset used in this model was taken from various online sources like social media articles, news clippings and magazine reports and devised using an HTML-based WebCrawler BeautifulSoup, comprising 3000 spam data in addition to 7000 non-spam data in total.

The PyDub Python package facilitates the AudioSegment Python module. To assess the performance of telecom fraud detection, we premeditated a few experiments to perceive it. This study calculated telecom fraud linked interchanges. We planned ten dialogues grounded taking place from the soundtracks on the Internet that are telecom fraud. We tested ten samples of the conversations that used clear English in addition to four conversations with unclear English. Among the calls, eight calls were distinguished as a scam. It states that the algorithm was able to detect 80% of fraudulent calls using English. Among the calls with unclear English, only one call was spotted as a scam. The foremost cause for this singularity is that the recognition accuracy rests on the eminence and quality of dialogue recognition. Therefore, as soon as the caller uses ambiguous language, the dialogue recognition algorithm is not able to return the precise transcript. However, in the future, through the advances in the power of speech recognition technology and the enhancement of recognition exactness, the uncovering accuracy of fraudulent calls by our model will burgeon.

Further, the efficiency of the LSTM and SVM algorithms was tested. The system was first tested on both the algorithms independently, and though the performance was appreciable, the independent usage of these algorithms triggered false positives occasionally. When used in tandem, the number of false positives triggered

Table 1 Comparison of the performance of the proposed method with other machine learning approaches

Search technique	Accuracy %	Precision %	Recall %	F-measure
Random forest	90.12	87.60	92.65	90.05
Decision tree	91.62	88.70	94.55	91.53
XGBoost	92.96	91.15	94.77	92.92
ScamBlk (LSTM + SVM)	97.08	96.53	97.64	96.32

had substantially reduced, thereby making the model much more efficient. In our experiments, the metrics used for evaluation are the F-Measure, recall, accuracy and precision. Recall is denoted as the percentage of the retrieved and significant true positive class to the overall sum of recovered true and false positive class and suitable in curtailing false negatives is the focus. The main attention is that the portion of true positive recovered besides valid toward the whole sum of relevant positives is called precision and suitable for curtailing false positives. Precision, recall, accuracy and F-Measure are defined in Eqs. (1)–(4), which are used to quantify the performance of the proposed approach.

$$\text{Recall\%} = \frac{\text{True sum of Positives}}{\text{True sum of Positives} + \text{False sum of Negatives}} \quad (1)$$

$$\text{Precision\%} = \frac{\text{True number of Positives}}{\text{True number of Positives} + \text{False number of Positives}} \quad (2)$$

$$\text{Accuracy\%} = \frac{\text{Precision} + \text{Recall}}{2} \quad (3)$$

$$\text{F-Measure\%} = \frac{2(\text{Precision} \times \text{Recall})}{(\text{Precision} + \text{Recall})} \quad (4)$$

Table 1 illustrates the performance exhibited by the proposed approach ScamBlk when compared with other machine learning algorithms. The algorithms used were random forest, decision tree and XGBoost algorithms with the proposed system using an ensemble approach involving the LSTM and SVM algorithms. All the models were tested on the same dataset in the same testing conditions.

The random forest produces the lowest recall, precision, F-Measure and accuracy. The lowest precision value is mainly because this algorithm uses statistics features such as Call Detail records and not the actual text, which decreases the accuracy of the model. Regarding the lack of actual text and NLP model makes the accuracy of 90.05 comparatively low. The decision tree algorithm's performance is also low due to the use of imbalanced data. For instance, if we have 50,000 spam datasets and 1000 non-spam datasets, it causes a bias toward spam. Due to this issue, decision tree yields low accuracy when compared. XGBoost is a boosting algorithm grounded on a gradient boosted decision tree algorithm. XGBoost presents a better-regularized

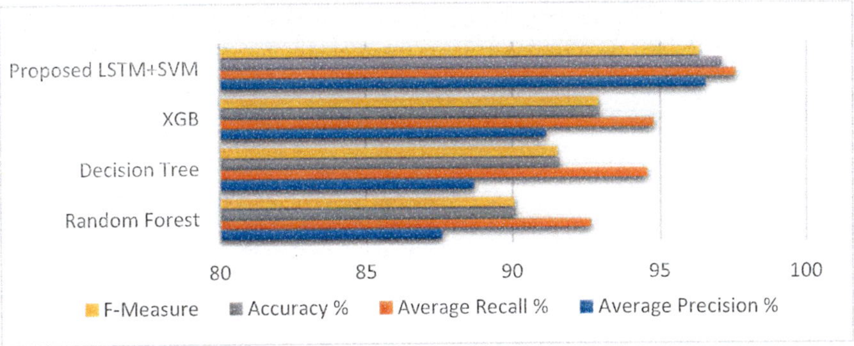

Fig. 2 Performance evaluation for different algorithms employed

technique to diminish overfitting. Decision tree uses a series of learning algorithms alone and a very high complexity-based XGBoost, making the approach computationally complex. However, infusing voice recognition and NLP in the proposed model is highly commendable, explaining why it has comparatively higher recall, precision, accuracy when related to other models.

Figure 2 depicts the results of the proposed ScamBlk system compared with the XGBoost, decision tree and random forest system, where a significant performance improvement can be noticed. The graph tells us that the proposed system has higher precision and accuracy. Hence, a method for Telecom Spam detection is offered by ScamBlk, which is an efficient and cost-effective approach. This is because this approach uses voice recognition and NLP infused with the machine learning model. LSTM + SVM is used for detecting spam content in call recording in addition to text data. Thereby, supplying the need for auxiliary data and background data as a result and voice recognition brings diversification of results in audio data.

5 Conclusion

Fraudulent or scam phone calls continue to pose a threat to the privacy and security of the user, and with newer malicious threats evolving every day related to such scams, effective detection becomes a challenging task. This work introduces a voice recognition-based NLP-infused machine learning framework (ScamBlk) for detecting scam calls. This method makes use of the content of the call, where audio data from the call is split into segments based on pauses and is subject to pre-processing. It is transcribed into the textual form, subject to text pre-processing and fed to an ensemble machine learning model employing the LSTM and SVM machine learning algorithm, which flags the call as a scam based on a set threshold. The approach involving this algorithm exhibited better performance in comparison to other baseline algorithms with high accuracy of 97.08% and was found to be superior in the task of detecting fraudulent telecommunication calls with respect to the other approaches. As a future work, we will continue to focus on voice cloning technology using AI and how cybercriminals can be trapped using our solutions from digital forensics and E-discovery point of view [16].

References

1. Q. Zhao, K. Chen, T. Li, Y. Yang, X. Wang, Detecting telecommunication fraud by understanding the contents of a call. Cybersecurity **1**(1), 1–12 (2018)
2. H. Li, X. Xu, C. Liu, T. Ren, K. Wu, X. Cao, et al., A machine learning approach to prevent malicious calls over telephony networks, in *2018 IEEE Symposium on Security and Privacy (SP)*, pp. 53–69. IEEE (2018, May)
3. Derakhshan, A., Harris, I. G., & Behzadi, M. (2021) Detecting telephone-based social engineering attacks using scam signatures, in *Proceedings of the 2021 ACM Workshop on Security and Privacy Analytics*, pp. 67–73
4. H.E. Bordjiba, E.B. Karbab, M. Debbabi, Data-driven approach for automatic telephony threat analysis and campaign detection. Digit. Investig. **24**, S131–S141 (2018)
5. J. Xing, M. Yu, S. Wang, Y. Zhang, Y. Ding, Automated fraudulent phone call recognition through deep learning, in *Wireless Communications and Mobile Computing, 2020*
6. I. Javed, K. Toumi, N. Crespi, N-Combat: a nuisance call combating framework for internet telephony, in *2018 17th IEEE International Conference on Trust, Security and Privacy in Computing and Communications/12th IEEE International Conference on Big Data Science and Engineering (TrustCom/BigDataSE)*, pp. 112–117. IEEE (2018)
7. Y. Li, D. Hou, A. Pan, Z. Gong, Demalc: a feature-rich machine learning framework for malicious call detection, in *Proceedings of the 2017 ACM on Conference on Information and Knowledge Management*, pp. 1559–1567 (2017)
8. X. Min, R. Lin, K-means algorithm: fraud detection based on signaling data, in *2018 IEEE World Congress on Services (SERVICES)*, pp. 21–22. IEEE (2018)
9. M. Balduzzi, P. Gupta, L. Gu, D. Gao, M. Ahamad, Mobipot: Understanding mobile telephony threats with honeycards, in *Proceedings of the 11th ACM on Asia Conference on Computer and Communications Security*, pp. 723–734 (2016)
10. M. Arafat, A. Qusef, G. Sammour, Detection of Wangiri telecommunication fraud using ensemble learning, in *2019 IEEE Jordan International Joint Conference on Electrical Engineering and Information Technology (JEEIT)*, pp. 330–335. IEEE (2019)
11. G. Vennila, M.S.K. Manikandan, M.N. Suresh, Detection and prevention of spam over Internet telephony in voice over internet protocol networks using Markov chain with incremental SVM. Int. J. Commun. Syst. **30**(11), e3255 (2017)
12. M. Swarnkar, N. Hubballi, SpamDetector: detecting spam callers in voice over internet protocol with graph anomalies. Secur. Priv. **2**(1), e54 (2019)
13. R.T. Pashiri, Y. Rostami, M. Mahrami, Spam detection through feature selection using artificial neural network and sine–cosine algorithm. Math. Sci. **14**(3), 193–199 (2020)
14. G. Vennila, M.S.K. Manikandan, M.N. Suresh, Dynamic voice spammers detection using Hidden Markov Model for voice over internet protocol network. Comput. Secur. **73**, 1–16 (2018)
15. Y. Tan, Q. Wang, G. Mi, Ensemble decision for spam detection using term space partition approach. IEEE Trans. Cybern. **50**(1), 297–309 (2018)
16. M. Sathiyanarayanan, O. Fadahunsi, Integrating digital forensics and digital discovery to improve e-mail communication analysis in organisations, in *Smart Computing Paradigms: New Progresses and Challenges*, pp. 187–193 (Springer, Singapore, 2020)

Personality BERT: A Transformer-Based Model for Personality Detection from Textual Data

Dipika Jain, Akshi Kumar, and Rohit Beniwal

Abstract Understanding personality type can aid in understanding people preferences and associated cognitive processes. Automated personality detection can commendably help NLP experts and psychoanalysts to identify the dominant or distinguishing qualities of a person. At its basic level, a personality is expressed through a person's temperament or emotional tone. Pertinent studies validate linguistic cues in written and spoken text as a coherent and consistent mode of assessing and interpreting personality. With the proliferation of social media applications, the psycholinguistic markers in user's online posts can facilitate comprehending variations in personalities. Transformer models have emerged as new generation NLP models and are already being implemented to benefit an array of NLP use cases. This research puts forward a transformer-based model for personality detection from textual data. The proposed personality BERT is a textual modality-specific deep neural model that fine-tunes a pretrained bidirectional representation for transformers (BERT) for the personality classification task. Kaggle's MBTI dataset is used to evaluate and validate the proposed model. An f1 score of 0.6945 is reported.

Keywords Personality · BERT · Text · Classification

1 Introduction

An individual's personality is a relatively distinct and a consistent pattern of thinking, feeling, and behaving. It entails how an individual affects others and how he/she understands and views himself/herself [1]. This general style of interacting with the

D. Jain · R. Beniwal
Department of Computer Science and Engineering, Delhi Technological University, New Delhi, India

A. Kumar (✉)
Department of Information Technology, Netaji Subhas University of Technology, New Delhi, India
e-mail: akshi.kumar@nsut.ac.in

A. K. Bashir et al. (eds.), *Proceedings of International Conference on Computing and Communication Networks*, Lecture Notes in Networks and Systems 394,
https://doi.org/10.1007/978-981-19-0604-6_48

world is primarily shaped by life experiences and inner psychological characteristics. Socio-cognitive processes contribute to learned behaviors that are central to one's personality [2]. The study of personality reveals three distinct aspects: Firstly that a personality reflects individual differences; secondly, personality is consistent and enduring, and lastly, that a personality can change. Personality can be studied through patterns of inner and outer measurable traits and the person-situation interaction. A good personality can help in better social and professional life, whereas a negative personality can lead to disorders like depressive, anti-social, and narcissistic personality disorders among others.

There are many personality theories which form the basis of various personality testing and assessment tools. The three most prominent ones include the Freudian theory, the Neo-Freudian theory, and the trait theory. The most popular trait theory is a quantitative approach to classify personality using a set of dominant characteristics or identifiable traits. Typically, traits are any distinguishing, relatively enduring way in which one individual differs from another and can be measured via behavioral indicators. Personality assessment using the trait theory includes techniques aimed to quantify representative signs and patterns of traits that an individual displays through various circumstances using these stable and tangible qualities. The traits can be captured through various verbal (voice, tone, text) and non-verbal (facial expressions, body language, physiological indicators) manifestations. The growing use of social media has boosted a virtual community to communicate individual ideas, sentiments, opinions, and emotions which reflect their attitude, behavior, and personality. Simultaneously, personality traits can be significantly deciphered from the trivial pieces of information people post on social media. That is, what an individual creates, consumes, likes/dislikes, and follows can partially reveal insights to understanding personality. A large body of research on personality detection on social media has focused on the use of trait theories such as the five-factor model-FFM and the Myers–Briggs type indicator (MBTI). The MBTI personality assessment developed in 1940's categorizes human behavior and personality into 16 distinct groups where traits are binary in nature, meaning that a person is categorized in an 'either this or that' position. The notion of MBTI is that every persona consists of four dimensions, and each dimension has two possibilities [3]. These dimensions and their corresponding possibilities are as follows: introversion (I)—extroversion (E), intuition (N)—sensing (S), thinking (T)—feeling (F), and judging (J)—perceiving (P). In total, there are 16 combinations, or 16 personality types, and are denoted using a four letter abbreviation. That is each person will have a personality abbreviated (ENFJ, INFJ, ENFP, INFP, ENTJ, INTJ, ENTP, INTP, ESFJ, ISFJ, ESFP, ISFP, ESTJ, ISTJ, ESTP, ISTP) from the combination of all four axis, i.e., someone who is extroverted, relies more on intuitions, thinking, and judging rather than perceiving will be labeled as an ENFJ.

Recently, computational psychology has revolutionized cognitive neuroscience and psychotherapy studies. Various modeling tools and natural language processing capabilities can be used to analyze the instantaneous behavior and traits using psycholinguistic markers from millions of online posts. Real-time analytics of personality traits has emerged as a vital market research and branding strategy that can facilitate personalization of Web, recommendation services, and intelligent empathetic conversational agents [4, 5]. Various modality-specific (textual, visual, audio, sensor-based psychophysiological, and brain signals) and multimodal (bi-modal and tri-modal) personality computing models have been reported in literature. This research puts forward a bidirectional representation for transformers-based (BERT) model for recognizing apparent personality traits from textual modality. BERT includes a transformer with multiple attention mechanisms to provide context of words in a text and a classification layer to the transformer output to detect the personality from textual data. It is one of the most popular neural architectures used for a wide variety of NLP tasks, and fine-tuning BERT allows to build a robust classification model to predict categories. The proposed personality BERT is textual modality-specific deep neural model that fine-tunes a pre-trained BERT for the personality classification task. The MBTI_kaggle dataset is used to evaluate and validate the proposed model.

The paper has five sections. The following Sect. 2 includes the relevant work in the domain. Section 3 describes the proposed personality BERT model followed by the results in Sect. 4 and the conclusion in Sect. 5.

2 Related Work

Pertinent literature has reported many approaches for recognizing personality traits from various modalities. For text modality [3–6], there has been work on various datasets, namely: MBTI_kaggle, PersonalityCafe, and myPersonality. Hernandez and Scott [6] used RNN on the MBTI Kaggle and reported the performance accuracy of 67.77%. Cui and Qi [7] reported a survey on various machine learning techniques for MBTI personality type prediction in text. Apart from MBTI dataset, the textual modality for personality traits has been studied on PersonalityCafe forums [3] which consists of 68,000 posts using BERT model and an accuracy of 0.479 with 30 epochs and 0.00001 as learning rate. Majumdar et al. [1] detected personality from input text by utilizing convolutional neural network. Salsabila et al. [8] implemented SVM along with BERT for big five personality detection of Twitter datasets. They also employed Linguistic Inquiry Word Count (LIWC) in the model. Bhavya et al. [9] also employed various machine learning techniques for personality detection. Li et al. [10] demonstrated the relation between personality traits and emotional behaviors. They used CNN-based multitask model on multiple famous personality and emotion datasets. Sun et al. [11] proposed a group-level personality detection method by using an unsupervised feature learning method. Lynn et al. [12] focused on message-level attention for personality detection. They used GRU with word-level and message-level attention on social media posts. Ren et al. [13] demonstrated the use and importance of sentiment in a BERT-based personality detection model on two datasets. Leonardi et al. [14] proposed a multilingual model by employing BERT on the myPersonality dataset.

3 The Proposed Personality BERT Model

Automatic personality recognition allows predicting how people will respond to certain situations and understanding the sorts of things they prefer and value. In the proposed personality BERT model, we have used Kaggle's MBTI text-based dataset which consists of the data from social media platforms. This consists of 8600 rows of data with data of sixteen different personality types, abbreviated from the personality keys. MBTI looks at our personal preferences across four dichotomies as given in Fig. 1.

The dataset is cleaned by removing links and punctuations and then split into training and test data [15]. Data processing and cleaning are initiated using the Python library, and then, the BERTTokenizer is then initialized to perform the personality classification task [16]. Transformer library is imported, and the cross-entropy is used to compute the loss which is the most important cost function used in the personality BERT model. BERT is mainly a family of transformer-encoder architecture which processes each token of input text with respect to the full context of all the tokens present and therefore called as the bidirectional encoder representation from transformers. The proposed 'personality BERT' model loads the pretrained BERT base model from the transformers library and proceeds by taking the first hidden state from BERT output (corresponding to CLS token) to feed it into a dense layer with 16 neurons and softmax activation. Figure 2 shows the architecture of the proposed model.

That is, the cleaned data pass through the transformer which further goes to the feed-forward neural network using softmax as the activation function. As it is

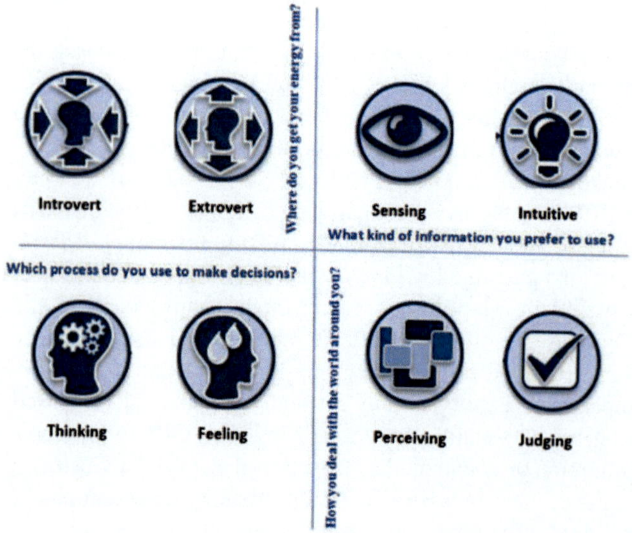

Personality Type	Dataset Count
INFJ	1470
INTP	1304
INTJ	1091
ENTP	685
ENFP	675
ISTP	337
ISFP	271
ENTJ	231
ISTJ	205
ENFJ	190
ISFJ	166
ESTP	89
ESFP	48
ESFJ	42

Fig. 1 MBTI's 8 preferences and dataset distribution

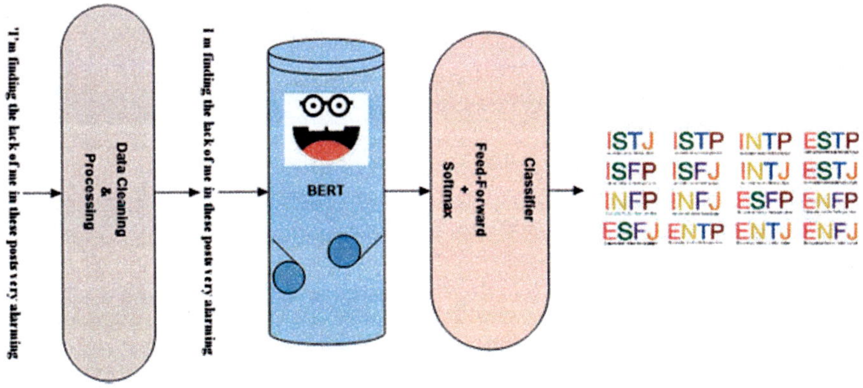

Fig. 2 Architecture of personality BERT model

a multiclass classification of the data, hence, the softmax function is used in the model. The intermediate results are converted into probabilities with the help of softmax activation function in the feed-forward neural network.

4 Results

The results obtained by using the personality BERT model are discussed in this section. It also mentions the hyperparameter settings for the BERT model. State-of-the-art (SOTA) comparison is also done in order to validate the work. Performance of the proposed model is assessed using accuracy, recall, precision, and f1 metrics. The dataset was divided in the ratio of 85:15 (where 85% was used for training the data and remaining 15% was used for testing purposes) using a ten-fold cross-validation technique. Table 1 shows the amount of data reserved for training and test per personality type.

Implementation was done using Jupyter Notebook and Python libraries (TensorFlow and Keras along with basic NumPy and pandas libraries). All the hyperparameter values are given in Table 2.

The training and test data performance in terms of loss, accuracy, f1, precision, and recall values is shown in Table 3.

As observed from Table 3, there is a huge incongruence in training and test results, and this overfitting is primarily due to the skewness in the dataset. Table 2 depicted the distribution of the dataset, and the class imbalance was quite evident. Though we corrected class imbalances, by tuning the hyperparameters using GridSearchCV from the Scikit-Learn package and generating stratified 5-folds, but this did not seem to improve the results marginally. The performance of the personality BERT model was also compared with the existing works as shown in Fig. 3. A superior performance is observed for the proposed model on the MBTI dataset.

Table 1 Personality type, data label, no. of post

Personality type	Data labels	Posts_split	
		Test	Train
INFJ	0	221	1249
ENTP	1	103	582
INTP	2	196	1108
INTJ	3	164	927
ENTJ	4	35	196
ENFJ	5	28	162
INFP	6	275	1557
ENFP	7	101	574
ISFP	8	41	230
ISTP	9	50	287
ISFJ	10	25	141
ISTJ	11	31	174
ESTP	12	13	76
ESFP	13	7	41
ESTJ	14	6	33
ESFJ	15	6	36

Table 2 Hyperparameters for BERT

Parameters	Values
Trainable parameters	335,158,288
Non-trainable parameters	0
Loss (cross-entropy)	1.5575
Max_length	1500
Learning rate	0.00001
Verbose	1
Epochs	20
Batch_Size	32

Table 3 BERT performance

Metric	Training	Test
Loss	0.0452	1.5094
Accuracy	0.9860	0.6945
F1	0.9857	0.6945
Precision	0.9868	0.7116
Recall	0.9848	0.6818

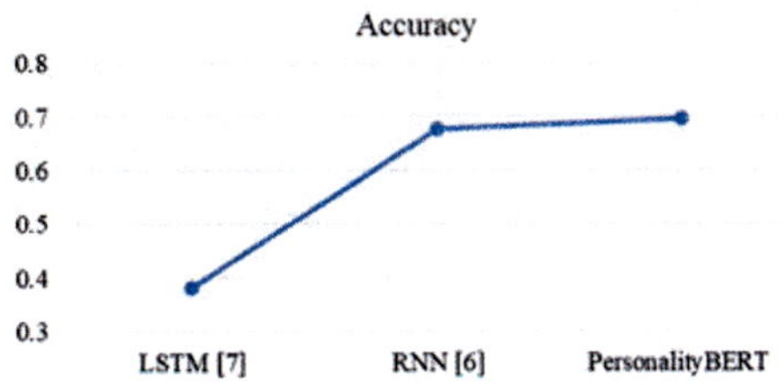

Fig. 3 Comparison with existing works on MBTI dataset

5 Conclusion

Personality computing aids from techniques intended to understand, predict, and identify human behavior. As personality is a distinctive trait which differentiates among individuals, their patterns, preferences and choices, automated personality detection finds many diverse applications from recommendation systems, job screening, psychological studies to consumer forecasting. Credibility and accuracy are the most important factors to determine different personalities, and it entails the creation of new models for more efficient, reliable, and accurate determination of personalities. The personality BERT model presented in this paper works in this direction and determines the different personalities from the text-based benchmark MBTI dataset by employing a transformer type of neural network architecture, BERT. The dataset is split into test and training, and the results are evaluated using f1, recall, precision, and accuracy scores. The results show an improvement over the existing works, but further, validation of results of all personalities and each personality type needs to be done. The dataset is highly skewed in favor of introverted types, and class imbalance is prominent. Future studies using bootstrapping or resampling techniques need to be done to balance this data.

References

1. N. Majumder, S. Poria, A. Gelbukh, E. Cambria, Deep learning-based document modeling for personality detection from text. IEEE Intell. Syst. **32**(2), 74–79 (2017)
2. A. Kumar, K. Srinivasan, W.H. Cheng, A.Y. Zomaya, Hybrid context enriched deep learning model for fine-grained sentiment analysis in textual and visual semiotic modality social data. Inform. Proces. Manage. **57**(1), 102141 (2020). https://doi.org/10.1016/j.ipm.2019.102141
3. S.S. Keh, I. Cheng, *Myers-Briggs Personality Classification and Personality-Specific Language Generation Using Pre-trained Language Models* (2019). ArXiv abs/1907.06333
4. Y. Hernández, C.A. Peña, A. Martínez, Model for personality detection based on text analysis, in *Advances in Computational Intelligence. MICAI 2018. Lecture Notes in Computer Science*, ed. I. Batyrshin, M. Martínez-Villaseñor, H. Ponce Espinosa, vol. 11289 (Springer, Cham, 2018). https://doi.org/10.1007/978-3-030-04497-8_17
5. Y. Mehta, N. Majumder, A. Gelbukh, E. Cambria, Recent trends in deep learning based personality detection. Artif. Intell. Rev. **53**(4), 2313–2339 (2020)
6. R.K. Hernandez, I. Scott, *Predicting Myers-Briggs Type Indicator with Text* (2017)
7. B. Cui, C. Qi, *Survey Analysis of Machine Learning Methods for Natural Language Processing for MBTI Personality Type Prediction* (2017)

8. G.D. Salsabila, E.B. Setiawan, Semantic approach for big five personality prediction on twitter. J. RESTI (Rekayasa Sistem Dan Teknologi Informasi) **5**(4), 680–687 (2021)
9. S. Bhavya, A.S. Pillai, G. Guazzaroni, Personality identification from social media using deep learning: a review, in *Soft Computing for Problem Solving*, pp. 523–534 (2020)
10. Y. Li, A. Kazameini, Y. Mehta, E. Cambria, *Multitask Learning for Emotion and Personality Detection* (2021). arXiv preprint arXiv:2101.02346
11. X. Sun, B. Liu, Q. Meng, J. Cao, J. Luo, H. Yin, Group-level personality detection based on text generated networks. World Wide Web **23**(3), 1887–1906 (2020)
12. V. Lynn, N. Balasubramanian, H.A. Schwartz, Hierarchical modeling for user personality prediction: the role of message-level attention, in *Proceedings of the 58th Annual Meeting of the Association for Computational Linguistics*, pp. 5306–5316 (2020, July)
13. Z. Ren, Q. Shen, X. Diao, H. Xu, A sentiment-aware deep learning approach for personality detection from text. Inform. Proces. Manage. **58**(3), 102532 (2021)
14. S. Leonardi, D. Monti, G. Rizzo, M. Morisio, Multilingual transformer-based personality traits estimation. Information **11**(4), 179 (2020)
15. A. Kumar, V. Hugo, C. Albuquerque, Sentiment analysis using XLM-R transformer and zero-shot transfer learning on resource-poor Indian language. ACM Trans. Asian Low-Resour. Lang. Inform. Proces. **20**(5), 1 (2021). Article No.: 90. https://doi.org/10.1145/3461764
16. M.P.S. Bhatia, S.R. Sangwan, Debunking online reputation rumours using hybrid of lexicon-based and machine learning techniques, in *Proceedings of First International Conference on Computing, Communications, and Cyber-Security (IC4S 2019)*, pp. 317–327. Springer, Berlin (2020)

A Bi-Level Stochastic Model with Averse Risk and Hidden Information for Cyber-Network Interdiction

MingChu Li, Wanyu Dong, Xiao Zheng, Anil Carie, and Yuan Tian

Abstract This paper proposes a method to enable a risk-averse and resource-constrained network defender to deploy security countermeasures in an optimal way to prevent multiple potential attackers with uncertain budgets. To solve the problem of information asymmetry between the attacker and the defender, a fake countermeasure (FC) is placed on the arc, and the situation of multiple attackers is also taken into consideration. This method is based on the risk aversion bi-level stochastic network interdiction model on the attack graph, which can easily map the path of attackers. Meanwhile, our method can minimize the weighted sum of all losses and minimize the risk of the defender's key nodes being destroyed. At the same time, in order to prevent the key node of the defender from being destroyed, the risk condition value measurement is taken into account in the stochastic programming model. We design a SA-CPLEX algorithm to provide a high-quality approximate optimal solution. And computational results suggest that our method provides better network interdiction decisions than traditional deterministic and risk-neutral models.

Keywords Stackelberg game · Averse risk · Hidden information · Bi-level programming · Cyber-security

M. Li · W. Dong (✉)
School of Software, Dalian University of Technology, Dalian, China
e-mail: dongwanyu@mail.dlut.edu.cn

M. Li
e-mail: mingchul@dlut.edu.cn

X. Zheng
School of Computer Science and Technology, Shandong University of Technology, Zibo, China
e-mail: xiao_zheng0910@163.com

A. Carie
School of Computer Science, VIT-AP, Amaravati, India

Y. Tian
School of Economics and Management, Dalian University of Technology, Dalian, China
e-mail: ytian@mail.dlut.edu.cn

A. K. Bashir et al. (eds.), *Proceedings of International Conference on Computing and Communication Networks*, Lecture Notes in Networks and Systems 394,
https://doi.org/10.1007/978-981-19-0604-6_49

1 Introduction

With the rapid development of information technology, people can browse a large number of websites through the Internet. The application scenarios of computer equipment have also expanded and penetrated into the public's access network technology and work [1–3]. However, while the computer brings more convenience, it also has certain security risks, causing some key information to be leaked and bringing certain economic losses. This paper studies the problem of a network defender to minimize worst-case damage by setting countermeasures against uncertain attacks. We propose a Stackelberg game between defenders and attackers, in which the defender not only can deploy true countermeasures (TCs) but also fake countermeasures (FCs). The deployment of FCs can be used to mislead the attacker's actions.

The goal of this research is to help the defender makes better use of the limited budget to protect the network from uncertain attacks. Therefore, it is necessary to establish a new interdiction model to formulate the risk aversion of the network defender under the uncertainty of the attackers' ability [4]. This paper establishes a defender-attacker stochastic Stackelberg game [5] model including risk aversion based on attack graph. Our stochastic network interdiction model can interdict multiple potential attackers with uncertain budgets. Compared with a model that considers a constant budget and a unique attacker, this modeling method is more representative of a realistic scenario. However, the traditional risk model stochastic programming usually takes the minimization of losses as their goal and does not take into account the risk of maximum loss scenarios. The risk aversion stochastic programming model minimizes the defender's expected loss and minimizing the risk of huge losses when the attackers' initial budget is uncertain.

We introduce a novel risk-averse bi-level stochastic network interdiction model based on attack graphs and use conditional risk value as risk measurement and customized accurate algorithm to solve the bi-level random network counter-measures model for risk aversion. This problem is defined as a bi-level stochastic network interdiction problem with risk aversion, the upper-level is the problem of the defender, and the lower-level is the problem of the attacker. In the upper-level model, the defender makes decisions without knowing the attacker's budget. While in the lower-level model, the attacker plans an attack route based on a known budget and a known interdiction strategy of the defender. And a simulated annealing algorithm based on the commercial solver CPLEX, namely, SA-CPLEX, is customized for our model to solve this NP-hard problem.

The contributions of this paper are as follows.

(1) We propose a risk-averse defender-attacker stochastic Stackelberg game model that merges fake countermeasures and multiple attackers with uncertain budgets.
(2) Conditional value-at-risk (CVaR) is involved in our model to measure the defender's risk performance with respect to the attackers' uncertain budgets.
(3) An effective algorithm is proposed to solve the resulting bi-level problem, which can provide an efficient solution for our model.

The remainder of this paper will be described in the following structure. In Sect. 2, we review the related works. The problem definition and formulation are presented in Sect. 3. In Sect. 4, we propose a heuristic algorithm and perform theoretical analysis on the proposed algorithm. We present the experimental results and analysis of the results in Sect. 5. Finally,we summarize our conclusions in Sect. 6.

2 Related Work

Attack graphs with different changes are widely used as a tool for network analysis, such as defensive tree [6], attack countermeasure tree [7], vulnerability dependency graph [8], etc. The way of network interdiction based on attack graphs to protect target nodes (key assets) is to remove a set of arcs or nodes from the attack graph. In the previous literature, attack graph network interdiction enhances network security by generating cut sets [9]. Khouzani et al. [10] studied the cyber-security defense problem using attack graphs to model a multi-stage attack. In addition to the mathematical model of attack graphs, some studies have proposed the use of traditional mathematical models to reduce the risk of network attacks. For example, Zheng et al. [11] allocated limited mitigation resources to increase the robustness of supply chain infrastructure information technology in cyber-attacks. A recent paper, Bhuiyan et al. [12] modeled multiple potential attackers, in which the attacker's actions are assumed to be absolutely unsuccessful if the defender deploys interdiction measures on the arc. But in real life, even if the defender installs defensive countermeasures, the attacker still has a certain chance to pass the arc. There are also studies that consider uncertainty in the bi-level network interdiction model, including the uncertainty of protection facilities to minimize the worst-case [13]. We have found that taking uncertainty into consideration has a positive direction for the completeness of the entire model.

In order to maximize the attackers' cost of the shortest path, Pay et al. [14] established a random shortest path network interdiction model. But in their network interdiction model, the huge risk posed by the attacker was not considered. In [15], the risk measure, i.e., conditional value-at-risk (CVaR), is incorporated into the location and protection problem. Furthermore, Lei et al. [16] studied stochastic flow interdiction problems using a risk-averse approach. As proved by Lei et al. [16], the model considering risk preferences can provide more robust solutions in comparison to the risk-neutral counterpart. In this regard, our paper also incorporates a risk measure to hedge against the huge risk.

3 Problem Definition and Formulation

This paper studies the stochastic Stackelberg game interaction between the defender and two or more attackers in the risk aversion network using the attack graph. As shown in Fig. 1, the node represents the attack state,and its set is N. Each node is represented in the form of N-D, where N is the value of each node, and D is the defender's loss when the head node is destroyed. The green node is the initial safety condition, the blue node is the key node, and the yellow is the transition node. The attacker's attack path consists of an arc from the initial node to the key node. In the case of NCs in Fig. 1, the attacker's optimal plan is to destroy the key node through the attack path $0 \to 2 \to 4$. The attacker destroys any node in the attack graph, and the defender has a certain loss. The attackers start from the green initial node, and their goal is to destroy the blue node to maximize the defender's loss. Once the target node is successfully attacked, it will be completely destroyed.

The arc between the two nodes represents the action of the attacker. Set of arcs with the tail pointing to node $A_t(i)$, which indicates the prerequisite for the attacker's action, and it is a necessary security condition that the attacker should break during the action. Set of arcs with the head pointing to node $A_h(i)$, which represents the post-condition of the attacker's action, which is the security condition that the attacker breaks after the action is successful. The value V of the arc between two nodes indicates the probability of a successful attack through this arc. We can calculate the loss when a node is destroyed as $V \times d$. Taking the attack path $0 \to 2 \to 4$ as an example, the expected maximum loss to the defender is 3.8584(= 0.79 × 0.04 × 5). Finding the optimal path by calculating and comparing the losses caused by different paths.

In our research, the defender-attacker stochastic Stackelberg game on the attack graph is modeled as a bi-level stochastic network interdiction problem with risk

Fig. 1 An example of attack graph

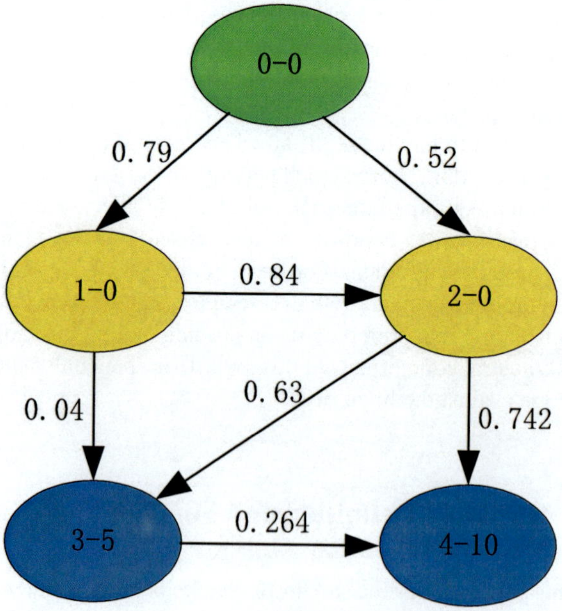

aversion. The upper-level indicates the problem of defender, and the lower-level represents the problem of attacker. Within the game, the defender first finds the arc where the attacker is most likely to attack without knowing the attackers' budget and spends a certain cost to install countermeasures (including TCs and FCs) on this arc within the deployment budget. The attacker's budget will only be obtained by the defender after they have completed the plan. b_d and $\bar{b}_d$ represent the defender's budget of deploying TCs and FCs, respectively. In order to optimize the defender model, some FCs will be placed in the path to mislead the attacker. It should be noted that there are also some paths in the graph that have no countermeasures (NCs). Minimizing expected losses is an important task for defenders while minimizing the huge risks caused by the loss of key assets.

In terms of the behavior of the defender, each attacker develops an attack plan based on their budget and destroys the key assets aiming to maximize the loss of the defender. Each attacker has an attack cost c_{ij}^{attack} when attacking through the arc, and their total cost should not exceed the budget. Intuitively, the FCs can be exposed or detected by the attacker, thus let o_{ij} be the exposure probability of FCs deployed on arc (i, j). If the attacker detects the FCs, the arc (i, j) equipped with FCs is the same as the one with NCs. Thus, according to both FCs and NCs, the attacker has the same probability to pass the arc, which is denoted by $\bar{p}_{ij}$. Also, let p_{ij} be the probability of passing the arc equipped with TCs.

In order to maximize the defender's loss, the attackers use a limited budget b_a to select the optimal attack plan in a given set of truncated arcs. The attackers start the attack from the vulnerable node and continue to penetrate the network through the transition node until the key asset (target node) is destroyed. If an attacker can break through one of the target nodes, the network defender will suffer losses. The attack path includes an arc from the initially vulnerable node N_I to the target node N_T. An attack plan consists of a combination of one or more attack arcs. Even if this arc can break through multiple target nodes in an attack strategy, the attacker only needs to successfully attack this arc once. In this way, the problem is transformed into a discrete optimization problem. In addition, once the target node is attacked, then it will be completely destroyed.

Decisions are made sequentially in bi-level stochastic programming [17]. The upper-level is to make decisions before the uncertainty is realized, and the lower-level is to make additional decisions after the uncertain parameters of each scenario are concretely realized. In the upper-level of our bi-level random programming model, the defender must make a decision to interdict even when the attacker's budget is not clear. In each case of the lower-level, each attacker specifies an attack plan with a known budget and knows the interdiction decision of the network defender. We use conditional risk value (CVaR) [18] as a risk indicator to measure risk aversion.

Our model will have to take into account the multiple attackers with uncertain budgets, where each attacker has a specific budget. The budgets of the defender and the attackers must be within their given limits. It is not certain which attacker the network defender will encounter, nor does it know the attacker's budget. However, according to the probability distribution of known parameter values, the defender can estimate the attacker's budget [19]. In order to simulate the uncertainty in the budget of multiple potential attackers, we consider a set of the limited number of scenarios S in the stochastic optimization problem. Each scenario represents an attacker with a specific budget. In the case of a limited budget, the defender chooses the best subset of arcs for deploying countermeasures to minimize the maximum loss in all scenarios (Table 1).

Table 1 Notation

Notation	Description
Sets	
N	Set of nodes
N_I	Set of initially nodes
N_T	Set of key nodes
A	Set of arcs
$A_t(i)$	Set of arcs with the tail pointing to node i
$A_h(i)$	Set of arcs with the head pointing to node i
S	Set of scenarios index by s
Parameters	
l_t	Loss resulting from breaching a key node $t \in N_T$
o_{ij}	Exposure probability of FCs deployed on arc (i, j)
p_{ij}	Probability of successful attack through the arc (i, j)
	equipped with TCs
$\bar{p}_{ij}$	Probability of successful attack through the arc (i, j)
	Equipped with FCs or NCs
p^s	Probability of scenario s
λ	The coefficient of risk
α	Confidence level
b_a	Attacker's budget
b_d	Defender's budget for deploying TCs
$\bar{b}_d$	Defender's budget for deploying NCs
c_{ij}^{attack}	Attack cost through arc (i, j)
c_{ij}^{d}	Cost of TCs on arc (i, j)
$\bar{c}_{ij}^{d}$	Cost of FCs on arc (i, j)
Decision variables	
x_{ij}	1 if TCs are deployed on arc (i, j), 0 otherwise
$\bar{x}_{ij}$	1 if FCs are deployed on arc (i, j), 0 otherwise
f_{ij}	1 if arc (i, j) is used for one or more attacks, 0 otherwise
z_i	Probability of node i being destroyed
y_{ij}	Product of z_i and f_{ij}
η	Upper-level variable (represents the value-at-risk, VaR)
v^s	Excess loss variable in scenario $s \in S$

3.1 Minimize the Disutility of Defender

The upper-level model is to minimize the disutility of defenders. The objective function (1) consists of two parts, where the first part calculates the expected minimum disutility of defenders in all scenarios. And the expected maximum loss in all scenarios is equal to the probability p^s of scenario s multiplied by the maximum total loss Q^s caused by the attackers to the defender in scenario s. The second part simulates the CVaR metric of huge loss risk. Constraints (2) and (3) limit the budget for deploying TCs and FCs, respectively. Constraints (4) prevent the TCs and FCs from being deployed at the same arc. Constraint (5) calculates the additional loss in all attack scenarios, and the excess loss variable in scenario s is greater than or equal to the maximum total loss minus the variables at the upper-level. Constraints (6) are binary requirements.

$$H = \min \sum_{s \in S} p^s Q^s(x, \bar{x}) + \lambda(\eta + \frac{1}{1-\alpha} \sum_{s \in S} p^s v^s) \tag{1}$$

$$\text{s.t.} \sum_{(i,j) \in A} c_{ij}^d x_{ij} \leq b_d \tag{2}$$

$$\sum_{(i,j) \in A} \bar{c}_{ij}^d \bar{x}_{ij} \leq \bar{b}_d \tag{3}$$

$$x_{ij} + \bar{x}_{ij} \leq 1 \quad \forall (i, j) \in A \tag{4}$$

$$Q^s(x, \bar{x}) - \eta \leq v^s \tag{5}$$

$$x_{ij}, \bar{x}_{ij} \in \{0, 1\} \quad \forall (i, j) \in A \tag{6}$$

$$\eta \in R \tag{7}$$

$$v^s \geq 0 \quad \forall s \in S \tag{8}$$

3.2 Maximize the Utility of Attacker

The lower-level model is to maximize the utility of the attacker. That is, the objective function (9) maximizes the loss caused by interdicting the target node multiplied by the probability of the target node t being destroyed. Constraint (10) limits that the total expenditure of the attack must be within their budget. Constraint (11) indicates that whether the attacker attacks the arc ij has a decisive influence on its success probability. If the attacker takes action, the probability of success is the product of the true attack probability, the fake attack probability, and the non-attack probability. Constraints (12) ensure that there is a higher probability that an attacker successfully destroyed node through an arc (i, j), and the probability of node j being attacked is less than or equal to the probability of successfully attacking through arc

ij.Constraints (13) indicate that only one attack is required on an arc ij.Constraints (14) indicate that if an arc is attacked one or more times, it will be 1, and if there is no attack, it will be zero.

$$Q^S(x, \bar{x}) = \max \sum_{t \in N_T} l_t z_t \tag{9}$$

$$s.t. \sum_{(i,j)\in A} c_{ij}^{\text{attack}} f_{ij} \leq b_a \tag{10}$$

$$\beta_{ij} = f_{ij} \cdot p_{ij}^{x_{ij}} \cdot \bar{p}_{ij}^{(1-o_{ij})\bar{x}_{ij}} \cdot \bar{p}_{ij}^{1-(x_{ij}+\bar{x}_{ij})} \tag{11}$$

$$z_j \leq \sum_{(i,j)\in A_e(j)} z_i \beta_{ij} \quad \forall j \in N/N_I \tag{12}$$

$$\sum_{i,j \in A_e(j)} f_{ij} \leq 1 \quad \forall j \in N/N_I \tag{13}$$

$$f_{ij} \in 0, 1 \quad \forall (i, j) \in A \tag{14}$$

$$0 \leq z_j \leq 1 \quad \forall j \in N \tag{15}$$

Constraints (12) are nonlinear; however, the only nonlinear terms are $z_i f_{ij}$. In this regard, we define the auxiliary variables w_{ij} to replace them. For each $(i, j) \in A$ and $i \in N/N_I$, a set of new constraints is added to the formulation to line $w_{ij} = z_i f_{ij}$.

$$w_{ij} \leq z_i \tag{16}$$

$$w_{ij} \leq f_{ij} \tag{17}$$

$$w_{ij} \geq 0 \tag{18}$$

$$w_{ij} \geq f_{ij} + z_i - 1 \tag{19}$$

4 Solution Approach

It is difficult to solve the bi-level linear problem using existing algorithms directly. Because the model is more complicated, and it is an NP-hard problem [20]. In the past few years, many studies have proposed the use of precise algorithms or hybrid heuristics to solve the bi-level optimization problem. For example, Shamekhi Amiri et al. [21] invented a global iterative search method, inferring the potential behavior of followers as a new constraint for each iteration in the leader problem. In this paper, we propose a heuristic solution algorithm, namely SA-CPLEX, where the SA algorithm is used to solve the defender problem in the upper-level, and the CPLEX solver is used to obtain the optimal attacker's strategy in the lower-level.

In the heuristic algorithm, one of the crucial parts is the representation of the solution [22]. The heuristic algorithm also acts alternately on the coding space and the solution space [23]. It is a way to find the best solution within an acceptable time. The generation of the neighborhood and the fast calculation of the objective function are the goals of this algorithm. In addition, it must ensure that it has access to the entire solution space. The value of the initial solution will have some impact on the performance of the heuristic algorithm. In order to give an initial solution to the defender problem, we use a randomly sized subset as the central node [24]. And we define and use a single operator to generate adjacent solutions. This operator is called "Swap" and is used to change the solution representation to an arc in the array. Four parameters, T_0, T_f, δ and $I_{\max}$ are used in the algorithm. Among them, T_0 represents the initial temperature, and T_f is the final temperature at which the SA process is stopped [25]. δ is used as the cooling rate parameter of the upper-level problem. $I_{\max}$ is the number of solutions generated by the algorithm at each temperature.

The detailed algorithm is given in Algorithm 1. As shown in Algorithm 1, the algorithm first generates the initial solution of the defender and then improves this initial solution through subsequent iterations. According to the given defender's strategy $(x, \bar{x})$, the attacker's objective value $Q(x, \bar{x})$ can be calculated according to (9). We fix the initial temperature as T_0 and use it as the initial parameter of the algorithm. $(x, \bar{x})_{\text{best}}$ represents the optimal solution found so far, and f_{best} represents its relative objective function value. For each temperature, we define ΔE as the difference between the newly obtained solution and the target of the existing solution, that is, $\Delta E = H(Q((x, \bar{x})')) - H(Q(x, \bar{x}))$. We repeat this cycle at most once at each temperature and use the optimal solution obtained so far. Then the temperature decrease to $T \leftarrow \delta \times T$ after each iteration. Repeat the training until the current temperature T is lower than the pre-specified final temperature T_f, the algorithm ends.

5 Experiments

All experiments were performed on a personal computer with a 2.90 GHz Core (TM) i7-10700 CPU AND 16GB RAM. We implement our proposed algorithm in Matlab 2020a and ILOG CPLEX 12.10 is applied to solve the attacker's problems optimally. We have conducted a lot of experiments so that the average result will not change too much (Table 2).

5.1 Parameter Setup

We use an attack graph with a node size of $|N|$ (=50) for numerical experiments, and the arc size is about $2.15 \times |N|$. Breach loss of the goal nodes is uniformly from (500, 150), while the budget of this random attacker is Weibull distribution (100, 200). The probability of attack success is uniformly from (0, 1). The probability of a successful attack through the arc (i, j) equipped with TCs and FCs is between 0 and 1. Moreover, three different level of confidence are also tested for the experiments,

Algorithm 1 SA-CPLEX ($T_0, T_f, \delta, I_{\max}$)

Generate a random initial solution $(x, \bar{x})$
$\forall s \in S$,Calculate $Q^S(x, \bar{x})$ using CPLEX
$Q(x, \bar{x}) = \sum_s Q^s(x, \bar{x})$
$T \leftarrow T_0$, $H_{\text{best}} \leftarrow H(Q(x, \bar{x}))$, $(x, \bar{x})_{\text{best}} \leftarrow (x, \bar{x})$, $I \leftarrow 0$
while $T > T_f$ **do**
  for $I < I_{\max}$ **do**
    Generate a new solution $(x, \bar{x})'$ based on $(x, \bar{x})$ using "Swap" operator
    $\forall s \in S$,Calculate $Q^S(x, \bar{x})'$ using CPLEX
    $Q((x, \bar{x})') = \sum_s Q^s((x, \bar{x})')$
    $\Delta E \leftarrow H(Q((x, \bar{x})')) - H(Q(x, \bar{x}))$
    if $\Delta E < 0$ **then**
      $(x, \bar{x}) \leftarrow (x, \bar{x})'$
      $H(Q(x, \bar{x})) \leftarrow H(Q((x, \bar{x})'))$
    else
      $\rho \leftarrow rand(0, 1)$
      if $\rho > e^{-|\Delta E|/T}$ **then**
        $(x, \bar{x}) \leftarrow (x, \bar{x})'$
        $H(Q(x, \bar{x})) \leftarrow H(Q((x, \bar{x})'))$
      end if
    end if
    if $H(Q(x, \bar{x})) < H_{\text{best}}(Q(x, \bar{x}))$ **then**
      $(x, \bar{x})_{\text{best}} \leftarrow (x, \bar{x})$, $H_{\text{best}} \leftarrow H(Q(x, \bar{x}))$
    end if
    $(x, \bar{x}) \leftarrow (x, \bar{x})_{\text{best}}$
    $I \leftarrow I + 1$
  end for
  $I \leftarrow 0$
  $T \leftarrow \delta \times T$
end while
return $(x, \bar{x})_{\text{best}}$, H_{best}

Table 2 Parameters and default values

Parameters	Values
Network size (nodes, $\|N\|$)	50
Arcs, $\|A\|$	$\approx 2.15 \times \|N\|$
Breach loss of the goal nodes	~uniform (500, 1500)
Defender's budget, b_d	150
Level of confidence, α	0.2, 0.5, 0.8
Risk coefficient, λ	2, 4, 8, 10
Random attacker budget, b	~weibull (50, 500)

i.e., $\alpha \in \{0.2, 0.5, 0.8\}$. We set the exposure probability of FCs deployed on arc (i, j) is in the range (0, 1). And Four different values of risk factors are also tested for the experiments, i.e., $\lambda \in \{2, 4, 8, 10\}$. Physical attacks or cyber-attacks on important infrastructure systems are also within the range that our attack graph can simulate [26].

5.2 Effects of Involving FCs

Figure 2 shows variation of mean-risk expected maximum loss (MREXPLoss) with and without FC budget. For maps of different sizes, we have different defender's budgets. The eventual experimental results showed that when the total budget remains the same, the more budget spent on FCs, the smaller the MREXPLoss. And as the total budget value increases, MREXPLoss becomes smaller. As the budget increases, the defender has sufficient budget to place countermeasures in more attack paths. As a result, the combination of various defensive countermeasures has increased, making it more difficult for attackers to attack. At the same time, it can be seen from the experiment that when there is a lot of total budgets, the defender can protect more attack paths. In other words, the defender can have more combinations of different attack paths.

5.3 Effects of the Probability of Exposure

Figure 3 shows the variation of MREXPLoss under different exposure probabilities of FCs. We can see from the experimental results that the greater the probability of FCs being exposed through the arc (i, j), the smaller the value of MREXPLoss. In other words, when the probability of FCs being exposed is very small, the FC can be well hidden. This will cause more interference to the attacker, which will cause the attacker to make more wrong decisions. Therefore, the defender can better interdict the attack and reduce some losses.

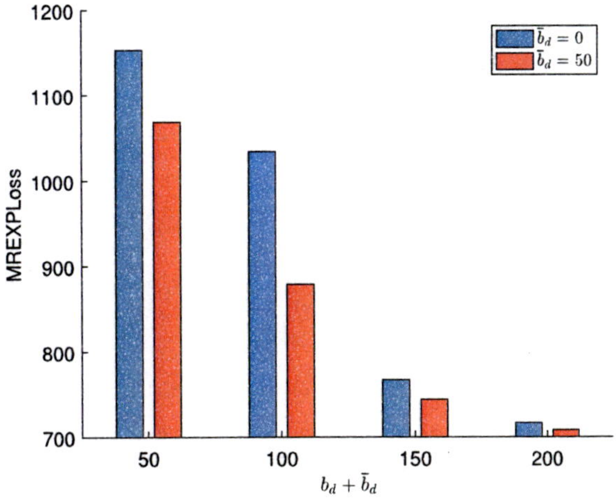

Fig. 2 Variation of b_d and $\bar{b}_d$ with number of total budgets. Other parameters are: $|N| = 50$, $p_{ij} = 0.5$, $\bar{p}_{ij} = 0.5$, $o = 0.75$, λ=0

Fig. 3 Variation of α with possibility of o_{ij}. Other parameters are: $|N| = 50$, $p_{ij} = 0.5$, $\bar{p}_{ij} = 0.5$, $\lambda = 0$

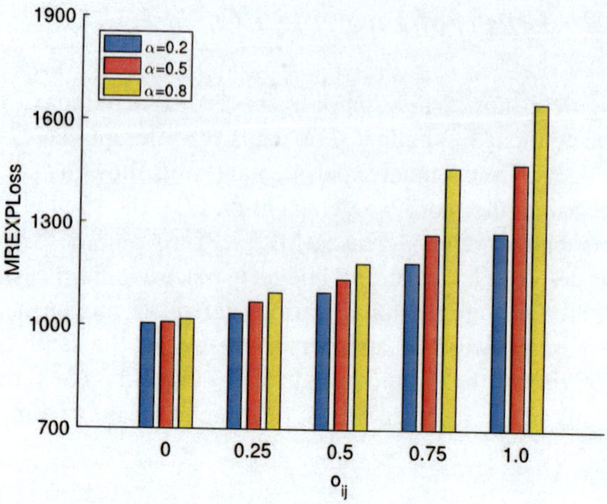

5.4 *Effects of the Probability of Successful Attack*

Figures 4 and 5 show the variation of MREXPLoss in the probability of successful attack through the arc (i, j) with TCs and FCs, respectively. It can be concluded from the experimental results that when the attacker's success probability to TCs increases, the defender is more vulnerable to attack. Similarly, when the attacker's probability of success in FCs increases, the attacker will be more likely to destroy these TCs. As the probability of being successfully attacked in TCs or FCs is higher, their loss is greater.

Fig. 4 Variation of α with possibility of p_{ij}. Other parameters are: $|N| = 50$, $\bar{p}_{ij} = 0.5$, $o = 0.75$, $\lambda = 0$

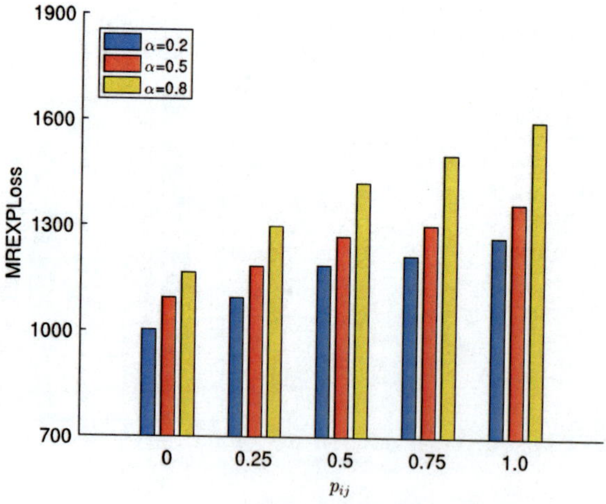

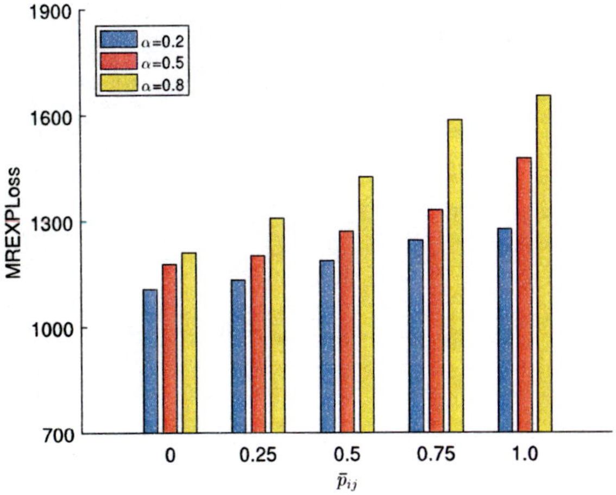

Fig. 5 Variation of α with possibility of $\bar{p}_{ij}$. Other parameters are:$|N| = 50$, $p_{ij} = 0.5$, $o = 0.75$, $\lambda = 0$

5.5 *Effects of the Budget of TCs and FCs*

Figure 6 shows the gap between the budget of TCs and FCs when the total budget remains the same. It can be seen from the experimental results that when the total budget value becomes larger, the constant real budget MREXPLoss is decreasing. As the total budget increases, MREXPLoss decreases at a higher rate. The defender can mislead the attacker by adjusting the ratio of the FCs budget to the TCs budget, thereby achieve a better protective effect.

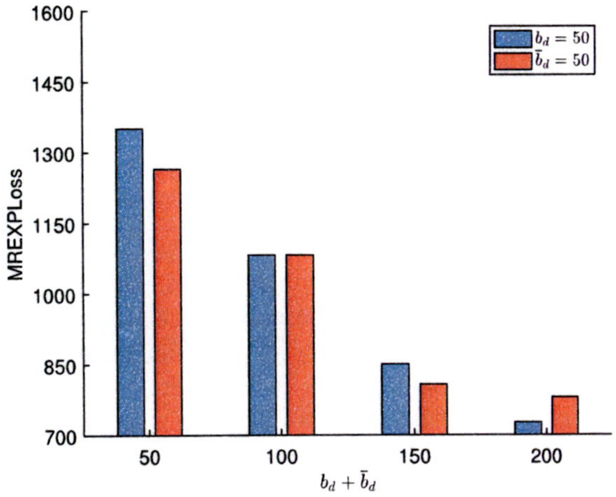

Fig. 6 Variation of $b_d = 0$ and $\bar{b}_d$ with possibility of total budgets. Other parameters are: $|N| = 50$, $p_{ij} = 0.5$, $\bar{p}_{ij} = 0.5$, $o = 0.75$, $\lambda = 0$

Fig. 7 Variation of α with possibility of λ. Other parameters are: $|N| = 50$, $p_{ij} = 0.5$, $\bar{p}_{ij} = 0.5$, $o = 0.75$

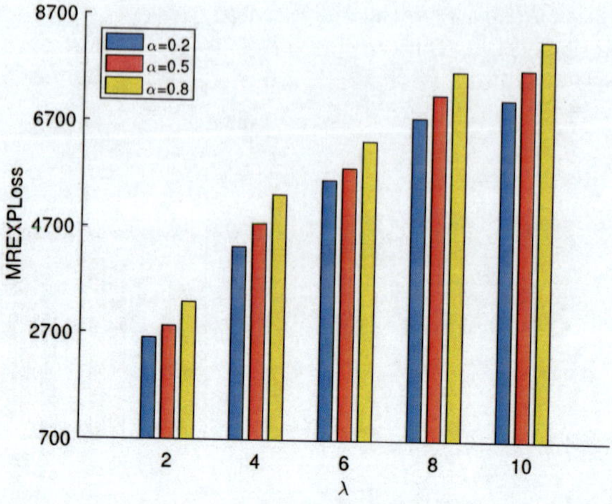

5.6 *Effects of Confidence Level and Risk Coefficient*

In our proposed model, two risk parameters, confidence level (α) and risk coefficient (λ) are our important parameter members. Figure 7 shows the variation of MREXPLoss with respect to λ under three different levels of α. As shown in the experimental results, we can clearly see the result of using risk metrics (CVaR) to minimize the losses caused by random attackers' budget cyber-attacks. The larger the value of α, the more concerned about the situation of major losses, and the more conservative the decision-making. To a certain extent, the goal of minimizing the expected value of the main loss scenario is also considered here. In other words, minimizing huge losses is not the only goal of our model. Because in this case, the optimal interdiction decision under risk-neutral preference also partially considers the minimization of large losses.

6 Conclusions

This article studies the problem of the best interdicting strategy from the perspective of the defender, where the defenders seek to minimize the risk of major losses. In addition, the budget uncertainty of multiple potential attackers and the fake countermeasures deployed by the defender are considered. Based on the extension of the traditional attack graph, we establish a risk aversion bi-level stochastic network interdiction model to formulate this problem. In our risk aversion model, our risk measure is CVaR. In response to this model, we developed a customized binary bi-level programming problem algorithm that combines randomness and risk aversion. Our model is closer to reality and considers more comprehensively for the defender. The experimental results show that the interdiction decision provided by our model is more robust than the traditional model. Successfully achieved the minimization of the huge loss risk caused by network attacks by avoiding risks. In the future, this paper can be easily applied to the security of the underwater wireless sensor networks [27], the gird monitoring systems [28] and the critical infrastructure systems [29].

Acknowledgements This work was support from the National Science Foundation of China (Grant No. 61877007), Fundamental Research Funds for the Central Universities (Grant No. DUT20GJ205), and the Shandong National Science Foundation of China (Grant No. ZR202103040468).

References

1. S. Noel, S. Jajodia, Optimal ids sensor placement and alert prioritization using attack graphs. J. Netw. Syst. Manage. **16**(3), 259–275 (2008)
2. S.B.H. Shah, F. Yin, I.U. Khan, Z. Chen, M. Zakarya, Collating and analysing state-of-the-art hierarchical routing protocols in WSN to increase network lifetime and conserve energy, in *Proceedings of the International Conference on Future Networks and Distributed Systems, ICFNDS 2017*, Cambridge, United Kingdom, July 19–20, 2017, ed. by M. Hammoudeh, R.M. Newman, ACM, 2017, pp. 42:1–42:10. [Online]. Available: https://doi.org/10.1145/3102304.3102346
3. S.B. Shah, C. Zhe, F. Yin, I.U. Khan, S. Begum, M. Faheem, F.A. Khan, 3d weighted centroid algorithm & rssi ranging model strategy for node localization in wsn based on smart devices. Sustain. Cities Soc. **39**, 298–308 (2018)
4. R. Zhang, Q. Zhu, Y. Hayel, A bi-level game approach to attack-aware cyber insurance of computer networks. IEEE J. Select. Areas Commun. **35**(3), 779–794 (2017)
5. S.A. Zonouz, H. Khurana, W.H. Sanders, T.M. Yardley, Rre: a game-theoretic intrusion response and recovery engine. IEEE Trans. Parallel Distrib. Syst. **25**(2), 395–406 (2014)
6. S. Bistarelli, F. Fioravanti, P. Peretti, Defense trees for economic evaluation of security investments, in *Proceedings of the The First International Conference on Availability, Reliability and Security, ARES 2006, The International Dependability Conference—Bridging Theory and Practice, April 20–22 2006, Vienna University of Technology, Austria*. IEEE Computer Society (2006), pp. 416–423 [Online]. Available: https://doi.org/10.1109/ARES.2006.46
7. S.D. Roy, S. Kundu, Performance of an adaptive power based cdma cognitive radio networks, in *IEEE Symposium on Industrial Electronics and Applications (ISIEA)*, pp. 28–33 (2010)
8. E. Serra, S. Jajodia, A. Pugliese, A. Rullo, V.S. Subrahmanian, Pareto-optimal adversarial defense of enterprise systems. ACM Trans. Inf. Syst. Secur. **17**(3), 11:1–11:39 (2015). Available: https://doi.org/10.1145/2699907
9. M. Alhomidi, M. Reed, Finding the minimum cut set in attack graphs using genetic algorithms, in *International Conference on Computer Applications Technology (ICCAT)*, pp. 1–6 (2013)
10. M.H.R. Khouzani, Z. Liu, P. Malacaria, Scalable min-max multi-objective cyber-security optimisation over probabilistic attack graphs. Eur. J. Oper. Res. **278**(3), 894–903 (2019)
11. K. Zheng, L.A. Albert, J.R. Luedtke, E. Towle, A budgeted maximum multiple coverage model for cybersecurity planning and management. IISE Trans. **51**(12), 1303–1317 (2019)
12. T.H. Bhuiyan, H.R. Medal, A.K. Nandi, M. Halappanavar, Risk-averse bi-level stochastic network interdiction model for cyber-security risk management. Int. J. Crit. Infrastructure Prot. **32**, 100408 (2021). https://doi.org/10.1016/j.ijcip.2021.100408
13. G. Oliva, R. Setola, M. Tesei, A stackelberg game-theoretical approach to maritime counter-piracy. IEEE Syst. J. **13**(1), 982–993 (2018)
14. B.S. Pay, J.R.W. Merrick, Y. Song, Stochastic network interdiction with incomplete preference. Networks **73**(1), 3–22 (2019). https://doi.org/10.1002/net.21831

15. S. Jalali, M. Seifbarghy, S.T.A. Niaki, A risk-averse location-protection problem under intentional facility disruptions: a modified hybrid decomposition algorithm. Transp. Res. Part E-logist. Transp. Review **114**, 196–219 (2018)
16. X. Lei, S. Shen, Y. Song, Stochastic maximum flow interdiction problems under heterogeneous risk preferences. Comput. Oper. Res. **90**, 97–109 (2018)
17. A.M.F. Fard, M. Hajiaghaei-Keshteli, A bi-objective partial interdiction problem considering different defensive systems with capacity expansion of facilities under imminent attacks. Appl. Soft Comput. **68**, 343–359 (2018)
18. G. Yu, J. Zhang, Multi-dual decomposition solution for risk-averse facility location problem. Transp. Res. Part E: Logist. Transp. Rev. **116**, 70–89 (2018)
19. A.K. Nandi, H.R. Medal, S. Vadlamani, Interdicting attack graphs to protect organizations from cyber attacks: a bi-level defender-attacker model. Comput. Oper. Res. **75**, 118–131 (2016)
20. Q. Li, M. Li, J. Gan, C. Guo, A game-theoretic approach for the location of terror response facilities with both disruption risk and hidden information. Int. Trans. Oper. Res. **28**(4), 1864–1889 (2021). https://doi.org/10.1111/itor.12900
21. A. Shamekhi Amiri, S.A. Torabi, R. Ghodsi, An iterative approach for a bi-level competitive supply chain network design problem under foresight competition and variable coverage. Transp. Res. Part E: Logist. Transp. Rev. **109**, 99–114 (2018). Available: https://www.sciencedirect.com/science/article/pii/S1366554517305483
22. N. Aliakbarian, F. Dehghanian, M. Salari, A bi-level programming model for protection of hierarchical facilities under imminent attacks. Comput. Oper. Res. **64**, 210–224 (2015)
23. R. Khanduzi, A.K. Sangaiah, A fast genetic algorithm for a critical protection problem in biomedical supply chain networks. Appl. Soft Comput. **75**, 162–179 (2019)
24. N. Ghaffarinasab, A. Motallebzadeh, Hub interdiction problem variants: models and metaheuristic solution algorithms. Eur. J. Oper. Res. **267**(2), 496–512 (2018)
25. S.-W. Lin, J.N. Gupta, K.-C. Ying, Z.-J. Lee, Using simulated annealing to schedule a flowshop manufacturing cell with sequence-dependent family setup times. Int. J. Prod. Res. **47**(12), 3205–3217 (2009)
26. P.J. Hawrylak, M. Haney, M. Papa, J. Hale, Using hybrid attack graphs to model cyber-physical attacks in the smart grid, in *2012 5th International Symposium on Resilient Control Systems* (2012), pp. 161–164
27. S.B.H. Shah, Z. Chen, S.H. Ahmed, F. Yin, M. Faheem, S. Begum, Depth based routing protocol using smart clustered sensor nodes in underwater WSN, in *Proceedings of the 2nd International Conference on Future Networks and Distributed Systems, Amman, Jordan, June 26–27, 2018*, ed. by A. Abuarqoub, B. Adebisi, M. Hammoudeh, S. Murad, M. Arioua. ACM (2018), pp. 53:1–53:7. Available: https://doi.org/10.1145/3231053.3231119
28. S.B.H. Shah, L. Wang, M.E. Haque, M.J. Islam, A. Carie, N. Kumar, Lifetime improvements of smart sensors maintenance protocol in prospect of iot-based rampal power plant, in *2020 16th International Conference on Mobility, Sensing and Networking (MSN)* (2020), pp. 260–267
29. Q. Li, M. Li, R. Zhang, J. Gan, A stochastic bilevel model for facility location-protection problem with the most likely interdiction strategy, in *Reliability Engineering & System Safety* (2021), pp. 1–50. Available: https://doi.org/10.1016/j.ress.2021.108005

Univariate Nonlinear VMs Instances Demand Forecasting for Optimized Cloud Resources Orchestration

Hamzaoui Ikhlasse, Duthil Benjamin, Courboulay Vincent, and Medromi Hicham

Abstract As the vast cloud traffic never stop growing in minutes, hours, and daily basis, autonomous and proactive cloud resources orchestration become a veritable obligation. This dominant trend is inciting to further seek for accurate forecasting models supporting multi-cloud scheduling levels decision-making. In this paper, we compare four nonlinear deep neural network models, namely: LSTM, GRU, and, respectively, their bidirectional variants: BiLSTM and BiGRU. Experimentation test scenarios demonstrated the performance of BiGRU models above other candidate models. This new developed version achieved preliminary until 0.0928 and 0.7823 of RMSE values, then 0.0729 and 0.6302 of MAE values, respectively for NASA and Amazon datasets. These findings hence prove the influence of the complex and nonparametric data on the accuracy of prediction values.

Keywords Cloud instances demand · Univariate time series forecasting · Deep neural networks · Bidirectional gated recurrent unit

H. Ikhlasse (✉) · M. Hicham
Research Foundation for Development and Innovation in Science and Engineering, 16469 Casablanca, Morocco
e-mail: ikhlasse.hamzaoui-etu@etu.univh2c.ma

M. Hicham
e-mail: hmderomi@yahoo.com

System Architecture Team (EAS), Engineering Research Laboratory (LRI), National High School of Electricity and Mechanic (ENSEM), Hassan II University, 8118 Casablanca, Morocco

H. Ikhlasse · D. Benjamin
EIGSI, La Rochelle, France
e-mail: duthil@eigsi.fr

D. Benjamin · C. Vincent
IT, Image, Interaction Laboratory (L3I), University of La Rochelle, La Rochelle, France
e-mail: vcourbou@univ-lr.fr

A. K. Bashir et al. (eds.), *Proceedings of International Conference on Computing and Communication Networks*, Lecture Notes in Networks and Systems 394,
https://doi.org/10.1007/978-981-19-0604-6_50

1 Introduction

As cloud industries' resources orchestration becomes increasingly challenging with cloud traffics bursting, accurate cloud instances demand forecasting becomes an obligation [1]. The demand for multimodal instances required by several users is a collection of time series vectors, forming either multivariate dependent variables, or single univariate instances time series. Multimodal instances correspond to various instances families with diverse resources criteria.

These multimodal cloud instances data typically incorporate numerous linear and nonlinear interdependencies with a wide spectrum of residual data. In turn, the presence of such complex features in a multi-VMs instances cloud environment involves in the first place the use of sophisticated models allowing to predict with great accuracy several shades of traffic demands and potential disruptive events [2].

Despite many existing time series forecasting models, the remaining daunting challenges are how to obtain highest precisions under complex data [3, 4]. In the case of AWS dataset [5], around 212 instance types with various resources' capacities may be provisioned in the form of multimodal use cases, including: compute-optimized, memory-optimized, storage-optimized, micro-instances, FPGAs, and GPUs instances. Figure 1a, b depicts the behavioral distribution of multimodal AWS instances requests observation, with respectively, the mean and median central distributions.

From the above distribution, one can obviously conclude that neither the mean (Fig. 1a) nor the median (Fig. 1b) tendencies cluster around fixed close central values. The central distribution then roughly discloses the complex and nonparametric nature of AWS instances request data. Deep learning techniques are inherently suitable to manifest the characteristics of such complex nonlinear cloud traffics data prediction. Motivating by above factors, the main paper contributions are mentioned as follows:

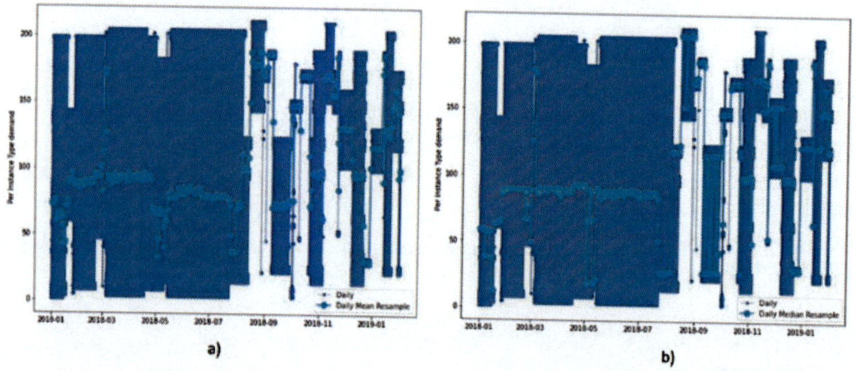

Fig. 1 AWS instances demand observations, with respectively, **a** the central mean distribution and **b** the central median distribution

- The paper compares two univariate time series datasets from Amazon and NASA in terms of demand forecasting accuracy, given the diverse nature of their data.
- Toward this end, both univariate datasets were used to evaluate four nonlinear deep neural networks variants, namely: the long short-term memory (LSTM), its improved version the gated recurrent unit (GRU), and their bidirectional variants: (BiLSTM) and (BiGRU).
- The proposed forecasting models were thereby evaluated under many test scenarios, to select the most adequate hyperparameters for preliminary precisions, then pave the way for future tangible improvements.

2 Literature Review

This section presents in a tabular scheme, the most complete and recent nonlinear cloud timeseries forecasting, by means of various deep neural networks models. Table 1 synthesizes literature studies aiming toward efficient cloud resources orchestration.

3 Methodology

The proposed AWS instances demand forecasting system (Fig. 2) starts with multimodal Amazon instances demand preprocessing, categorized by the aforementioned instances families. For the sake of simplicity, we only performed prediction on the compute-optimized AWS instances family. Using the both AWS and NASA univariate datasets, this paper intents to validate the most prediction-accurate deep neural network models among LSTMs, GRUs, and their bidirectional versions. Once data preprocessed, we partitioned the whole datasets into training, validation, and testing sets. Primary experimental tests prompted us to choose as input time series windows' sizes: six hours for AWS compute-optimized instances time series and one hour for NASA time series. Thereafter, the four adopted models were created using four layers among which two hidden layers and one input layer, each with 100 neural units and ReLU as activation function. The models' parameters were fitted using the mean squared error (MSE) and Adam optimizer, by means of validation losses and training losses. The training process was performed initially using 20 epochs, 10 patience, and various batch sizes. The transformed and generated predictions were eventually evaluated on test sets, using the root mean squared error (*RMSE*) and the mean absolute error (*MAE*) metrics.

Table 1 Literature review on recent and most accurate DNNs prediction studies

Study/year	Adopted DNNs models	Paper contributions	Implementation tool	Adopted dataset	Accuracy intervals
2020 [6]	Online multi-resource feed-forward neural network (OM-FNN)	Forecasting (CPU and memory) demands for future applications, using, respectively, prediction windows sizes of (5, 10, 20, 30, and 60 min)	Python version 3 on Intel Xeon Silver	Google Cluster dataset	RMSE: [0.0198–0.00025]
2019 [7]	Deep Belief Network (DBN)	Predicting cloud services response time for increased performance	Python	14,400 created experimental data, using nine statistical distributions	RMSE: [0.05–0.0159]
2021 [8]	Evolutionary Quantum Neural Network (EQNN)	Encoding workload data into qubits and propagating them through the network to predict cloud workload, using prediction windows sizes of (5, 30, and 60 min)	Python version 3.7 on Intel Xeon Silver	– Google Cluster Data – PlanetLab VMs traces – NASA traces – Grid archive workload	RMSE: [0.00960–0.00089]
2019 [9]	Multi-Layer Perceptron (MLP)	Predicting workload demand, using a prediction windows sizes from 1 to 60 min intervals	Python 2.7 on Google Colab Nvidia Tesla K80 GPU	NASA Web traces	RMSE: [544.08–16.26]
	Gated Recurrent Unit (GRU)				RMSE: [521.66–16.51]
2020 [10]	Extreme Learning Machines (ELM)	Predicting cloud workload for time intervals of 1, 10, 20, 30, 60 min and 1 day	Python dual Intel Core i5-3230 M	Google cluster	MPE: [0.003–0.002]

(continued)

Table 1 (continued)

Study/year	Adopted DNNs models	Paper contributions	Implementation tool	Adopted dataset	Accuracy intervals
2019 [11]	Auto Encoder with one class SVM (AE-SVM)	Cloud network intrusion detection	Python	– NSL-KDD – UNSW-NB15 – The campus network traffic records (Palo Alto system log)	Accuracy: [91.7–96.1%]
2021 [12]	Multi-layered LSTM	Multiple VMs workload classification prediction	Python Google Colab Nvidia Tesla K80 GPU	Grid Workload Archive (GWA)	Accuracy: [85–88%]
2017 [13]	Bidirectional Univariate-(BLSTM-U)	Predicting cloud resources usage using univariate and multivariate features on many validation steps sizes	Theano Nvidia Tesla K80 GPU	Google cluster traces	RMSE: [0.0358–0.0115]
	Multivariate (BLSTM-M)				RMSE: [0.0255–0.0095]

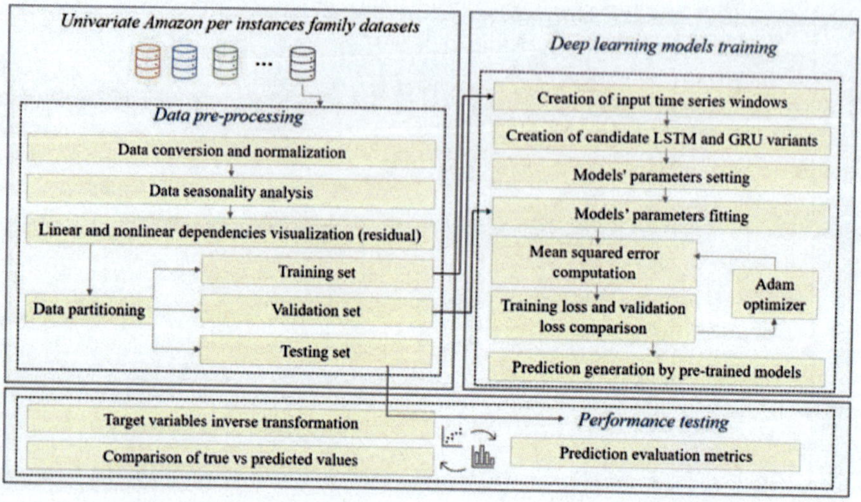

Fig. 2 AWS univariate instances demand time series forecasting system

4 Results Performance Analysis

In this section, we first describe the main experimental setups along with a description of adopted datasets. Afterward, we provide a performance analyses of obtained prediction results.

4.1 Experimentation Setup and Dataset Analysis

The implementations of the four deep neural networks models were conducted on Python 3.8, using TensorFlow and Keras library, on the top of a DELL G5 15 intel core 7 machine with Nvidia GeForce RTX2070 GPU and 16 Go of RAM. The adopted extended AWS dataset [14] includes one million multimodal instances requests with their resource capabilities. As mentioned earlier, we chose to perform the prediction on only one instance family: the compute-optimized, from which we collected a sample of one hundred thousand instances. In order to evaluate the influence of diverse time series sequences' shades on predictions accuracies, we also performed prediction on the univariate NASA dataset [15] of about thirty thousand instances.

Figure 3 plots the quantile–quantile (Q–Q) graphs of, respectively, per minute resampled Amazon (Fig. 3a) and NASA datasets (Fig. 3c) and per hour resampled Amazon (Fig. 3b) and NASA datasets (Fig. 3d). One can obviously deduce from entire plotted quantiles that neither Amazon nor NASA datasets follow normal distributions since sample quantiles data clearly deviate nonlinearly from the red line. However, the presence of a wide spectrum of residual data is more demonstrated in Amazon dataset.

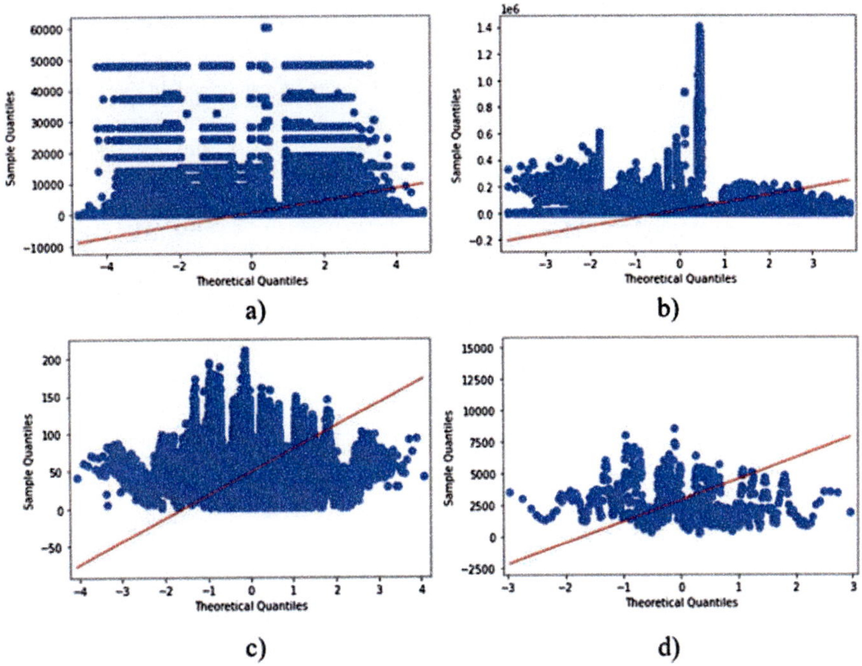

Fig. 3 Q–Q plot of **a** per minute sampled AWS instances, **b** per hour sampled AWS instances, **c** per minute sampled NASA time series, and **d** per hour sampled NASA time series

4.2 Prediction Results

As previously stated, we have created input windows of 6 and 1 h, respectively, for both Amazon and NASA datasets based on their sizes. The goal is to predict future data windows of the same size, distributed per minutes. Table 2 introduces the metrics' evaluation results for the four DNNs models, along both adopted datasets, and according to 3 batch sizes scenarios (64, 128, and 256). The first noticed statement is the influence of batch and training data sizes on the training time in second. Indeed, when the training data size is large enough, larger batches lead to less training time (AWS case), and vice versa (NASA case). It should also be noted that this time increases within the bidirectional variants, since the training is performed in a double direction. Regarding models' performance, the BiGRU and LSTM achieved the best prediction results in case of AWS dataset. Then, the BiGRU and BiLSTM performed well on NASA dataset. Though the most so far accurate DNNs model is the BiGRU version, which did not exceed an RMSE value of 2.1764. Figure 4 depicts the four models' prediction values, including the most precise ones provided by the BiGRU.

Table 2 Performance evaluation metrics' results for the four DNNs models under both univariate AWS and NASA datasets

Dataset	DNNs model	Batch size	Training time in (s)	RMSE	MAE
Univariate AWS instances	GRU	64	854	2.8588	2.2539
		128	1211	3.7418	2.9159
		256	305	3.2802	2.5316
	BiGRU	64	1391	**0.7823**	**0.6302**
		128	1832	2.1764	1.7573
		256	663	1.5135	1.1876
	LSTM	64	1021	**0.8988**	**0.7134**
		128	585	3.3215	2.5825
		256	342	3.6021	2.6637
	BiLSTM	64	2027	4.5845	3.5859
		128	1282	4.1107	3.0570
		256	804	4.0884	3.0984
Univariate NASA http workload	GRU	64	277	0.1476	0.1181
		128	223	0.1467	0.1161
		256	109	2.5591	1.9682
	BiGRU	64	522	**0.0928**	**0.0729**
		128	444	0.1597	0.1240
		256	279	0.2006	0.1526
	LSTM	64	278	0.3148	0.2420
		128	294	0.2403	0.1857
		256	159	1.9222	1.3254
	BiLSTM	64	790	0.2524	0.2330
		128	1091	**0.0935**	**0.0777**
		256	277	1.7796	1.4045

Bold indicates RMSE and MAE values of the most accurate prediction results

5 Conclusion

Cloud provider industries continually sake nowadays toward preserving their elasticity and quality of services, due to the growing sudden changes in various cloud demand rates. In an attempt to accurately predict future cloud instances demand, we compare in this paper four DNNs variants under two univariate time series datasets. In the entire experimentation test scenarios, the BiGRU achieved the less RMSE and MAE values of, respectively, 0.0928 and 0.0729. Nevertheless, the prediction results of AWS instances were less accurate than NASA forecasting results, owing to the strong nonlinear dependency found in AWS time series. As future perspectives, we strive to forecast other existing AWS instances families demand with their

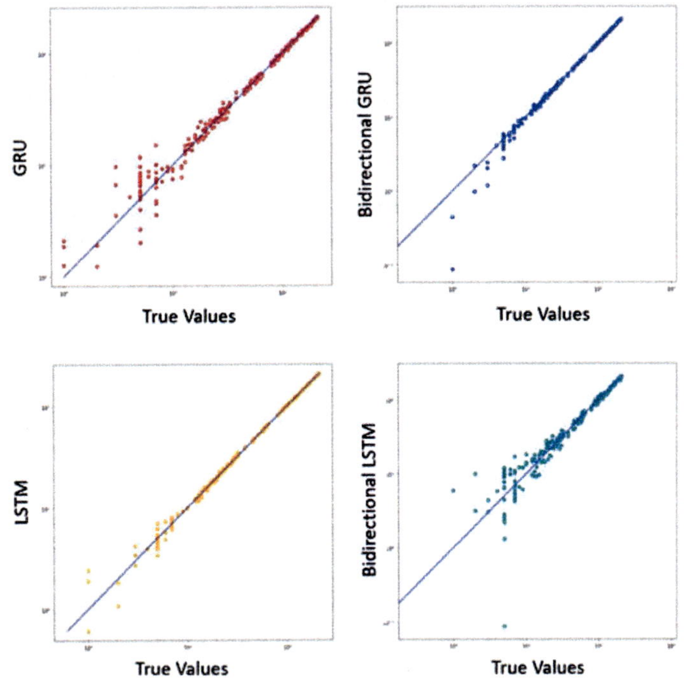

Fig. 4 Comparison between predicted and true values for the four DNNs models deployed using AWS instances (Best BiGRU scenario, where RMSE and MAE reached 0.7823 and 0.6302, respectively)

resource occupancies. Toward higher performance, we plan to develop an improved extended BiGRU model using additional test scenarios and further layers of noise and nonlinear residual reductions.

References

1. I. Hamzaoui, G. Bourgeois, B. Duthil et al., Parallel, proactive and power efficient virtual network embedding in a green and distributed SD-ODCN architecture. IEEE Access **9**, 39344–39362 (2021). https://doi.org/10.1109/ACCESS.2021.3063708
2. H. Ikhlasse, D. Benjamin, C. Vincent, M. Hicham, An overall statistical analysis of AI tools deployed in cloud computing and networking systems, in *2020 5th International Conference on Cloud Computing and Artificial Intelligence: Technologies and Applications (CloudTech)*. IEEE, pp 1–7 (2020)
3. S. Taherizadeh, M. Grobelnik, Key influencing factors of the Kubernetes auto-scaler for computing-intensive microservice-native cloud-based applications. Adv. Eng. Softw. **140**, 102734 (2020). https://doi.org/10.1016/j.advengsoft.2019.102734
4. I. Hamzaoui, B. Duthil, V. Courboulay, H. Medromi, A survey on the current challenges of energy-efficient cloud resources management. SN Comput. Sci. **1**, 1–28 (2020). https://doi.org/10.1007/s42979-020-0078-9

5. AWS Amazon EC2 Instance Comparison. https://instances.vantage.sh/. Accessed 7 Jun 2021
6. D. Saxena, A.K. Singh, A proactive autoscaling and energy-efficient VM allocation framework using online multi-resource neural network for cloud data center. Neurocomputing **426**, 248–264 (2021). https://doi.org/10.1016/J.NEUCOM.2020.08.076
7. Y. Gao, B. Zhang, S. Wang, A. Ma, DBN based cloud service response time prediction method. Int. Conf. Adv. Commun. Technol. ICACT 2019-February 42–46 (2019). https://doi.org/10.23919/ICACT.2019.8701922
8. A.K. Singh, D. Saxena, J. Kumar, V. Gupta, A quantum approach towards the adaptive prediction of cloud workloads. IEEE Trans. Parallel Distrib. Syst. **32**, 2893–2905 (2021). https://doi.org/10.1109/TPDS.2021.3079341
9. D.F. Kirchoff, M. Xavier, J. Mastella, C.A.F. De Rose, A preliminary study of machine learning workload prediction techniques for cloud applications, in *Proceedings of 27th Euromicro Int Conf Parallel, Distrib Network-Based Process PDP 2019*, pp. 222–227 (2019). https://doi.org/10.1109/EMPDP.2019.8671604
10. J. Kumar, A.K. Singh, Decomposition based cloud resource demand prediction using extreme learning machines. J. Netw. Syst. Manag. **284**(28), 1775–1793 (2020). https://doi.org/10.1007/S10922-020-09557-6
11. Y.F. Hsu, Z.Y. He, Y. Tarutani, M. Matsuoka, Toward an online network intrusion detection system based on ensemble learning. IEEE Int. Conf. Cloud Comput. CLOUD **2019**, 174–178 (2019). https://doi.org/10.1109/CLOUD.2019.00037
12. P. Bhagtya, S. Raghavan, K.D. Chandraseakran, Workload classification in multi-vm cloud environment using deep neural network model. Proc. ACM Symp. Appl. Comput. (2021). https://doi.org/10.1145/3412841.3442068
13. S. Gupta, D.A. Dinesh, Resource usage prediction of cloud workloads using deep bidirectional long short term memory networks, in *IEEE International Conference on Advanced Networks and Telecommunications Systems ANTS 2017*, pp. 1–6 (2018). https://doi.org/10.1109/ANTS.2017.8384098
14. Kaggle AWS EC2 Pricing Data|Kaggle. https://www.kaggle.com/akashsarda/aws-ec2-pricing-data. Accessed 7 Aug 2021
15. GitHub GitHub—dionatrafk/workload_prediction. https://github.com/dionatrafk/workload_prediction. Accessed 20 Aug 2021

A Study on Audio Quality of Experience for WebRTC-Based Communication Applications

M. Ugur Seker and H. Hakan Kilinc

Abstract Users of communication applications always want uninterrupted and high-quality sound. It is possible to provide these requests by determining the quality experienced by the end-user. Under different network conditions, it is important to identify the effects of parameters and possible scenarios in order to select the most optimal Opus codec configuration in real-time. In this way, it is possible to establish an adaptive configuration structure. This study focuses on the effects of Opus audio codec within an application that uses the WebRTC library. When the call started, test scenarios were created to investigate the effects of changing Opus parameters, and the results were discussed.

1 Introduction

Many massive communication applications such as Whatsapp Messenger, Facebook Messenger, Google Meet/Hangout, Discord, Amazon Chime, GotoMeeting, Kandy use WebRTC technology [1–3]. Audio and video quality is very significant in these applications. In terms of audio, Quality of Experience (QoE) guarantees that you deliver audio that will delight end users. Changes in bitrates affect the quality of the sound, and the higher the bitrate is the better the sound.

Web Real-Time Communication (WebRTC) technology can use many audio codec components; however, it encourages the use of the Opus audio codec component. Opus is an open, completely royalty-free, versatile audio codec component designed to handle a wide variety of audio applications including voice over IP, videoconferencing, in-game chat, musical performances, and is standardized with RFC6716 [4]. Opus is an adaptive codec, and it decreases or increases the sound quality according to the condition of the bandwidth. It is scalable from low bitrate narrowband speech at 6 kilobit per second (kbit/s) to very high-quality stereo music at 510 kbit/s. Details

M. Ugur Seker · H. Hakan Kilinc (✉)
Orion Innovation Turkey, Istanbul, Turkey
e-mail: hakan.kilinc@orioninc.com

M. Ugur Seker
e-mail: ugur.seker@orioninc.com

A. K. Bashir et al. (eds.), *Proceedings of International Conference on Computing and Communication Networks*, Lecture Notes in Networks and Systems 394,
https://doi.org/10.1007/978-981-19-0604-6_51

of media encoding such as signal sampling rate, frame size, and timing are defined in a Real-time Transport Protocol (RTP) payload format. Format information of RTP data is transmitted to endpoints via protocols such as Session Description Protocol (SDP) or Extensible Messaging and Presence Protocol (XMPP). RTP payload format and technical parameters for Opus are specified in RFC7587 [5].

In this study, our main goal is to investigate the effects of Opus parameters on bit rate, sound quality, and data usage size in the WebRTC-based communication applications. Thus, the way to create an adaptive codec configuration structure that can be used in different network conditions will be opened. For this, we use a software development platform called the MobileSDK library developed in-house. The MobileSDK includes the video and audio communication features of the WebRTC library, enables its users to develop communication applications for mobile systems, and is used in well-known communication products in the world.

In the next section, a literature review is given. In Sect. 3, information about the features and parameters of the Opus audio codec is given. In Sect. 4, the test environment and the effects of parameters are discussed. In the last section, studies are summarized and concluded.

2 Related Work

There are studies in the literature that conduct Opus usage tests and analyzes within WebRTC.

Alahmadi [6] defined the equipment distortion factor and packet loss robustness factor for all bitrates and operating modes supported by the Opus codec in speech applications in his doctoral thesis. These factors are used to evaluate the quality experienced by users in WebRTC-based applications. He also proposed an adaptive bitrate switching algorithm for speech applications in the WebRTC-based communication applications.

Garcia et al. [7] conducted a detailed study on the QoE evaluation and performance of WebRTC-based applications. They analyzed WebRTC topologies that affect QoE, proposed key performance indicators to predict QoE in WebRTC, and systematically reviewed QoE applied to WebRTC applications. Two years before these studies, Garcia et al. [8] proposed a methodology and test tool to obtain reliable and statistical both network-dependent QoS (Quality of Service) measurements and media-dependent QoE indicators of WebRTC-based applications and infrastructures.

A research team from the Leipzig University of Telecommunications has many similar publications on the performance of Opus codec in WebRTC. Some of these are as follows.

Meszaros et al. [9] explained the difficulties in quality assessment methods in WebRTC-based audio calls and presented a quality measurement tool called QuARTCS for the evaluation of these calls. Meszaros and Trojahn describe this work in more detail in their bachelor thesis [10].

Jokisch and Maruschke [11] evaluated audio and speech quality using 81 audio files and showed the effects of transcoding. Jokisch et al. [12] realized the instrumental quality assessment of Opus-coded speech within the WebRTC-based application using the POLQA testbed and the AQuA tools. They also tested the Opus with mixed audio and music signals and achieved similar results to the quality performance of the Opus codec in WebRTC with the quality provided by standalone coding.

Maruschke et al. [13] investigated the Opus codec behavior and showed that Opus properties can be configured by changing the Session Description Protocol (SDP) parameters in the WebRTC.

In this study, we have performed real-world tests to achieve the least cost and best performance by changing SDP messages. Unlike the studies given in the literature, we carried out our tests by focusing on bitrate, data usage, and sound quality.

3 Opus Audio Codec Features

Opus [4] has signal sampling rates from 8 kilohertz (kHz) (narrowband) to 48 kHz (fullband). It supports both Constant Bit Rate (CBR) and Variable Bit Rate (VBR) from narrowband to fullband. In addition, it has speech&music support and mono&stereo support. According to RFC7587 [5], recommended sampling rate values for Opus with respect to bandwidth are shown in Table 1.

In Opus, two different modes can be selected as speech and music mode. Speech mode allows efficient encoding of audio signals at lower bit rates, while music mode optimizes audio signals at medium and higher bit rates. By default, WebRTC uses fullband speech mode as the band, unless stereo is selected. The recommended bitrates in kbit/s according to bandwidth and mode are as follows;

- Narrowband speech mode: 8–12 kbit/s
- Wideband speech mode: 16–20 kbit/s
- Fullband speech mode: 28–40 kbit/s
- Fullband mono music mode: 48–64 kbit/s
- Fullband stereo music mode: 64–128 kbit/s

Table 1 Audio bandwidth and sampling rates for Opus

Bandwidth name	Audio bandwidth (kHz)	Sampling rate (kHz)
Narrowband	0–4	8
Mediumband	0–6	12
Wideband	0–8	16
Super wideband	0–12	24
Fullband	0–20	48

Opus has multiple configurable features that can affect on bitrate and audio quality. These features are VBR, CBR, discontinuous transmission (DTX) and comfort noise (CN), forward error correction (FEC), stereo operation, packetization time (Frame size), and sampling rates.

According to RFC7587, there are ten optional parameters for Opus audio codec. In our tests, we will examine the effects of parameters written in bold on bitrates and sound quality. These parameters are **MaxPlaybackRate**, Sprop-MaxCaptureRate, **Ptime**, MaxPtime, **MaxAverageBitrate**, **Stereo**, Sprop-Stereo, CBR or VBR, **UseinbandFEC**, **UseDTX**.

4 Experiment

4.1 Testing Environment

We used the WebRTC-based mobile software development platform (MobileSDK) to perform the tests. We developed this platform for communications platform as a service (CPaaS) environments. It enables service providers to provide services in their own communication infrastructure to application developers via application programming interface (API) and software development kits (SDK) as in Fig. 1.

Other important notes about the test environment are as follows;

- Samsung Note 5 and Iphone 6S phones with applications developed with MobileSDK are used. Samsung Note 5 is caller and Iphone 6S is callee.

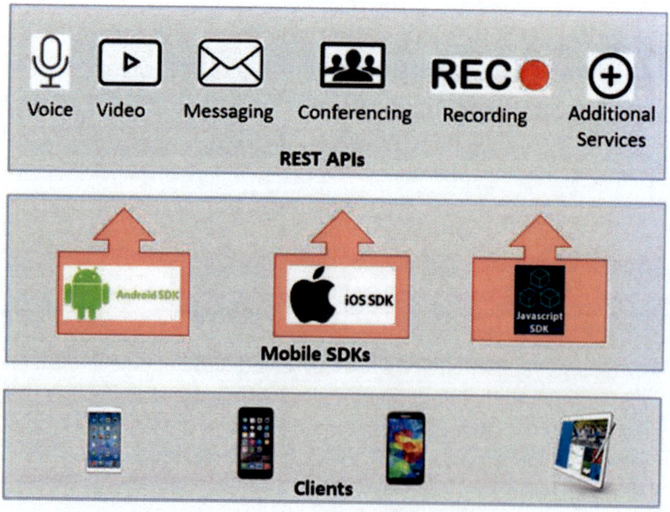

Fig. 1 Mobile SDKs reference model

- Data is collected using RTPStatistics feature of MobileSDK for mobile only tests.
- A Web client is used for DTX testing.
- All parameters will be in *fmtp* line of Opus audio codec in SDP except MaxPtime and Ptime.
- MaxPtime and Ptime values effect all codec.

4.2 Testing Scenarios

As the methodology of the test, we want to examine the sent and received bitrate changes. Bitrate is the number of bits, or data processed per unit of time. Changes in bitrates affect the quality of the sound, and the higher the bitrate means the better the sound.

4.2.1 Effect of Parameters

The effect of MaxPlaybackRate values on bitrate is shown in Fig. 2. We assume the receiver to have no limitations according to RFC7587. This parameter can be any value between 8 kHz and 48 kHz. The default value is 24 kHz and the total bitrate is 87 kbit/s. Using 8 kHz MaxPlaybackRate can reduce the total bitrate usage by 58 kbit/s. When we look at the sent and received bitrates, the quality of the sound obtained does not decrease.

According to RFC7587, the default value for Ptime is 20 millisecond (ms). There is no default value for Ptime in SDP. As shown in Fig. 3, there is a decrease in bitrate up to 60 ms. It does not make a difference in bitrate after 60 ms. Increasing MaxPtime alone does not make a difference in bitrate.

In Fig. 4, the effect of MaxAvarageBitrate values on bitrate is shown, and decreasing this value decreases the total bitrate continuously. If no value is specified, Opus

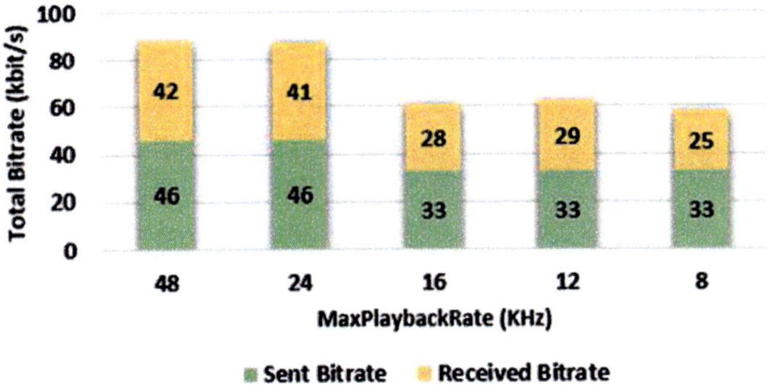

Fig. 2 The effect of MaxPlaybackRate values on bitrate

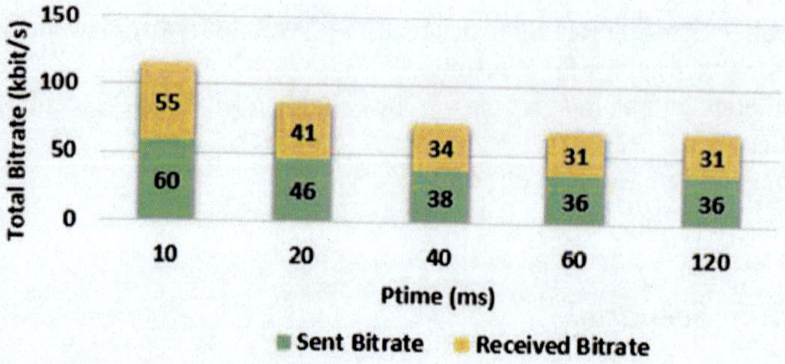

Fig. 3 The effect of Ptime values on bitrate

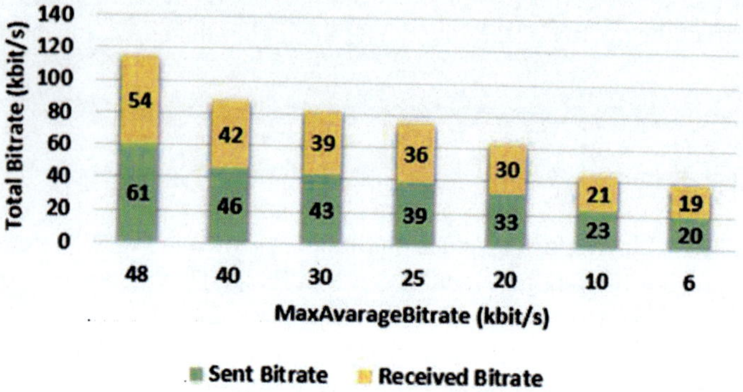

Fig. 4 The effect of MaxAvarageBitrate values on bitrate

chooses it according to the desired mode. The bitrate value for the fullband speech mode is 28–40 kbit/s. It has been observed that 40 kbit/s is the default in WebRTC. Using 10 kbit/s can reduce the total bitrate usage to 44 kbit/s, which is good sound quality up to 10 kbit/s.

Figure 5 shows the case whereby default the stereo and DTX values are enabled and the FEC value is disabled. Enabling the stereo parameter nearly doubled the bitrate usage. Enabling DTX reduces the total bitrate to 52 kbit/s while disabling FEC reduces the total bitrate by only 2 kbit/s.

4.2.2 Effect of Parameter Combinations

In all case scenarios in Table 2 and Fig. 6, DTX was enabled, MaxPtime = 60 ms and MinPtime = 10 ms were accepted, and other parameters were changed according to the scenarios. In the evaluation, the most important criterion was sound quality and also data usage during the 1-min call. In Scenario-1, the default values of the parameters in WebRTC are used. The data usage of 900 kilobytes (kB) is remarkable,

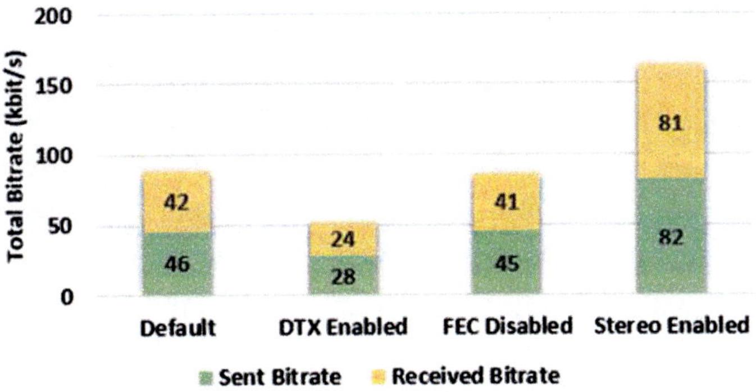

Fig. 5 The effect of Stereo, DTX and FEC values on bitrate

Table 2 Test scenarios and results

No	Max playback rate (KHz)	Max average bitrate (kbit/s)	Ptime (ms)	FEC	Average data usage (kB)	Audio quality
1	24	40	20	Enable	900	Good
2	12	10	40	Enable	350	Good
3	8	10	60	Enable	310	Good
4	8	8	60	Enable	270	Rare sound interrupts
5	8	6	60	Enable	260	Frequent sound interrupts
6	8	10	60	Disable	300	Good
7	16	20	60	Enable	450	Good
8	8	10	20	Enable	420	Good

and we accept the sound quality as good. Namely, the speeches on both sides were clearly heard.

One of the striking points in the scenarios is the 4th and 5th scenarios where the MaxAvarageBitrate value is low. In these scenarios, a serious deterioration in sound quality has been detected. On the other hand, although the total bitrate is low in Scenario-3 and Scenario-6, the sound quality is good. When we evaluate the values in both Table 2 and Fig. 6 together, we see the direct proportion between total bitrate, data usage, and sound quality.

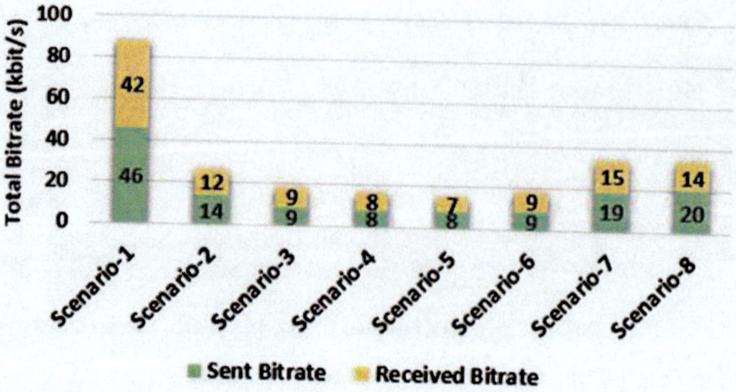

Fig. 6 The effect of scenarios on bitrate

5 Conclusion

Nowadays, in the design of communication tools that are used extensively, it is necessary to consider, test, and analyze as wide a range of scenarios as possible in order to establish an adaptable structure according to the connection quality.

In this study, in the WebRTC-based communication applications, test scenarios that will reveal the effects of Opus parameters on bit rate, sound quality, and data size have been realized and analyzed using the MobileSDK library. We determined scenarios that the bitrate and data usage is low, and the sound quality is good at the same time. This study can be extended with test cases about parameter changes in the middle of the call.

In future studies, we will study adaptive codec configuration algorithms and their performances in real-time and the mid-call event, considering the experiences in test scenarios and also security requirements.

References

1. T. Levent-Levi, *10 Massive Applications Using WebRTC* (2017), https://bloggeek.me/massive-applications-using-webrtc/
2. P. Hancke, *What's up with WhatsApp and WebRTC?* (2015), https://webrtchacks.com/whats-up-with-whatsapp-and-webrtc/
3. Kandy Cloud Communications Platform as a Service (CPaaS) (2021). https://www.kandy.io/
4. JM. Valin, K. Vos, ***RFC6716: Definition of the Opus Audio Code*** (2012). https://datatracker.ietf.org/doc/html/rfc6716
5. J. Spittka, K. Vos, J.M. Valin, *RFC7587: RTP Payload Format for the Opus Speech and Audio Codec* (2015). https://datatracker.ietf.org/doc/html/rfc7587
6. M. Alahmadi, *An adaptive bitrate switching algorithm for speech applications in the context of WebRTC*. Ph.D. Thesis, National University of Ireland (2020)
7. B. Garcia, M. Gallego, F. Gortázar, A. Bertolino, Understanding and estimating quality of experience in WebRTC applications. Computing, **101**, 1585–1607 (2019)

8. B. Garcia, F. Gortázar, L. López-Fernández, M. Gallego, M.P. Diaz, WebRTC Testing: challenges and Practical Solutions. IEEE Commun. Stand. Mag. **1**, 36–42 (2017)
9. M. Meszaros, F. Trojahn, M. Maruschke, O. Jokisch, Quartcs: A tool enabling end-to-any speech quality assessment of webrtc-based calls, in *International Conference on Speech and Computer (SPECOM 2018)* (Springer, Leipzig, 2018), pp. 408–418
10. M. Meszaros, F. Trojahn, *Definition and Analysis of WebRTC Performance Parameters as well as Conception and Realization of an End-to-End Audio Quality Monitoring Solution for WebRTC-Based "immmr" Call Scenarios*. Master Thesis, Leipzig University of Telecommunications (2017)
11. O. Jokisch, M. Maruschke, Audio and speech coding/transcoding in web real-timecommunication, in*International Symposium on Human Life Design (HLD 2016)* (Kanazawa, Japan, 2016)
12. O. Jokisch, M. Maruschke, M. Meszaros, V. Iaroshenko, Audio and speech quality survey of the Opus codec in web real-time communication, *In Proceedings of ESSV-27th Conference of Electronic Signal Processing*, Germany (2016), pp. 254–262
13. M. Maruschke, O. Jokisch, M. Meszaros, V. Iaroshenko, Review of the Opus codec in a WebRTC scenario for audio and speech communication, in *Proceedings of 17th International SPECOM Conference*, Greece (2015)

AVR Technologies in Sustainable Tourism: A Bibliometric Review

Sandeep Kumar Dey, Vo Viet Hung, Huynh Thai Hoc, and Quynh Giao Ngoc Pham

Abstract An extensive, PRISMA-led bibliometric analysis of scientific literature from the last 20 years was conducted to extract future research agenda in the niche field of sustainable tourism and the application of virtual reality+ technologies (extended, mixed, hyper, and augmented). The field being investigated contains significant opportunities through meta-analysis methods like linear–logarithmic transformations, coupling clustering, and network analysis. Astringent keyword inclusion rule ensured that the most pertinent scientific literature entered the analysis. The authors used critical content analysis parameters like Cohen's kappa to include the best fit papers. This bibliometric analysis convenes the need to focus on emerging areas like cyber-neurotics, telekinesis, cyber-optics, and gamification to provide impetus to sustainable consumption and production in the tourism and travel industry. The paper expresses the paradigm shift in research topics as the world enters the COVID-19-induced pandemic and its impact on future research endeavours.

Keywords Bibliometric review · Virtual reality · Sustainable tourism · Augmented reality

S. K. Dey (✉) · V. V. Hung · H. T. Hoc
Faculty of Management and Economics, Tomas Bata University in Zlin, Mostni 5139, Zlin 76001, Czech Republic
e-mail: dey@utb.cz

V. V. Hung
e-mail: vo_van@utb.cz; vvo@utb.cz

H. T. Hoc
e-mail: huynh_thai@utb.cz

Q. G. N. Pham
Faculty of Multimedia Communications, Tomas Bata University in Zlin, Univerzitní 2431, Zlin 76001, Czech Republic
e-mail: qpham@utb.cz

A. K. Bashir et al. (eds.), *Proceedings of International Conference on Computing and Communication Networks*, Lecture Notes in Networks and Systems 394,
https://doi.org/10.1007/978-981-19-0604-6_52

1 Introduction

Sustainability has become a part of the academic lexicon for the past three decades. Earlier studies in the discipline involved a producer-based narrative which transformed into a significantly consumer-based one in the recent past [1]. Consumers are more eager to modify their consumption lifestyles by showing eco-friendliness and willingness to pay for green items to minimize environmental pollution as environmental conservation becomes more prominent [2].

Predominant sustainable practices in tourism that are in focus include wastage reduction [3], energy conservation [4], alternative fuels [5], and green human resources [6]. In essence, sustainability is the coexistence of the biosphere and human civilization, with the core idea being to improve both current and future potential to meet human needs. According to [7] and Neuhofer et al. [8] the realm of sustainability not only confines itself to the minimization of negative impacts on the environment but also includes other assessments that concentrate on the economy, society, and ecology. Tourism academicians and industry practitioners observe that digitalization has integrated capacities from information and communication technologies, giving rise to novel technologies in the sector [7].

Stankov and Gretzel [9] have called this trend evolution of the tourism industry under the wide ambit of Industry 4.0, which has facilitated the emergence of disruptive technologies like virtual/augmented/mixed and hyper realities, IoT, AI-computer vision, and advanced human–computer interactions (HCI). Various examples of this paradigm shift include autonomous agents and things being used in hospitality establishments [10] and humanoid service robots at food and beverage businesses [11, 12]. Satta et al. [13] are of the notion that eco-friendly initiatives signify a key focus of innovation for tourist enterprises and destinations in redesigning their approaches to reduce the environmental impact and manage relationships with stakeholders, especially in the new normal situation sustainability in tourism has become a central focus [14].

This is aligned with the UN Sustainable Development Goals, where one of the objectives emphasizes encouraging sustainable industrialization and promoting innovation [15]. In the present-day scenario, VR capacities have been used widely in the tourism and hospitality industry ranging from destination marketing [16], accommodation sales [17], event and festival management and design [18], training of personnel [19], and new sensory museology [20]. Given the above premises, virtual reality/augmented reality can play a major role in propagating sustainable tourism [21].

The current COVID-19 has ushered in radical changes in the way academia looks at sustainable tourism. The narrative towards an integrated sustainable tourism development environment is more pronounced now [22]. Under the umbrella of Industry 4.0, disruptive technologies combine virtual reality technology with advanced immersive haptic capabilities resulting in a plethora of alternative simulated spaces like augmented, hyper, and mixed realities are seen as a bricolage of the sustainable future of the tourism industry [23].

Taking note of all the significant developments in the sphere of VR in sustainable tourism, we are encouraged to conduct a bibliometric review. The objectives of this study are (i) acquire and process state-of-the-art researches in the field of virtual realities+ (augmented/hyper/mixed/extended) in the context of sustainable tourism; (ii) investigate the future trends of the research into the field.

The following sections are organized: Sect. 2 shows research methodology, Sect. 3 presents Discussion, and Sect. 4 conveys future research.

2 Research Methodology

Draper et al. [24] define bibliometric analysis as a structured approach that reviews published articles and other literature of scientific nature. It involves micro and deep text analysis of the literature to uncover trends and projections for future research [25]. Bibliometric research is a complex field that considers journal ranking, the assessment of research quality, bibliometric literature review, and the analysis of trends and patterns [26].

For the purpose of conducting a bibliometric analysis, two judges from the authoring team were commissioned to analyse 100 papers obtained after inclusion keywords were entered to filter the literature load. An inter-rater framework was installed with two reviewers from the authoring team who have experience in the teaching of sustainable tourism and virtual reality at the university level. They were commissioned to select the most pertinent papers from the $n = 100$ papers selected for further analysis. Cohen's kappa coefficient (κ) is a statistic that is used to assess qualitative (categorical) item inter-rater reliability (and also intra-rater reliability). It is typically believed to be a more reliable statistic than simple agreement estimates since it considers the potential of agreement occurring by chance [27]. The outcome of the inter-rater intervention extracted 33 papers, thereby achieving a *kappa* statistic of 0.75, which is robust [28]. The following table illustrates the kappa statistic or κ derivation.

This section is divided into stages: 1. Data collection, 2. Analysis.

2.1 Data Collection

The research uses the leading scientific directory called the Web of science to identify pertinent literature in the field of sustainable tourism and the use of VR+ technologies from the last 20 years (2002–2021). Researchers have vehemently used the Web of science database for tasks like citation and co-occurrence analysis, author network, and path mapping [24]. All papers in the English language, including peer-reviewed academic journal articles, book chapters, conference proceedings, were analysed for this study.

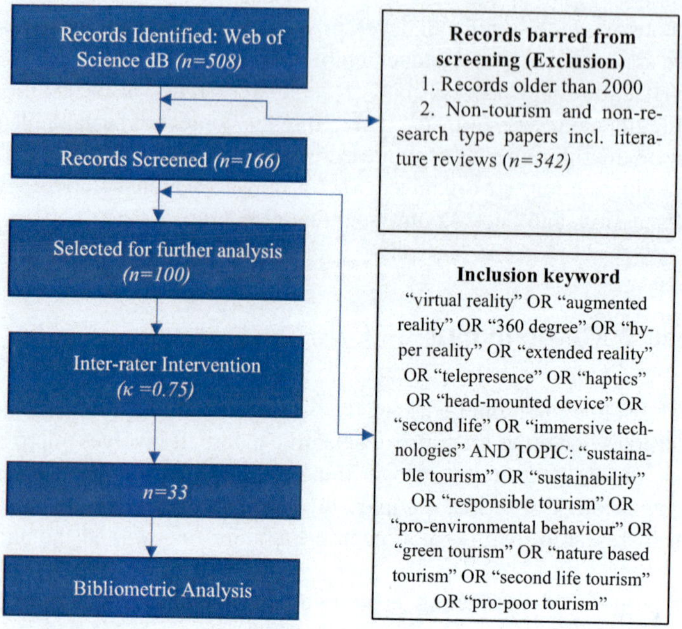

Fig. 1 Bibliometric analysis framework

Data collection took place in the month of June 2021. Prospective terms were extracted and treated with BOOLEAN values (AND, OR). Firstly, general keywords were involved in the search; they were basically "virtual reality" AND "sustainable tourism". The search result returned 508 publications, which were scrutinized for thematic patterns using NVIVO software, and standard features were extracted that were established in the form of keywords. Secondly, we use a PRISMA-led framework to guide our bibliometric review of the literature. The following figure presents the framework that provides the researchers with the required operating protocol to conduct the analysis (Fig. 1).

2.2 Analysis

Figure 2 displays pattern analysis of investigations having high impacts done in the field of VR+ and sustainable tourism in the last 20 years. It appears that one of the pioneering works in the field under investigation was by Guttentag [29] who investigated the utilization of wearable technologies in tourism and travel research. Furthermore, through the high impact works of [30–34] and [35], the topic of VR in tourism has expanded to include potent research areas like climate change and cultural sustainability.

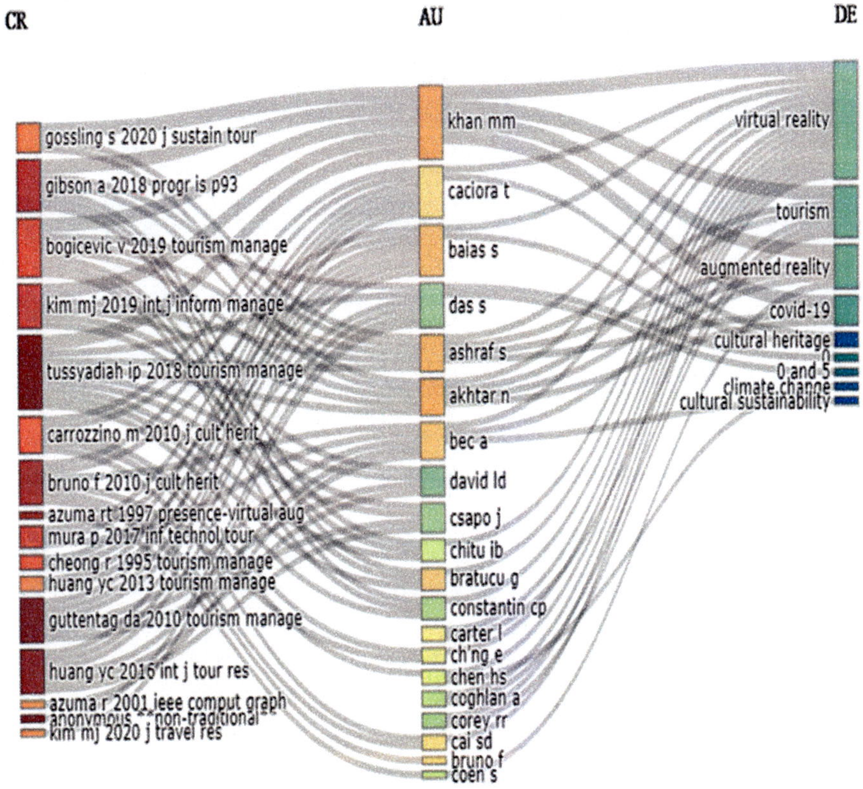

Fig. 2 Pattern analysis. (*Source* biblioshiny)

A closer inspection of the pattern analysis presented via Fig. 2 will reveal that themes pertaining to the concept of sustainability in the context of VR+ is nascent. Also, to be noted is the emergence of "COVID-19"-related topics in the pattern diagram post the pandemic. This pattern is mainly due to a discourse on the notion of mass tourism and digital tourism in the new normal by authors [36]. Among all of these advancements in the virtual tourism domain, the sustainability aspect remains a grey area for research.

Publications in this particular field have leaped in the years following 2018. It can be evaluated from the below-given Fig. 3 that the pandemic ushered in a new and revamped interest in the field of VR+ technologies in tourism.

Figure 4 provides a glimpse of the evolution of themes in the world of VR+ technologies in tourism and travel research. From the below-given figure, it might be discerned that themes in this particular field of study are moving from motifs like "experience" to core areas like "tourism". This may imply that the interest towards the sustainable dimension will surface from 2021 onwards as researchers will focus on VR+ interventions to gauge and capture tourists' dynamics with wide-spread interest in the area [37]

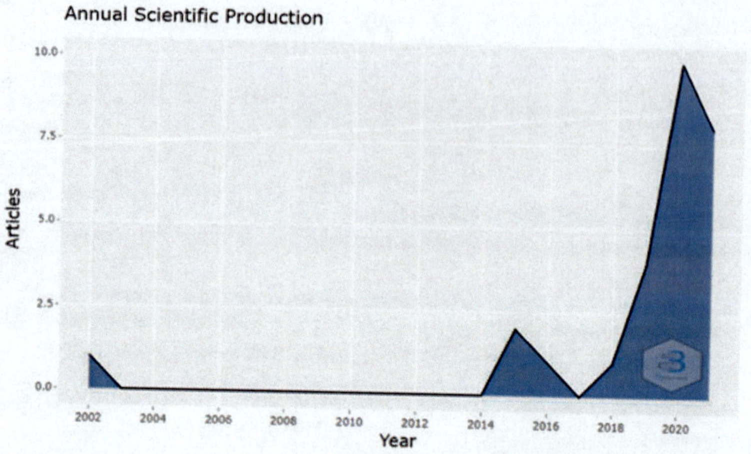

Fig. 3 Annual scientific production in the last 20 years. (Source: biblioshiny)

Fig. 4 Thematic evolution last 20 years. (*Source* biblioshiny)

3 Discussion

The linear logarithmic modulation of the abstract content (given in Fig. 2) observed a particular network of scientific works which have been extensively studied, and a list of probable research questions have been developed, which is given hereunder Table 1. The research paper clustering was based on network analysis of the authors' co-occurrence fields.

Furthermore, it is observed from a scoping review of literature in the field of VR+ technologies in tourism that were found eligible for this bibliometric review ($n =$

Table 1 Cohen's kappa derivation (*Source* Author's Own)

		Reviewer 1			Kappa statistic
		Accept	Reject	Total accepted/Rejected	
Reviewer 2	Accept	33	5	38	
	Reject	7	55	62	
	Total	40	60	100	
Observed agreement					0.88
Agreement by chance					0.52
Cohen's kappa					0.75

33), 75% of the literature used classical conditioning models like S–O-R (stimulus-organism-response), the remaining either deployed theory of planned behaviour (marketing-related papers), and two papers were qualitative investigations. An evaluation of the research questions developed from content analysis of the 33 literature reveals that the futuristic research inquiries dwell upon a comprehensive mix of experimentations and mixed-method studies. The authors conducted a scoping review of the significant literature with a kappa score of >0.80 from the thematic cluster analysis and identified the following research questions given in Table 1.

4 Conclusion and Future Research Agenda

Figure 5 presents the most relevant sources of scientific literature. It is observed that acute studies on tourists and VR interventions are basic themes. Topics pertaining to augmented reality and the experience of VR+ technologies remain unchanged through the years 2001–2020. With the onset of the COVID-19 pandemic, there is an emergence of impact studies in the field of VR+ technologies and Industry 4.0 capacities.

Studies that revolve around VR and intention to use are gradually declining. In reference to probable research questions given in Table 1 and thematic analysis of Fig. 5, the future of VR+ disruptions in the field of sustainable research will probably centre on the following themes given in Table 2. The researchers are recommended to dwell on the emerging research areas in the field of VR+ technologies in sustainable tourism. The patterns also observe that museums have widely used VR/AR technologies to promote heritage (Table 3).

The bibliometric analysis of 33 different scientific articles has redeemed the fact that investigations on the application of VR+ technologies like XR/AR/MR/HR in sustainable tourism research is still nascent. Future directions should focus on the broad thematic aspects given in Table 2 above. Furthermore, deliberations in innovative scientific methods like eye-tracking and neural traffic analysis should be considered to progress the current understanding of simulated and immersive environments on human beings.

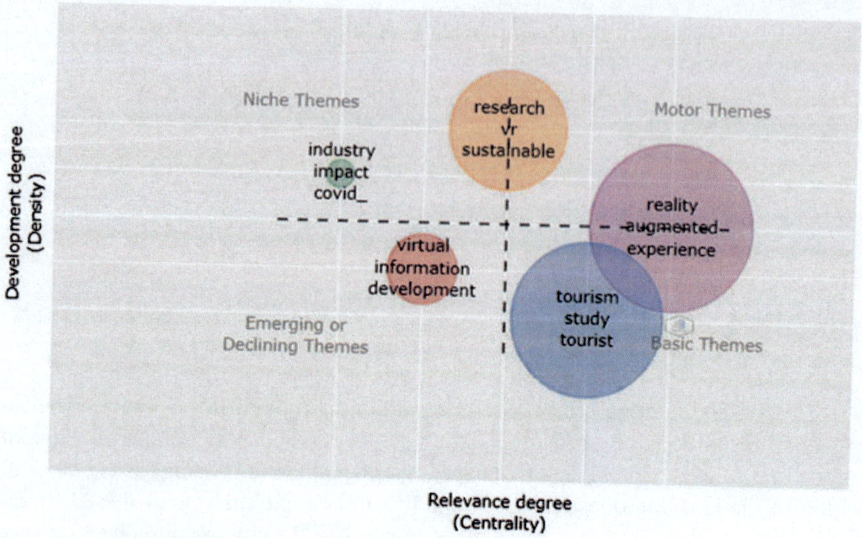

Fig. 5 Most relevant sources of scientific literature. (*Source* Biblioshiny App)

Table 2 Potential research questions from thematic clustering

Research paper cluster	Sources	Probable research questions
1	[30, 33, 38, 39]	RQ1: What will be the role of flow state in a cross-cohort environment RQ2: How does VR entice a stimulus–response towards tourism promotion RQ3: Can haptic response moderate the mental imagery formation in the tourism experience RQ4: What is the actual behavioural outcome of the VR intervention
2	[40–42]	RQ1: What is the difference between AR and VR interventions in the attitudes of tourists RQ2: What is the effect of telepresence on the attitudes of tourists RQ3: What is the efficacy of an embodied virtual agent in moderating involvement RQ4: What is the role of AR applications in "last chance" tourism

Table 3 Potential emerging topics for future research

Citation	Themes	Sub themes
[43–45]	User experience (UX)	Cyber-optics, haptic response, teleflow, cyberkinesis, embodiment and "avatarization", telepresence
[46]	Information development	Storytelling and immersive gamification
[47, 48]	COVID-19 impact	Last chance tourism, museology, and accessible tourism

Lest we forget, ethical and moral dimensions should be in harmony with research investigations as the question of human subjects has always obstructed various interventions in the field of behavioural science, especially human–computer interactions.

Acknowledgements This project has been financed within OP RDE project Junior grants of Tomas Bata University in Zlín, reg. No. CZ.02.2.69/0.0/19_073/0016941.

Appendix

An extract of the most relevant key words from the abstracts that were scrutinized for further analysis reveal that while "virtual reality vr" was the most frequent,

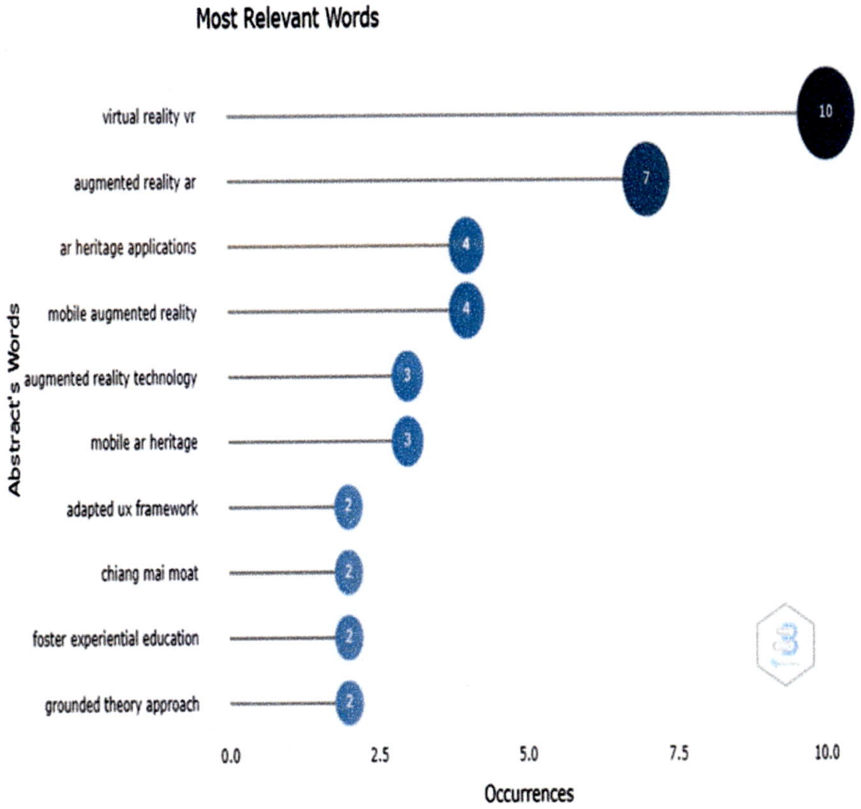

Fig. 6 Most relevant words identified in the bibliometric analysis. (*Source* biblioshiny)

other words of interest include "mobile ar", "adapted ux framework" and "experiential education". This further implies that VR+ technologies not only trigger certain behaviour but have an educational effect as well (Fig. 6).

References

1. R. Buckley, Sustainable tourism: research and reality. Ann. Tourism Res. **39**(2), 528–546. https://doi.org/10.1016/j.annals.2012.02.003
2. P. He, Y. He, F. Xu, Evolutionary analysis of sustainable tourism. Ann. Tour. Res. **69**, 76–89 (2018). https://doi.org/10.1016/j.annals.2018.02.002
3. G. Yfantidou, M. Matarazzo, The future of sustainable tourism in developing countries. Sustain. Dev. John Wiley & Sons Ltd. **25**(6):459–466 (2017). https://doi.org/10.1002/SD.1655
4. L. He, J. Zha, H.A. Loo, How to improve tourism energy efficiency to achieve sustainable tourism: evidence from China. Routledge **23**(1), 1–16, (2019). https://doi.org/10.1080/13683500.2018.1564737
5. G. Miller et al., Public understanding of sustainable tourism. Anna. Tourism Res. Pergamon **37**(3), 627–645 (2010). https://doi.org/10.1016/J.ANNALS.2009.12.002
6. Q.A. Nisar et al., Green human resource management practices and environmental performance in Malaysian green hotels: the role of green intellectual capital and pro-environmental behavior. J. Cleaner Production. Elsevier 311, p. 127504. https://doi.org/10.1016/J.JCLEPRO.2021.127504
7. C. Van Winkle, J. Bueddefeld, Information and communication technology in event management (2020). https://doi.org/10.1007/978-3-030-05324-6_86-1
8. B. Neuhofer, K. Celuch, T.L. To, Experience design and the dimensions of transformative festival experiences. Int. J. Contemp. Hospitality Manage. Emerald Publishing Limited, **32**(9), 2881–2901. https://doi.org/10.1108/IJCHM-01-2020-0008
9. U. Stankov, U. Gretzel, Tourism 4.0 technologies and tourist experiences: a human-centered design perspective. In *Information Technology and Tourism* (Springer Science and Business Media, Deutschland GmbH) **22**(3), 477–488. https://doi.org/10.1007/S40558-020-00186-Y
10. S.A. Cohen, D. Hopkins, Autonomous vehicles and the future of urban tourism. Annal. Tourism Res. Pergamon **74,** 33–42 (2019). https://doi.org/10.1016/J.ANNALS.2018.10.009
11. J. Murphy, U. Gretzel, J. Pesonen, Marketing robot services in hospitality and tourism: the role of anthropomorphism. Routledge **36**(7), 784–795 (2019). https://doi.org/10.1080/10548408.2019.1571983
12. I.P. Tussyadiah, S. Park, Consumer evaluation of hotel service robots, in *Information and Communication Technologies in Tourism 2018* (Springer, Cham, 2018), pp. 308–320. https://doi.org/10.1007/978-3-319-72923-7_24
13. G. Satta, R. Spinelli, F. Parola, Is tourism going green? A literature review on green innovation for sustainable tourism. Tourism Analysis. Cognizant Commun. Corporation **24**(3), 265–280. https://doi.org/10.3727/108354219X15511864843803
14. P. Brouder et al., 'Reflections and discussions: tourism matters in the new normal post COVID-19 Routledge **22**(3), 735–746 (2020). https://doi.org/10.1080/14616688.2020.1770325
15. United Nations, *The 2030 Agenda for sustainable development, transforming our world: The 2030 Agenda for sustainable development* (2016). https://doi.org/10.1201/b20466-7
16. A. McFee et al., The effects of virtual reality on destination image formation, in *Information and Communication Technologies in Tourism 2019* (Springer, Cham, 2019), pp. 107–119. https://doi.org/10.1007/978-3-030-05940-8_9
17. X.Y. Leung, J. Lyu, B. Bai, A fad or the future? Examining the effectiveness of virtual reality advertising in the hotel industry. Int. J. Hospitality Manage.Pergamon **88**, 102391. https://doi.org/10.1016/J.IJHM.2019.102391

18. J. Xu et al., A development of stage event management system using virtual reality technique. Int. Soc. Optics Photonics **11049**, 110492Y (2019). https://doi.org/10.1117/12.2522171
19. B. Tracey, M.P. Swart (Nellie) (2020) Training and development research in tourism and hospitality: a perspective paper Tourism Rev. Emerald Publishing Limited **75**(1), 256–259. https://doi.org/10.1108/TR-06-2019-0206
20. D. Romanek, B. Lynch, Touch and the value of objects handling: Final conclusions for a new sensory museology, in *Touch in Museums: Policy and Practice in Object Handling* (Routledge, 2020), pp. 275–286. https://doi.org/10.4324/9781003135616-28
21. J. Dewailly, Sustainable tourist space: From reality to virtual reality? Tourism Geographies Taylor & Francis Group **1**(1), 41–55 (2007). https://doi.org/10.1080/14616689908721293
22. M. Cristofaro, L. Leoni, S. Baiocco, Promoting co-evolutionary adaptations for sustainable tourism: the "alpine convention" case. Tourism Plann. Dev. Routledge **17**(3), 275–294 (2020). https://doi.org/10.1080/21568316.2019.1600162
23. S. Kask, *Virtual Reality in Support of Sustainable Tourism: Experiences from Eastern Europe*. (Estonian University of Life Sciences, 2018). Available at: https://click.endnote.com/viewer?doi=10.15159%2Femu.35&token=WzE0ODA4OTYsIjEwLjE1MTU5L2VtdS4zNSJd.HCEdYO6_VHQfeh2Z0sBp8-H98mI
24. J. Draper, L. Young Thomas, G.G. Fenich, Event management research over the past 12 years: what are the current trends in research methods data collection data analysis procedures and event types?. J. Convention Event Tourism **19**(1), 3–24 (2018). https://doi.org/10.1080/15470148.2017.1404533
25. F. Mehraliyev et al., A state-of-the-art review of smart tourism research. J. Travel Tourism Mark. Routledge 37(1):78–91. https://doi.org/10.1080/10548408.2020.1712309
26. L.Y. Leong et al., Tourism research progress—a bibliometric analysis of tourism review publications. Tourism Rev. Emerald Publishing Limited **76**(1), 1–26 (2020). https://doi.org/10.1108/TR-11-2019-0449
27. M.L. McHugh, Interrater reliability: The kappa statistic. Biochemia Med. Editorial Office, 22(3), 276–282 (2012). https://doi.org/10.11613/BM.2012.031
28. J.R. Landis, G.G. Koch, The measurement of observer agreement for categorical data. Biometrics. JSTOR **33**(1), 159 (1977). https://doi.org/10.2307/2529310
29. D.A. Guttentag, Virtual reality: applications and implications for tourism. Tour. Manage. **31**(5), 637–651 (2010). https://doi.org/10.1016/j.tourman.2009.07.003
30. V. Bogicevic et al., 'Virtual reality presence as a preamble of tourism experience: The role of mental imagery', *Tourism Management*. Pergamon **74**, 55–64 (2019). https://doi.org/10.1016/J.TOURMAN.2019.02.009
31. M. Carrozzino, M. Bergamasco, Beyond virtual museums: Experiencing immersive virtual reality in real museums. J. Cult. Heritage. Elsevier Masson **11**(4), 452–458 (2010). https://doi.org/10.1016/J.CULHER.2010.04.001
32. Y.C. Huang et al., Exploring the implications of virtual reality technology in tourism marketing: an integrated research framework. Int. J. Tourism Res. John Wiley and Sons Ltd, **18**(2), 116–128. https://doi.org/10.1002/jtr.2038
33. M.J. Kim, C.M. Hall, A hedonic motivation model in virtual reality tourism: Comparing visitors and non-visitors. Int. J. Inf. Manage. Pergamon **46**, 236–249 (2019). https://doi.org/10.1016/J.IJINFOMGT.2018.11.016
34. I.P. Tussyadiah et al. (2018a) Virtual reality, presence, and attitude change: empirical evidence from tourism. *Tourism Manage.* Pergamon, **66**, 140–154. https://doi.org/10.1016/J.TOURMAN.2017.12.003
35. M.J. Kim et al., Relationships between lifestyle of health and sustainability and healthy food choices for seniors. Int. J. Contemp. Hospitality Manage. Emerald Group Publishing Limited **25**(4), 558–576. https://doi.org/10.1108/09596111311322925
36. N. Akhtar et al., Post-COVID 19 tourism: will digital tourism replace mass tourism? Sustainability **13**(5352), (2021). Multidisciplinary Digital Publishing Institute, **13**(10), 5352. https://doi.org/10.3390/SU13105352

37. S.M.C. Loureiro, J. Guerreiro, F. Ali, 20 years of research on virtual reality and augmented reality in tourism context: A text-mining approach. Tourism Manage. Pergamon **77**, 104028 (2020). https://doi.org/10.1016/J.TOURMAN.2019.104028
38. A. Gibson, M. O'Rawe, Virtual reality as a travel promotional tool: Insights from a consumer travel fair, in *Augmented Reality and Virtual Reality* eds by A. Gibson, M. O'Rawe. 1st edn. (Springer, Cham, Dublin), pp. 93–107 (2018). https://doi.org/10.1007/978-3-319-64027-3_7
39. S. Gössling, Technology, ICT and tourism: from big data to the big picture. J. Sustain. Tourism. Routledge **29**(5), 849–858 (2020). https://doi.org/10.1080/09669582.2020.1865387
40. F. Bruno et al., From 3D reconstruction to virtual reality: A complete methodology for digital archaeological exhibition. J. Cult. Heritage. Elsevier Masson **11**(1), 42–49 (2010). https://doi.org/10.1016/J.CULHER.2009.02.006
41. M. Carrozzino et al., Comparing different storytelling approaches for virtual guides in digital immersive museums, in *Lecture Notes in Computer Science (including subseries Lecture Notes in Artificial Intelligence and Lecture Notes in Bioinformatics).* (Springer, Cham, 2018), 10851 LNCS, pp. 292–302. https://doi.org/10.1007/978-3-319-95282-6_22
42. I.P. Tussyadiah et al. (2018b) Virtual reality, presence, and attitude change: Empirical evidence from tourism. Tourism Manage. Elsevier Ltd, **66,** 140–154. https://doi.org/10.1016/j.tourman.2017.12.003
43. L. Men et al., The impact of transitions on user experience in virtual reality, in *Proceedings—IEEE Virtual Reality.* (IEEE Computer Society, 2017), pp. 285–286. https://doi.org/10.1109/VR.2017.7892288
44. C. Wienrich et al. (2018) Assessing user experience in virtual reality—a comparison of different measurements, in *Lecture Notes in Computer Science (including subseries Lecture Notes in Artificial Intelligence and Lecture Notes in Bioinformatics)* (Springer, Cham, 2018), 10918 LNCS, pp 573–589. https://doi.org/10.1007/978-3-319-91797-9_41
45. X. Xia, W. Wu, User experience of virtual reality interfaces based on cognitive load. (Springer, Cham, 2021), pp. 340–347. https://doi.org/10.1007/978-3-030-80091-8_40
46. S. Farra et al., Storyboard Development for Virtual Reality Simulation. Clin. Simul. Nurs. Elsevier **12**(9), 392–399 (2016). https://doi.org/10.1016/J.ECNS.2016.04.002
47. D. Sarkady, L. Neuburger, R. Egger, Virtual reality as a travel substitution tool during COVID-19, in *Information and Communication Technologies in Tourism 2021* (Springer, Cham, 2021), pp. 452–463. https://doi.org/10.1007/978-3-030-65785-7_44
48. A.F. Schiopu et al., Virus tinged? Exploring the facets of virtual reality use in tourism as a result of the COVID-19 pandemic. Telematics Inform. Pergamon **60,** 101575 (2021). https://doi.org/10.1016/J.TELE.2021.101575

A Memory-Based Particle Swarm Optimization for Parameter Identification of Lorenz Chaotic System

Rizk M. Rizk-Allah, M. A. Farag, Mahmoud H. Barghout, and Aboul Ella Hassanien

Abstract A novel modified version of particle swarm optimization (PSO) is introduced in this paper to estimate the parameters of the chaotic Lorenz system. The parameters estimation of the Lorenz system is modeled as a multidimensional problem and solved by the proposed algorithm, a memory-based particle swarm optimization (MbPSO) algorithm. In MbPSO, two new terms are added to the standard PSO to vary the population direction and enhance search capability. Firstly, the impact of parameter configuration on MbPSO is studied. After that, the parameter estimation problem is solved. The performance of the proposed MbPSO is compared with other meta-heuristic algorithms in terms of parameter accuracy and convergence speed. According to the results, linking the memory of each particle to the memory of other particles has a very significant effect on the proposed algorithm compared to the original PSO. Briefly, the MbPSO algorithm is a successful and powerful optimization algorithm for parameter estimation of chaotic systems with accurate performance.

Keywords Chaotic system · Lorenz system · Parameter estimation · Particle swarm optimization

1 Introduction

Nonlinear dynamics, especially the study of chaotic systems (CSs), have been increasing interest in various fields, including science, engineering, communication, biomedical, finance, administration, and other areas [1]. The chaotic system,

R. M. Rizk-Allah (✉) · M. A. Farag · M. H. Barghout
Faculty of Engineering, Menoufia University, Shebin El-Kom, Egypt
e-mail: rizk_masoud@yahoo.com

A. E. Hassanien
Faculty of Computers and Information, Cairo University, Giza, Egypt
URL: http://www.egyptscience.net

Scientific Research Group in Egypt, Giza, Egypt

A. K. Bashir et al. (eds.), *Proceedings of International Conference on Computing and Communication Networks*, Lecture Notes in Networks and Systems 394,
https://doi.org/10.1007/978-981-19-0604-6_53

commonly known as the butterfly effect, means unpredictable, erratic motions, and a complex behavior governed by deterministic laws. The chaotic system that is extremely responsive to the initial conditions and parameters [2]. Chaotic systems with low-order deterministic models could enhance interpretation and description of phenomena (recorded data) of complicated real-world systems. The machinery condition, electrical circuits [3], and subjects of diagnosis of human health status [4] are some of the recent applications of chaotic systems. Many chaotic systems simulate real-world systems such as Chen, Lorenz, Newton–Leipnik, Volta, Rossler [5], and Lotka–Volterra [6].

In recent years, the control and synchronization of CSs have been intensively studied in multiple subjects [7]. Most of the recommended techniques only work according to the conditions of CSs that are defined in advance. Nevertheless, the parameters may be very difficult to identify in the actual world due to the CSs complexity. Parameter estimation of CSs has great value in nonlinear science. Control theory and signal processing have drawn great attention in different research subjects and can be investigated as a multidimensional optimization problem (multi-DOP) [2].

In the last decade, parameter identification of CSs has attracted a lot of attention in the literature [8]. Different types of traditional techniques have been improved to investigate these problems [9]. The most common approaches utilized to express the problem of parameters estimation of CSs as a multi-DOP are metaheuristic algorithms such as PSO [10], the genetic algorithm (GA) [11], and differential evolution algorithm (DE) [12]. They seem to be an efficient alternative to traditional methods because they do not require the gradient of the objective function, are not responsive to the starting point, and seldom get trapped in local optima [13].

The investigation of the parameter identification problems has a significant history, with a special focus on the Lorenz system. Lorenz introduced this system in 1963 while observing atmospheric convection [14], and it is considered the first chaotic attractor of a three-dimensional self-government system. The parameters are not easy to acquire due to the unstable dynamic behavior of CSs. So in recent years, metaheuristic algorithms have been popular and frequently used by the researchers for their efficient performance in a solution of optimization problems [15–17]; various metaheuristic algorithms employed for parameters identification of this system have been proposed by minimizing synchronization error such as GA [11], PSO [10], improved particle swarm optimization (IPSO) [18], a drifting particle swarm optimization (DPSO) [19], a quantum-behaved particle swarm optimization (QPSO) [20], PSO with dynamic inertia weight PSO (DIW-PSO) [21], and chaotic ant swarm (CAS) [5, 22, 23]. It is noted that PSO and its modifications are the most frequently used approaches for parameter estimation of CSs.

PSO was proposed in 1995 by Kennedy and Eberhart [24] as a population-based algorithm inspired by the social behavior of swarms in nature such as bird flock and fish school. The algorithm seeks out the best solution via sharing the historical and social knowledge between the candidates (particles) throughout the global solution space. PSO algorithm has been used as an important optimization technique in various applications due to its simple implementation, few parameters to adjust, and fast convergence rate. On the other hand, PSO demonstrates some shortages. It is possible to get stuck in a local optimum in another sense, premature convergence, and the convergence rate reduces significantly in the later stages of evolution [25–27]. Therefore, many variants of the standard PSO approach have been proposed for different applications to make the most of the merits, improve the search capability and overcome the deficiencies such as immune PSO (IPSO) [28], continuous trait-based PSO (CTB-PSO) [29], and the hybrid GA-PSO [30, 31].

In this paper, parameters estimation of chaotic Lorenz system is modeled as a multi-DOP and solved based on a modified version of PSO by adding two terms to the velocity update formula. The modified algorithm is called memory-based particle swarm optimization (MbPSO). This modification of its role diversifies and enhances search capability. This is the first research of MbPSO for parameters estimation of the Lorenz system that the authors are aware of. The comparisons of the results acquired by other techniques show this algorithm's efficiency, effectiveness, and robustness.

The main contributions of this paper can be summarized as follows:

- The chaotic system is modeled, generally, and the Lorenz system is illustrated clearly.
- A modified version of PSO is proposed called MbPSO.
- The proposed algorithm (MbPSO) is applied for the Lorenz system to estimate its parameters.
- Comparisons are made between the proposed algorithm (MbPSO), modified PSO (PSO+), original PSO (programmed algorithms for this study), and other algorithms used in the literature.

The remainder of this paper is organized as follows: Section 2 introduces the problem formulation regarding the Lorenz system. Section 3 reviews the basic principles of PSO. Section 4 introduces the proposed MbPSO algorithm, and the parameter configuration for the proposed algorithm is illustrated in Sect. 5. The numerical simulation, and comparisons are presented in Sect. 6. Finally, this paper is concluded in Sect. 7.

2 Problem Formulation

Considering the n-dimensional chaotic system modeled by the ordinary nonlinear differential equation as below (ONLDE):

$$\dot{X} = f(X, X_o, \theta_o) \tag{1}$$

where $X = (x_1, x_2, \ldots, x_n)^T \in R^n$ indicates the state vector, x_0 indicates the initial state and $\theta_0 = (\theta_{1o}, \theta_{2o}, \ldots, \theta_{do})^T$ is a set of original parameters. During calculating the parameters, assume the chaotic system structure is known in advance, and hence, the approximated system can be defined as follows:

$$\dot{Y} = f(Y, X_o, \theta) \tag{2}$$

where $Y = (y_1, y_2, ..., y_n)^T \in R^n$ indicates the state vector, and $\theta = (\theta_1, \theta_2, ..., \theta_d)^T$ is a set of calculated parameters. Thus, the problem of the parameter estimation can be described as the following optimization problem:

$$\text{Min } J = \frac{1}{M}\sum_{k=1}^{M} X_k - Y_k^2 \text{ by searching suitable } \theta^* \tag{3}$$

where the length of data used for parameters estimation is defined by M. The state vectors of the original and the estimated systems at time k ($k = 1,2, ..., M$) are denoted by Xk and Yk, respectively.

The parameters estimation for CSs is obviously expressed as a problem of multi-DOP, where the vector of decision is θ, and minimization of J is the objective function of the optimization problem. The parameters estimation problem of CSs based on optimization techniques can be demonstrated by Fig. 1. The parameters are not easy to acquire due to the unstable dynamic behavior of CS. Furthermore, multiple variables always exist in the problem, and multiple local optima solutions mislead the algorithm in the search zone of J, so conventional techniques are very easy to trap in local optima, and the global optimal parameters are difficult to attain. Thus, Lorenz system was selected as a CS model to test the performance of the proposed algorithm. Lorenz system [14] has an unpredictable complex dynamic behavior and exhibits infinite erratic periodic motions with extremely dependent initial conditions and parameters. The Lorenz model is employed as an equivalent model about the behavior of the atmosphere because it replicates some of the features of large-scale weather patterns, such as the simulation of system behavior, variations in the predictability of local events, and various time scales. The equations of the Lorenz system are system of three ONLDE resulting from a simple form of the fluid convection between two horizontal plates that is called Rayleigh Bénard problem. Lorenz system designated three time-dependent variables: convection motion intensity is represented by x, the temperature differential between increasing and decreasing currents is denoted by y, and z is the deviation from linearity in the vertical temperature profile [32]. The

Fig. 1 The overall scheme for optimization concept of parameters estimation in CSs

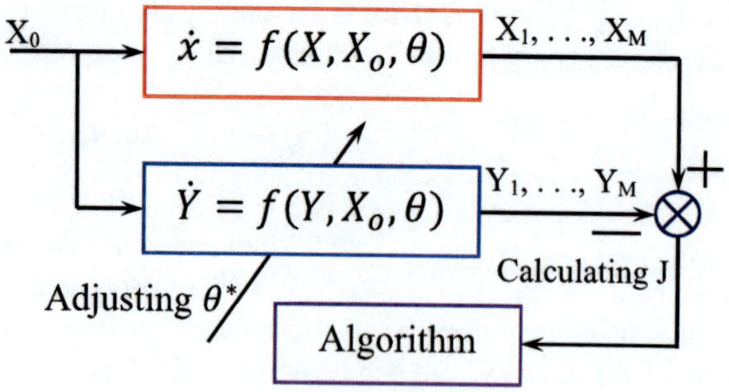

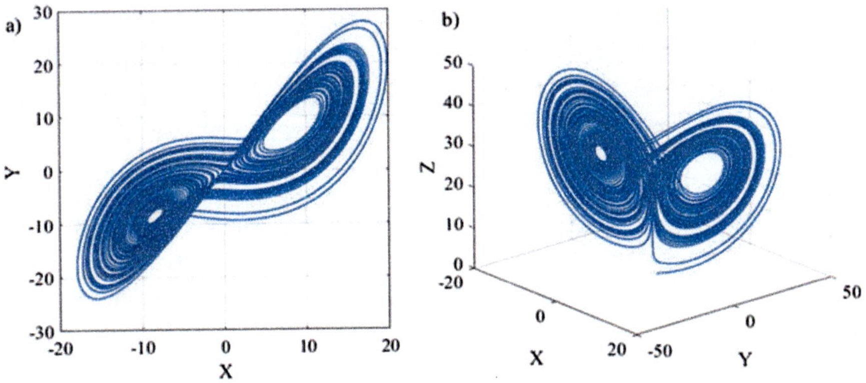

Fig. 2 The behavior of Lorenz system with $A = 10$, $B = 28$, $C = 8/3$ and initial conditions ($X0 = 1$, $Y0 = 0$, $Z0 = 0$) for evolving time = 100 s. **a)** the projection of Lorenz attractor into x–y plane; **b)** the Lorenz attractor in three-dimensional space

following model is a mathematical representation of the Lorenz system:

$$\left\{\begin{array}{l} \dot{x} = A(y - x) \\ \dot{y} = Bx - xy - y \\ \dot{z} = xy - Cz \end{array}\right. \quad (4)$$

where $\dot{x} = \mathrm{d}x/\mathrm{d}t$, $\dot{y} = \mathrm{d}y/\mathrm{d}t$, $\dot{z} = \mathrm{d}z/\mathrm{d}t$, A is defined as the Prandtl number, B ($B = R_a/R_c$) is defined as the Rayleigh number over the critical Rayleigh number, and C gives the size approximated by the system of the region [32]. All the parameters are positive numbers; for example, Ref. [33] demonstrates the behavior of a chaotic attractor with initial conditions ($X0 = 1$, $Y0 = 0$, $Z0 = 0$) and $A = 10$, $B = 28$, and $C = 8/3$. For a complete study of this system, see [33] (Fig. 2).

3 Particle Swarm Optimization (PSO)

PSO is considered as an evolutionary algorithm that is based on individual improvement in addition to collaboration and competition in the population. It depends on the simulation of simplified social models, such as the swarm theory: fish schooling, bird flocking [24]. PSO has a highly obvious, straightforward, and easy-to-implement theoretical structure.

PSO initializes randomly with a swarm of birds (particles) over the searching space. The particles search with a certain velocity and find the global optimum after several iterations. The main idea is to adjust the trajectory of each particle to its own best position and to the best particle of the swarm at each iteration. At each iteration, the velocity vector of each particle is affected by its inertia, its best position, and the position of the best particle. Then, each particle moves to a new position. Assume that, the location and the velocity of particle i in the d-dimensional search space can be expressed as $X_i = \left[x_{i,1}, x_{i,2}, \ldots, x_{i,d}\right]^T$ and $V_i = \left[v_{i,1}, v_{i,2}, \ldots, v_{i,d}\right]^T$, respectively. The particle fitness can be estimated according to the objective function of the optimization problem. The best position visited previously of particle i is denoted by the personal best position (pbest). It can be expressed as $P_i(t) = \left[p_{i,1}(t), p_{i,2}(t), \ldots, p_{i,d}(t)\right]^T$. The position of the best particle of the swarm is defined by the global best position (gbest) $P_g(t) = \left[p_{g,1}(t), p_{g,2}(t), \ldots, p_{g,d}(t)\right]^T$. At each step, the velocity and position of each particle is determined as follows:

$$v_{i,j}(t+1) = v_{i,j}(t) + c_1 r_1 \left(p_{i,j}(t) - x_{i,j}(t)\right) + c_2 r_2 \left(p_{g,j}(t) - x_{i,j}(t)\right), \quad j = 1, 2, \ldots, d \tag{5}$$

$$x_{i,j}(t+1) = x_{i,j}(t) + v_{i,j}(t+1), \quad j = 1, 2, \ldots, d \tag{6}$$

where ω, indicates to the inertia weight, controls the effect of previous particle velocity on its current one. $C1$ and $C2$ are the cognitive and collective acceleration coefficients, respectively, which are balance the effects of self-awareness and social awareness on particle movement toward the target and adjust the step size. $r1$ and $r2$ are two independent uniform distributed random constants within the range of [0, 1].

In PSO, Eq. (5) implies that the particle's new velocity is updated based on its previous velocity and the distance of its present position from both the global best position of the entire swarm and its best historical position. A typical approach is to constrain the values of the elements of the vector vi to a range between [Vi, min, Vi, max] to control excessive wandering of particles outside of the search space [Xi, min, Xi, max]. The particle then moves toward a new position in accordance with Eq. (6). The procedure is repeated until a termination condition determined is met [34].

4 Memory-Based Particle Swarm Optimization (MbPSO)

This section presents a new modified version of PSO called memory-based particle swarm optimization (MbPSO). The suggested algorithm (MbPSO) is a proposed improvement for the particle velocity updating equation to determine the parameters of Lorenz system. Although the position updating equation in MbPSO is identical to the standard PSO, the equation of the velocity updating is improved by the addition of two new terms, as illustrated below:

$$v_{i,j}(t+1) = \Psi\left(\omega v_{i,j}(t) + c_1 r_1 \left(\text{pbest}_{i,j}(t) - x_{i,j}(t)\right) + c_2 r_2 \left(\text{gbest}_j(t) - x_{i,j}(t)\right) + c_3 r_3 \left(\text{pbest}_{i,j}(\text{rand}) - x_{i,j}(t)\right) + c_4 r_4 \left(v_{i,j}(\text{rand})\right)\right) \tag{7}$$

$$x_{i,j}(t+1) = x_{i,j}(t) + v_{i,j}(t+1),$$
$$\forall i \in \text{number of particles}, \forall j \in \text{problem dimension, and } \forall t \in \text{number of iterations} \quad (8)$$

where ω is defined by the inertia weight. $r1$, $r2$, $r3$, and $r4$ are random numbers within [0, 1]. $C1$, $C2$, $C3$, and $C4$ denote the acceleration constants that control the effect of each term in the update process. pbest denotes the particle's best solution, while gbest denotes the best solution reached by all particles. Knowing that the acceleration constants pull each particle to its best individual and global locations. As a result, small values cause short movements toward target regions, whereas high values cause abrupt movements. High values of the inertia weight, on the other hand, result in a broad search across the search space, whereas low values result in a more focused search [34].

In the original PSO, the first term, "inertial part" is represented as $\omega v_{i,j}(t)$ which is utilized as a search skill factor by using $\omega = \omega_{\max} - t(\omega_{\max} - \omega_{\min})/t_{\max}$, where $\omega_{\min}$ and $\omega_{\max}$ are the values of minimum and maximum inertia weight, respectively, and $t_{\max}$ refers to the maximum number of iterations [35]. The second term "cognitive part" is represented as $c_1 r_1 \left(\text{pbest}_{i,j}(t) - x_{i,j}(t)\right)$ that is defined as the distance between the particle's current position and the particle's most well-known position, and it means that the particle's decision will be influenced by its previous experiences. The third term "social part" is represented as $c_2 r_2 \left(\text{gbest}_j(t) - x_{i,j}(t)\right)$ which is defined as the distance between the particle's current position and its best neighborhood position, and it means that the particle's decision is influenced by the rest of the particles [34]. From this idea, we added two new terms. The fourth term, "random self-cognition," is represented as $c_3 r_3 \left(\text{pbest}_{i,j}(\text{rand}) - x_{i,j}(t)\right)$ which relates the particle's position to the best position of random particles, which in fact let each particle exploit the memory of other particles and allow the particles to randomly share their knowledge during the updating process; it results in a stronger interrelation between the whole swarm, higher chance in convergence toward good solutions, and higher diversity of the search process. The fifth term, which is represented as $c_4 r_4 v_{i,j}(\text{rand})$, includes a random velocity that leads to increasing the potential of space exploration and preventing the MbPSO from being stuck in local optima. The new terms are inspired by [36].

The researchers utilize many additional terminologies with different meanings (e.g., [23, 37]). Still, this paper incorporates modifications to diversify the population search direction and enhance the swarm's search capability. In original PSO, gbest is utilized to enhance convergence characteristics, but the drawback of using gbest is decreasing the diversity of the population that results in local minima. Thus, adding two new terms increases search capability, providing new knowledge and more information to the population. It can guide the particles to a better position, and the attraction of gbest position to local optima in the search space is reduced. Figure 3 depicts a graphical view of the updating process for a particle's position

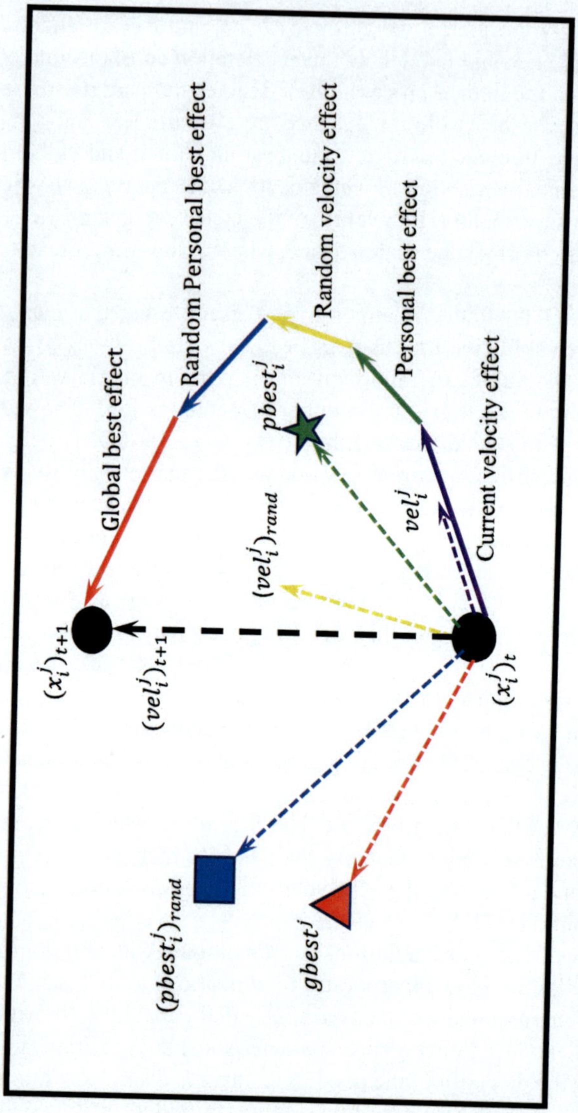

Fig. 3 Illustration of the particle's update equations in MbPSO

and velocity in the MbPSO algorithm. Also, the overall procedures of MbPSO are demonstrated in Fig. 4.

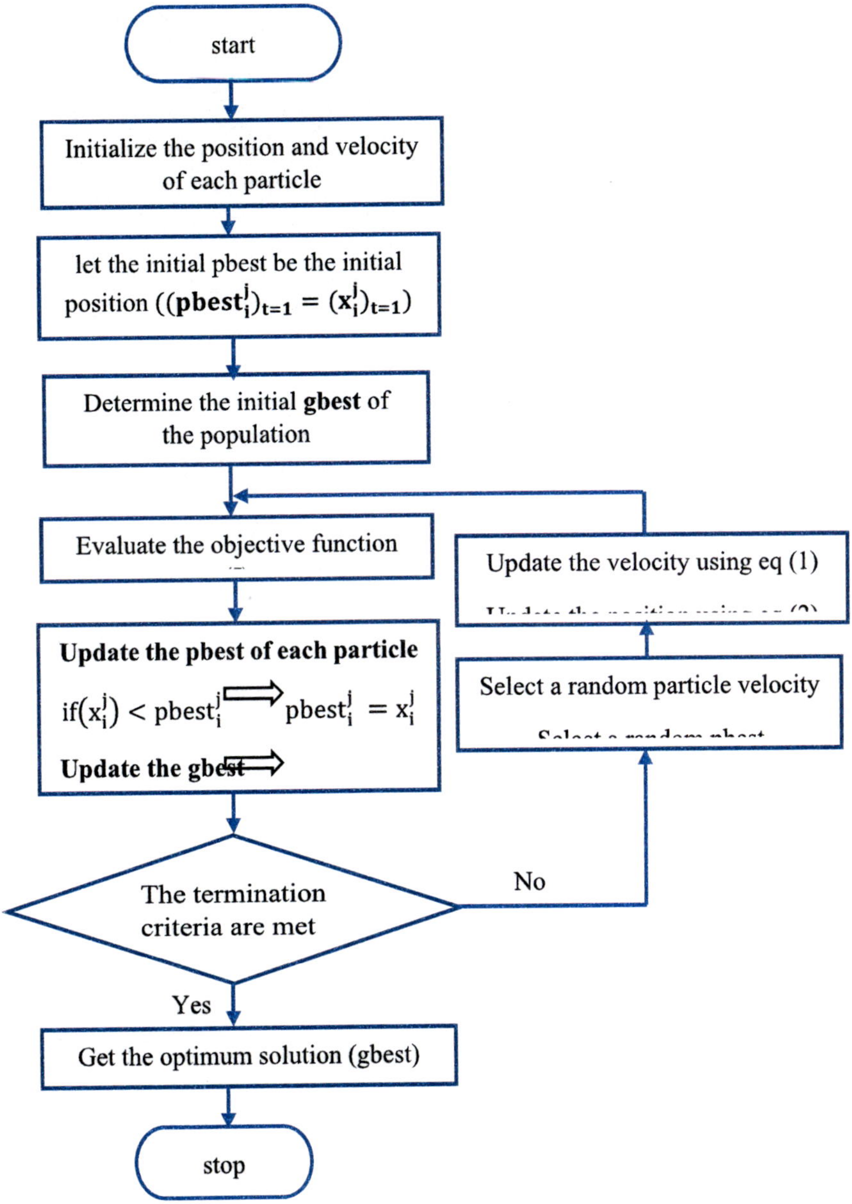

Fig. 4 Flowchart of the proposed algorithm (MbPSO)

5 Parameter Configuration for the Proposed Algorithm (MbPSO)

To make the best use of the proposed algorithm, the range of the parameter values are taken from the literature [36], and the trial and error method was used through the implementation on Lorenz system for configuring the parameters of the proposed algorithm during the operation of minimizing J. Thus, a population of 100 particles (p) was used; an inertia weight ω was established as linearly reducing from the peak value of 0.9 to a minimum value of 0.4. The acceleration constants were used as follows $C1 = 1.5$, $C2 = 2$, and $C3 = 3$ [38]. As different values of parameter $C4$ have a significant impact on the proposed algorithm, the various effects of $C4$ were studied for values under one; based on the literature, $C4$ was studied without including the constriction factor, and Fig. 5 shows the various effect of $C4$. For this study, the constant $C4 = 0.3$ introduces a better influence on the convergence than the others (the procedures for $C4$ are based on the coefficient mentioned above with other terms in [23]); in another aspect, the constriction factor has an extremely great effect on the proposed algorithm; following $C4$ is set to 0.3, the effect of constriction factor was analyzed from 0.2 to 0.8. Figure 6 shows the various effect of the constriction factor on the convergence and demonstrates that the constant $\psi = 0.4$ presents the best convergence. The maximum generation number was set to $t_{\max} = 100$ and $t_{\max} = 200$, and this is illustrated in the section of discussion and results. All simulations were implemented 20 times. Table 1 displays the parameters used in the compared algorithms (PSO, PSO+, and MbPSO).

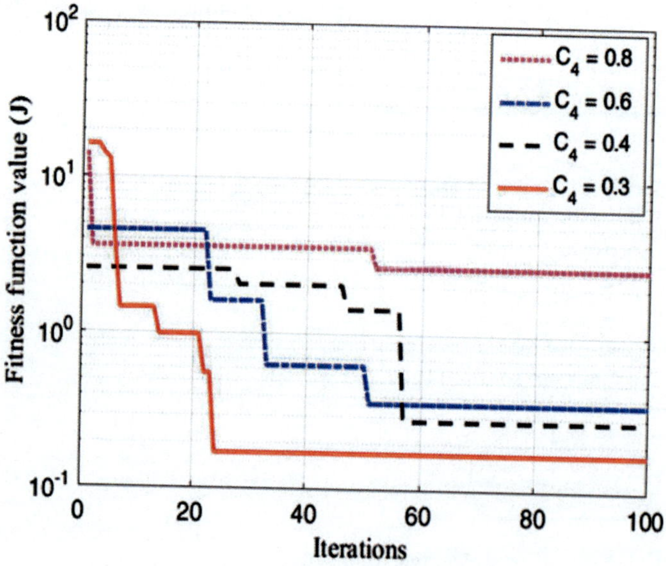

Fig. 5 Convergence graph. Specifying of the best value of constant C_4

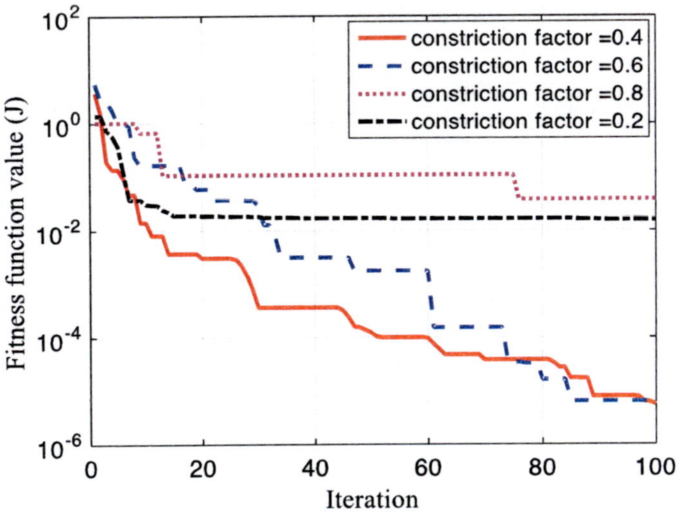

Fig. 6 Convergence graph. Specifying of the best value of constriction factor Ψ

Table 1 Parameter configuration for all algorithms used

Algorithm	Parameter configuration
PSO	$p = 100$, $\omega_{\max} = 0.9$, $\omega_{\min} = 0.4$, $c_1 = c_2 = 2$
PSO+	$p = 100$, $\omega_{\max} = 0.9$, $\omega_{\min} = 0.4$, $c_1 = c_2 = 1.5$, $c_3 = 0.4$
MbPSO	$p = 100$, $\Psi = 0.4$, $\omega_{\max} = 0.9$, $\omega_{\min} = 0.4$, $c_1 = 1.5$, $c_2 = 2$, $c_3 = 3$, $c_4 = 0.3$

6 Results and Discussion

The proposed algorithm MbPSO is implemented to estimate the parameters of the Lorenz chaotic system (A, B, and C). Figure 7 and Fig. 8 show the convergence characteristic of the fitness function J for MbPSO, PSO, and PSO + . These figures display that the value of J reduces to zero rapidly, which means MbPSO can converge to the global optimum very fast.

Furthermore, MbPSO is compared with a particle swarm optimization (PSO) [34] and a modified PSO (PSO+) [6] to test the performance of the proposed algorithm. The comparison was made for two cases: the first case, the maximum number of iterations was set to 100, the range of the estimated parameters is ($8 < A < 12, 20 < B < 30, 2 < C < 3$).

Table 2 compared the error and standard deviation (St.D.) values found by PSO, PSO+, and MbPSO. As shown in Table 2, MbPSO has fast convergence and accurate performance. Figure 7 illustrated that MbPSO has high performance and fast convergence, but PSO and PSO+ had almost the same performance. The second case, when the maximum generation number was set to 200 and the range of the estimated parameters increased ($0 < A < 20, 0 < B < 50, 0 < C < 5$), the performance of

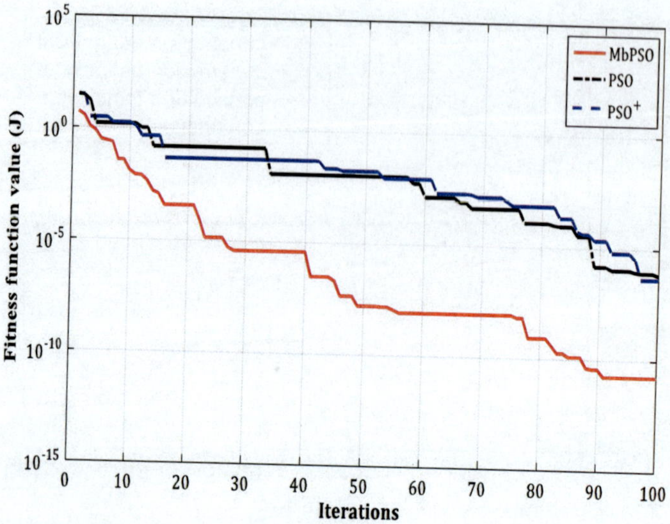

Fig. 7 Convergence characteristic of objective function (J) for three algorithms PSO, PSO+, and MbPSO (100 iterations)

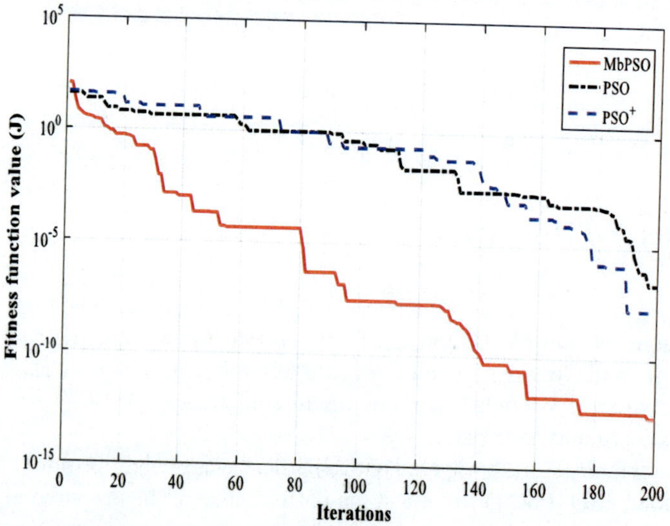

Fig. 8 Convergence characteristic of the objective function (J) for three algorithms PSO, PSO+, and MbPSO (200 iterations)

the modified algorithm (PSO+ and MbPSO) enhanced. Table 3 compared the error and St.D. values of PSO, PSO+ , and MbPSO. After the modifications of simulation conditions, MbPSO has the highest performance and the fastest convergence rate, and PSO+ has the second-best performance, as shown in Fig. 8.

Table 2 Comparison of convergence and statistical results of PSO, PSO+, and MbPSO (100 iterations)

Algorithm	MSE(J)	St.D		
	Best	Average	Worst	
PSO	1.2836E−06	1.2836E−06	3.7944E−04	1.5743E−04
PSO+	7.5670E−07	4.5711E−05	1.4774E−04	5.9247E−05
MbPSO	1.7666E−11	8.4102E−04	7.5000E−03	1.9000E−03

Table 3 Comparison of convergence and statistical results of PSO, PSO+, and MbPSO (200 iterations)

Algorithm	MSE(J)	St.D		
	Best	Average	Worst	
PSO	2.3475E−07	4.5743E−06	1.1863E−05	5.0764E−06
PSO+	1.6007E−08	1.5318E−05	5.3659E−05	2.3867E−04
MbPSO	2.5447E−13	1.4715E−04	5.5921E−04	2.5709E−05

Other comparisons can be made between the undefined parameter values A', B', and C' found by the MbPSO and the values found by with PSO and PSO+ from fitness function performance. The convergence rates of the parameters for the three-dimensional Lorenz system using MbPSO, PSO, and PSO+ are shown in Fig. 9. All estimated parameters found by MbPSO are very similar to the real values in all simulations, as shown in Fig. 9. This result demonstrates that the trajectories of the estimated parameters converge to their real values asymptotically. Therefore, MbPSO converges much faster than PSO and PSO+ . To get a full picture of our estimates, the statistical results found in estimating the parameters A', B', and C' using MbPSO, PSO, and PSO+ are summarized in Table 4. As shown in this table, the best results found by MbPSO are better than those obtained by PSO and the others. Additionally, the estimated parameter values found by MbPSO are still very similar to the real values of the original parameters.

On the other side, the abilities of MbPSO are compared to other metaheuristic techniques solved the same problem. Table 4 shows the comparison between the proposed algorithm and: (DE) [12], PSO [2], and a hybrid swarm intelligence algorithm (PSO–ACO) [23] in parameters estimation of Lorenz system. In general, MbPSO has accurate results better than the best results determined by these metaheuristic techniques (Table 5).

7 Conclusion

In this work, parameters identification for chaotic systems is modeled as a multi-dimensional optimization problem. The problem is solved using a modified PSO

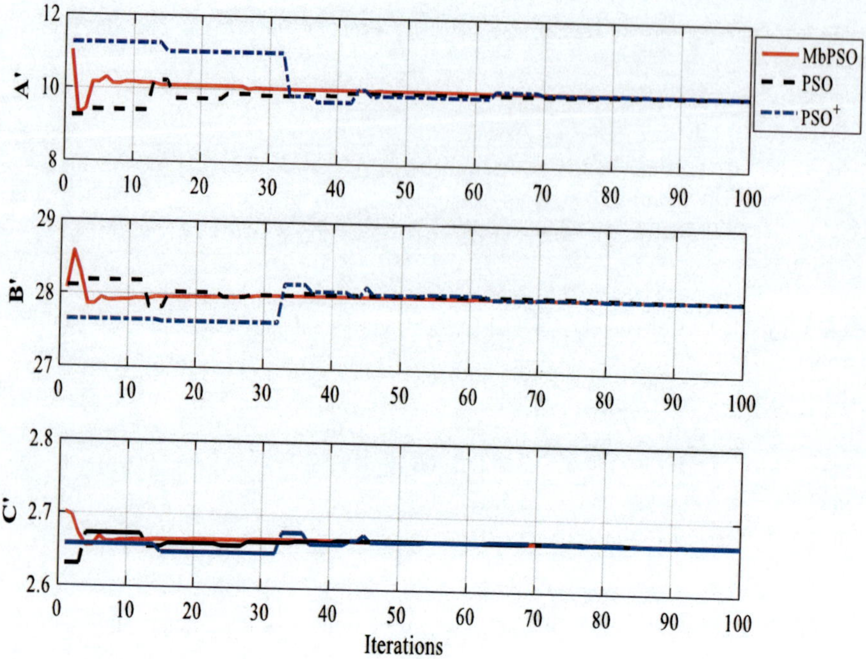

Fig. 9 Lorenz system searching process for A', B', and C' using PSO, PSO+ , and MbPSO

Table 4 Statistical results found by PSO, PSO+ , and MbPSO for the estimated parameters

Statistical results	Algorithms	Parameters		
		A'	B'	C'
Best	MbPSO	10.0000	28.0000	2.6667
	PSO+	10.0003	27.999	2.6667
	PSO	10.0009	27.9995	2.6666
Average	MbPSO	9.9981	28.0009	2.6667
	PSO+	10.0039	27.9985	2.6666
	PSO	9.9976	28.0011	2.6667
Worst	MbPSO	9.8567	28.0641	2.6689
	PSO+	10.0155	27.9938	2.6666
	PSO	9.9912	28.0001	2.6662

version called a memory-based particle swarm optimization (MbPSO). Lorenz system is selected to test the performance of MbPSO. In the proposed algorithm, MbPSO, two new terms are added to the standard PSO to diversify and enhance search capability. Comparisons are made between the proposed algorithm MbPSO, the original PSO, a modified PSO (PSO+), and other algorithms published in the literature. The comparisons and results demonstrate that the suggested algorithm

Table 5 Results obtained by many parameter estimation algorithms and the proposed MbPSO

Parameter	PSO [2]		DE [12]		PSO–ACO [23]		MbPSO (this work)		
	Average	Best	Average	Best	Average	Best	Average	Best	Worst
A′	10.0184	9.9953	10.0101	10.0001	10.0005	10.000	9.9981	10.00000	9.8567
B′	27.9934	28.0071	28.0000	28.0000	28.0011	28.0000	28.0009	28.00000	28.0641
C′	2.6663	2.6670	2.6667	2.6667	2.6673	2.6666	2.6667	2.66667	2.6689
MSE(J)	4.18E + 0	4.86E − 2	2.00E − 7	2.00E − 7	1.20E − 5	1.03E − 6	1.53E-6	1.76E-11	7.5E-3

is an effective and useful arithmetic method for parameter identification of chaotic systems, especially the Lorenz system, with high efficiency, fast convergence process, and accurate performance.

References

1. S.H. Strogatz, Nonlinear dynamics and chaos: with applications to physics, biology, chemistry, and engineering, 1st pbk (2000)
2. Q. He, L. Wang, B. Liu, Parameter estimation for chaotic systems by particle swarm optimization. Chaos Solitons Fractals **34**, 654–661 (2007). https://doi.org/10.1016/j.chaos.2006.03.079
3. H.-G. Ma, X.-F. Zhu, J.-F. Xu, M.-S. Ai, Circuit state analysis using chaotic signal excitation. J. Fr. Inst. **345**, 75–86 (2008)
4. W. Klonowski, From conformons to human brains: an informal overview of nonlinear dynamics and its applications in biomedicine. Nonlinear Biomed. Phys. **1**, 5 (2007). https://doi.org/10.1186/1753-4631-1-5
5. Y. Mousavi, A. Alfi, Fractional calculus-based firefly algorithm applied to parameter estimation of chaotic systems. Chaos Solitons Fractals **114**, 202–215 (2018). https://doi.org/10.1016/j.chaos.2018.07.004
6. J.A. Lazzús, P. Vega-Jorquera, C.H. López-Caraballo, et al., Parameter estimation of a generalized Lotka–Volterra system using a modified PSO algorithm. Appl. Soft Comput. J. **96,** 106606 (2020). https://doi.org/10.1016/j.asoc.2020.106606
7. E. Elabbasy, H. Agiza, M. El-Dessoky, Global synchronization criterion and adaptive synchronization for new chaotic system. Chaos. Chaos Solitons Fractals **23**, 1299–1309 (2005)
8. Y. Liu, W.K.S. Tang, Modified dynamic minimization algorithm for parameter estimation of chaotic system from a time series. Nonlinear Dyn **66**, 213–229 (2011). https://doi.org/10.1007/s11071-010-9922-0
9. A. Alireza, Ò. Acta Autom. Sin. 37 (2011)
10. B. Samanta, C. Nataraj, Particle swarm optimization for chaotic system parameter estimation, in *2009 IEEE Swarm Intelligence Symposium SIS 2009*, pp. 74–80. https://doi.org/10.1109/SIS.2009.4937847
11. D. Dai, X. Ma, F. Li, Y. You, An approach of parameter estimation for a chaotic system based on genetic algorithm. Acta Phys. Sin **11**, 2459–2462 (2002)
12. B. Peng, B. Liu, F.-Y.Y. Zhang, L. Wang, Differential evolution algorithm-based parameter estimation for chaotic systems. Chaos Solitons Fractals **39**, 2110–2118 (2009). https://doi.org/10.1016/j.chaos.2007.06.084
13. A. ALFI, PSO with adaptive mutation and inertia weight and its application in parameter estimation of dynamic systems. Acta Autom. Sin. 37:541–549 (2011). https://doi.org/10.1016/s1874-1029(11)60205-x
14. E.N. Lorenz, Deterministic nonperiodic flow. J. Atmos. Sci. **20**, 130–141 (1963)
15. R.M. Rizk-Allah, A.E. Hassanien, M. Elhoseny, M. Gunasekaran, A new binary salp swarm algorithm: development and application for optimization tasks. Neural Comput. Appl. **31**, 1641–1663 (2019)
16. R.M. Rizk-Allah, A.E. Hassanien, New binary bat algorithm for solving 0–1 knapsack problem. Complex Intell. Syst. **4**, 31–53 (2018)
17. A.E. Hassanien, R.M. Rizk-Allah, M. Elhoseny, A hybrid crow search algorithm based on rough searching scheme for solving engineering optimization problems. J. Ambient Intell. Humaniz Comput. First Online: 25 June 2018
18. H. Modares, A. Alfi, M.-M. Fateh, Parameter identification of chaotic dynamic systems through an improved particle swarm optimization. Expert Syst. Appl. **37**, 3714–3720 (2010)

19. J. Sun, J. Zhao, X. Wu et al., Parameter estimation for chaotic systems with a drift particle swarm optimization method. Phys. Lett. A **374**, 2816–2822 (2010)
20. K. Yang, K. Maginu, H. Nomura, Parameters identification of chaotic systems by quantum-behaved particle swarm optimization. Int. J. Comput. Math. **86**, 2225–2235 (2009)
21. A. Alfi, Particle swarm optimization algorithm with dynamic inertia weight for online parameter identification applied to Lorenz chaotic system. Int. J. Innov. Comput. Inf. Control **8**, 1191–1203 (2012)
22. L. Li, Y. Yang, H. Peng, X. Wang, An optimization method inspired by "chaotic" ant behavior. Int. J. Bifurc Chaos **16**, 2351–2364 (2006)
23. J.A. Lazzús, M. Rivera, C.H. López-Caraballo, Parameter estimation of Lorenz chaotic system using a hybrid swarm intelligence algorithm. Phys. Lett. A **380**, 1164–1171 (2016)
24. J. Kennedy, R. Eberhart, Particle swarm optimization, in *Proceedings of ICNN'95-International Conference on Neural Networks* (IEEE, 1995), pp. 1942–1948
25. R.M. Rizk-Allah, A. Slowik, A.E. Hassanien, Hybridization of grey wolf optimizer and crow search algorithm based on dynamic fuzzy learning strategy for large-scale optimization. IEEE Access **8**, 161593–161611 (2020). https://doi.org/10.1109/ACCESS.2020.3021693
26. R.M. Rizk-Allah, A.E. Hassanien, Locomotion-based hybrid salp swarm algorithm for parameter estimation of fuzzy representation-based photovoltaic modules. J. Mod. Power Syst. Clean Energy (2020)
27. R.M. Rizk-Allah, Hybridizing sine cosine algorithm with multi-orthogonal search strategy for engineering design problems. J. Comput. Des. Eng. **5**, 249–273 (2018)
28. C. Lin, Y. Liu, C. Lee, An efficient neural fuzzy network based on immune particle swarm optimization for prediction and control applications. Int. J. Innov. Comput. Inf. Control **4**, 1711–1722 (2008)
29. E. Keedwell, M. Morley, D. Croft, Continuous trait-based particle swarm optimisation (CTB-PSO). in *International Conference on Swarm Intelligence* (Springer, 2012), pp. 342–343
30. A. Gálvez, A. Iglesias, A new iterative mutually coupled hybrid GA–PSO approach for curve fitting in manufacturing. Appl. Soft Comput. **13**, 1491–1504 (2013)
31. I. Jain, V.K. Jain, R. Jain, Correlation feature selection based improved-binary particle swarm optimization for gene selection and cancer classification. Appl. Soft. Comput. **62**, 203–215 (2018)
32. X.-J. Wu, S.-L. Shen, Chaos in the fractional-order Lorenz system. Int. J. Comput. Math. **86**, 1274–1282 (2009)
33. I. Pan, S. Das, Evolving chaos: Identifying new attractors of the generalised Lorenz family. Appl. Math. Model **57**, 391–405 (2018). https://doi.org/10.1016/j.apm.2018.01.015
34. J. Kennedy, R. Eberhart, Y. Shi, *Swarm Intelligence* (Morgan Kaufman, San Francisco, 2001)
35. R.C. Eberhart, Y. Shi, Comparing inertia weights and constriction factors in particle swarm optimization, in: *Proceedings of the 2000 Congress on Evolutionary Computation. CEC00 (Cat. No.00TH8512),* vol. 1 (2000), pp. 84–88
36. D. Sedighizadeh, E. Masehian, M. Sedighizadeh, H. Akbaripour, GEPSO: A new generalized particle swarm optimization algorithm. Math. Comput. Simul. **179**, 194–212 (2021)
37. J.A. Lazzús, Hybrid particle swarm-ant colony algorithm to describe the phase equilibrium of systems containing supercritical fluids with ionic liquids. Commun. Comput. Phys. **14**, 107–125 (2013)
38. A. Rezaee Jordehi, J. Jasni, Parameter selection in particle swarm optimisation: a survey. J. Exp. Theor. Artif. Intell. **25**, 527–542 (2013). https://doi.org/10.1080/0952813X.2013.782348

Author Index

A. K. Bashir et al. (eds.), *Proceedings of International Conference on Computing and Communication Networks*, Lecture Notes in Networks and Systems 394,
https://doi.org/10.1007/978-981-19-0604-6